Lecture Notes in Computer Science   9658

Commenced Publication in 1973
Founding and Former Series Editors:
Gerhard Goos, Juris Hartmanis, and Jan van Leeuwen

More information about this series at http://www.springer.com/series/7409

Bin Cui · Nan Zhang · Jianliang Xu
Xiang Lian · Dexi Liu (Eds.)

Web-Age Information Management

17th International Conference, WAIM 2016
Nanchang, China, June 3–5, 2016
Proceedings, Part I

 Springer

Editors
Bin Cui
Peking University
Beijing
China

Nan Zhang
The George Washington University
Washington, D.C.
USA

Jianliang Xu
Hong Kong Baptist University
Kowloon Tong, Hong Kong
SAR China

Xiang Lian
University of Texas Rio Grande Valley
Edinburg, TX
USA

Dexi Liu
Jiangxi University of Finance and
  Economics
Nanchang
China

ISSN 0302-9743   ISSN 1611-3349   (electronic)
Lecture Notes in Computer Science
ISBN 978-3-319-39936-2   ISBN 978-3-319-39937-9   (eBook)
DOI 10.1007/978-3-319-39937-9

Library of Congress Control Number: 2016940123

LNCS Sublibrary: SL3 – Information Systems and Applications, incl. Internet/Web, and HCI

Printed on acid-free paper

This Springer imprint is published by Springer Nature
The registered company is Springer International Publishing AG Switzerland

Preface

This volume contains the proceedings of the 17th International Conference on Web-Age Information Management (WAIM), held during June 3–5, 2016, in Nanchang, Jiangxi, China. As a flagship conference in the Asia-Pacific region focusing on the research, development, and applications of Web information management, its success has been witnessed through the previous conference series that were held in Shanghai (2000), Xi'an (2001), Beijing (2002), Chengdu (2003), Dalian (2004), Hangzhou (2005), Hong Kong (2006), Huangshan (2007), Zhangjiajie (2008), Suzhou (2009), Jiuzhaigou (2010), Wuhan (2011), Harbin (2012), Beidahe (2013), Macau (2014), and Qingdao (2015). With the fast development of Web-related technologies, we expect that WAIM will become an increasingly popular forum to bring together outstanding researchers in this field from all over the world.

This high-quality program would not have been possible without the authors who chose WAIM for disseminating their contributions. Out of 249 submissions to the research track and 17 to the demonstration track, the conference accepted 80 research papers and eight demonstrations. The contributed papers address a wide range of topics, such as big data analytics, data mining, query processing and optimization, security, privacy, trust, recommender systems, spatial databases, information retrieval and Web search, information extraction and integration, data and information quality, distributed and cloud computing, among others.

The technical program of WAIM 2016 also included two keynote talks by Profs. Beng Chin Ooi (National University of Singapore) and Yanchun Zhang (Victoria University, Australia), as well as three talks in the Distinguished Young Lecturer Series by Profs. Tingjian Ge (University of Massachusetts at Lowell), Hua Lu (Aalborg University), and Haibo Hu (Hong Kong Polytechnic University). We are immensely grateful to these distinguished guests for their invaluable contributions to the conference program.

A conference like WAIM can only succeed as a team effort. We are deeply thankful to the Program Committee members and the reviewers for their invaluable efforts. Special thanks to the local Organizing Committee headed by Guoqiong Liao and Xiaobing Mao. Many thanks also go to our workshop co-chairs (Shaoxu Song and Yongxin Tong), proceedings co-chairs (Xiang Lian and Dexi Liu), DYL co-chairs (Hong Gao and Weiyi Meng), demo co-chairs (Xiping Liu and Yi Yu), publicity co-chairs (Ye Yuan, Hua Lu, and Chengkai Li), registration chair (Yong Yang), and finance chair (Bo Shen). Last but not least, we wish to express our gratitude for the hard work of our webmaster (Bo Yang), and for our sponsors who generously supported the smooth running of our conference.

We hope you enjoy the proceedings WAIM 2016!

June 2016

Zhanhuai Li
Sang Kyun Cha
Changxuan Wan
Bin Cui
Nan Zhang
Jianliang Xu

Organization

Organizing Committee

Honor Chair

Qiao Wang   Jiangxi University of Finance and Economics, China

General Co-chairs

Zhanhuai Li   Northwestern Polytechnical University, China
Sang Kyun Cha   Seoul National University, Korea
Changxuan Wan   Jiangxi University of Finance and Economics, China

PC Co-chairs

Bin Cui   Peking University, China
Nan Zhang   George Washington University, USA
Jianliang Xu   Hong Kong Baptist University, SAR China

Workshop Co-chairs

Shaoxu Song   Tsinghua University, China
Yongxin Tong   Beihang University, China

Proceedings Co-chairs

Xiang Lian   The University of Texas Rio Grande Valley, USA
Dexi Liu   Jiangxi University of Finance and Economics, China

DYL Series Co-chairs (Distinguished Young Lecturer)

Hong Gao   Harbin Institute of Technology, China
Weiyi Meng   SUNY Binghamton, USA

Demo Co-chairs

Xiping Liu   Jiangxi University of Finance and Economics, China
Yi Yu   National Institute of Informatics, Japan

Publicity Co-chairs

Ye Yuan   Northeastern University, China
Hua Lu   Aalborg University, Denmark
Chengkai Li   The University of Texas at Arlington, USA

Local Organization Co-chairs

Xiaobing Mao   Jiangxi University of Finance and Economics, China
Guoqiong Liao   Jiangxi University of Finance and Economics, China

Registration Chair

Yong Yang   Jiangxi University of Finance and Economics, China

Finance Chair

Bo Shen   Jiangxi University of Finance and Economics, China

Web Chair

Bo Yang   Jiangxi University of Finance and Economics, China

Steering Committee Liaison

Weiyi Meng   SUNY Binghamton, USA

CCF DBS Liaison

Xiaofeng Meng   Renmin University of China, China

Program Committee

Alex Thomo   University of Victoria, Canada
Anirban Mondal   Xerox Research Centre India, India
Baihua Zheng   Singapore Management University, Singapore
Baoning Niu   Taiyuan University of Technology, China
Byron Choi   Hong Kong Baptist University, SAR China
Carson Leung   University of Manitoba, Canada
Ce Zhang   Stanford University, USA
Chengkai Li   The University of Texas at Arlington, USA
Chih-Hua Tai   National Taipei University, China
Cuiping Li   Renmin University of China, China
David Cheung   The University of Hong Kong, SAR China
Dejing Dou   University of Oregon, USA
De-Nian Yang   Academia Sinica, Taiwan, China

Dongxiang Zhang	National University of Singapore, Singapore
Feida Zhu	Singapore Management University, Singapore
Feifei Li	University of Utah, USA
Fuzhen Zhuang	ICT, Chinese Academy of Sciences, China
Gang Chen	Zhejiang University, China
Gao Cong	Nanyang Technological University, Singapore
Giovanna Guerrini	Università di Genova, Italy
Guohui Li	Huazhong University of Science and Technology, China
Guoliang Li	Tsinghua University, China
Guoqiong Liao	Jiangxi University of Finance and Economics, China
Haibo Hu	Hong Kong Polytechnic University, SAR China
Hailong Sun	Beihang University, China
Hiroaki Ohshima	Kyoto University, Japan
Hongyan Liu	Tsinghua University, China
Hongzhi Wang	Harbin Institue of Technology, China
Hongzhi Yin	The University of Queensland, Australia
Hua Lu	Aalborg University, Denmark
Jae-Gil Lee	Korea Advanced Institute of Science and Technology, Korea
Jeffrey Xu Yu	Chinese University of Hong Kong, SAR China
Jiaheng Lu	Renmin University of China
Jianbin Qin	The University of New South Wales, Australia
Jianbin Huang	Xidian University, China
Jiannan Wang	University of Berkerley, USA
Jie Shao	University of Electronic Science and Technology of China, China
Jinchuan Chen	Renmin University of China, China
Jingfeng Guo	Yanshan University, China
Jiun-Long Huang	National Chiao Tung University, Taiwan, China
Jizhou Luo	Harbin Institue of Technology, China
Ju Fan	National University of Singapore, Singapore
Junfeng Zhou	Yanshan University, China
Junjie Yao	East China Normal University, China
Ke Yi	Hong Kong University of Science and Technology, SAR China
Kun Ren	Yale University, USA
Kyuseok Shim	Seoul National University, South Korea
Lei Zou	Peking University, China
Leong Hou U	University of Macau, SAR China
Lianghuai Yang	Zhejiang University of Technology, China
Lidan Shou	Zhejiang University, China
Lili Jiang	Max Planck Institute for Informatics, Germany
Ling Chen	University of Technology, Sydney, Australia
Luke Huan	University of Kansas, USA
Man Lung Yiu	Hong Kong Polytechnic University, SAR China
Muhammad Cheema	Monash University, Australia

Peiquan Jin	University of Science and Technology of China, China
Peng Wang	Fudan University, China
Qi Liu	University of Science and Technology of China, China
Qiang Wei	Tsinghua University, China
Qingzhong Li	Shandong University, China
Qinmin Hu	East China Normal University, China
Quan Zou	Xiamen University, China
Richong Zhang	Beihang University, China
Rui Zhang	The University of Melbourne, Australia
Rui Chen	Samsung Research America, USA
Saravanan Thirumuruganathan	Qatar Computing Research Institute, Qatar
Senjuti Basu Roy	University of Washington, USA
Shengli Wu	Jiangsu University, China
Shimin Chen	Chinese Academy of Sciences, China
Shinsuke Nakajima	Kyoto Sangyo University, Japan
Shuai Ma	Beihang University, China
Sourav Bhowmick	National Taiwan University, China
Takahiro Hara	Osaka University, Japan
Taketoshi Ushiama	Kyushu University, Japan
Tingjian Ge	University of Massachusetts Lowell, USA
Wang-Chien Lee	Penn State University, USA
Wei Wang	University of New South Wales, Australia
Weiwei Sun	Fudan University, China
Weiwei Ni	Southeast University, China
Wen-Chih Peng	National Chiao Tung University, Taiwan, China
Wenjie Zhang	University of New South Wales, Australia
Wolf-Tilo Balke	TU-Braunschweig, Germany
Wookey Lee	Inha University, Korea
Xiang Lian	The University of Texas Rio Grande Valley, USA
Xiangliang Zhang	King Abdullah University of Science and Technology, Saudi Arabia
Xiaochun Yang	Northeast University, China
Xiaofeng Meng	Renmin University of China, China
Xiaohui Yu	Shandong University, China
Xiaokui Xiao	Nanyang Technological University, Singapore
Xifeng Yan	University of California at Santa Barbara, USA
Xin Lin	East China Normal University, China
Xin Cao	Queen's University Belfast, UK
Xin Wang	Tianjin University, China
Xingquan Zhu	Florida Atlantic University, USA
Xuanjing Huang	Fudan University, China
Yafei Li	Henan University of Economics and Law, China
Yang Liu	Shandong University, China
Yanghua Xiao	Fudan University, China
Yang-Sae Moon	Kangwon National University, Korea

Contents – Part I

Recommender Systems

Contents – Part II

Social Media

Big Data Analytics

Distributed and Cloud Computing

Demo Papers

Data Mining

More Efficient Algorithm for Mining Frequent Patterns with Multiple Minimum Supports

Wensheng Gan[1], Jerry Chun-Wei Lin[1(✉)], Philippe Fournier-Viger[2],
and Han-Chieh Chao[1,3]

[1] School of Computer Science and Technology,
Harbin Institute of Technology Shenzhen Graduate School, Shenzhen, China
wsgan001@gmail.com, jerrylin@ieee.org
[2] School of Natural Sciences and Humanities,
Harbin Institute of Technology Shenzhen Graduate School, Shenzhen, China
philfv@hitsz.edu.cn
[3] Department of Computer Science and Information Engineering,
National Dong Hwa University, Hualien County, Taiwan
hcc@ndhu.edu.tw

Abstract. Frequent pattern mining (FPM) is an important data mining task, having numerous applications. However, an important limitation of traditional FPM algorithms, is that they rely on a single minimum support threshold to identify frequent patterns (FPs). As a solution, several algorithms have been proposed to mine FPs using multiple minimum supports. Nevertheless, a crucial problem is that these algorithms generally consume a large amount of memory and have long execution times. In this paper, we address this issue by introducing a novel algorithm named efficient discovery of Frequent Patterns with Multiple minimum supports from the Enumeration-tree (FP-ME). The proposed algorithm discovers FPs using a novel Set-Enumeration-tree structure with Multiple minimum supports (ME-tree), and employs a novel *sorted downward closure (SDC) property* of FPs with multiple minimum supports. The proposed algorithm directly discovers FPs from the ME-tree without generating candidates. Furthermore, an improved algorithms, named FP-ME$_{DiffSet}$, is also proposed based on the *DiffSet* concept, to further increase mining performance. Substantial experiments on real-life datasets show that the proposed approaches not only avoid the "rare item problem", but also efficiently and effectively discover the complete set of FPs in transactional databases.

Keywords: Frequent patterns · Multiple minimum supports · Sorted downward closure property · Set-enumeration-tree · DiffSet

1 Introduction

In the process of knowledge discovery in database (KDD) [2,3], many approaches have been proposed to discover more useful and invaluable information from

© Springer International Publishing Switzerland 2016
B. Cui et al. (Eds.): WAIM 2016, Part I, LNCS 9658, pp. 3–16, 2016.
DOI: 10.1007/978-3-319-39937-9_1

huge databases. Among them, frequent pattern mining (FPM) and association rule mining (ARM) [2–4] have been extensively studied. Most studies in FPM or ARM focus on developing efficient algorithms to mine frequent patterns (FPs) in a transactional database, in which the occurrence frequency of each itemset is no less than a specified number of customer transactions w.r.t. the user-specified minimum support threshold (called *minsup*). However, they suffer from an important limitation, which is to utilize a single minimum support threshold as the measure to discover the complete set of FPs. Using a single threshold to assess the occurrence frequencies of all items in a database is inadequate since each item is different and should not all be treated the same. Hence, it is hard to carry out a fair measurement of the frequencies of itemsets using a single minimum support when mining FPs.

For market basket analysis, a traditional FPM algorithm may discover many itemsets that are frequent but generate a low profit and fail to discover itemsets that are rare but generate a high profit. For example, clothes i.e., {*shirt, tie, trousers, suits*} occur much more frequent than {*diamond*} in a supermarket, and both have positive contribution to increase the profit amount. If the value of *minsup* is set too high, though the rule {*shirt, tie ⇒ trousers*} can be found, we would never find the rule {*shirt, tie ⇒ diamond*}. To find the second rule, we need to set the *minsup* very low. However, this will cause lots of meaningless rules to be found at the same time. This is the co-called "rare item problem" [7]. To address this issue, the problem of frequent pattern mining with multiple minimum supports (FP-MMS) has been studied. Liu et al. [7] introduced the problem of FP-MMS and proposed the MSApriori algorithm by extending the level-wise Apriori algorithm. The goal of FP-MMS is to discover the useful set of itemsets that are "frequent" for the users. It allows the users to free set multiple minimum support thresholds instead of an uniform minimum support threshold to reflect different natures and frequencies of all items. Some approaches have been designed for the mining task of FP-MMS, such as MSApriori [7], CFP-growth [11], CFP-growth++ [10], etc. The state-of-the-art CFP-growth++ was proposed by extending the FP-growth [4] approach to mine FPs from a condensed CFP-tree structure. However, the mining efficiency of them is still a major problem. Since the previous studies of FP-MMS still suffer the time-consuming and memory usage problems, it is thus quite challenging and critically important to design an efficient algorithm to solve this problem. In this paper, we propose a novel mining framework named mining frequent patterns from the Set-enumeration-tree with multiple minimum supports to address this important research gap. Major contributions are summarized as follows:

- Being different from the Apriori-like and FP-growth-based approaches, we propose a novel algorithm for directly extracting FPs with multiple minimum supports from the Set-enumeration-tree (abbreviated as FP-ME). It allows the user to specify multiple minimum support thresholds to reflect different natures and frequencies of items. This increases the applicability of FPM to real-life situations.

- Based on the proposed Set-Enumeration-tree with Multiple minimum supports (ME-tree), a new *sorted downward closure (SDC) property* of FPs in ME-tree holds. Therefore, the baseline algorithm FP-ME can directly discover FPs by spanning the ME-tree with the *SDC property*, without the candidate generate-and-test, thus greatly reduce the running time and memory consumption.
- The *DiffSet* concept is further extend in the improved FP-ME$_{DiffSet}$ algorithm to greatly speed up the process of mining FPs.
- Extensive experiments show that the proposed FP-ME algorithm is more efficient than the state-of-the-art CFP-growth++ algorithm for mining FPs in terms of runtime, effectiveness of prune strategies and scalability, especially the improved algorithm considerably outperforms the baseline algorithm.

2   Related Work

Up to now, many approaches have been extensively developed to mine FPs. Since only a single *minsup* value is used for the whole database, the model of ARM implicitly assumes that all the items in the database have similar occurrence frequencies. However, as most of the real-world databases are non-uniform in nature, mining FPs with a single *minsup* (or *mincof*) constraint leads to the following problems: (i) If *minsup* is set too high, we will not find the patterns involving rare items. (ii) In order to find the patterns that involve both frequent and rare items, we have to set *minsup* very low. However, this may cause combinatorial explosion, producing many meaningless patterns.

The problem of frequent pattern mining with multiple minimum support thresholds has been extensively studied, and algorithms such as MSApriori [7], CFP-growth [11], CFP-growth++ [10], REMMAR [8] and FQSP-MMS [6] have been proposed, among others [9]. MSApriori extends the well-known Apriori algorithm to mine FPs or ARs by considering multiple minimum support thresholds [7]. The major idea of MSApriori is that by assigning specific minimum item support (*MIS*) values to each item, rare ARs can be discovered without generating a large number of meaningless rules. MSApriori mines ARs in a level-wise manner but suffers from the problem of "pattern explosion" since it relies on a generate-and-test approach. Lee et al. [9] then proposed a fuzzy mining algorithm for discovering useful fuzzy association rules with multiple minimum supports by using maximum constraints. An improved tree-based algorithm named CFP-growth [11] was then proposed to directly mine frequent itemsets with multiple thresholds using the pattern growth method based on a new MIS-Tree structure. An enhanced version of CFP-growth named CFP-growth++ [10] was also proposed, it employs *LMS* (least minimum support) instead of *MIN* to reduce the search space and improve performance. *LMS* is the least *MIS* value amongst all *MIS* values of frequent items. Moreover, three improved strategies were also presented to reduce the search space and runtime. However, it is still too time-consuming and memory cost.

Table 1. An example database

TID	Transaction
T_1	a, c, d
T_2	a, d, e
T_3	b, c
T_4	a, c, e
T_5	a, b, c, d, e
T_6	b, d
T_7	a, b, c, e
T_8	b, c, d
T_9	c, d, e
T_{10}	a, c, d

Table 2. Derived FPs

Itemset	$MIS \times |D|$	sup	Itemset	$MIS \times |D|$	sup
(a)	4	6	(ae)	4	4
(b)	5	5	(bc)	3	4
(c)	3	8	(cd)	3	5
(d)	6	7	(ce)	3	4
(ac)	3	5	(acd)	3	3
(ad)	4	4	(ace)	3	3

3 Preliminaries and Problem Statement

Let $I = \{i_1, i_2, \ldots, i_m\}$ be a finite set of m distinct items. A transactional database $D = \{T_1, T_2, \ldots, T_n\}$, where each transaction $T_q \in D$ is a subset of I, contains several items with their purchase quantities $q(i_j, T_q)$, and has an unique identifier, TID. A corresponding multiple minimum supports table, $MMS\text{-}table = \{ms_1, ms_2, \ldots, ms_m\}$, indicates the user-specified minimum support value ms_j of each item i_j. A set of k distinct items $X = \{i_1, i_2, \ldots, i_k\}$ such that $X \subseteq I$ is said to be a k-itemset. An itemset X is said to be contained in a transaction T_q if $X \subseteq T_q$. An example database is shown in Table 1. It consists of 10 transactions and 5 items, denoted from (a) to (e), respectively. For example, transaction T_1 contains items a, c and d. The minimum support value of each item, denoted as ms, is defined and as shows the $MMS\text{-}table = \{ms(a)\colon 40\%;$ $ms(b)\colon 50\%; ms(c)\colon 30\%; ms(d)\colon 60\%; ms(e)\colon 100\%\}$.

Definition 1. The number of transactions that contains an itemset is known as the occurrence frequency of that itemset. This is also called the support count of the itemset. The support of an itemset X, denoted by $sup(X)$, is the number of transactions containing X w.r.t. $X \subseteq T_q$.

Definition 2. The minimum support threshold of an item i_j in a database D, which is related to $minsup$, is redenoted as $ms(i_j)$ in this paper. A structure called $MMS\text{-}table$ indicates the minimum support threshold of each item in D and is defined as: $MMS\text{-}table = \{ms_1, ms_2, \ldots, ms_m\}$.

Definition 3. The minimum item support value of a k-itemset $X = \{i_1, i_2, \ldots, i_k\}$ in D is denoted as $MIS(X)$, and defined as the smallest ms value for items in X, that is: $MIS(X) = min\{ms(i_j)|i_j \in X\}$.

For example, $MIS(a) = min\{ms(a)\} = 40\%$, $MIS(ae) = min\{ms(a),$ $ms(e)\} = min\{40\%, 100\%\} = 40\%$, and $MIS(ace) = min\{ms(a), ms(c),$ $ms(e)\} = min\{40\%, 30\%, 100\%\} = 30\%$. The extended model enables the user to simultaneously specify high *minsup* for a pattern containing only frequent items and low *minsup* for a pattern containing rare items. Thus, efficiently addressing the "rare item problem".

Definition 4. An itemset X in a database D is called a frequent pattern (FP) iff its support count is no less than the minimum itemset support value of X, such that $sup(X) \geq MIS(X) \times |D|$.

Definition 5. Let be a transactional database D ($|D| = n$) and a *MMS-table*, which defines the minimum support thresholds $\{ms_1, ms_2, \ldots, ms_m\}$ of each item in D. The problem of mining FPs from D with multiple minimum supports (FP-MMS) is to find all itemsets X having a support no less than $MIS(X) \times |D|$.

For the running example, the derived complete set of FPs is shown in Table 2.

4 Proposed FP-ME Algorithm for FP-MMS

4.1 Proposed ME-Tree

Based on the previous studies, the search space of mining frequent patterns with multiple minimum supports can be represented as a lattice structure [12] or a Set-enumeration tree [12], both of them for the running example are respectively shown in Fig. 1. Note that the well-known downward closure property in association rule mining does not hold for FP-MMS. For example, the item (e) is not a FP but its supersets (ae) and (ace) are FPs in the running example. To solve this problem, Liu et al. [7] proposed a concept called *sorted closure property*, which assumes that all items within an itemset are sorted in increasing order of their minimum supports.

Property 1. If a sorted k-itemset $\{i_1, i_2, \ldots, i_k\}$, for $k \geq 2$ and $MIS(i_1) \leq MIS(i_2) \leq \ldots \leq MIS(i_k)$, is frequent, then all of its sorted subsets with $k - 1$ items are frequent, except for the subset $\{i_2, i_3, \ldots, i_k\}$ [6].

Definition 6 (Total order $\prec$ on items). Without loss of generality, assume that items in each transaction of a database are sorted according to the lexicographic order. Furthermore, assume that the total order $\prec$ on items in the designed ME-tree is the ascending order of items MIS values.

Definition 7 (Set-enumeration-tree with multiple minimum supports). A ME-tree is a sorted Set-enumeration tree using the defined total order $\prec$ on items.

Definition 8 (Extension nodes in the ME-tree). In the designed ME-tree with the total order $\prec$, all child nodes of any tree node are called its extension nodes.

Since $MIS(c) < MIS(a) < MIS(b) < MIS(d) < MIS(e)$, the total order $\prec$ on items in the ME-tree is $c \prec a \prec b \prec d \prec e$. The ME-tree for the running example is illustrated in Fig. 2, and the following lemmas are obtained.

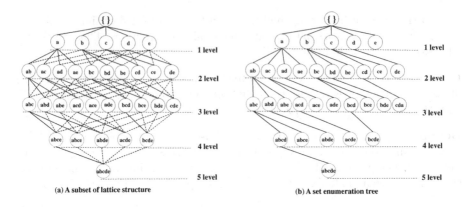

(a) A subset of lattice structure

(b) A set enumeration tree

Fig. 1. The search space presentation of FP-MMS.

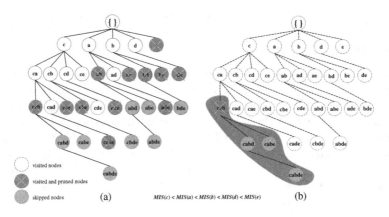

Fig. 2. Applied pruning strategies in the ME-tree.

Lemma 1. *The complete search space of the proposed FP-MMS algorithm can be represented by a ME-tree where items are sorted according to the ascending order of the MIS values on items.*

Lemma 2. *The support of a node in the ME-tree is no less than the support of any of its child nodes (extension nodes).*

Proof. Let X^k be a node in the ME-tree containing X items, and let X^{k-1} be any parent node of X^k containing $(k-1)$ items. It is straightforward from the well-known Apriori property that $sup(X^k) \leq sup(X^{k-1})$, this lemma can be proven.

Lemma 3. *The MIS of a node in the ME-tree equals to the MIS of any of its child nodes (extension nodes).*

4.2   Proposed Pruning Strategies

It is important to note that the *sorted downward closure (SDC) property* of FPs in the ME-tree can only guarantee partial anti-monotonicity for FPs, but not general anti-monotonicity. In other words, the *SDC property* holds for any extensions (child nodes) of a given node, but it may not hold for any supersets of that node. Thus, if the *SDC property* of FPs is used to determine if all supersets of an itemset should be explored, some FPs may not be found. For instance, in the running example database, the item (e) is not a FP since $sup(e) = 5$ (<10), while its supersets (ae), (ce) and (ace) are FPs, as shown in Table 2. It is thus incorrect to directly determine the FPs based only on the proposed *SDC property*. In MSApriori and CFP-growth++, it was shown that the MIN/LMS concept can guarantee the global anti-monotonicity of frequent patterns with multiple minimum supports and ensure the completeness of the set of derived FPs. To address this problem, we further adopt the MIN/LMS concept in the proposed FP-ME algorithm.

Definition 9 (Least minimum support, *LMS*). The least minimum support (LMS) refers to the lowest *minsup* of all frequent patterns. Therefore, the LMS in a database is always equal to the lowest MIS value among all frequent items. Thus, the LMS is equal to the lowest value in the *MMS-table* and is defined as $min\{ms(i_1), ms(i_2), \ldots, ms(i_m)\}$, where m is the total number of items in a database.

For example, the LMS of Table 2 is calculated as $LMS = min\{ms(a), ms(b), ms(c), ms(d), ms(e)\} = min\{40\%, 50\%, 30\%, 60\%, 100\%\} = 30\%$.

Property 2. If $X = \{i_1, i_2, \ldots, i_k\} \subseteq I$, where $1 \le k \le n$, is a pattern such that $sup(X) < LMS$, then $sup(X) < min\{MIS(i_1), MIS(i_2), \ldots, MIS(i_k)\}$, it never could be a frequent pattern.

Property 3. If X and Y are two patterns such that $X \subset Y$ and $sup(X) < LMS$, then $sup(Y) < LMS$. It indicates that LMS guarantees the global anti-monotonicity of frequent patterns with multiple minimum supports.

Proof. Let be an itemset X such that X is a subset of Y. Thus, $sup(Y) \le sup(X)$. The relationship $sup(Y) \le sup(X) < LMS$ holds.

Note that the set of 1-items which having $sup(X) \ge LMS$ is denoted as $LMS\text{-}FP^1$, the following theorem can be obtained.

Theorem 1. *Assume that 1-itemsets which having a MIS lower than LMS are discarded and that the sorted downward closure (SDC) property is applied. We have that if an itemset is not a $LMS\text{-}FP^1$, then it is not a FP as well as all its supersets.*

Proof. Let X^{k-1} be a $(k-1)$-itemset and its superset k-itemset is denoted as X^k. Since $X^{k-1} \subseteq X^k$, (1) For a $LMS\text{-}FP^1$, $sup(X) \ge LMS$; (2) Since items are sorted by ascending order of MIS values, $sup(X^{k-1}) \ge sup(X^k)$

and $MIS(X^{k-1}) = MIS(X^k) = min\{MIS(i_1), MIS(i_2), ..., MIS(i_m)\} = MIS(i_1)$. Thus, if X^{k-1} is not a $LMS\text{-}FP^1$ ($MIS(X^{k-1}) < LMS$), none of its supersets are FPs.

Based on the designed ME-tree, the above lemmas and theorems ensure that all FPs are included in the extensions of the set of $LMS\text{-}FP^1$ and thus that we can safely discard the other itemsets. Thus, the designed *sorted downward closure (SDC) property* in ME-tree can guarantee the completeness and correctness of the proposed FP-MMS framework. These properties facilitate to use LMS as a constraint to reduce the search space. Based on the above analysis, it can be shown that the LMS plays a significant role in the FP-ME algorithm, it not only ensures the completeness of the set of derived FPs, but also can be used to prune the search space.

Strategy 1. *In the designed ME-tree using the total order $\prec$, if a node X has a support value less than the LMS, then any nodes which contains X w.r.t. all supersets of X can be directly pruned, X would not be used to explored in the ME-tree.*

Lemma 4. *Given a transactional database D and multiple minimum support threshold $MIS(i_k)$ of each item i_k, the constructed ME-tree contains the complete information about frequent patterns in D.*

Proof. In the construction process of ME-tree, each transaction in D can be mapped to one path in the ME-tree whenever necessary. And according to Theorems 1 and 2, all promising itemsets, the $LMS\text{-}FP^1$, their information in each transaction is completely stored in the ME-tree based on the total order $\prec$. Notice that we retained those infrequent items with supports no less than LMS w.r.t. the $LMS\text{-}FP^1$ in the ME-tree because these items their supersets may be frequent.

Strategy 2. *Let be the designed ME-tree using the MIS-ascending order of items. If a node X has a support value less than its MIS w.r.t. the MIS of its prefix level-2 node, then any extension of X w.r.t. all child nodes of X can be directly pruned.*

For example, the effect of **Pruning Strategy 1** and **Pruning Strategy 2** in the running example are respectively shown in Fig. 2(a) and (b). The itemset (cab) is not considered to be a FP since $sup(cab)$ ($= 2 < MIS(cab) \times |D|$). By applying the **Pruning Strategy 2**, all the child nodes of itemset (cab) are not considered to be FPs since their support values are always no greater than those of (cab). Hence, the child nodes $(cabd)$, $(cabe)$ and $(cabde)$ (the shaded nodes in Fig. 2) are guaranteed to be uninteresting and can be pruned safely.

4.3   Proposed FP-ME Algorithm

Note that the top-down traversal strategy is adopted in the ME-tree. As shown in Algorithm 1, the FP-ME algorithm first scans the database to calculate the MIS

value of each single item, as well as the LMS, then discover the set of I^* w.r.t. LMS-FP^1 (Lines 1 to 2). It is important to notice that here the derived patterns are the set of LMS-FP^1 but not the set of FPs, the reason had been mentioned before. The discovered LMS-FP^1 are then sorted in MIS-ascending in order to construct their TID-Sets by scanning database again, thus forming the final set of TID-Sets of all item $i \in I^*$ (Lines 3 to 4). Afterwards each 1-itemset X in the set of I^* is processed in designed order $\prec$ to find FPs from the ME-tree, using the constructed TID-Sets without multiple rescaning the database (Line 5, FP-Spanning procedure). As mentioned before, the spanning miner mechanism in the FP-ME is different form the generate-and-test algorithm and the pattern-growth approaches. The main idea of the FP-Spanning procedure (c.f. Algorithm 2) is that for each 1-itemset X_a, extensions of X_a are recursively explored using a depth-first search. Moreover, each itemset encountered during that search is evaluated to determine if it is a defined FP (Lines 2 to 4). Note that the depth search is only performed for an itemset X_a if the support of X_a (which is directly obtained from the relevant TID-Sets by calculating its TIDs) is larger than or equal to $MIS(X_a) \times |D|$ (Lines 3 to 4). Simultaneously, the TID-Sets of the

---

Input: D $(n = |D|)$, MMS-$table$.
Output: The set of frequent patterns (FPs).
1 scan MMS-$table$ to calculate the MIS value of each single item and put into the set of $MISArray$, as well as the LMS among them ;
2 scan D to find $I^* \leftarrow \{i \in I | sup(i) \geq LMS\}$, w.r.t. the LMS-FP^1 ;
3 sort the set I^* in MIS ascending order $\prec$;
4 scan D once again to construct the TID-Set of each $i \in I^*$ such as $i.TidSet$;
5 call **FP-Spanning**$(\phi, I^*, MISArray)$;
6 return FPs;

Algorithm 1: FP-ME algorithm

Input: X: an itemset, $extensionsOfX$: a set of all extensions of X, $MISArray$: an array containing the minimum item support thresholds of all items.
Input: The set of frequent patterns (FPs).
1 **for** each itemset $X_a \in extensionsOfX$ **do**
2 $\quad$ calculate the $sup(X_a)$ and $MIS(X_a)$ from the built structure of X_a;
3 $\quad$ **if** $sup(X_a) \geq MIS(X_a) \times |D|$ **then**
4 $\quad\quad$ $FPs \leftarrow FPs \cup X_a$;
5 $\quad\quad$ $extensionsOfX_a \leftarrow \emptyset$;
6 $\quad\quad$ **for** each itemset $X_b \in extensionsOfX_a$ such that b after a **do**
7 $\quad\quad\quad$ $X_{ab} \leftarrow X_a \cup X_b$;
8 $\quad\quad\quad$ calculate the $X_{ab}.TidSet$ by merging $X_a.TidSet$ and $X_b.TidSet$;
9 $\quad\quad\quad$ $extensionsOfX_a \leftarrow extensionsOfX_a \cup X_{ab}$;
10 $\quad\quad$ call **FP-Spanning**$(X_a, extensionsOfX_a, MIS(X_a))$;
11 return FPs;

Algorithm 2: FP-Spanning Procedure

extensions of X_a is built (Lines 6 to 9). The above process is recursively executed until all items in $LMS\text{-}FP^1$ have been processed (Line 10).

4.4   Improved Algorithm with the *DiffSet* Strategy

In [13], a novel vertical data representation called *DiffSet* was presented, that only keeps track of differences in the TIDs of a candidate pattern from its generating frequent patterns. It had been shown that *DiffSet* drastically cut down the size of memory required to store intermediate results. Furthermore, we incorporated the *DiffSet* into previous vertical mining method FP-ME, significantly increasing the mining performance, and denoted this improved algorithm as FP-ME$_{\text{DiffSet}}$, for fast mining FPs. Details of the concept *DiffSet* and its construction in the FP-ME$_{\text{DiffSet}}$ algorithm were skipped here due to the space limitation.

5   Experimental Results

In this section, substantial experiments were conducted to verify the effectiveness and efficiency of the proposed baseline FP-ME algorithm (FP-ME$_{\text{baseline}}$) and the improved FP-ME$_{\text{DiffSet}}$ algorithm. Note that there are several studies have been done previously on the topic of mining FPs with multiple minimum supports, and it had been shown that the CFP-growth++ [10] significantly outperforms the MSApriori [7] and CFP-growth [11]; the state-of-the-art CFP-growth++ was thus executed to derive FPs, which can provide a benchmark to verify the efficiency of the proposed two algorithms. All algorithms are implemented in Java language and performed on a personal computer with an Intel Core i5-3460 dual-core processor and 4 GB of RAM and running the 32-bit Microsoft Windows 7 operating system. And experiments are conducted on two datasets named kosarak [1] and BMSPOS [1]. The source code of the MSApriori and CFP-growth++ algorithms can be download from the SPMF data mining library [5]. Note that the front 100 K transactions in kosarak are selected in the experiments.

Furthermore, the discussed method in [7] is adopt in the proposed FP-ME algorithm to automatically assign multiple item supports to items. The methodology is as follows: $MIS(i_j) = max[\beta \times f(i_j), LMS]$, where β is a constant used to set the *MIS* values of items as a function of their frequency (or support). To ensure the randomness and equipment diversity, β was set in the [0.0, 1.0] interval for the datasets. The parameter *LMS* is the user-specified least minimum itemset support allowed, and $f(i_j)$ is the frequency (or support) of an item i_j. Note that if β is set to zero, then a single *LMS* value will be used for all items, and this will be equivalent to traditional FPM. If $\beta = 1$ and $f(i_j) \geq LMS$, then $MIS(i_j) = f(i_j)$.

5.1   Execution Time

For the conducted experiments, the parameter β was randomly set to a fixed number of each item. Fig. 3 shows the runtime of the algorithms under various

LMS with a fixed β and under various β with a fixed *LMS* for different datasets. From Fig. 3, it can be observed that both the FP-ME$_{baseline}$ and the improved FP-ME$_{DiffSet}$ algorithms outperform the CFP-growth++ algorithm on the two datasets under various *LMS* with a fixed β or under a fixed *LMS* with various β. Moreover, we can see that FP-ME$_{DiffSet}$ has in general the best performance among them. It is reasonable since CFP-growth++ consists of three phases. It first scans the database once to construct a global MIS-Tree. All nodes in the Header-table of a MIS-Tree are sorted by the order of descending MIS values. Then, the MIS-Tree is restructured to reduce the search space by four pruning techniques. At last, CFP-growth++ recursively mines the tree by creating projected trees to generate all desired itemsets. This process is too time-consuming, thus it performs worse than the two proposed FP-ME algorithms. In contrast, the proposed algorithms utilize the "structuring when mining" property, and apply two pruning strategies to early prune unpromising itemsets and search space in ME-tree, which can avoid the costly join operations of a huge number of unpromising patterns. Moreover, the FP-ME$_{DiffSet}$ algorithm applies the *Diff-Set* to quickly calculate the TID-Set, thus greatly reducing the computations than the baseline algorithm.

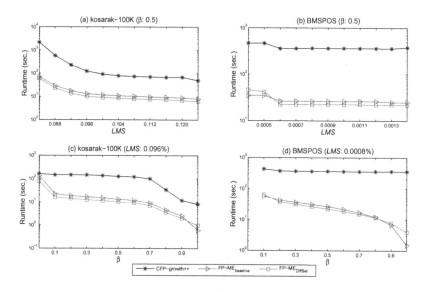

Fig. 3. Execution time.

5.2   Memory Usage

With the same test parameters, we also assessed the memory consumption of the compared algorithms. Memory measurements were done using the Java API. Note that the peak memory consumption of each algorithm was recorded for all datasets. Results are shown in Fig. 4. It can be clearly seen that the proposed algorithms, both FP-ME$_{baseline}$ and the improved FP-ME$_{DiffSet}$, require less memory compared to the state-of-the-art CFP-growth++ algorithm under various parameters on the two datasets, and even up to 8 times as shown in Fig. 4(a). Specifically, the two ME-tree-based approaches require nearly constant memory under various parameter values on the all datasets. The memory usage of the generation-and-test CFP-growth++ algorithm dramatically increases as LMS or β decreases, while the memory usage of the proposed algorithms remain stable. Besides, the improved FP-ME$_{DiffSet}$ always consumes little more memory than that of FP-ME$_{baseline}$. This result is reasonable since the vertical data structure $DiffSet$ is adopted to keep track of differences in the TIDs of a prefix pattern from its generating FPs.

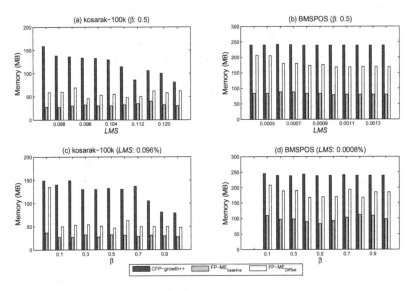

Fig. 4. Memory usage. (Color figure online)

5.3 Scalability Analysis

The scalability of the proposed methods are further evaluated by performing experiments when the kosarak dataset size was varied from 100 K to 500 K, by increments of 100 K each time. Figure 5 shows the runtime and memory usage for the three compared algorithms when the LMS and β were set to 0.001 and 0.5, respectively. It can be observed that the runtime of all compared algorithms is linear increased along with the increasing of dataset size $|X|$. The performance of P-ME$_{baseline}$ and the improved FP-ME$_{DiffSet}$ significantly scale better than that of CFP-growth++. With the increasing of the size of dataset, the runtime of FP-ME$_{baseline}$ is close to that of FP-ME$_{DiffSet}$. Specially, the gap of runtime among them grows wider with the increasing of dataset size. It also can be clearly seen that the proposed two algorithms require less memory compared to the state-of-the-art CFP-growth++ algorithm in a wide range of dataset size. From the observed results of the scalability test, it can be concluded that the proposed algorithms are more scalable than the previous algorithms.

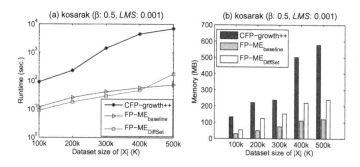

Fig. 5. Scalability of compared algorithms. (Color figure online)

6   Conclusion

In this paper, we proposed a novel FP-MMS framework namely FP-ME for mining frequent patterns with multiple minimum supports. Based on the designed Set-enumeration-tree with multiple minimum supports (ME-tree), a novel *sorted downward closure (SDC) property* was proposed. In order to guarantee the completeness of derived results, the *LMS* concept was extend in FP-ME to mine the FPs. Different from the generate-and-test approach, FP-ME can directly discover FPs by spanning the ME-tree with two pruning strategies. In addition, an improved algorithm by adopting the *DiffSet* concept is further developed to speed up the mining process by reducing the cost of database scans and pruning search space. From the experiments, it can be found that the proposed two algorithms significantly outperform the state-of-the-art CFP-growth++ algorithm in terms of execution time, memory usage and scalability. Specifically, the improved algorithm outperforms the baseline algorithm.

Acknowledgment. This research was partially supported by the National Natural Science Foundation of China (NSFC) under grant No. 61503092 and by the Tencent Project under grant CCF-TencentRAGR20140114.

References

1. Frequent itemset mining dataset repository. http://fimi.ua.ac.be/data/
2. Agrawal, R., Imielinski, T., Swami, A.: Database mining: A performance perspective. IEEE Trans. Knowl. Data Eng. **5**, 914–925 (1993)
3. Agrawal, R., Srikant, R.: Fast algorithms for mining association rules in large databases. In: The International Conference on Very Large Data Bases, pp. 487–499 (1994)
4. Han, J., Pei, J., Yin, Y., Mao, R.: Mining frequent patterns without candidate generation: A frequent-pattern tree approach. Data Min. Knowl. Disc. **8**(1), 53–87 (2004)
5. Fournier-Viger, P., Gomariz, A., Soltani, A., Gueniche, T., Wu, C.W., Tseng, V.S.: SPMF: A java open-source pattern mining library. J. Mach. Learn. Res. **15**, 3389–3393 (2014)

6. Huang, T.C.K.: Discovery of fuzzy quantitative sequential patterns with multiple minimum supports and adjustable membership functions. Inf. Sci. **222**, 126–146 (2013)
7. Liu, B., Hsu, W., Ma, Y.: Mining association rules with multiple minimum supports. In: ACM SIGKDD International Conference on Knowledge Discovery and Data Mining, pp. 337–341 (1999)
8. Liu, Y.C., Cheng, C.P., Tseng, V.S.: Discovering relational-based association rules with multiple minimum supports on microarray datasets. Bioinformatics **27**(22), 3142–3148 (2011)
9. Lee, Y.C., Hong, T.P., Wang, T.C.: Mining fuzzy multiple-level association rules under multiple minimum supports. In: IEEE International Conference on Systems, Man and Cybernetics, pp. 4112–4117 (2006)
10. Kiran, R.U., Reddy, P.K.: Novel techniques to reduce search space in multiple minimum supports-based frequent pattern mining algorithms. In: ACM International Conference on Extending Database Technology, pp. 11–20 (2011)
11. Hu, Y.H., Chen, Y.L.: Mining association rules with multiple minimum supports: A new mining algorithm and a support tuning mechanism. Decis. Support Syst. **42**(1), 1–24 (2006)
12. Rymon, R.: Search through systematic set enumeration. Technical reports (CIS), vol. 297 (1992)
13. Zaki, M.J., Gouda, K.: Fast vertical mining using diffsets. In: ACM SIGKDD International Conference on Knowledge Discovery and Data Mining, pp. 326–335 (2003)

Efficient Mining of Uncertain Data for High-Utility Itemsets

Jerry Chun-Wei Lin[1]([✉]), Wensheng Gan[1], Philippe Fournier-Viger[2],
Tzung-Pei Hong[3,4], and Vincent S. Tseng[5]

[1] School of Computer Science and Technology,
Harbin Institute of Technology Shenzhen Graduate School, Shenzhen, China
jerrylin@ieee.org, wsgan001@gmail.com
[2] School of Natural Sciences and Humanities,
Harbin Institute of Technology Shenzhen Graduate School, Shenzhen, China
phifv@hitsz.edu.cn
[3] Department of Computer Science and Information Engineering,
National University of Kaohsiung, Kaohsiung, Taiwan
tphong@nuk.edu.tw
[4] Department of Computer Science and Engineering,
National Sun Yat-sen University, Kaohsiung, Taiwan
[5] Department of Computer Science, National Chiao Tung University,
Hsinchu, Taiwan
vtseng@cs.nctu.edu.tw

Abstract. High-utility itemset mining (HUIM) is emerging as an important research topic in data mining. Most algorithms for HUIM can only handle precise data, however, uncertainty that are embedded in big data which collected from experimental measurements or noisy sensors in real-life applications. In this paper, an efficient algorithm, namely Mining Uncertain data for High-Utility Itemsets (MUHUI), is proposed to efficiently discover potential high-utility itemsets (PHUIs) from uncertain data. Based on the probability-utility-list (PU-list) structure, the MUHUI algorithm directly mine PHUIs without candidate generation and can reduce the construction of PU-lists for numerous unpromising itemsets by using several efficient pruning strategies, thus greatly improving the mining performance. Extensive experiments both on real-life and synthetic datasets proved that the proposed algorithm significantly outperforms the state-of-the-art PHUI-List algorithm in terms of efficiency and scalability, especially, the MUHUI algorithm scales well on large-scale uncertain datasets for mining PHUIs.

Keywords: Data mining · Uncertainty · High-utility itemset · PU-list · Pruning strategies

1 Introduction

Knowledge Discovery in Database (KDD) aims at finding the meaningful and useful information from the amounts of mass data [4,10,11]. Depending on

© Springer International Publishing Switzerland 2016
B. Cui et al. (Eds.): WAIM 2016, Part I, LNCS 9658, pp. 17–30, 2016.
DOI: 10.1007/978-3-319-39937-9_2

different requirements in various domains and applications, frequent itemset mining (FIM) [11] and association rule mining (ARM) [4] are the important and fundamental issues in KDD. Instead of traditional FIM and ARM, high-utility itemset mining (HUIM) [8,21] incorporates both quantity and profit values of an item/set to measure how "useful" an item or itemset is. An itemset is defined as a high-utility itemset (HUI) if its utility value in a database is no less than a user-specified minimum utility count. In general, the "utility" of an item/set can be presented as the user-specified factor in real-life applications, i.e., weight, cost, risk, or unit profit. Chan et al. [8] first proposed the concept of HUIM. Yao et al. [21] then defined a unified framework for mining high-utility itemsets (HUIs). The goal of HUIM is to identify the rare items or itemsets in the transactions, but it can bring valuable profits for the retailers or managers. HUIM serves as a critical role in data analysis and has been widely utilized to discover knowledge and mine valuable information. Many approaches developed in HUIM have been extensively studied, such as Two-Phase [15], IHUP [6], UP-growth [18], UP-growth+ [19], HUI-Miner [14], FHM [17], among others. In real-world situations, the mass data may be uncertainly collected from the incomplete data sources, such as wireless sensor network, RFID, GPS, or WiFi systems [2,3], and the size of the collected uncertain data is very large. Traditional data mining technologies for handling precise data cannot be directly, however, applied to the incomplete or inaccurate data for discovering the required information on business service.

In real-life applications, utility and probability are two different measures for an object (e.g., an useful pattern). The utility is a semantic measure (how "utility" of a pattern is based on the user's priori knowledge and goals), while probability is an objective measure (the probability of a pattern is an objective existence) [10]. Up to now, most algorithms of HUIM have been extensively developed to handle precise data, which are not suitable to mine the data with uncertainty. It may be useless or misleading if the discovered results of HUIs with low existential probability. To the best of our knowledge, the PHUIM framework [13] is the first work to address the issue of mining HUIs from uncertain data. However, the task of mining HUIs from uncertain data, especially large-scale uncertain data, remains very costly in terms of execution time. Therefore, it is a non-trivial task and an important challenge to design more efficient algorithms to solve the limitation. In this paper, an efficient mining model namely Mining large-scale Uncertain data for High-Utility Itemsets (MUHUI) is proposed to effectively discover the Potential High-Utility Itemsets (PHUIs). Major contributions of this paper are summarized as follows:

- Fewer studies on HUIM have addressed the mining of HUIs from uncertain data by taking into account both the semantics-based utility measure and objective probability measure. In this paper, an efficient MUHUI algorithm is designed to successfully and directly mine the HUIs from uncertain databases without candidate generation.
- Based on the utility and probability properties, several pruning strategies are developed to efficiently and early prune the search space and the unpromising

itemsets. Thus, the search space for mining HUIs can be greatly reduced, and the mining performance can be significantly improved.

– Extensive experiments show that the proposed algorithm is more efficient than the state-of-the-art PHUI-List algorithm for mining uncertain HUIs in terms of runtime, effectiveness of prune strategies and scalability, especially MUHUI scales well than PHUI-List in large-scale uncertain databases.

2   Related Works

HUIM is different from FIM and ARM, which incorporates the measurements of local transaction utility (occur quantity) and external utility (unit profit) to discover the profitable itemsets from the quantitative databases. The HUIM was first proposed by Chan et al. [8]. Yao et al. [21] then defined a strict unified framework for HUIM. Since the downward closure property of ARM does no longer hold for HUIM, Liu et al. [15] designed the TWU model to maintain the transaction-weighted downward closure (TWDC) property, which can be used to greatly reduce the number of unpromising candidates for mining HUIs in a level-wise mechanism. Several tree-based approaches for mining HUIs such as IHUP [6], UP-growth [18], and UP-growth+ [19] have been extensively studied. Based on these pattern-growth approaches, more computations are still required to generate and keep the huge number of discovered candidates for mining the actual HUIs.

To solve the above limitations of traditional HUIM, the HUI-Miner algorithm [14] was proposed to directly mine HUIs to avoid the multiple database scans without candidate generation based on the designed utility-list structure. The FHM algorithm [17] was further proposed to enhance the performance of HUI-Miner by analyzing the co-occurrences among 2-itemsets. Instead of traditional HUIM, the variants of HUIM have been also extended and developed [12,20]. The development of other algorithms for HUIM is still in progress, but most of them are processed to handle precise data, the PHUIM framework [13] is the only work which focuses on mining high-utility itemsets on uncertain data.

3   Preliminaries and Problem Statement

3.1   Preliminaries

Based on the mentioned reason in [13], the tuple uncertainty model [3,7] and the expected support-based model [9] are also adopted in the proposed algorithm. Let $I = \{i_1, i_2, \ldots, i_m\}$ be a finite set of m distinct items in an uncertain quantitative database $D = \{T_1, T_2, \ldots, T_n\}$, where each transaction $T_q \in D$ is a subset of I, contains several items with their purchase quantities $q(i_j, T_q)$, and has an unique identifier, TID. In addition, each transaction has a unique probability of existence $p(T_q)$, which indicates that T_q exists in D with probability $p(T_q)$ based on a tuple uncertainly model. A corresponding profit table,

Table 1. An example database

TID	Transaction	Probability
T_1	$(A,1); (C,3); (D,4)$	0.95
T_2	$(B,1); (C,1); (D,2)$	0.85
T_3	$(A,2); (B,2); (E,3)$	0.85
T_4	$(C,2); (E,2)$	0.5
T_5	$(B,1); (D,2); (E,2)$	0.75
T_6	$(A,1); (C,2); (D,1)$	0.7
T_7	$(A,3); (C,1); (D,3); (E,4)$	0.45
T_8	$(B,1); (C,4); (E,1)$	0.36
T_9	$(B,3); (D,5)$	0.81
T_{10}	$(D,5); (E,2)$	0.6

Table 2. Derived PHUIs

Itemset	Utility	Pro
(C)	130	5.26
(E)	70	4.61
(AC)	90	2.75
(CD)	80	3.55
(CE)	105	2.51
(DE)	50	2.40
(ACD)	98	2.75

$ptable = \{pr_1, pr_2, \ldots, pr_m\}$, in which pr_j is the profit value of an item i_j, is created. A k-itemset X, denoted as X^k, is a set of k distinct items $\{i_1, i_2, \ldots, i_k\}$. Given two user-specified thresholds, the minimum utility threshold (ε) and the minimum potential probability threshold (μ).

Table 1 shows an example of a tuple uncertainty quantitative database. The profit table is defined as $\{pr(A): 6; pr(B): 3; pr(C): 10; pr(D): 1; pr(E): 5\}$, and the two thresholds are respectively set at $\varepsilon \ (= 15\%)$ and $\mu \ (= 18\%)$.

Definition 1. The utility of an item i_j in a transaction T_q is denoted as $u(i_j, T_q)$ and defined as: $u(i_j, T_q) = q(i_j, T_q) \times pr(i_j)$.

For example, $u(A, T_1) = q(A, T_1) \times pr(A) = (1 \times 6) = 6$.

Definition 2. The probability of an itemset X occurring in T_q is denoted as $p(X, T_q)$, which can be defined as: $p(X, T_q) = p(T_q)$, where $p(T_q)$ is the corresponding probability of T_q.

For example, $p(A, T_1) = p(T_1) = 0.95$, and $p(AD, T_1) = p(T_1) = 0.95$.

Definition 3. The utility of an itemset X in transaction T_q is denoted as $u(X, T_q)$, which can be defined as: $u(X, T_q) = \sum\limits_{i_j \in X \wedge X \subseteq T_q} u(i_j, T_q)$.

For example, the utility of (AD) in T_1 is calculated as $u(AD, T_1) = u(A, T_1) + u(D, T_1) = q(A, T_1) \times pr(A) + q(D, T_1) \times pr(D) = (1 \times 6) + (4 \times 1) = 10$.

Definition 4. The utility of an itemset X in D is denoted as $u(X)$, which can be defined as: $u(X) = \sum_{X \subseteq T_q \wedge T_q \in D} u(X, T_q)$.

Definition 5. An itemset X is defined as a HUI if its utility value is no less than the minimum utility count as: $u(X) \geq \varepsilon \times TU$, in which TU is the total utility of the be processed database.

Definition 6. The potential probability of an itemset X in D is denoted as $Pro(X)$, which can be defined as: $Pro(X) = \sum_{X \subseteq T_q \wedge T_q \in D} p(X, T_q)$.

Definition 7. An itemset X in an uncertain database D is defined as a potential high-utility itemset (PHUI) if it satisfies the following two conditions: (1) X is a HUI w.r.t. $u(X) \geq \varepsilon \times TU$; (2) $Pro(X) \geq \mu \times |D|$. A desired PHUI indicates the itemset has both high potential probability and high utility value.

Problem Statement: Given an uncertain database D with total utility is TU, the minimum utility threshold and the minimum potential probability threshold are respectively set as ε and μ. The problem of potential high-utility itemset mining (PHUIM) from uncertain data is to mine PHUIs whose utilities are larger than or equal to $(\varepsilon \times TU)$, and its potential probability is larger than or equal to $(\mu \times |D|)$.

4   Proposed MUHUI Algorithm for Mining HUIs

Alougth the PHUI-List algorithm has better performance compared to the upper-bound-based PHUI-UP algorithm [13], however, it explores the search space of itemsets by generating itemsets, and a costly join operation of probability-utility-list (PU-list) has to be performed recursively to evaluate the probability and utility information of each itemset. By utilizing the PU-list structure, a more efficient MUHUI algorithm is proposed here to improve the performance for efficiently mining PHUIs.

4.1   The PU-list Structure

The PU-list structure [13] is a new vertical data structure, it incorporates the probability and utility properties to keep necessary information from uncertain data in terms of TID information, probability, utility, and remaining utility information. Let an itemset X and a transaction (or itemset) T such that $X \subseteq T$, the set of all items from T that are not in X is denoted as $T \backslash X$, and the set of all the items appearing after X in T is denoted as T/X. Thus, $T/X \subseteq T \backslash X$. For example, consider $X = \{CD\}$ and transaction T_7 in Table 1, $T_7 \backslash X = \{AE\}$, and $T_7/X = \{E\}$.

Definition 8 (Probability-Utility-list, PU-list). The PU-list of an itemset X in a database is denoted as $X.PUL$. It contains an entry (element) for each transaction T_q where X appears ($X \in T_q \subseteq D$). An element consists of four fields: (1) the **tid** of X in T_q ($X \subseteq T_q \in D$); (2) the probabilities of X in T_q (**prob**); (3) the utilities of X in T_q (**iu**); and (4) the remaining utilities of X in T_q (**ru**), in which **ru** is defined as $X.ru(T_q) = \sum_{i_j \in (T_q/X)} u(i_j, T_q)$.

Therefore, all necessary information from uncertain data can be compressed into the designed PU-list structure without losing any useful information. Thanks to the property of PU-list, the probability and utility information of the longer

k-itemset can be built by joining its parent node and uncle node, i.e., $(k\text{-}1)$-itemset. The join operation can be easily done without rescanning the database. The construction procedure of the PU-list is recursively processed if it is necessary to determine the k-itemsets in the search space, details of the construction can be referred to [13]. Note that it is necessary to initially construct the PU-list of the complete set of HTWPUI[1] [13] as the input for the later recursive process. The PU-list is constructed in TWU ascending order as $(B \prec A \prec D \prec E \prec C)$, which is shown in Fig. 1.

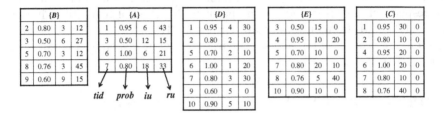

Fig. 1. Constructed PU-list structure of HTWPUI[1].

Definition 9. The sum of the utilities and remaining utilities of an itemset X in D, denoted as $X.IU$ and $X.RU$, respectively, which can be defined as:

$$X.IU = \sum_{X \subseteq T_q \wedge T_q \in D} (X.iu), X.RU = \sum_{X \subseteq T_q \wedge T_q \in D} (X.ru). \tag{1}$$

4.2 Search Space and Properties

Based on the PU-list structure, the search space of the proposed MUHUI algorithm can be represented as the Set-enumeration tree by the TWU values of the 1-items in the set of HTWPUI[1] in ascending order, as shown in Fig. 2 (left). Based on the constructed Set-enumeration tree, the following lemmas can be obtained.

Lemma 1. *The sum of all the probabilities of any node in the Set-enumeration tree is greater than or equal to the sum of all the probabilities of any of its child nodes.*

Proof. Assume a $(k\text{-}1)$-itemset w.r.t. a node in the Set-enumeration tree be $X^{k-1}(k \le 2)$, and any of its child nodes be denoted as X^k. Since $p(X^k, T_q) = p(T_q)$ for any transaction T_q in D, it can be found that: $\frac{p(X^k, T_q)}{p(X^{k-1}, T_q)} = \frac{p(T_q)}{p(T_q)} = 1$. Since X^{k-1} is subset of X^k, the $TIDs$ of X^k is the subset of the $TIDs$ of X^{k-1}, thus,

$$Pro(X^k) = \sum_{X^k \subseteq T_q \wedge T_q \in D} p(X^k, T_q) \le \sum_{X^{k-1} \subseteq T_q \wedge T_q \in D} p(X^{k-1}, T_q) = Pro(X^{k-1}).$$

Lemma 2. *For any node X in the Set-enumeration tree, the sum of X.IU and X.RU is greater than or equal to the sum of all the utilities of any one of its child nodes.*

Proof. From [13], this lemma holds.

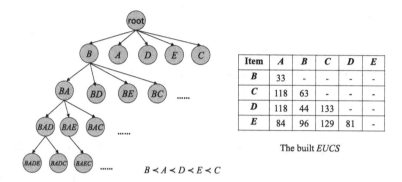

Item	A	B	C	D	E
B	33	-	-	-	-
C	118	63	-	-	-
D	118	44	133	-	-
E	84	96	129	81	-

The built *EUCS*

$$B \prec A \prec D \prec E \prec C$$

Fig. 2. Constructed Set-enumeration tree and *EUCS*.

4.3 Proposed Pruning Strategies

Based on the PU-list and the properties of probability and utility, five efficient pruning strategies are designed in MUHUI to early prune unpromising itemsets.

Theorem 1 (Downward Closure Property of HTWPUI). *Let X^k and X^{k-1} be the HTWPUI from uncertain databases, and $X^{k-1} \subseteq X^k$. The $TWU(X^{k-1}) \geq TWU(X^k)$ and $Pro(X^{k-1}) \geq Pro(X^k)$.*

Proof. Let X^{k-1} be a $(k-1)$-itemset and its superset k-itemset is denoted as X^k, then

$$TWU(X^k) = \sum_{X^k \subseteq T_q \wedge T_q \in D} tu(T_q) \leq \sum_{X^{k-1} \subseteq T_q \wedge T_q \in D} tu(T_q) = TWU(X^{k-1}).$$

From Lemma 1, it can be found that $Pro(X^{k-1}) \geq Pro(X^k)$. Therefore, if X^k is a HTWPUI, any its subset X^{k-1} is also a HTWPUI.

Theorem 2 (PHUIs $\subseteq$ HTWPUIs). *The transaction-weighted probability and utilization downward closure (TWPUDC) property ensures that PHUIs $\subseteq$ HTWPUIs, which indicates that if an itemset is not a HTWPUI, then none of its supersets will be PHUIs [13].*

By utilizing the TWPUDC property, we only need to construct the PU-list for those promising itemsets w.r.t. the HTWPUIs. Furthermore, the following several pruning strategies are proposed in MUHUI to speed up the computations.

Strategy 1. *After the first database scan, we can obtain the TWU and probability value of each 1-item in database. If the TWU of a 1-item i (TWU(i)) and the sum of all the probabilities of i (Pro(i)) do not satisfy the two conditions of HTWPUI, this item can be directly pruned, and none of its supersets is a desired PHUI.*

Strategy 2. *When traversing the Set-enumeration tree based on a depth-first search strategy, if the sum of all the probabilities of a tree node X w.r.t. Pro(X) in its constructed PU-list is less than the minimum potential probability, then none of the child nodes of this node is a desired PHUI.*

Strategy 3. *When traversing the Set-enumeration tree based on a depth-first search strategy, if the sum of X.IU and X.RU of any node X is less than the minimum utility count, any of its child node is not a PHUI, they can be regarded as irrelevant and be pruned directly.*

Theorem 3 (Estimated Utility Co-occurrence Pruning strategy, EUCP). *If the TWU of 2-itemset is less than the minimum utility count, any superset of this 2-itemset is not a HTWUI and would not be a HUI either [17].*

To effectively apply the EUCP strategy, a structure named Estimated Utility Co-occurrence Structure (*EUCS*) [17] is built in the proposed algorithm. It is a matrix that stores the *TWU* values of the 2-itemsets, as shown in Fig. 2 (right). Note that *EUCS* is built in the first database scan after discovering the $HTWPUI^1$.

Strategy 4. *Let X be an itemset (node) encountered during the depth-first search of the Set-enumeration tree. If the TWU of a 2-itemset $Y \subseteq X$ according to the constructed EUCS is less than the minimum utility threshold, X is not a HTWPUI and would not be a PHUI; none of its child nodes is a PHUI. The construction of the PU-lists of X and its children is unnecessary to be performed.*

Strategy 5. *Let X be an itemset (node) encountered during the depth-first search of the Set-enumeration tree. After constructing the PU-list of an itemset, if X.PUL is empty or the Pro(X) value is less than the minimum probability threshold, X is not a PHUI, and none of X its child nodes is a PHUI. The construction of the PU-lists for the child nodes of X is unnecessary to be performed.*

4.4   Proposed MUHUI Algorithm

As shown in the main procedure of MUHUI (Algorithm 1), it first scans the uncertain database to calculate the $TWU(i)$ and $Pro(i)$ values of each item $i \in I$ (Line 1), and then find the set of I^* w.r.t. $HTWPUI^1$ (Line 2, pruning Strategy 1). After sort I^* in TWU ascending order (Line 3), the MUHUI algorithm scans D again to construct the PU-list for each 1-item $i \in I^*$, and build the *EUCS* structure (Line 4). After that, recursively using a depth-first search procedure PHUI-Search (Line 5) to mine PHUIs. Details of the PHUI-Search procedure are shown in Algorithm 2.

Input: D, *ptable*, ε, μ.
Output: The set of potential high-utility itemsets (PHUIs).
1 scan D to calculate the $TWU(i)$ and $Pro(i)$ of each item $i \in I$;
2 find $I^* \leftarrow \{i \in I | TWU(i) \geq TU \times \varepsilon \wedge Pro(i) \geq |D| \times \mu\}$;   /* **strategy 1***/ ;
3 sort I^* in TWU ascending order $\prec$;
4 scan D once again to construct the PU-list of each $i \in I^*$ and build the $EUCS$;
5 call **PHUI-Search**$(\phi, I^*, EUCS, \varepsilon, \mu)$;
6 return $PHUIs$;

Algorithm 1. MUHUI algorithm

The PHUI-Search procedure takes the inputs as: X, $extendOfX$, $EUCS$, ε and μ. Firstly, each itemset X_a is determined to directly produce the PHUIs (Lines 2 to 4). Two pruning strategies, both Strategy 2 and Strategy 3, are applied to further determine whether its extensions satisfy the PHUI conditions for executing the later depth-first search (Line 5). Before executes the PU-list construction procedure to build PU-lists for itemsets with prefix itemset, the MUHUI algorithm utilizes the EUCP strategy to check whether those itemsets are need to build PU-list or not (Line 8, pruning Strategy 4). If X_a is promising, the $Construct(X, X_a, X_b)$ is executed to construct a set of PU-list of all 1-extensions of itemset X_a (w.r.t. $extendOfXa$) (Lines 8 to 12). Note that each constructed X_{ab} is a 1-extension of itemset X_a (Line 10), if the built PU-list of X_{ab} satisfies the pruning Strategy 5, X_{ab} should be put into the set of $extendOfXa$ for executing the later depth-first search; otherwise, X_{ab} would be directly pruned (Lines 11 to 12). The designed PHUI-Search procedure is recursively processed to mine PHUIs (Line 13). Based on the above pruning strategies, the designed MUHUI algorithm can prune the itemsets with lower potential probability and utility count early, without constructing their PU-lists of extensions.

5 Experiments

We performed extensive experiments to evaluate the proposed MUHUI algorithm. Note that the PHUI-UP and PHUI-List algorithms are the first work focus on mining uncertain data for HUIs, and the state-of-the-art PHUI-List algorithm significantly outperforms the PHUI-UP algorithm [13]. Only the PHUI-UP and PHUI-List algorithms for mining PHUIs are compared against the proposed MUHUI algorithm, in terms of runtime, memory usage, the effect of different pruning strategies, and scalability. Note that the MUHUI2 algorithm adopts the all designed pruning strategies, the MUHUI1 algorithm does not use the pruning Strategy 5.

In order to perform a fair comparison, all algorithms used in the experiments are also implemented in Java language and performed on a personal computer with 4 GB of RAM and running the 32-bit Microsoft Windows 7 operating system. And experiments are also conducted on four datasets including both real-world datasets (foodmart [16], accident [1], and retail [1]), and synthetic dataset

---

Input: X: an itemset, $extendOfX$: a set of all extensions of X, $EUCS$, ε, μ.
Input: The set of potential high-utility itemsets (PHUIs)
1 **for** *each itemset* $X_a \in extendOfX$ **do**
2   obtain the $X_a.IU$, $X_a.RU$ and $Pro(X_a)$ values from the built $X_a.PUL$;
3   **if** $X_a.IU \geq TU \times \varepsilon \wedge Pro(X) \geq |D| \times \mu$ **then**
4          $PHUIs \leftarrow PHUIs \cup X_a$;

5   **if** $X_a.IU + X_a.RU \geq TU \times \varepsilon \wedge Pro(X_a) \geq |D| \times \mu$ **then**
6          $extendOfX_a \leftarrow \emptyset$;
7   **for** *each itemset* $X_b \in extendOfX_a$ *such that* X_b *after* X_a **do**
8   **if** $\exists\, TWU(ab) \in EUCS \wedge TWU(ab) \geq TU \times \varepsilon$ **then**
9                  $X_{ab} \leftarrow X_a \cup X_b$;
10                  $X_{ab}.PUL \leftarrow construct(X, X_a, X_b)$;
11   **if** $X_{ab}.PUL \neq \emptyset \wedge Pro(X_{ab}) \geq |D| \times \mu$ **then**
12                      $extendOfX_a \leftarrow extendOfX_a \cup X_{ab}$;

13   call **PHUI-Search**$(X_a, extendOfX_a, EUCS, \varepsilon, \mu)$;

14 **return** PHUIs;

Algorithm 2. PHUI-Search Procedure

(T10I4D100K) generated using the IBM Quest Synthetic Data Generator [5]. Both the quantity (internal) and profit (external) values are assigned to the items in the test datasets by the simulation method in previous studies [13,15,19] except for foodmart dataset. In addition, due to the tuple uncertainty property, each transaction in these datasets is randomly assigned a unique probability value in the range of (0.0, 1.0].

5.1 Execution Time

Experiments are compared under varied minimum utility thresholds (abbreviated as MUs) with the fixed minimum potential probability threshold (abbreviated as MP). The runtime results under varied MUs with a fixed MP are shown in Fig. 3.

From Fig. 3, it can be observed that the runtime of all the algorithms is decreased along with the increasing of MU. In particular, the proposed MUHUI algorithm is generally up to almost one or two orders of magnitude faster than the PHUI-UP algorithm, and also outperforms the state-of-the-art PHUI-List algorithm on all datasets. It is reasonable since the upper-bound-based generate-and-test mechanism has worse results than the vertical PU-list-based approaches. Besides, the MUHUI algorithm uses five pruning strategies to early prune unpromising itemsets and search space, which can avoid the costly join operations of a huge number of PU-lists for mining PHUIs, but the PHUI-List. When the MU is set quite low, longer patterns of HTWPUIs are first discovered by the PHUI-UP algorithm, and thus more computations are needed to process with the generate-and-test mechanism, especially in a dense dataset. While the

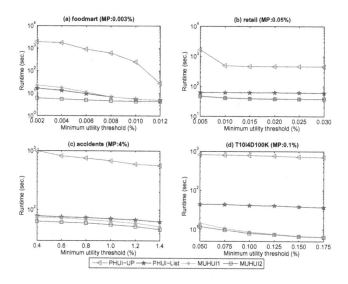

Fig. 3. Runtime under varied MUs with a fixed MP (Color figure online).

PHUI-List, MUHUI1 and MUHUI2 algorithms directly determine the PHUIs from the Set-enumeration tree without candidate generation in a level-wise way, it can effectively avoid the time-consuming dataset scan. Moreover, the MUHUI algorithm applies five pruning strategies to early prune the unpromising items, thus greatly reducing the computations than the PHUI-List algorithm.

5.2   Memory Usage

The memory usage of the compared algorithms were evaluated by using the Java API. In the same way, the performance was tested and examined under varied MUs with a fixed MP. From Fig. 4, it can be clearly seen that the proposed MUHUI algorithm requires less memory usage compared to PHUI-UP, but consumes little more memory than the state-of-the-art PHUI-List algorithm except for the retail dataset. Specially, the memory usage of the two PU-list-based algorithms, PHUI-List and MUHUI, change smoothly under varied parameters in four datasets. This performance is somehow similar to the conclusion given as the above analysis of runtime. This result is reasonable since both PHUI-List and MUHUI are PU-list-based algorithms, they can easily prune unpromising itemsets with the help of actual utilities and remaining utilities. The reason why a little more memory than PHUI-List always consumes for MUHUI is that it has to spend the extra memory cost for storing the additional *EUCS* data structure. Hence, the memory usage of the MUHUI algorithm is somehow similar to that of PHUI-List.

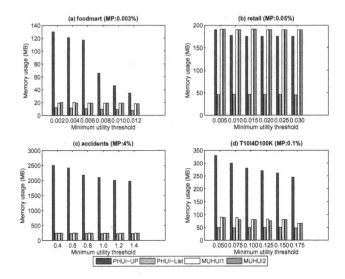

Fig. 4. Memory usage under varied MUs with a fixed MP. (Color figure online)

5.3 Scalability Analysis

As shown in Fig. 5, the scalability of the four algorithms is compared in the synthetic dataset T10I4N4KD$|X|$K with different scales, which is set MP = 0.05%, MU = 0.1%, and $|X|$ is set varying from 100 to 500. It can be observed that the runtime of all compared algorithms is linear increased along with the increasing of dataset size $|X|$. The performance of MUHUI1 and MUHUI2 are, however, relatively stable to the variations of $|X|$. With the increasing of the size of dataset, the time of MUHUI1 is close to that of MUHUI2, but significantly faster than that of PHUI-List. Specially, the gap of runtime among them grows wider with the increasing of dataset size. With the increasing of dataset size, the running time of algorithms is linearly increasing as well, and the proposed MUHUI algorithm scales well on large-scale dataset. Figure 5(b) shows the memory usages of four algorithms which indicates the linearity in term of dataset size. In addition, we can find that the memory usage of the two PU-list-based algorithms is steady increasing than that of PHUI-UP.

To evaluated the effect of proposed different pruning strategies, the number of visited nodes in the search space are further compared, as can be observed in Fig. 5(c). Note that the number of visited nodes in the PHUI-List, MUHUI1 and MUHUI2 algorithms are denoted as N_1, N_2, and N_3, respectively. It shows that the number of the visited nodes of MUHUI1 and MUHUI2 are quite less than that of PHUI-List.

6 Conclusion

In this paper, an efficient algorithm called Mining of Uncertain data for High-Utility Itemsets (MUHUI) is proposed to consider the mining of those itemsets

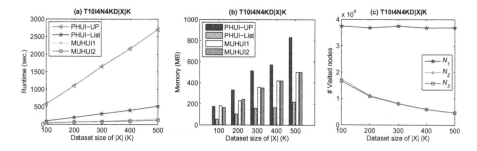

Fig. 5. Scalability of compared algorithms. (Color figure online)

with high-utility and high probability. The MUHUI algorithm is developed by proposing several efficient pruning strategies, and to improve the performance. Based on the PU-list structure, MUHUI utilizes the properties of probability and utility. Several efficient pruning strategies are also designed to speed up the computations, which can effectively avoid the construction of PU-lists of the huge number of unpromising itemsets. Substantial experiments both on real-life and synthetic datasets show that the proposed algorithm consumes a little more memory than the PHUI-List algorithm, but has great performance in terms of runtime, the number of visited nodes in the enumeration tree, and scalability compared to the past works. Specifically, the MUHUI algorithm is more scalable than the PHUI-UP and PHUI-List algorithms on large-scale uncertain databases.

Acknowledgment. This research was partially supported by the National Natural Science Foundation of China (NSFC) under grant No. 61503092 and by the Tencent Project under grant CCF-TencentRAGR20140114.

References

1. Frequent itemset mining dataset repository. http://fimi.ua.ac.be/data/
2. Aggarwal, C.C.: Managing and mining uncertain Data (2010)
3. Aggarwal, C.C., Yu, P.S.: A survey of uncertain data algorithms and applications. IEEE Trans. Knowl. Data Eng. **21**(5), 609–623 (2009)
4. Agrawal, R., Srikant, R.: Fast algorithms for mining association rules in large databases. In: The International Conference on Very Large Data Bases, pp. 487–499 (1994)
5. Agrawal, R., Srikant, R.: Quest synthetic data generator. http://www.Almaden. ibm.com/cs/quest/syndata.html
6. Ahmed, C.F., Tanbeer, S.K., Jeong, B.S., Le, Y.K.: Efficient tree structures for high utility pattern mining in incremental databases. IEEE Trans. Knowl. Data Eng. **21**(12), 1708–1721 (2009)
7. Bernecker, T., Kriegel, H.P., Renz, M., Verhein, F., Zuefl, A.: Probabilistic frequent itemset mining in uncertain databases. In: ACM SIGKDD International Conference on Knowledge Discovery and Data Mining, pp. 119–128 (2009)

8. Chan, R., Yang, Q., Shen, Y.D.: Mining high utility itemsets. In: IEEE International Conference on Data Mining, pp. 19–26 (2003)
9. Chui, C.-K., Kao, B., Hung, E.: Mining frequent itemsets from uncertain data. In: Zhou, Z.-H., Li, H., Yang, Q. (eds.) PAKDD 2007. LNCS (LNAI), vol. 4426, pp. 47–58. Springer, Heidelberg (2007)
10. Geng, L., Hamilton, H.J.: Interestingness measures for data mining: a survey. ACM Comput. Surv. **38** (2006)
11. Han, J., Pei, J., Yin, Y., Mao, R.: Mining frequent patterns without candidate generation: a frequent-pattern tree approach. Data Min. Knowl. Disc. **8**(1), 53–87 (2004)
12. Lin, J.C.W., Gan, W., Hong, T.P., Tseng, V.S.: Efficient algorithms for mining up-to-date high-utility patterns. Adv. Eng. Inform. **29**(3), 648–661 (2015)
13. Lin, J.C.W., Gan, W., Fournier-Viger, P., Hong, T.P., Tseng, V.S.: Mining potential high-utility itemsets over uncertain databases. In: ACM ASE BigData & Social Informatics, p. 25 (2015)
14. Liu, M., Qu, J.: Mining high utility itemsets without candidate generation. In: ACM International Conference on Information and Knowledge Management, pp. 55–64 (2012)
15. Liu, Y., Liao, W., Choudhary, A.K.: A two-phase algorithm for fast discovery of high utility itemsets. In: Ho, T.-B., Cheung, D., Liu, H. (eds.) PAKDD 2005. LNCS (LNAI), vol. 3518, pp. 689–695. Springer, Heidelberg (2005)
16. Microsoft: Example Database foodmart of Microsoft Analysis Services. http://msdn.microsoft.com/en-us/library/aa217032(SQL.80).aspx
17. Fournier-Viger, P., Wu, C.-W., Zida, S., Tseng, V.S.: FHM: faster high-utility itemset mining using estimated utility co-occurrence pruning. In: Andreasen, T., Christiansen, H., Cubero, J.-C., Raś, Z.W. (eds.) ISMIS 2014. LNCS, vol. 8502, pp. 83–92. Springer, Heidelberg (2014)
18. Tseng, V.S., Wu, C.W., Shie, B.E., Yu, P.S.: UP-growth: an efficient algorithm for high utility itemset mining. In: ACM SIGKDD International Conference on Knowledge Discovery and Data Mining, pp. 253–262 (2010)
19. Tseng, V.S., Shie, B.E., Wu, C.W., Yu, P.S.: Efficient algorithms for mining high utility itemsets from transactional databases. IEEE Trans. Knowl. Data Eng. **25**(8), 1772–1786 (2013)
20. Wu, C.W., Shie, B.E., Tseng, V.S., Yu, P.S.: Mining top-k high utility itemsets. In: ACM SIGKDD International Conference on Knowledge Discovery and Data Mining, pp. 78–86 (2012)
21. Yao, H., Hamilton, H.J., Butz, C.J.: A foundational approach to mining itemset utilities from databases. In: SIAM International Conference on Data Mining, pp. 211–225 (2004)

An Improved HMM Model for Sensing Data Predicting in WSN

Zeyu Zhang, Bailong Deng, Siding Chen, and Li Li$^{(\boxtimes)}$

School of Computer and Information Science, Southwest University,
Chongqing 400715, China
lily@swu.edu.cn

Abstract. Wireless sensor networks (WSN) have been employed in numerous fields of real world applications. Data failure and noise reduction still remain tough unsolved problems for WSN. Predicting methods for data recovery by empirical treatment, mostly based on statistics has been studied exclusively. Machine learning models can greatly enhance the predicting performance. In this paper, an improved HMM is proposed for multi-step predicting of wireless sensing data given historical data. The proposed model is based on clustering of wireless sensing data and multi-step predicting is accordingly accomplished for different varying patterns using HMM whose parameters are optimized by Particle Swarm Optimization (PSO). We evaluate our model on two real wireless sensing datasets, and comparison between Naive Bayesian, Grey System, BP Neural Networks and traditional HMMs are conducted. The experimental results show that our proposed model can provide higher accuracy in sensing data predicting. This proposed model is promising in the fields of agriculture, industry and other domains, in which the sensing data usually contains various varying patterns.

Keywords: HMM model · Sensing data predicting · k-means algorithm · PSO algorithm

1 Introduction

Nowadays, wireless sensor network (WSN) has been widely applied in numerous fields of human life. Due to the convenient features exemplified by wireless communication and self-organization, typical applications of WSN include environmental surveillance, industrial detection, health care monitoring, etc. However, with the applications expanding, representative drawbacks like data failure impede the further development of WSN.

Data failure is inevitable since it is caused in numerous reasons such as electromagnetic interference, environmental hazards, hardware malfunctioning, etc. [1]. Penetration of WSN into many sensitive fields may turn data failure into disasters. Thus various solutions for WSN data failure are explored, in which predicting stands as representative. The objectives of WSN data predicting

© Springer International Publishing Switzerland 2016
B. Cui et al. (Eds.): WAIM 2016, Part I, LNCS 9658, pp. 31–42, 2016.
DOI: 10.1007/978-3-319-39937-9_3

generally contain: (1) recovering received failure data; (2) forecasting prospective data [2,3].

Sensing data acquired in continuous time, therefore predicting of sensing data equals time series predicting [3]. There are many methods to predict time series. Among them, machine learning proved efficient for time series predicting with outstanding capability of generalization. In many typical situations like stock series predicting or human activities predicting, diverse mature machine learning methods already exist. But because of the uncertainty and different varying patterns in sensing data, inadequate choice of machine learning model can lead to poor performances [2].

In this paper, we proposed an improved HMM model to satisfy the demand of sensing data predicting. This model contains two specific improvements: (1) clustering sensing data into different groups based on different varying patterns; (2) seeking out the optimal parameter setting with intelligent algorithm for the model.

The rest of this paper is organized as follows. In Sect. 2, different typical predicting methods for sensing data is presented. Section 3 introduces our main improvements on HMM. Experimental results are shown in Sect. 4 and conclusion are given in Sect. 5.

2   Related Work

Detailed research work from different perspectives for sensing data predicting has been published. A comprehensive survey of existing data mining techniques and their multi-level classification scheme for sensing data predicting are given in [4], and an adaptive data mining framework of WSNs for future research is also proposed. [5] presents a predicting method based on neural network and extreme learning machine (ELM) algorithm. The results show the neural network trained by ELM reduces time-cost efficiently with barely loss of prediction accuracy. In [6], a WSN system is constructed in big data centers where stays many servers. An online predicting of steep environmental change is provided to decrease energy waste. This event predicting method indicates the very feasibility of machine learning in sensing data analysing and predicting. [7] predicts sensing data for finding potential data failure with time series analytic method ARIMA, the results demonstrates the close relationship between sensing data and time series. Saini A, etc. in [8] introduces strategies for predicting outliers in WSN with different filtering methods like Kalman Filter.

In this paper, the popular mature machine learning model HMM is used as the main predicting method. Various research related HMM finished in recent years. Jialing Li, etc. use HMM to predict the time series of human behaviors based on the historical habits in [9]. In [10], predicting of hardware failure is conducted by HMM, the modeling procedure of hardware failure events shows typicality in HMM predicting. [11] describes a prediciton method with HMM and dynamic principal component analysis (DPCA). The experimental results based on a bearing test bed show the plausibility and effectiveness of the proposed method. HMM based predicting is shown feasible on financial time series in [12]. The proposed model is to make a Maximum a Posteriori decision over all the possible stock values for the next day. In WSN, [13] predicts link quality with HMM. This predicting approach is realized by considering the next varying trend of specific links. In 2014, [14] introduces an adaptive HMM combing k-means to predict the learning style of specific student and then gives the student learning suggestions. [15] demonstrates a self-adaptive HMM whose parameters are optimized by intelligent optimizing algorithms. And the experimental results comparing with other models show the outperformance of the proposed self-adaptive HMM.

3   Methodology

The randomness and diversity of data varying in WSN causes difficulties in analyzing, feature extracting and predicting. Therefore traditional predicting model can hardly be competent to overcome sensing data failure problem. In this study, we use a improved HMM to analyze and predict sensing data by historical data. This section will introduce the main methods for wireless sensing data predicting in our model.

3.1   Hidden Markov Model (HMM)

As a dynamic model, HMM is specially designed to account for random sequences. In HMM, two finite sets of states are contained as variables: *observable state variables* and *hidden state variables*. The sequences of observable variables present directly observable discrete-time process of states in the system. Hidden state variables stand for the unobservable state of the system, which emit observable variables with different probabilities. A typical formal description of HMM is:

$$\lambda = \{s, o, T, C, \pi\} \tag{1}$$

- $\lambda = \{$A specific setting of HMM parameters.$\}$
- $s = \{$Number of (hidden) states that is manually set.$\}$
- $o = \{$Number of observable states, which is also set by user.$\}$
- $T = \{$Transition matrix: each element t_{ij} represents the probability that transiting from current state s_i to s_j in the next time step.$\}$
- $C = \{$Confusion matrix: whose element c_{ij} represents the probability that the system observes o_i in the hidden state s_j.$\}$
- $\pi = \{$Represents initial probability for each state.$\}$

Apparently, parameter setting λ has significant impact of HMM performance. In this study, intelligent optimizing is conducted and for parameter setting λ, especially for parameter s since its unobservability stands for an obstacle in improving HMM performance.

3.2 Sensing Data Preprocessing

In our approach, original decimal sensing data will be mapped into five different monitor states. Each monitor state represent a specific interval that stays
between two distinct thresholds that determined by existing professional knowledge. Every threshold indicates a *"critical value"* in sensing data varying curve,
thus the monitor states can explicitly reflect the current situation of monitored
object with barely data loss.

In Fig. 1, a sample of data varying curve in agricultural WSN is illustrated.
Figure 2 shows an instance of agricultural sensing data mapping, and a typical
mapping rule based on profession knowledge is shown in Table 1 as an instance.

Utilizing data mapping, a process of sensing data varying can be described
as a time series of monitor states. Therefore, a *finite-state machine* is competent
to completely represent the sensing data varying in WSNs. Table 1 presents the
statistics of 1.5 million agricultural sensing data, each row includes one-step
transiting probabilities to every potential state from current state. For example,
the second row whose first element is "1" depicts the transiting probability of
"$s_1 \to s_1$", "$s_1 \to s_2$", "$s_1 \to s_3$"..., respectively. According to Table 1, the *5-state
machine* will be realized as the basic structure and the monitor states will be
the input of the HMM based model as observations.

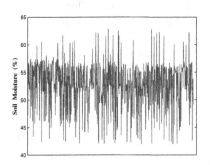

Fig. 1. Sample of sensing data varying

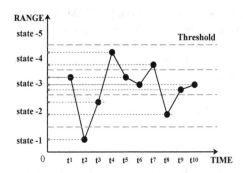

Fig. 2. Illustration of sensing data
mapping

3.3 Sensing Data Clustering

Due to the diversity of data varying patterns in WSN, accuracy of prediction
can hardly improves using single model with constant parameter setting. Thus,
we apply *k-means algorithm* to cluster the whole sensing dataset into different
smaller subsets on the basis of different data varying patterns.

K-means is an efficient clustering algorithm that aiming to find natural
groups data. In *k-means*, preparation like feature extraction or rule-making is not
indispensable. The main idea of *k-means* clustering is to minimize the distances
between the points and the cluster center in each clusters. Formula 2 shows the

Table 1. Transiting probability matrix of agricultural dataset

$State_t$ \ $State_{t+1}$	1	2	3	4	5
1	0.2222	0.6803	0.0332	0.0326	0.0317
2	0.0350	0.4450	0.3500	0.1050	0.0650
3	0.0845	0.1296	0.6225	0.0958	0.0676
4	0.0762	0.1238	0.3714	0.3810	0.0476
5	0.0454	0.1022	0.2614	0.3401	0.2509

k-means objective function in this study: M and N indicate the number of clusters and the number of data in specific cluster, c_i represents the center of i_{th} cluster, $x_j^{(i)}$ indicates the j_{th} data sequence is contained by the i_{th} cluster.

Clustering results of the two sensing datasets used in this study are shown in Figs. 3 and 4. Considering the computation cost by excessive clusters, the best clustering result is the point on the curve that is the nearest to the origin of coordinates. Aiming to calculate the distance between coordinate origin and the points on the result curve, Formula 3 shows the normalization of two axises to erase impact of different dimensions

$$minimize[\frac{1}{M}\sum_{i=1}^{M}(\frac{1}{N}\sum_{j=1}^{N}|x_j^{(i)} - c_i|^2)] \tag{2}$$

$$x' = \frac{x - x_{min}}{x_{max} - x_{min}} \tag{3}$$

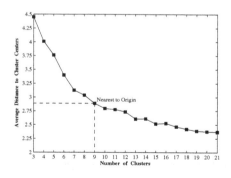

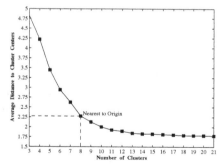

Fig. 3. Clustering in agricultural dataset

Fig. 4. Clustering in industrial dataset

According to Formulas 2 and 3, the most possible number of data varying patterns is 9 in agricultural dataset and 8 in industrial dataset. In our model, various HMM parameter settings λ^{c_t} are built for each data varying pattern c_t. All the parameter settings for one dataset are then combined together into a HMM library to provide precise prediction of sensing data.

3.4 Parameter Optimizing

Aiming to reach higher prediction accuracy, optimizing of HMM parameters is very necessary since some of the parameters are initiated randomly. Among various available optimizing strategies, Particle Swarm Optimization (PSO) algorithm is chosen for optimizing our proposed model. PSO is a stochastic optimizing method inspired by social behavior of bird flocking or fish schooling. In many cases, PSO shows commendable performance with simpler rules comparing with Genetic Algorithm.

In our approach, we focus on the parameter of hidden states number in HMM, namely s in λ. Each particle in PSO symbolize a candidate parameter, and all the particles move around the solution space. If one particle locally finds the best known position, the others will be guided there. The best solution will be located eventually after finite iterations. By PSO, the parameter settings managed by HMM library will be optimazed for each data varying pattern.

The core procedures of PSO in this study are introduced following. "[]" indicates vector variables. Formula 4 shows the top priority for velocity decision. The main strategy of particle velocity updating is described in Formula 5. And Formula 6 updates the position for every particle in searching space.

$$v \in [-V_{max}, V_{max}] \tag{4}$$

$$\begin{aligned} v[] = w_0 \times v[] &+ w_1 \times r_1 \times (locBest[] - curPos[]) \\ &+ w_2 \times r_2 \times (gloBest[] - curPos[]) \end{aligned} \tag{5}$$

$$curPos[] = curPos[] + v[] \tag{6}$$

- $v = \{$A manually set variable indicates the limit of velocity.$\}$
- $V_{max} = \{$Superior limit of velocity, which is manually set.$\}$
- $locBest[] = \{$Keep records of individual optimal values for each particle.$\}$
- $gloBest[] = \{$Records the global optimal value.$\}$
- $curPos[] = \{$Memory space for the current positions of each particle.$\}$
- $w_0 = \{$A nonnegative constant represents of particles motion inertia.$\}$
- $w_1, w_2 = \{$Indicate constant weight of acceleration for $locBest[]$ and $gloBest[]$, respectively.$\}$
- $r_1, r_2 = \{$Two random values between [0,1] for tuning of velocity updating.$\}$

3.5   Predicting Method

By sensing data clustering and parameter optimizing, HMM library is constructed, which contains a series of trained and optimized parameter settings λ^{c_t}. The sensing data multi-step predicting considering historical data thus conduct by following steps:

(1) Seek out the optimal parameter setting λ from HMM library with *Forward Algorithm* for given observation sequence $O = \{o_1, o_2, ..., o_t\}$.
(2) According to the explicit λ, the data varying pattern p_O and the most probable hidden states sequence $S = \{s_1, s_2, ..., s_t\}$ of given O are obtained.
(3) Variable $l(O|S, \lambda)$ is then introduced to discover the similar historical data sequence for O. $l(O|S, \lambda)$ is a lilelihood value describes the relationship between given O and discovered S under the explicit λ. Formula 7 describes the definition of $l(O|S, \lambda)$.

$$l(O|S, \lambda) = \prod_{i=1}^{t} p(d_i|s_i, \lambda) \tag{7}$$

(4) With Formula 7, a specific sequence $H = \{h_1, h_2, ..., h_t\}$ that has the highest similarity to O will be found from all the historical data under the data varying pattern p_O.
(5) By obtaining of H, the subsequent data $H_s = \{h_{t+1}, h_{t+2}, ..., h_{t+n}\}$ is decided. Varying trends in H_s are then applied as references for predicting subsequent unknown data of O.

As a crucial part of this predicting method, the main goal of *Forward Algorithm* is to find the most possible λ that generates given observation sequence $O = \{o_1, o_2, ..., o_t\}$. Formal description of *Forward Algorithm* is shown in Formula 8, in which λ^{c_t} indicates a parameter setting in HMM library L. If the given O obtained the highest probability on the parameter setting λ^{c_1}, O belongs to cluster c_1 then the varying pattern of O is accordingly determined.

$$\theta' = \arg \max_{\lambda^{c_t} \in L} P(O|\theta(h)) \tag{8}$$

4   Experiments

4.1   Model Architecture

Figure 5 illustrates the architecture of the proposed model. Four main procedures are contained in the model: *Preprocessing, Training, Optimizing* and *Predicting*. In preprocessing, raw sensing data is firstly converted to monitor states by given mapping rules. Then a finite-state machine is obtained to describe sensing data varying, which is also used to construct HMM model as basic structure in training phase. Another task in training procedure is to build HMM library consists of various parameter settings to match different varying patterns of sensing data.

Optimizing procedure is in charge of seeking out the optimal parameter settings to improve prediction accuracy further. With the obtaining of optimal HMM library, multi-step predicting on the basis of historical data is conducted in the procedure of predicting.

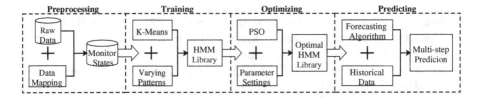

Fig. 5. General representation of the proposed model

4.2 Datasets

Two real wireless sensing datasets respectively acquired from orange orchard and manufactory are used to evaluate the performances of the proposed model. Both of the datasets have the same allocation rates: 60 % for model training, 20 % for model validation and 20 % for testing.

- **Agricultural Dataset.** A comprehensive dataset includes over 6 million data of soil moisture, temperature and many other types of wireless sensing data acquired from an orange orchard in Chongqing. Experiments mainly concentrates on the data of soil moisture.
- **Industrial Dataset.** A special dataset consists of 3,3 million data of temperature of industrial devices, which is provided by Chongqing Small and Medium Enterprise Bureau.

4.3 Evaluation Metrics

Three different metrics are used for evaluation in this paper: *Prediction Correct Rate (PCR)*, *Average Prediction Error (APE)* and *Varying Trends Prediction Correct Rate (VT-PCR)*. Definitions of each metric and variables are explained following:

- **Prediction Correct Rate (PCR)** indicates the correct rate of completely correct predicting results. Definition of PCR is shown in Formula 9. N represents the data scale of testing. p_i and r_i indicate the predicting result and the real observation. $\odot$ indicates calculation of *exclusive NOR*.

$$\frac{1}{N} \sum_{i=1}^{N} (p_i \odot r_i) \times 100\,\% \tag{9}$$

- **Varying Trends Prediction Correct Rate (VT-PCR)** describes the correct rate of varying trend predicting results. In Formula 10, r_{i-1} indicates the previous real observation of the current. Thus $(p_i - p_{i-1})$ is prediction of sensing data varying trend while $(r_i - r_{i-1})$ is the real varying trend. Specially, p_0 and r_0 represent the last element of observation sequence.

$$\frac{1}{N} \sum_{i=1}^{N} [\frac{(p_i - p_{i-1})}{|p_i - p_{i-1}|} \odot \frac{(r_i - r_{i-1})}{|r_i - r_{i-1}|}] \times 100\,\% \tag{10}$$

- **Multi-step Average Prediction Error (MAPE)** reflects the overall performance of the proposed model in multi-step predicting. In Formula 11, M is the number of multi-step predicting results which is also equal to the number of input data sequences, S represents step number of predicting that is set to 4 in our study, p_{ij} indicates a single output of multi-step predicting while r_{ij} indicates single real observation. As an example, $M = 100$ means 100 monitor state sequences input into the model, while $S = 4$ means the trained model will predict the subsequent 4 data for each sequence as output.

$$\frac{1}{M} \sum_{i=1}^{M} (\frac{1}{S} \sum_{j=1}^{S} |p_{ij} - r_{ij}|) \times 100\,\% \tag{11}$$

4.4 Results

In following, a series of predicting results comparing the proposed model with several other typical machine learning models are shown. The contrast models in this study include: *Grey System, Naive Bayes, BP Neural Network, Traditional HMM* and *improved HMM with only k-means (K-HMM)*.

Figures 6 and 7 show the prediction correct rate (PCR) that is given by Formula 9 in both datasets. With uncertainty increases, PCR decreases gradually whatever the model is. However, the proposed model named as KO-HMM in figures, generates higher PCR than any other models in each phase of multi-step predicting.

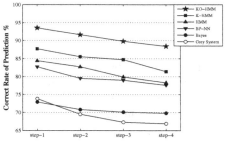
Fig. 6. Prediction of agricultural data

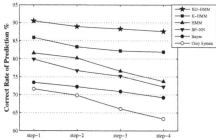
Fig. 7. Prediction of industrial data

Predicting of sensing data trends is valuable in reflecting the status of specific system. Figures 8 and 9 indicate the overall PCR and VT-PCR generated by each model according to Formulas 9 and 10. Because of the relatively loose definition, VT-PCR obtained by each model is higher than PCR. As can be seen from Figs. 8 and 9, the proposed model has definite advantages in predicting accracy comparing with other models.

Figures 10 and 11 demonstrate the average error of multi-step prediction. The evaluation metric MAPE is defined in Formula 11. The least MAPEs on two datasets indicate the capability of the proposed model to provide precise prediction of unknown sequences.

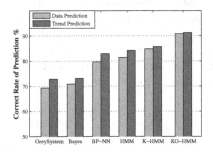

Fig. 8. Agricultural data trend prediction

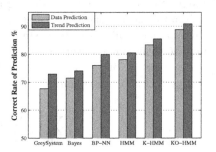

Fig. 9. Industrial data trend prediction

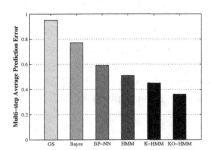

Fig. 10. MAPE in Agricultural data

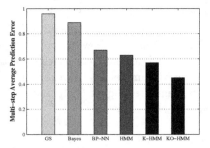

Fig. 11. MAPE in Industrial data

- Evaluations on different datesets and comparisons with other model verified the effectiveness of the proposed model in wireless sensing data predicting.
- Construction of HMM library based on k-means algorithm shows positive effect on sensing data predicting. Therefore clustering in HMM based models proved effective for predicting.
- Optimizing HMM library with PSO obviously enhances the capability of our model in unknown data predicting. Necessity of parameter optimizing in the proposed model is directly testified.

- With precise prediction of sensing data and their varying trends, the proposed model can: (1) significantly strengthen the robustness of WSN by providing efficient data recovery; (2) give instructions for subsequent jobs with prospective prediction.
- Experiments of multi-step sensing data predicting shows promising results for sequences forecasting using our proposed model.
- Generalization capacity of the proposed model is shown from the experimental results on different datasets. The potential effects of this model in other similar fields is revealed.

5 Conclusion

In this paper, we first analyze the characteristics of agricultural and industrial sensing data from real datasets. For improving accuracy of sensing data prediction, k-means algorithm is conducted to cluster sensing data according to different varying patterns. Then a HMM library contains a series of parameter settings are constructed, in which every parameter setting is corresponding to a specific varying pattern of sensing data. PSO algorithm is then introduced to optimize HMM parameter settings in the HMM library. Comparative analyses of the performance of our proposed model show obvious enhancement on prediction accuracy, and efficient feasibility for general predicting of random sequence, especially for the data including various varying patterns.

In the future, we will consider how to upgrade the existing HMM library realtimely to achieve precise prediction for new varying pattern. Moreover, how to cluster different data more appropriately will be investigated further.

Acknowledgment. This work is supported by Natural Science Foundations of China (No. 61170192), National High-tech R&D Program of China (No. 2013AA013801).

References

1. Kamal, A.R.M., Bleakley, C.J., Dobon, S.: Failure detection in wireless sensor networks: A sequence-based dynamic approach. ACM Trans. Sensor Netw. **10**(2), 35 (2014)
2. Wu, M., Tan, L., Xiong, N.: Data prediction, compression, and recovery in clustered wireless sensor networks for environmental monitoring applications. Inf. Sci. **329**, 800–818 (2016)
3. Aggarwal, C.C.: Managing and mining sensor data (2013)
4. Mahmood, A., Shi, K., Khatoon, S., Xiao, M.: Data mining techniques for wireless sensor networks: A survey. Int. J. Distrib. Sensor Netw. **2013**, 24 (2013)
5. Liu, Q., Zhang, Y.Y., Shen, J., Xiao, B., Linge, N.: A wsn-based prediction model of microclimate in a greenhouse using an extreme learning approach. In: Advanced Communication Technology, pp. 133–137 (2015)
6. Vuppala, S.K., Ghosh, A., Patil, K.A., Padmanabh, K.: A scalable WSN based data center monitoring solution with probabilistic event prediction. In: Advanced Information Networking and Applications (2012)

7. Kharade, S.S., Khiani, S.: Fault prediction and relay node placement in wireless sensor network-a survey. Int. J. Sci. Res. **3**(10), 702–704 (2014)
8. Saini, A., Sharma, K.K., Dalal, S.: A survey on outlier detection in wsn. Int. J. Res. Aspects Eng. Manage. **1**(2), 69–72 (2014)
9. Li, J., Li, L., Wu, Y., Chen, S.: An improved recommender based on hidden Markov model. In: Anthony, P., Ishizuka, M., Lukose, D. (eds.) PRICAI 2012. LNCS, vol. 7458, pp. 870–874. Springer, Heidelberg (2012)
10. Teoh, T.T., Cho, S.Y., Nguwi, Y.Y.: Hidden markov model for hard-drive failure detection, pp. 3–8 (2012)
11. Jianbo, Y.: Health condition monitoring of machines based on hidden markov model and contribution analysis. IEEE Trans. Instrum. Meas. **61**(8), 2200–2211 (2012)
12. Gupta, A., Dhingra, B.: Stock market prediction using hidden Markov models. Engineering and Systems, pp. 1–4 (2012)
13. de Araújo, G.M., Kaiser, J., Becker, L.B.: An optimized markov model to predict link quality in mobile wireless sensor networks. In: Computers and Communications, pp. 000307–000312 (2012)
14. Deeb, B., Hassan, Z., Beseiso, M.: An adaptive HMM based approach for improving e-learning methods (2014)
15. Liu, L., Luo, D., Liu, M., Zhong, J., Wei, Y., Sun, L.: A self-adaptive hidden markov model for emotion classification in chinese microblogs. Mathematical Problems in Engineering (2015)

eXtreme Gradient Boosting for Identifying Individual Users Across Different Digital Devices

Rongwei Song, Siding Chen, Bailong Deng, and Li Li$^{(\boxtimes)}$

School of Computer and Information Science, Southwest University,
Chongqing 400715, China
srw123@email.swu.edu.cn, sidingchan@gmail.com, microbye64@gmail.com,
lily@swu.edu.cn

Abstract. With the increasing popularity of tablets, smartphones and other mobile electronic devices, it is not uncommon for users to complete online tasks through different electronic devices. Identifying individual users across different digital devices is now becoming a hot research topic. Methods based on name, email and other demographic information have received much attention. However, it is often difficult to obtain a complete set of information. In this paper, we use a probabilistic approach for cross-device identity issue and focus on comparing different algorithms. We conduct an in-depth study and expand the attribute of data through the study of the relationship between attributes. Dummy variables are introduced to improve the efficiency of the models. Experimental results on four datasets (released by ICDM Challenge) show that the eXtreme Gradient Boosting can consistently and significantly outperform other algorithms on both accuracy and F1-score. It also consistently provides a better performance compared to the methods we used in ICDM Challenge (We took part in the ICDM 2015 Challenge, and achieved a moderate score ranking use C4.5 and BP model), and achieves a better comprehensive evaluation ranking.

1 Introduction

It is very common for one user to have multiple Internet devices, typical representatives are computers and mobile phones. While these devices are not directly connected together, they have more or less common behaviors, due to the fact that they often belong to the same user. Users can access through different Internet devices to accomplish the same task, as consumers move across devices to complete online tasks [1], their identifier becomes fragmented. However, it isn't always able to discern when activity on different devices for marketers who are hoping to target them with meaningful messages, recommendations, and customized experiences.

A common method for bridging these device identifiers involves relying on deterministic identifiers [2], such as names, email addresses, phone numbers, or other personal information. This kind of methods has been used for online harassment and offline violence among young people [3], cross-device recognition,

© Springer International Publishing Switzerland 2016
B. Cui et al. (Eds.): WAIM 2016, Part I, LNCS 9658, pp. 43–54, 2016.
DOI: 10.1007/978-3-319-39937-9_4

and so on. However, we can not get these personal information. It is difficult to identify individual users across different digital devices relying on deterministic identifiers.

A better solution to the problem is to infer multiple device identifiers belonging to the same user using a probabilistic approach, and to optimize the level of accuracy of the prediction. It is necessary to train a classifier for each device [4]. However, it is very costly and time-consuming to label enough training data for each device. Thus, we are interested in exploring whether it is technically possible to select some attributes to represent the data, and feasible in time.

Motivated by above observation, we use the device table data to generate positive cases and negative cases, and combine them with cookie table to generate data for training and testing. Then we compare eight models on accuracy and F1-score, including several models based on tree. We conduct experiments on four types of datasets. Experimental results show that XGBoost model can outperform baseline methods significantly and consistently in terms of accuracy and F1-score. Finally, we discuss the parameter of XGBoost model.

The rest of this paper is organized as follows. In Sect. 2, we briefly review several representative works related to the problem. In Sect. 3, we present XGBoost model in detail. In Sect. 4, we report experimental results on different models. In Sect. 5, we conclude this paper.

2   Related Work

In this section, we briefly review representative works related to user identification and gradient boosting model.

There are some works on user identification. For example, Carmagnola and Cena [5] describe the conceptualization and implementation of a framework that provides a common base for user identification for cross-system personalisation among web-based user-adaptive systems. Guna et al. [6] present an intuitive, implicit, gesture based identification system suited for applications such as the user login to home multimedia services, with less strict security requirements. However, these methods are used to identify users based on specific knowledge, it does not take into account the identification problem when users have a variety of devices.

The concept of gradient boosting is proposed by Friedman [7], he present "TreeBoost" model, and design method to train it. Then, Friedman [8] introduce bagging trick to gradient boosting. Friedman et al. [9] use second-order statistics for tree splitting to solve the problem of training speed. Johnson and Zhang [10] propose to do fully corrective step, as well as regularizing the tree complexity. However, the efficient of these methods are not very good in terms of time and space.

XGBoost model has done a great optimization compare to the traditional gradient boosting model, it has a very good efficiency in time and space. Our work is to identify individual users across different digital devices, XGBoost model is more accurate and suitable for the problem compared with the state-of-the-art methods.

3   XGBoost Model

In this section, we will present XGBoost model[1] in detail. In addition, we will introduce the accelerate optimization algorithm to solve the model.

3.1   Notations

The model in supervised learning usually refers to the mathematical structure of how to make the prediction y_i given x_i. First, we introduce several notations that will be used in our model. Denote $\{X_i \in R^{N_i \times D}, y_i \in R^{N_i \times 1}\}$ as the $i - th$ labeled samples in datasets, where N_i is the number of labeled samples and D is the number of attributes. $x_i^j \in R^{D \times 1}$ is the transpose of the $j - th$ row of X_i, and y_i is its label. In this paper we focus on binary classification, i.e., classifying a correspondence into positive or negative, thus $y_i \in \{0, 1\}$. F is a functional space of the set of all possible CARTs. w is the vector of scores on leaves, q is a function assigning each data point to the corresponding leaf and T is the number of leaves. γ is the gamma parameter and λ present L2 regularization term on weights in the model.

3.2   Model

Given a set of labeled training data, our goal is to train a general classifier for identifying cross-device user, XGBoost model in this paper is as follows.

Tree Ensemble. XGBoost model is tree ensembles. The tree ensemble model is a set of classification and regression trees (CART). Mathematically, we can write the model in the form

$$\hat{y}_i = \sum_{k=1}^{K} f_k(x_i), \quad f_k \in F \tag{1}$$

where K is the number of trees, f is a function in the functional space F, and F is the set of all possible CARTs. Therefore the objective to optimize can be written as

$$obj(\Theta) = \sum_{i}^{n} l(y_i, \hat{y}_i) + \sum_{k=1}^{K} \Omega(f_k) \tag{2}$$

where $l(y_i, \hat{y}_i)$ is the training loss function, and $\Omega(f_k)$ is the regularization term, the goal of XGBoost model is to minimize $obj(\Theta)$.

Additive Training. It is not easy to train all the trees at once. Instead, we use an additive strategy: fix what we have learned, add one new tree at a time.

---

[1] http://xgboost.readthedocs.org/en/latest/index.html.

We note the prediction value at step t by $\hat{y}_i^{(t)}$, so we have

$$\hat{y}_i^{(t)} = \sum_{k=1}^{t} f_k(x_i) = \hat{y}_i^{(t-1)} + f_t(x_i) \tag{3}$$

We use MSE(mean squared error) as our loss function, it becomes the following form.

$$
\begin{aligned}
obj^{(t)} &= \sum_{i=1}^{n} \left(y_i - (\hat{y}_i^{(t-1)} + f_t(x_i))\right)^2 + \sum_{i=1}^{t} \Omega(f_i) \\
&= \sum_{i=1}^{n} [2(\hat{y}_i^{(t-1)} - y_i)f_t(x_i) + f_t(x_i)^2] + \Omega(f_t) + constant
\end{aligned}
\tag{4}
$$

then we take the Taylor expansion of the loss function up to the second order

$$obj^{(t)} = \sum_{i=1}^{n} [l(y_i, \hat{y}_i^{(t-1)}) + g_i f_t(x_i) + \frac{1}{2} h_i f_t^2(x_i)] + \Omega(f_t) + constant \tag{5}$$

where the g_i and h_i are defined as

$$g_i = \partial_{\hat{y}_i^{(t-1)}} l(y_i, \hat{y}_i^{(t-1)})$$

$$h_i = \partial^2_{\hat{y}_i^{(t-1)}} l(y_i, \hat{y}_i^{(t-1)})$$

after we remove all the constants, the specific objective at step t becomes

$$obj^{(t)} = \sum_{i=1}^{n} [g_i f_t(x_i) + \frac{1}{2} h_i f_t^2(x_i)] + \Omega(f_t) \tag{6}$$

This becomes our optimization goal for the new tree. One important advantage of this definition is that it only depends on g_i and h_i. This is how xgboost can support custom loss functions. We can optimize every loss function, including logistic regression, weighted logistic regression, using exactly the same solver that takes g_i and h_i as input.

Model Complexity. We need to define the complexity of the tree $\Omega(f)$. In order to do so, we first refine the definition of the tree $f(x)$ as

$$f_t(x) = w_{q(x)}, w \in R^T, q : R^d \rightarrow \{1, 2, \cdots, T\} \tag{7}$$

here w is the vector of scores on leaves, q is a function assigning each data point to the corresponding leaf and T is the number of leaves. In XGBoost, we define the complexity as

$$\Omega(f) = \gamma T + \frac{1}{2} \lambda \sum_{j=1}^{T} w_j^2 \tag{8}$$

3.3 Structure Score

After reformalizing the tree model, we can write the objective value with the $t - th$ tree as:

$$obj^{(t)} \approx \sum_{i=1}^{n} [g_i w_{q(x_i)} + \tfrac{1}{2} h_i w^2_{q(x_i)}] + \gamma T + \tfrac{1}{2}\lambda \sum_{j=1}^{T} w_j^2$$
$$= \sum_{j=1}^{T} [(\sum_{i \in I_j} g_i)w_j + \tfrac{1}{2}(\sum_{i \in I_j} h_i + \lambda)w_j^2] + \gamma T \qquad (9)$$

where $I_j = \{i|q(x_i) = j\}$ is the set of indices of data points assigned to the $j - th$ leaf. We further compress the expression by defining $G_j = \sum_{i \in I_j} g_i$ and $H_j = \sum_{i \in I_j} h_i$:

$$obj^{(t)} = \sum_{j=1}^{T} [G_j w_j + \frac{1}{2}(H_j + \lambda)w_j^2] + \gamma T \qquad (10)$$

in this equation w_j is independent to each other, the form $G_j w_j + \frac{1}{2}(H_j + \lambda)w_j^2$ is quadratic and the best w_j for a given structure $q(x)$ and the best objective reduction we can get is:

$$w_j^* = -\frac{G_j}{H_j + \lambda}$$
$$obj^* = -\frac{1}{2}\sum_{j=1}^{T} \frac{G_j^2}{H_j + \lambda} + \gamma T \qquad (11)$$

The last equation measures how good a tree structure $q(x)$ is.

3.4 Learn the Tree Structure

Now that we have a way to measure how good a tree is, ideally we would enumerate all possible trees and pick the best one. In practice it is intractable, so we will try to optimize one level of the tree at a time. Specifically we try to split a leaf into two leaves, and the score it gains is

$$Gain = \frac{1}{2}[\frac{G_L^2}{H_L + \lambda} + \frac{G_R^2}{H_R + \lambda} - \frac{(G_L + G_R)^2}{H_L + H_R + \lambda}] - \gamma \qquad (12)$$

This formula can be decomposed as (1) the score on the new left leaf, (2) the score on the new right leaf (3), the score on the original leaf, (4) regularization on the additional leaf. We can see an important fact here: if the gain is smaller than γ, we would do better not to add that branch. This is exactly the pruning techniques in tree based models. By using the principles of supervised learning, we can naturally come up with the reason these techniques work.

For real valued data, we usually want to search for an optimal split. To efficiently do so, we place all the instances in sorted order. Then a left to right scan is sufficient to calculate the structure score of all possible split solutions, and we can find the best split efficiently.

4 Experiments

In this section we use XGBoost model and other algorithms to do the experiment, and report the experimental results. We first introduce the real-world user networks device datasets used in our experiments. Then we explore the variable importance of the dataset. After that we compare our method against the state-of-the-art algorithms, and show the results in the figure. Finally, we try to find out the best parameter setting on the performance of XGBoost model.

4.1 Datasets

The datasets is from the ICDM 2015 challenge: Drawbridge Cross-Device Connections[2]. we combine the dev_train_basic and cookie_all_basic files to generate the datasets of our experimental. The whole dataset has 23 kinds of attributes, for example, the attribute of country represent where the data is collected, the device_type attribute represent the type of the device. And there is some anonymous attributes, it is unfortunate that we do not know what it means. We also expand the property of the datasets, the samecountry attribute is generated by the difference between the country attribute of the dev_train_basic and cookie_all_basic files. Be the same, the same_co attribute is generated by the difference between the anonymous_c0 attribute of the dev_train_basic and cookie_all_basic files. The last column represent the data is positive case or negative case, represented by 0 or 1. Each data line represents one example of the dataset.

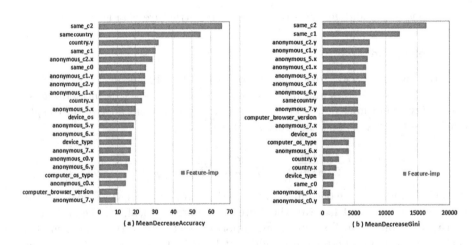

Fig. 1. Feature importance

Several preprocessing steps were taken before experiments. Figure 1 shows the feature importance of the data sets. we use the Mean Decrease Accuracy

---

[2] https://www.kaggle.com/c/icdm-2015-drawbridge-cross-device-connections.

and Mean Decrease Gini metric standards to get the feature importance of the data sets. As you can see, from Fig. 1(a), the feature of same_c2, samecountry are more important than the other features, from Fig. 1(b), the feature of same_c2, same_c1 are more important. So we choose the same_c1, same_c2 and samecountry as the important feature. The factor variable to be a number of feature is Pointless, and some models can not deal with factor variable. We put some factor variable convert into dummy variables to enhance the generalization ability of the datasets. In order to facilitate data processing and faster convergence, We also conducted a normalized data processing.

Table 1. The four types of datasets

Data Sources	Rows	Columns	Positive	Negative
data-all	360042	23	180018	180024
data-all*	94201	23	49234	44967
data-imp	360042	4	180018	180024
data-imp*	94201	4	49234	44967

Table 1 shows the detailed statistics of our real world datasets. We process the data collection into four types. Firstly, depending on how much the data sets selected properties, we divide the data sets into data-all and data-imp, the data-imp means that we only select the more important properties, not all. Then we also divide it into data-all* and data-imp* depend on whether deal with the unknown value. Besides, we also show the number of the data sets, such as row numbers, column numbers. The Positive and Negative represent the number of positive cases and negative cases. During the experiment, we intercepted 70 percent of the data as a training set, and did 10-fold cross validation, the other as the valid set.

4.2   Baseline Algorithms

To show the effectiveness of XGBoost model, we compare our method against the following state-of-the-art algorithms:

- **Decision Tree(DT):** The decision tree is a predictive model which maps observations about an item to conclusions about the item's target value. Decision tree learning is a method commonly used in data mining [11].
- **C4.5:** C4.5 is an algorithm used to generate a decision tree developed by Ross Quinlan [12]. C4.5 is an extension of Quinlan's earlier ID3 algorithm [13].
- **Gradient Boosting Model(GBM):** Gradient boosting [14] is a machine learning technique for regression and classification problems, which produces a prediction model in the form of an ensemble of weak prediction models, typically decision trees.

- **Random Forrest(RF):** In machine learning, Random Forrest [15] is a classifier which comprising a plurality of decision trees,and its output category is decided by the plural of individual tree.
- **Support Vector Machine(SVM):** support vector machines (SVMs, also support vector networks [16]) are supervised learning models with associated learning algorithms that analyze data and recognize patterns, used for classification and regression analysis.
- **Error Back Propagation(BP):** Error Back Propagation algorithm [17] is a common method of training artificial neural networks used in conjunction with an optimization method such as gradient descent.
- **k-Nearest Neighbors(k-NN):** The k-Nearest Neighbors algorithm is a non-parametric method used for classification and regression [18]. In both cases, the input consists of the k closest training examples in the feature space. The output depends on whether k-NN is used for classification or regression

4.3   Configuration

All experiments run on Macbook Pro 13-inch(Intel Core i55257U, 8G). The main parameters of XGBoost are as follows: max.depth = 5, eta = 0.1, ite = 300, gamma = 0, max_delta_step = 1, objective = "binary:logistic".

Table 2. The accuracy of the eight models on four types of datasets

Model	Data-all	Data-imp	Data-allnoNA	Data-impnoNA	Average
DecisionTree	0.6490	0.6358	0.6956	0.6937	0.6685
C4.5	0.6428	0.6500	0.7105	0.6937	0.6743
GBM	0.6679	0.6489	0.7104	0.7016	0.6822
RandomForrest	0.6798	0.6508	0.7247	**0.7039**	0.6898
SVM	0.5048	0.6490	0.6453	0.6457	0.6112
BP	0.6345	0.6339	0.6912	0.6899	0.6624
k-NN	0.5909	0.6009	0.6147	0.6614	0.6170
XGBoost	**0.7067**	**0.6594**	**0.7369**	0.7032	**0.7016**

4.4   Comparison of Different Models

In this section we conducted experiments to compare the performance of XGBoost model with several state-of-the-art methods.

The experimental results of baseline methods on the four types of datasets are summarized in Tables 2 and 3 respectively. Table 2 shows the accuracy of all the model on the four types of datasets, and Table 3 shows the F1-score.

From Tables 2 and 3 we can see that XGBoost model performs best on all four types of datasets. Although XGBoost model is not as good as the RandomForrest

Table 3. The F1-score of the eight models on four types of datasets

Model	Data-all	Data-imp	Data-allnoNA	Data-impnoNA	Average
DecisionTree	0.5028	0.4789	0.5906	0.5871	0.5399
C4.5	0.5909	0.5131	0.6645	0.5871	0.5889
GBM	0.5582	0.5171	0.6165	0.6003	0.5730
RandomForrest	0.6398	0.5157	0.6932	0.6158	0.6161
SVM	0.0282	0.5309	0.6726	0.6442	0.4690
BP	0.4751	0.4742	0.5988	0.5830	0.5328
k-NN	0.5969	0.5670	0.6473	**0.6732**	0.6211
XGBoost	**0.6754**	**0.5800**	**0.7185**	0.6516	**0.6564**

and k-NN model in accuracy and F1-score, but XGBoost model far ahead of other models in Average value, both in the accuracy and F1-score. In all of the datasets, we can find that all the models perform best on the data-all*, no matter the evaluation is accuracy or F1-score. This can be explained that the data-all* is cleaned the best. It deal with the unknown value, and retain more information than the data-imp*, so it is better for classification model to predict. All the models perform the worst on the data-imp because of the lack of information and unknown value. We can also see that XGBoost model performs far more stable on four types of datasets than other models. So XGBoost model has generalization capability on the four types of datasets and it is generic.

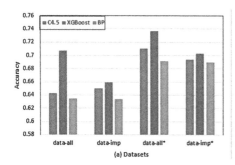 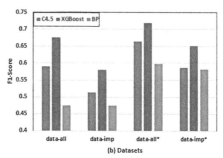

Fig. 2. The comparison between the three models on the accuracy and the F1-score

We also conduct experiments to compare the performance of XGBoost model with C4.5 and BP model. We use C4.5 and BP model to solve the ICDM 2015 Challenge. The experimental results are shown in Fig. 2. The results prove the performance improvement of XGBoost model. We compare the accuracy and F1-score. From Fig. 2 we can see that XGBoost model performs best compare to the other models on the four types of datasets, both in the accuracy and F1-score.

Furthermore, From Fig. 2(a) we can find that when use data-all, XGBoost model improves much more than other datasets in the accuracy. The same can also be found from Fig. 2(b) in the F1-score. It prove that XGBoost model can deal with comprehensive data consistently. The experimental results validate that XGBoost model is more suitable for the ICDM 2015 Challenge than the model what we used, and help to achieve a better comprehensive evaluation ranking.

4.5   Parameter Analysis

In this section we conducted experiments to explore the influence of parameter setting on the performance of XGBoost model. There are four important parameters in the model, i.e., ite, gamma, max_delta_step, max.depth. ite, max.depth are used to control the model iterations and maximum depth of a tree respectively. gamma are used to minimize loss reduction required to make a further partition on a leaf node of the tree. max_delta_step is a value that we allow each trees weight estimation to be. The experimental results are summarized in Fig. 3(a), (d), (b) and (c) respectively. Here we use the results on data-imp* for illustration and the objective parameter used in XGBoost model is "binary:logistic".

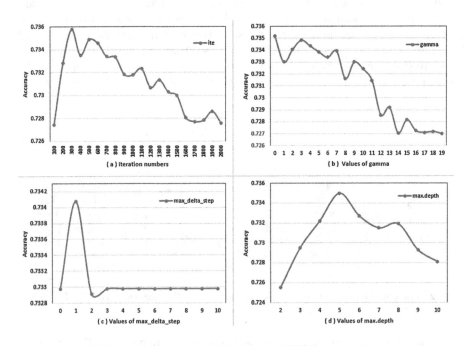

Fig. 3. Comparison of parameters of XGBoost model

From Fig. 3(a) we can see that as the values of ite grow from 100, the performance of XGBoost model first increases quickly then decreases slowly, and we find that when the ite is equal to 300, the model perform best on the datasets.

This is because that when the values of ite is too small, the relations between the datasets properties are not fully exploited. Thus the performance of XGBoost model improves quickly as the ite increase from 100. However, when the values of ite is too big, the relations between the datasets properties are overemphasized and the labeled training data is not fully respected. Thus the performance starts to decrease.

From Fig. 3(d), a moderate value is most suitable for max.depth. When the value of max.depth is too small, the Depth of the tree of XGBoost model is not enough, so the accuracy rate is not good. However, when the values of max.depth is too big, the model is easy to overfit the training data. From Fig. 3(b), we can see that with the increase of the gamma, the accuracy is a downward trend, and the best value is 0. From Fig. 3(c), the values of max_delta_step is not stable. According to the experience value, we can find that when the value of max_delta_step taken 1, the performance of XGBoost model will achieve the best.

5   Conclusions and Future Work

The contribution of this paper is three-fold. First, we conduct a more in-depth study in identifying individual users across different digital devices, and solve that problem with XGBoost model. Second, we fully compare the accuracy and F1-score of eight models in the four datasets, and we discuss the optimal parameters of XGBoost model. Third, our experiments on the four data sets show that XGBoost model is superiority and promising.

In the future, we will focus on the same user behavior between different digital devices and the improvement of XGBoost model.

Acknowledgments. This work is supported by Natural Science Foundations of China (No.61170192), National High-tech R&D Program of China (No. 2013AA013801).L Li is the corresponding author for the paper.

References

1. Wells, J.D., Fuerst, W.L., Palmer, J.W.: Designing consumer interfaces for experiential tasks: an empirical investigation. Eur. J. Inf. Syst. **14**(3), 273–287 (2005)
2. Setoguchi, S., Zhu, Y., Jalbert, J.J., et al.: Validity of deterministic record linkage using multiple indirect personal identifiers linking a large registry to claims data. Circ. Cardiovas. Qual. Outcomes **7**(3), 475–480 (2014)
3. Ojanen, T.T., Boonmongkon, P., Samakkeekarom, R., et al.: Investigating online harassment and offline violence among young people in Thailand: methodological approaches, lessons learned. Cult. Health sex. **16**(9), 1097–1112 (2014)
4. Gueth, P., Dauvergne, D., Freud, N., et al.: Machine learning-based patient specific prompt-gamma dose monitoring in proton therapy. Phys. Med. Biol. **58**(13), 4563 (2013)
5. Carmagnola, F., Cena, F.: User identification for cross-system personalisation. Inf. Sci. **179**(1), 16–32 (2009)

6. Guna, J., Stojmenova, E., Lugmayr, A., et al.: User identification approach based on simple gestures. Multimedia Tools Appl. **71**(1), 179–194 (2014)
7. Friedman, J.H.: Greedy function approximation: a gradient boosting machine. Ann. stat. 1189–1232 (2001)
8. Friedman, J.H.: Stochastic gradient boosting. Comput. Stat. Data Anal. **38**(4), 367–378 (2002)
9. Friedman, J., Hastie, T., Tibshirani, R.: Additive logistic regression: a statistical view of boosting (with discussion and a rejoinder by the authors). Ann. Stat. **28**(2), 337–407 (2000)
10. Johnson, R., Zhang, T.: Learning nonlinear functions using regularized greedy forest. arXiv preprint arXiv:1109.0887 (2011)
11. Rokach, L., Maimon, O.: Data mining with decision trees: theory and applications. World Scientific Pub Co Inc. (2008). ISBN 978-9812771711
12. Quinlan, J.R.: C4. 5: programs for machine learning. Elsevier (2014)
13. Quinlan, J.R.: Induction of decision trees. Mach. Learn. **1**(1), 81–106 (1986)
14. Breiman, L.: Arcing the edge. Technical Report 486, Statistics Department. University of California at Berkeley (1997)
15. Breiman, L.: Arcing classifier (with discussion and a rejoinder by the author). Ann. Stat. **26**(3), 801–849 (1998)
16. Cortes, C., Vapnik, V.: Support-vector networks. Mach. Learn. **20**(3), 273–297 (1995)
17. Rumelhart, D.E., Hinton, G.E., Williams, R.J.: Learning representations by back-propagating errors. Cogn. Model. **5**, 3 (1988)
18. Altman, N.S.: An introduction to kernel and nearest-neighbor nonparametric regression. Am. Stat. **46**(3), 175–185 (1992)

Two-Phase Mining for Frequent Closed Episodes

Guoqiong Liao[1(✉)], Xiaoting Yang[1], Sihong Xie[2], Philip S. Yu[2],
and Changxuan Wan[1]

[1] School of Information Technology, Jiangxi University of Finance and Economics,
Nanchang, China
liaoguoqiong@163.com, 523338968@qq.com, wanchangxuan@263.net
[2] Department of Computer Science, University of Illinois at Chicago, Chicago,
IL, USA
xiesihong1@gmail.com, psyu@uic.edu

Abstract. The concept of episodes was introduced for discovering the useful and interesting temporal patterns from the sequential data. Over the years, many episode mining strategies have been suggested, which can be roughly classified into two classes: Apriori-based breadth-first algorithms and projection-based depth-first algorithms. As we know, both kinds of algorithms are level-wise pattern growth methods, so that they have higher computational overhead due to level-wise growth iteration. In addition, their mining time will increase with the increase of sequence length. In the paper, we propose a novel two-phase strategy to discover frequent closed episodes. That is, in phase I, we present a level-wise shrinking mechanism, based on maximal duration episodes, to find the candidate frequent closed episodes from the episodes with the same 2-neighboring episode prefix, and in phase II, we compare the candidates with different prefixes to discover the final frequent closed episodes. The advantage of the suggested mining strategy is it can reduce mining time due to narrowing episode mapping range when doing closure judgment. Experiments on simulated and real datasets demonstrate that the suggested strategy is effective and efficient.

1 Introduction

The concept of episodes was first introduced for discovering the useful and interesting temporal patterns from sequential data [1]. Over the years, many episode mining strategies have been proposed for various applications, such as alarm sequence analysis in telecommunication networks [2], user-behavior prediction for web applications [3], and stock trend analysis for security companies [4], etc.

In general, the existing episode mining methods can be classified into Apriori-based breadth-first algorithms [2,5,6,9–13] and projection-based depth-first algorithms [4,7,8,14,15]. [2] presented two Apriori-based frequent episode mining methods, WINEPI and MINEPI. [5,6] proposed two kinds of probabilistic frequent episode mining methods from uncertain sequences. [9] proposed a new notion for episode frequency based on the non-overlapped occurrences, and presented an efficient counting algorithm using finite state automata. [10] presented

© Springer International Publishing Switzerland 2016
B. Cui et al. (Eds.): WAIM 2016, Part I, LNCS 9658, pp. 55–66, 2016.
DOI: 10.1007/978-3-319-39937-9_5

two fast non-overlapped occurrences counting algorithms for the serial and parallel episodes. In terms of episode mining in complex sequences, [11] proposed two frequent episode mining algorithms, MINEPI+ and EMMA. [12] proposed Clo-episode to mine the closed serial episodes following a breadth-first search order and integrating the pruning techniques with a prefix tree. [13] introduced the concept of strict episodes and defined instance-closed episodes. [4] used an event tree structure to represent the sets of event types with paths and nodes, and to mine the frequent episodes by a recursive procedure. [7] presented an algorithm that can find all frequent partite episodes satisfying a partite constraint in an input sequence. [8] proposed two episode mining approaches based on an Episode Prefix Tree (EPT) and a Position Pairs Set (PPS). [14] extended the definition of an episode in order to be able to represent the cases where events often occur simultaneously. It proposed an efficient depth-first search algorithm using a directed acyclic graph. Based on the rule of minimal and non-overlapping, [15] presented a frequent closed episode mining method FCEMine. In addition, [16] incorporated the concept of utility mining into episode mining and proposed a framework for mining high utility episodes in complex event sequences. [17] proposed an algorithm named MESELO for online frequent episode mining. [18] proposed a new quality measure for episodes based on minimal windows.

As we know, both the breadth-first and the depth-first algorithms are level-wise growth methods, i.e., mining the longer episodes by growing from the shorter. The main shortcoming is they have higher computational overhead due to level-wise growth iteration. In addition, their mining time will increase with the increase of sequence length. Focusing on these problems, in the paper we intend to study a novel two-phase strategy for mining frequent closed episodes. That is, in phase I, we present a level-wise shrinking mechanism, based on maximal duration episodes, to find the candidate frequent closed episodes with the same 2-neighboring episode prefix, and in phase II, we compare the candidates with different prefixes to discover the final frequent closed episodes.

The rest of the paper is organized as follows. We put forward the theoretical foundation of level-wise shrinking mechanism based on the maximal duration episodes in Sect. 2. In Sect. 3, we discuss the details of the two-phase strategy. Section 4 is performance evaluation. We conclude the paper at the end.

2   Theoretical Foundation of Level-Wise Shrinking

In this section, we discuss the theoretical foundation of level-wise shrinking mechanism based on maximal duration episodes.

2.1   Preliminary

Definition 1 (Event sequence). *Let E be a set of event types, an event is defined by a event-time pair (e, t) where $e \in E$ and t is the time at which e occurs. An ordered event sequence is denoted as $S = < (e_1, t_1), (e_2, t_2), \ldots, (e_n, t_n) >$, such that $\forall i \in \{1, \ldots, n\}$, $e_i \in E$, and $t_i < t_j$, $1 \leq i < j \leq n$.*

Definition 2 (Episode). *Let maxwin be a user-specified maximal duration constraint,* $P_i =< (e_{i1}, t_{i1}), (e_{i2}, t_{i2}), \ldots, (e_{ik}, t_{ik}) >$ *is an episode in sequence* S, *if and only if* P_i *is a subsequence of* S, *and* $t_{ik} - t_{i1} \le maxwin$. *For simplicity,* P_i *is also represented as* $< e_{i1}, e_{i2}, \ldots, e_{ik} >$, *where the time field in each event-time pair is omitted.*

Definition 3 (Frequent episode). *Let* $sup(P)$ *be the support of* P, *and minsup be a user-specified minimum support threshold,* P *is a frequent episode, if and only if* $sup(P) \ge minsup$.

Definition 4 (Closed episode). P_i *is a closed episode in* S, *if and only if there is no any episode* $P_j \in S$, $P_i \subseteq P_j$ *and* $sup(P_i) = sup(P_j)$.

Definition 5 (Frequent Closed episodes). *If* P *is both a frequent and closed episode,* P *is called a frequent closed episode.*

2.2 Candidate Episodes and Index Points

According to the definition of episodes, if the difference of the occurrence time of a pair of adjacent events in a sequence is more than *maxwin*, they cannot be the elements of any episode. We make use of this feature to extract adjacent frequent event pair as candidate episodes.

Definition 6 (Frequent 2-neighboring episodes, F-2NE). *Let* S *be a sequence, and* $< (e_j, t_j), (e_{j+1}, t_{j+1}) >$ *be an adjacent frequent event pair in* S, $< (e_j, t_j), (e_{j+1}, t_{j+1}) >$ *is a frequent 2-neighboring episode, if and only if* $t_{j+1} - t_j \le maxwin$ *and* $sup(< (e_j, t_j), (e_{j+1}, t_{j+1}) >) \ge minsup$.

Definition 7 (Index point). *Let* $< e_j, e_{j+1} >$ *be an F-2NE in sequence* S, $< t_j, sn_j >$ *is its index point, where* t_j *is the occurring time of* e_j, *and* sn_j *is the serial number of* e_j *in* S.

Figure 1 is a sequence example $S =< EBCDEBCDEBCDE >$. Supposed *minsup* of 3 and *maxwin* of 4, $< EB >$, $< BC >$, $< CD >$ and $< DE >$ are the F-2NEs in S, and their index points are listed in Table 1. For example, the index point $< 1, 1 >$ of EB represents the first event E occurs in S at time 1, and its serial number is 1.

Fig. 1. An example of sequence S

In order to avoid scanning the sequence again after getting the F-2NEs, we record all index points of a sequence into an inverted list called InxF2NE, indexed by F-2NEs.

Table 1. The index points in S

F-2NE	Support	Index points
$< EB >$	3	$< 1, 1 >, < 7, 5 >, < 13, 9 >$
$< BC >$	3	$< 2, 2 >, < 8, 6 >, < 14, 10 >$
$< CD >$	3	$< 3, 3 >, < 9, 7 >, < 15, 11 >$
$< DE >$	3	$< 4, 4 >, < 10, 8 >, < 16, 12 >$

The purpose of extracting the F-2NE is to clean out the non-frequent 2-neighboring episodes from a sequence. By the rule that if an episode isn't frequent, it is certain that all of its super episodes aren't frequent. Thus, we use the collection of the F-2NEs as the set of candidate episodes ($CanES$), which extraction procedure is shown as Algorithm 1.

Algorithm 1. CanES_Generation()

INPUT: S; $minsup$; $maxwin$;

OUTPUT: $CanES$;

BEGIN

1. $FreEv = \phi$; /*initialize the set of frequent events*/
2. $2NES = \phi$; /*initialize the set of 2-neighboring episodes*/
3. $CanES = \phi$; /*initialize the set of candidate episodes*/
4. $FreEv$=Get-FrequentEvent(S); /*Scan S and get all frequent events*/
5. for each event e_i in S
6.   if $e_i \notin FreEv$
7.   remove e_i from S /*clean out the non-frequent events*/
8.   else /*generate the 2-neighboring episodes
9.   if $(e_{i+1} \in FreEv) \vee ((t_{j+1} - t_j) \leq maxwin)$
10.   if $< e_j, e_{j+1} > \notin 2NES$
11. $2NES = 2NES \cup < e_j, e_{j+1} >$;
12. $sup(< e_j, e_{j+1} >) = 1$;
13.   InxF2NE-Insert($< e_j, e_{j+1} >$);
14.   else
15. $sup(< e_j, e_{j+1} >) + +$;
16.   InxF2NE-Update($< e_j, e_{j+1} >$);
17.   end if
18.   end if
19. end for;
20. for each $X \in 2NES$ do /*generate the CanES*/
21.   if $(sup(X) \geq minsup)$
22. $CanES = CanES \cup X$
23.   else
24.   InxF2NE-Delete($< e_j, e_{j+1} >$)
25.   end if
26. end for;
27. return $CanES$,

END

2.3   Maximal Duration Episodes

Since there is a time duration constraint for each episode, we extend the F-2NEs into the maximal episodes by *maxwin* as far as possible.

Definition 8 (F-2NE prefix). *Let* $\varphi = < e_{i1}, e_{i2} >$ *be a F-2NE, φ is called a F-2NE prefix of* $P_i = < e_{i1}, e_{i2}, \ldots, e_{ik} > (k \geq 2)$. *For the sake of simplicity, we use "prefix" to represent the F-2NE prefix in the paper.*

Definition 9 (Maximal   duration   episode,   MaxDE). *Let* $P_i = < e_{i1}, e_{i2}, \ldots, e_{ik} > (k \geq 2)$ *be a maximal duration episode in* S, *if and only if* $t_{ik} - t_{i1} \leq maxwin$, *and there is no super episode* $P_i' = < e_{i1}, e_{i2}, \ldots, e_{ik}, e_{i,k+1} > (k \geq 3)$ *in* S, $t_{i,k+1} - t_{i1} \leq maxwin$.

For *minsup* of 3 and *maxwin* of 4, Fig. 2 shows the maximum duration episodes extended from different prefixes in S (not including the F-2NEs).

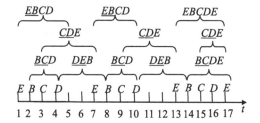

Fig. 2. The generation of the maximum duration episodes

The prefixes and their corresponding MaxDEs of S are listed in Table 2, where the underlined events mean the prefixes, and the numbers in parentheses are the supports. For example, $EBCDE(1)$ represents that the prefix of the maximum duration episodes $EBCDE$ is EB and its support is 1.

Table 2. The MaxDEs in S

F-2NE	MaxDEs
$EB(3)$	$\underline{EB}CDE(1)$, $\underline{EB}CD(2)$
$BC(3)$	$\underline{BC}DE(2)$, $\underline{BC}D(1)$
$CD(3)$	$\underline{CD}E(3)$
$DE(3)$	$\underline{DE}B(2)$

The construction procedure of MaxDEs is shown in Algorithm 2. It first gets all index points of an F-2NE from the index. Then, for each index point, it determines the start position p and the maximal event occurring time $tmax$ in the

MaxDE. Finally, it starts from the second event after p, and judges the occurring time of the after events in the sequence. If the time isn't more than $tmax$, the event should be included in the MaxDE. Otherwise, the generation procedure terminates.

Algorithm 2. MaxDE_Construction()

INPUT: $\varphi = < e_j, e_{j+1} >$; $maxwin$;

OUTPUT: $\varphi.MaxDES$; /*The set of maximal duration episodes of φ*/

BEGIN

1. $\varphi.MaxDES = \phi$;
2. Ψ=InxF2NE-Query(φ); /* query the index points of φ from InxF2NE*/
3. for each index point $x \in \Psi$
4. $t_{start}=x.t_j$; /*get the occurring time of e_j*/
5. $p = x.sn_j$; //get the serial number of e_j;
6. $t_{max} = t_{start} + maxwin$;
7. $p=p+2$; /*start from the second event after p*/
8. $tmp =< e_j, e_{j+1} >$;
9.   while $t_p \leq t_{max}$
10. $tmp = tmp + e_p$; /*add e_p into the temporary episode*/
11. $p=p+1$;
12.   end while
13. $\varphi.MaxDES = \varphi.MaxDES \cup tmp$
14. end for
15. return $\varphi.MaxDES$

END

2.4   Candidate Frequent Closed Episodes

Based on the maximal duration serial episodes, we can get the candidate frequent closed episodes in sequences.

Definition 10 (Episodes with the same prefix). $P_i =< e_{i1}, e_{i2}, \ldots, e_{in} >$ and $P_j =< e_{j1}, e_{j2}, \ldots, e_{jm} >$, $m \leq n$, are two different episodes with the same prefix in a sequence, if and only if $e_{i1} = e_{j1}$, and $e_{i2} = e_{j2}$.

Definition 11 (Super episode and subepisode with the same prefix). $P_j = < e_{j1}, e_{j2}, \ldots, e_{jm} >$ is a subepisode of $P_i =< e_{i1}, e_{i2}, \ldots, e_{in} >$ $(m < n)$ with the same prefix (i.e., P_i is the super episode of P_j with the same prefix), if and only if $e_{i1} = e_{j1}$, and $e_{i2} = e_{j2}$, and there is m integers, $1 \leq i_3 < i_4 < \ldots < i_m \leq n$, $\forall k(3 \leq k \leq m)$, $e_{ik} = e_{jk}$, denoted as $P_j \preceq P_i$.

Definition 12 (Candidate frequent closed episode, CanFCE). P is a candidate frequent closed episode based on prefix φ, if and only if $sup(P) \geq minsup$, and there is no P' with the same prefix, $P \preceq P'$ and $sup(P) = sup(P')$.

Property 1. If P is a maximal duration episode and $sup(P) \geq minsup$, P is a candidate frequent closed episode based on its prefix.

Proof: (1) If there is another maximal duration episode P' with prefix φ and $P \preceq P'$, the total support of P should be the sum of $sup(P)$ and $sup(P')$. Thus, P cannot be closed by P'.

(2) If there is no any maximal duration episode P' with prefix φ and $P \preceq P'$, there is no any super episode with the same prefix which can close P. □

By Definition 12, if $sup(P) \geq minsup$, P is a candidate frequent closed episode based on φ. Property 1 shows, when doing closure relation judgment between two maximal duration episodes, we only need to determine closure relations between the maximal duration episodes and their subepisodes, instead of checking the whole sequences. This property is the premise that we can do level-wise shrinking processing.

Property 2. *Let P be a maximal duration episode and φ be its prefix, if $sup(P) = sup(\varphi)$, all subepisodes with the same prefix of P are closed by P.*

Proof: Obviously, if $sup(P) = sup(\varphi)$, φ cannot be the prefix of any other maximal duration episode in the sequence.

Therefore, all P's subepisodes with prefix φ have the same support as P, so that they are closed by P. □

Thus, if a maximal duration episode has the same support as its prefix, it isn't necessary to further judge the closure relationships between the episode and its subepisodes. This is an important termination condition of level-wise shrinking mechanism.

The candidate frequent closed episodes can eliminate the redundant episodes among the episodes with the same prefix. Thus, we can redefine the frequent closed episodes.

Proposition 1. *P is a frequent closed episode, if and only if there is no any candidate frequent closed episode P' with different prefix in the sequence, $P \subseteq P'$ and $sup(P) = sup(P')$.*

3   Two-Phase Mining Strategy

Differing from the level-wise pattern growth algorithms, the two-phase mining strategy is based on the idea of "divide and conquer", namely, to decompose the whole sequence into multiple maximal subsequences, and adopts the level-wise shrinking mechanism to discover the frequent closed episodes.

3.1   Phase I Closure: Generation of Candidate Frequent Closed Episodes

The task of Phase I closure is to compare the MaxDEs with the same prefix in a sequence, to discover the candidate frequent closed episodes.

By Property 1, if the support of a maximal duration episode isn't less than the support threshold, it can be directly determined as a candidate frequent closed episode. By Property 2, if the support of a maximal duration episode is equal to that of its F-2NE prefix, all of its subepisodes with the same prefix are closed by it. Otherwise, we need to compare the closure relationships between the maximal duration episodes and their subepisodes.

Assumed X is a maximal duration episode with length L, and φ is its prefix, the shrinking procedure works like this:

- *Step 1.* Get all X's length-$(L-1)$ subepisodes with prefix φ.
- *Step 2.* For each length-$(L-1)$ subepisode, match it with all maximal duration episodes with prefix φ except X.
- *Step 3.* Compute the support of each length-$(L-1)$ subepisode. If the support isn't less than *minsup* and not equal to $sup(X)$, the subepisode is determined as a candidate frequent closed episode. Otherwise, as a non candidate frequent closed episode.
- *Step 4.* If the support of the length-$(L-1)$ subepisode is equal to that of its F-2NE prefix (termination condition A) or its length is 2(termination condition B), stop shrinking. Otherwise, shrink the length-$(L-1)$ subepisode into length-$(L-2)$ subepisodes, and so on, until one of the termination conditions is held.

The shrinking algorithm based on the maximal duration episodes is shown as Algorithm 3.

Algorithm 3. Shrink_processing()
INPUT: φ; X;
OUTPUT: Ω; /*The set of CCFE of X*/
BEGIN
1. L=Get-Length(X); /* get the length of X */
2. Z=Get-Subepisode(X,φ,L-1); /*get all L-1 sub-episodes with prefix φ*/
3. for each $Y \in Z$ do
4.   if $sup(Y) = sup(\varphi) \vee$ the length of Y is 2
5. $\Omega = \Omega \cup Y$
6.   else
7.   if $sup(Y) \neq sup(X) \wedge sup(Y) \geq minsup$
8. $\Omega = \Omega \cup Y$
9.   end if
10.   Shrink_processing(Y)
11.   end if
12.end for
13.return Ω
END

3.2   Phase II Closure: Mining the Frequent Closed Episodes

The task of Phase II closure is to compare the candidate frequent closed episodes in the sequence, and find the final frequent closed episodes.

Since the closure relations only occur between the episodes with the same support, we only need to compare the closure relationships among the episodes with different prefixes which have the same support after obtaining the candidate closed episodes with the same prefix. For the space limitation, we don't introduce the details of the algorithm of in Phase II.

For example, $\underline{EBC}D(3)$, $\underline{BC}D(3)$, $\underline{CD}E(3)$, and $\underline{DE}(3)$ are the candidate frequent closed episodes in S, and their supports are the same, so it needs to match them with each other. Since $\underline{BC}D$ and $\underline{DE}$ are the subepisodes of $\underline{EBC}D$ and $\underline{CD}E$ respectively, they are closed by their super episodes. Thus, $\underline{EBC}D$ and $\underline{CD}E$ are the final frequent closed episodes in S.

4   Performance Evaluation

This section tests and analyses the performance of the suggested Two-phase Episode Mining strategy, denoted as 2PEM.

All experiments are performed both on the simulated and real datasets: (1) The simulated dataset T10I4D100K is generated from the IBM Almaden Quest (http://fimi.ua.ac.be/data/), which contains 100000 sequences; (2) The real dataset is the retail records from an anonymous retail store in Belgium (http://fimi.ua.ac.be/data/), which contains 88162 sequences. We give a timestamp to each event randomly according to event occurrence order.

The performance measures include running time and memory overhead, and the selected comparison methods are the breadth-first algorithm Clo_episode [12] and the depth-first algorithm FCEMine [15]. All experiments are performed on 2.5 GHZ Intel Core I5 PC with 2 GB memory and in Windows 7 environment.

Figure 3 is the running time comparison under different support thresholds with $maxwin = 10$. It can be found the running time of 2PEM is obviously less than the others. The reason is 2PEM can quickly generate the maximal duration episodes using index, and the two-phase strategy can reduce the number of closure operations greatly.

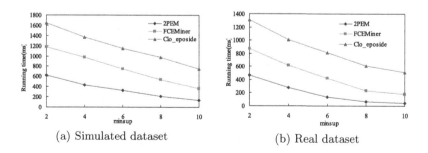

(a) Simulated dataset   (b) Real dataset

Fig. 3. The running time comparison under different support thresholds

Figure 4 is memory overhead comparison of the three methods under different support thresholds with $maxwin = 10$. As shown in the figures, the memory overhead of the 2PEM method is slightly higher than those of other methods when *minsup* is less than 4. This is because when the support threshold is smaller, it needs more memory space to store the F-2NEs and index information. But when the support is up to 8, the memory required by Clo_episode and FCEMine become larger than 2PEM.

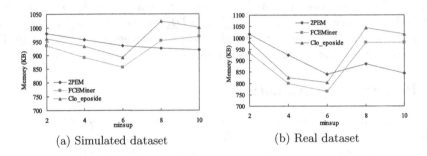

(a) Simulated dataset   (b) Real dataset

Fig. 4. The memory comparison under different support thresholds

Figure 5 is the running time comparison of the three methods under different *maxwin* with *minsup* = 4. Overall, the running time of all methods rises with the increase of *maxwin*. Similarly, the time of 2PEM is less than those of other two methods, and the increase ratio of 2PEM is also less than the other methods. This is because the sequence search time of 2PEM doesn't change with the change of *maxwin*.

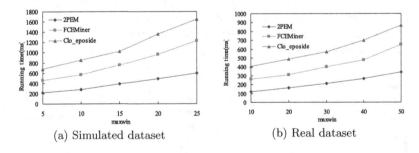

(a) Simulated dataset   (b) Real dataset

Fig. 5. The running time comparison under different duration constraints

Figure 6 is memory overhead under different *maxwin* with *minsup* = 4. It can be found, the memory overhead of three methods is relatively stable, and the space required by 2PEM is slightly higher than the others.

Figure 7 is the running time comparison of the three methods under different sequence lengths (5000–25000) with *minsup* = 4 and *maxwin* = 10. It can be seen, as the growth of sequence lengths, the growth of the running time of 2PEM

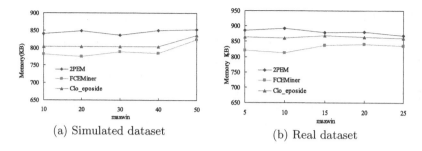

Fig. 6. The memory overhead comparison under different duration constraints

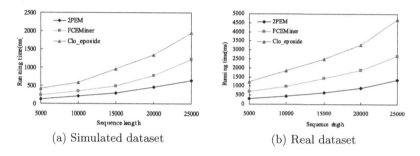

Fig. 7. The running time comparison under different sequence lengths

is relatively stable. This is because the number of loop iterations of Clo_episode and FCEMine increase exponentially with the increase of sequence length.

Figure 8 is memory overhead comparison under different lengths with $minsup = 4$ and $maxwin = 9$. We can see from the figure, the memory sizes of the three methods increase with the increase of sequence length, and the memory space of 2PEM is slightly higher than the others. This is because the space storing the maximal duration episodes for 2PEM increases with the increase of sequence length.

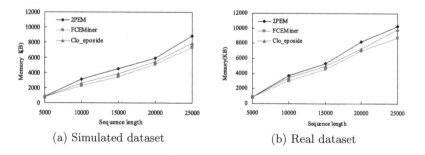

Fig. 8. The memory comparison under different sequence lengths

5   Conclusions

Episode mining is an important branch of sequential pattern mining. In this paper, we propose a novel two-phase mining strategy to mine the frequent closed episodes based on the maximal duration episodes. Experiments verified that the proposed strategy has better mining effectiveness and efficiency.

How to integrate the level-wise episode shrinking framework with the parallel episode mining and complex episode mining is our future work.

References

1. Mannila, H., Toivonen, H., Verkamo, A.I.: Discovering frequent episodes in sequences (Extended Abstract). In: Proceedings of KDD 1995, pp. 210–215 (1995)
2. Mannila, H., Toivonen, H., Verkamo, A.I.: Discovery of frequent episodes in event sequences. Data Min. Knowl. Disc. **1**(3), 259–289 (1997)
3. Laxman, S., Tankasali, V., White, R.: Stream prediction using a generative model based on frequent episodes in event sequences. In: Proceedings of KDD 2008, pp. 453–461 (2008)
4. Ng, A., Fu, A.W.C.: Mining frequent episodes for relating financial events and stock trends. In: Proceedings of PAKDD 2003, pp. 27–39 (2003)
5. Wan, L., Chen, L., Zhang, C.: Mining frequent serial episodes over uncertain sequence data. In: Proceedings of EDBT 2013, pp. 215–226 (2013)
6. Wan, L., Chen, L., Zhang, C.: Mining dependent frequent serial episodes from uncertain sequence data. In: Proceedings of ICDM 2013, pp. 1211–1216 (2013)
7. Katoh, T., Tago, S., Asai, T., Morikawa, H., Shigezumi, J., Inakoshi, H.: Mining frequent partite episodes with partwise constraints. In: Appice, A., Ceci, M., Loglisci, C., Manco, G., Masciari, E., Ras, Z.W. (eds.) NFMCP 2013. LNCS, vol. 8399, pp. 117–131. Springer, Heidelberg (2014)
8. Ma, X., Pang, H., Tan, L.: Finding constrained frequent episodes using minimal occurrences. In: Proceedings of ICDM 2004, pp. 471–474 (2004)
9. Laxman, S., Sastry, P.S., Unnikrishnan, K.P.: Discovering frequent episodes and learning hidden markov models: A formal connection. IEEE Trans. Knowl. Data Eng. **17**(11), 1505–1517 (2005)
10. Laxman, S., Sastry, P., Unnikrishnan, K.: A fast algorithm for finding frequent episodes in event streams. In: Proceedings of KDD 2007, pp. 410–419 (2007)
11. Huang, K., Chang, C.: Efficient mining of frequent episodes from complex sequences. Inf. Syst. **33**(1), 96–114 (2008)
12. Zhou, W., Liu, H., Cheng, H.: Mining closed episodes from event sequences efficiently. In: Proceedings of PAKDD 2010, pp. 310–318 (2010)
13. Tatti, N., Cule, B.: Mining closed strict episodes. In: Proceedings of ICDM 2010, pp. 501–510 (2010)
14. Tatti, N., Cule, B.: Mining closed episodes with simultaneous events. In: Proceedings of KDD 2011, pp. 1172–1180 (2011)
15. Zhu, H., Wang, W., Shi, B.: Frequent closed episode mining based on minimal and non-overlaping occurrence. J. Comput. Res. Dev. **50**(4), 852–860 (2013)
16. Wu, C., Lin, Y., Yu, P.S., Tseng, V.S.: Mining high utility episodes in complex event sequences. In: Proceedings of KDD 2013, pp. 536–544 (2013)
17. Ao, X., Luo, P., Li, C., Zhuang, F., He, Q.: Online frequent episode mining. In: Proceedings of ICDE 2015, pp. 891–902 (2015)
18. Tatti, N.: Discovering episodes with compact minimal windows. Data Min. Knowl. Disc. **28**(4), 1046–1077 (2014)

Effectively Updating High Utility Co-location Patterns in Evolving Spatial Databases

Xiaoxuan Wang, Lizhen Wang$^{(\boxtimes)}$, Junli Lu, and Lihua Zhou

Department of Information Science and Engineering,
Yunnan University, No. 2 Cuihu North Road,
Wuhua District, Kunming City, Yunnan Province, China
wangxiaoxuan1037@163.com, lzhwang2005@126.com

Abstract. Spatial co-location mining has been used for discovering spatial feature sets which show frequent association relationships based on the spatial neighborhood. In spatial high utility co-location mining, we should consider the utility as a measure of interests, by considering the different value of individual instance that belongs to different feature. This paper presents a problem of updating high utility co-locations on evolving spatial databases which are updated with fresh data at some areas. Updating spatial patterns is a complicated process in that fresh data increase the new neighbor relationships. The increasing of neighbors can affect the result of high utility co-location mining. This paper proposes an algorithm for efficiently updating high utility co-locations and evaluates the algorithm by experiments.

Keywords: Spatial high utility mining · Co-location patterns · Incremental update

1 Introduction

With the spatial data explosive growth and spatial databases being used widely, discovering spatial knowledge is more important and necessary. Spatial data mining is used to discover interesting and previously unknown, but potentially useful patterns from spatial databases [1, 2]. Spatial data mining is a task that more difficult than traditional transaction itemset mining, because there are complex spatial data types/features and confusing spatial relationships.

A spatial co-location pattern is made of a set of features whose instances are frequently appeared and located in a nearby area. Examples of frequently appeared together of features include Symbiotic species such as West Nile Virus often appears in region with many mosquito, and the traffic problem like the relationships between the traffic jam and car accident. In communication, spatial co-location patterns can be used to deal with the service requested by mobile users in nearby region [8].

With respect to the classic co-location mining, this paper considers the high utility and updating data two factors. The importance of patterns as utility is considered in high utility pattern mining on transaction databases, which is the same in spatial database. Many spatial co-location pattern mining works [3, 4] are mainly used to mine the frequent patterns, and the importance of different features is same. However,

© Springer International Publishing Switzerland 2016
B. Cui et al. (Eds.): WAIM 2016, Part I, LNCS 9658, pp. 67–81, 2016.
DOI: 10.1007/978-3-319-39937-9_6

in many spatial databases, the importance of different features is different. These approaches which focus on the participation ratio ignore some interesting but low frequency patterns. At the same time, the previous works assume that all data is available and the spatial database is static. However, there are many applications, such as climate monitoring, collected their data periodically and continually. For example, daily climate measurement values are collected at every 0.5 degree grid of the globe [8]. For keeping the interesting of analysis, discovered the importance of features should be considered and the patterns should be updated, with respect to the most recent database.

Compared with updating frequent co-location patterns, updating spatial high utility co-location patterns faced more challenges. Updating of databases mean there are data points increased or deleted. Unlike the transaction databases, if a new data point insert into a spatial database, it can make neighbor relationships with exiting data points as well as other new data points on the continuous space [8]. The all neighbor relationships need to be re-calculated. And due to the changed of neighbor relationships, the utility of patterns may be changed, which means the pattern utilities also need to be re-calculated. The task of updating the high utility co-location patterns is complex, and just re-calculating all of the patterns is not enough.

Our contributions are as follows:

First, we proposed a basic algorithm for mining spatial high utility patterns in an incremental database, which based on join-based co-location mining algorithm [3]. The basic algorithm is complete but not efficient. Second, we proposed a PUCP (Part Update of Co-location Patterns) algorithm for effectively updating the high utility co-location patterns with the increasing of spatial data points, which just need notice the changed patterns. The PUCP algorithm is efficient and can help us achieve the expected result.

Related works are given in Sect. 2. Section 3 describes the basic concepts of co-location mining and high utility co-location mining, while the proposed algorithm and new concepts are presented in Sect. 4. Experimental results are presented in Sect. 5. Finally, the conclusion and future work are given in Sect. 6.

2   Related Work

High Utility Itemst Mining: TThe theoretical model and definitions of high utility itemset mining were proposed in [5], which is called expected utility. Every item in the itemsets is associated with an additional value, called internal utility which is quantity of the item. The UMining algorithm was proposed in [6]. In [12], C.F.Ahmed proposed a very useful measure, called frequency affinity. UP-Growth method proposed in [13] enhances the mining performance in utility mining through maintaining the information of high utility itemsets by UP-tree. [10, 11] mainly researched how to mine high utility sequential patterns. [7] presented a two-phase algorithm to prune down the number of candidates and precisely obtain the complete set of high utility itemsets.

Spatial Co-location Pattern Mining: Spatial co-location pattern mining [1, 4] discovers spatial feature sets which appear frequency association relationships based on the spatial neighborhood. Co-location pattern mining was first discussed in [1] which mines association based on spatial relationships. Previous works on co-location pattern mining have presented different approaches, Huang et al. [3] proposed statistically meaningful interest measures and a join-based mining algorithm. Join-less algorithm for co-location mining was proposed in [4], which using instances-lookup scheme to find neighbor relationships. High utility co-location mining was discussed in [21].

Incremental Patterns Mining: Incremental pattern mining focuses on the dynamic nature of databases, it needs to update the patterns when added fresh data. An incremental mining algorithm for high utility patterns on the transaction database was proposed in [14]. [15–19] are mainly about solving the problem of incrementally mining frequent itemsets. Algorithms of mining spatial prevalent co-location patterns in an incremental database were proposed in [8, 9]. In [22], Lu et al.proposed a new algorithm for mining co-locations on spatial database which are constantly changed with add and disappeared data. [20] presented an efficient algorithm UWEP for updating large itemsets when new transactions are added to the set of old transactions.

3 High Utility Co-location Mining

In this section, we will introduce the basic concepts of high utility co-location mining and the problem definition.

3.1 Basic Concepts

3.1.1 Basic Concepts of Co-location Pattern Mining

Given a set of spatial instances S, and a set of spatial features F, the **neighbor relationship R** is a Euclidean distance metric with its threshold value d. In the Fig. 1(a), each instance denoted by the feature type and a numeric id value e.g. a_1. If two instances satisfy the neighbor relationship R, they link by the black line. A **co-location pattern c** is a subset of spatial features, $c \subseteq F$, whose instances frequently form cliques under the neighboring relationship R.

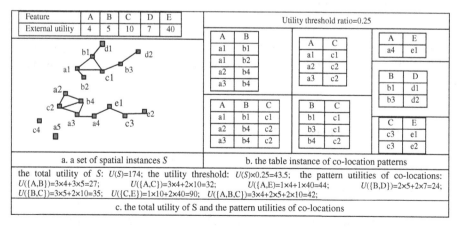

c. the total utility of S and the pattern utilities of co-locations

the total utility of S: $U(S)=174$; the utility threshold: $U(S) \times 0.25=43.5$; the pattern utilities of co-locations:
$U(\{A,B\})=3 \times 4+3 \times 5=27$; $U(\{A,C\})=3 \times 4+2 \times 10=32$; $U(\{A,E\})=1 \times 4+1 \times 40=44$; $U(\{B,D\})=2 \times 5+2 \times 7=24$;
$U(\{B,C\})=3 \times 5+2 \times 10=35$; $U(\{C,E\})=1 \times 10+2 \times 40=90$; $U(\{A,B,C\})=3 \times 4+2 \times 5+2 \times 10=42$;

Fig. 1. (a) a set of spatial instance S; (b) the table instances of co-locations; (c) the total utility of S and the patterns' utilities

A co-location c is called a k-co-location if the number of features in c is k. Such as, $c = \{A,B,C\}$ is a 3-co-location. A **Co-location Instance** I is a set of instances, $I \subseteq S$, which contains the instances of all the features in the co-location c, they satisfy the neighbor relationship R and form the clique relationship. For example, $I = \{a_1,b_1,c_1\}$ in Fig. 1(b) is a co-location instance of $\{A,B,C\}$. A co-location instance is called a **Row Instance**. The set of all row instances of c is the **table instance** of c, such as the Fig. 1(b), denoted as $T(c)$.

3.1.2   Basic Concepts of High Utility Co-location Pattern Mining

The **external utility** is used to describe the importance of different features. Given the feature set of a spatial database $F = \{f_1, f_2, ..., f_n\}$, we denote the external utility of f_i as v (f_i). The **internal utility** of f_i is the quantity of distinct instances of f_i appearing in $T(c)$. We denote it as $q(f_i, c) = |\prod_{f_i} T(c)|$, where $\prod$ is the projection operation. For example, in the Fig. 1(b), for co-location $\{A,B\}$, the internal utility of A: $q(A,\{A,B\}) = |\prod_A T(\{A,B\})| = |\{a_1,a_2,a_3\}| = 3$. Given a co-location $c = \{f_1, f_2, ..., f_k\}$, we can get the external utility and the internal utility of any feature f_i in c, so we define the product of the external utility and the internal utility of f_i as the **utility of f_i in c**, denoted as u $(f_i, c) = q(f_i, c) \times v(f_i)$. In the Fig. 1(c), $u(A,\{A,B\}) = q(A,\{A,B\}) \times v(A) = 3 \times 4$. The **pattern utility of c** is the sum of the utility of all features in c, i.e., $U(c) = \sum\limits_{f_i \in c} u(f_i, c)$.

For example, the pattern utility of $\{A,B\}$: $U(\{A,B\}) = u(A,\{A,B\}) + u(B,\{A, B\}) = 27$. **The total utility of database** S is the sum of the utilities of all features, we denoted it as $U(s) = \sum\limits_{f_i \in S} u(f_i, \{f_i\})$, $u(f_i, \{f_i\})$ is the product of the number of the instances of f_i appearing in S and the external utility of f_i. In Fig. 1, $u\{A,\{A\}\} = 5 \times 4$, and $U(S) = u\{A,\{A\}\} + u\{B,\{B\}\} + u\{C,\{C\}\} + u\{D,\{D\}\} + u\{E,\{E\}\} = 174$. At the same time, we need a threshold to measure the interesting of patterns, so we come up with the following definitions:

Definition 1 (utility threshold ratio). *The utility threshold ratio is a constant, which is specified by users, between 0 and 1, denoted as $\xi (0 \leq \xi \leq 1)$.*

Definition 2 (utility threshold). *The total utility of S is U(S). We define the utility threshold of S as the product of the total utility of S and the utility threshold ratio, denoted as Minutility(S) = U(S) $\times$ ξ.*

Given a co-location c, if $U(c) \geq Minutility(S)$, c is called a high utility pattern. In Fig. 1, if the threshold ratio is 0.25. The total utility of S is $U(S) = 174$. According to Definition 2, $Minutility(S) = 174 \times 0.25$.

3.2   The Problem of Evolving Databases

Previous researches mainly focus on the static spatial databases. However, if given an evolving database which is updated with fresh data, many problems occur. Like the databases in Fig. 2, the green points represent the original data and the red points are the fresh data that added into the original database. We can see that from Fig. 2(b), the fresh data are just added into the part1 and part2, so only the neighbor relationships in part1 and part2 are changed. However, these changes may affect the high utility co-locations of the update spatial database. In the following examples, we can find that the change of the high utility set of the spatial database. From Fig. 3(a) and (b), the original instances set $S1$ has been added some new instances. Obviously, the total utility of $S1$ and utility threshold has been changed because the utility of some features in $S1$ have been increased.

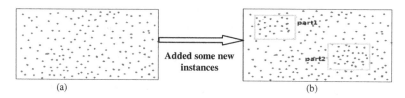

(a)   (b)

Fig. 2. (a) an original spatial database; (b) added some new instances into the original database

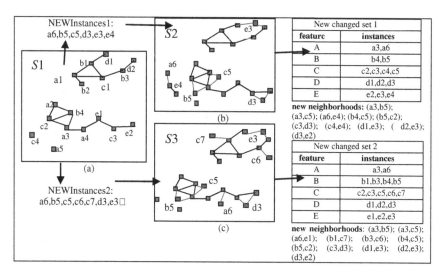

Fig. 3. (a) the original spatial instances set $S1$; (b) the new spatial instances set $S2$; (c) the new spatial instances set $S3$.

We also find some co-location patterns' utilities have been changed. For example, the pattern utility of {D,E} is 0 in $S1$ of Fig. 3(a), and {D,E} is not the high utility pattern. But in $S2$ of Fig. 3(b), {D,E} is a high utility pattern when d_3, e_3 has been added into $S1$, and the utility of {D,E} $U(\{D,E\}) = 101$. For the original database, we have found the result of high utility patterns. If using a naive algorithm to discover the new high utility patterns, we must re-scan and re-calculate the whole new spatial database. So we are faced with the following problem:

How to effectively update spatial high utility co-location patterns by using the known original high utility patterns and computing the changed patterns?

Thus, we propose a new algorithm to deal with the problem. In Sect. 4, we will describe our algorithm in detail.

4 Incremental Mining for High Utility Co-locations

4.1 Basic Algorithm

Given an original spatial database, some new data points and a utility threshold, we first should calculate all new relationships when inserted new data points into original database. Then co-location candidates are generated, the high utility patterns will be selected and some low utility co-location patterns will be filtered. Finally, the all high utility patterns satisfying the utility threshold are generated (see Fig. 4). Using the basic algorithm, we must re-calculate many existed neighbor relationships and re-calculate many patterns which the utility is unchanged. Thus, we propose another algorithm that just needs to calculate the changed patterns. In the following subsections, we discuss our method in detail.

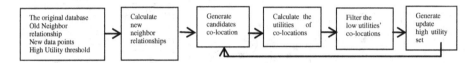

Fig. 4. The Framework of Basic Algorithm

4.2 Four Cases of Changed Co-location Patterns

Let $S_{old} = \{o_1,..o_n\}$ be a set of old data points in a spatial database and $S_{new} = \{o_{n+1},.. o_{n+m}\}$ be a set of new data points added in the database. Let $S_{updated}$ be all data points in the updated database, $S_{updated} = S_{old} \cup S_{new}$.

Definition 3 (changed instance). *Given an old instance $o \in S_{old}$, if $\exists oi \in Snew$ and $R (o,o_i) = true$, we call the old instance o as a changed instance.*

For example, In Fig. 3(b), $a_3 \in S_{old}$, $b_5 \in S_{new}$ and $R(a_3,b_5) = $ true, so a_3 is a changed instance.

Definition 4 (new-changed set). *New-changed set is an instances' set which contains S_{new} and all of changed instances.*

For example, In Fig. 3(b), the new-changed set of S2 is $\{a_3,a_6,b_4,b_5,c_2,c_3,c_4,c_5,d_1,d_2, d_3,e_2,e_3,e_4\}$.

According to Definition 2, we can get the utility of new-changed set *U(new-changed set)*, and the utility threshold of new-changed set *Minutility(new-changed set)*, and the high utility co-locations in new-changed set. In considering an original database and the new-changed set, there are four cases may occur (see Fig. 5):

> *Case 1: A co-location is high utility both in original database and new-changed set;*
> *Case 2: A co-location is high utility in original database but not high utility in new-changed set;*

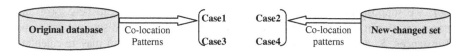

Fig. 5. Four cases of changed co-location patterns

Case 3: *A co-location is not high utility in original database but high utility in new-changed set;*
Case 4: *A co-location is not high utility both in original database and in new-changed set;*

Co-locations in *Case 1* are high utility in both original database and new-changed set, but the update utilities of them may less than the update utility threshold. Similarly, the pattern utilities of co-locations in *Case 2* ~ *Case 4*, may be changed. So we need use some methods to re-calculate their utilities and determine whether they are the high utility patterns. So we proposed a Part Update of Co-location Pattern (PUCP) algorithm.

4.3   Description of PUCP Algorithm

4.3.1   Related Definitions and Lemmas

Definition 5 (instance set). *Given the table instance of c, T(c). Instance set means getting all different instances belong to different features in T(c), and it is defined as:*

$$S(c) = \bigcup_{fi \in c} \prod_{fi} T(c)$$

e.g., in original database $S1$ of Fig. 3(a), $S_{s1}(\{A,E\}) = \{a_4,e_1\}$.

Lemma 1. *Let $U(S_{old})$ be the total utility of the original database, $U(new\text{-}changed\ set)$ is the total utility of the new-changed set, $U(S_{updated})$is the total utility of the update database. There must be $U(S_{updated}) \leq U(S_{old}) + U(new\text{-}changed\ set)$ and Minutility $(S_{old}) + Minutility(new\text{-}changed\ set) \geq Minutility(S_{updated})$.*

Proof: New-changed set contains some old instances, the total utility of $S_{updated}$ must less than $U(S_{old}) + U(new\text{-}changed\ set)$ which re-calculate some same instances. At the same time, because the utility threshold is equal to the product of the threshold ratio and the total utility of database according to Definition 2, $Minutility(S_{old}) + Minutility$ $(New\text{-}changed\ set) \geq Minutility(S_{updated})$.

For example, $Minutility(S1) + Minutility(new\text{-}changed\ set\ of$ $S2) = 49.75.$

$U(S1) + U(new\text{-}changed\ set\ of\ S2) > U(S_{updated}) = 280.$

Lemma 2. *Let $U_{old}(c)$ be the utility of co-location c in original database, $U_{nc}(c)$ be the utility of c in new-changed set, and let $U_{updated}(c)$ be the utility of c in the updated database, then $U_{old}(c) + U_{nc}(c) \geq U_{updated}(c)$ holds.*

Proof: we suppose the instance set of c in the original database to be $S_{old}(c)$, the instance set of c in the new-changed set to be $S_{nc}(c)$. New-changed set contains some old instances, $S_{old}(c) \cap S_{nc}(c)$ may not be empty, $U_{old}(c) + U_{nc}(c)$ may re-calculate some old instance value, so $U_{old}(c) + U_{nc}(c) \geq U_{updated}(c)$.

For example, $S_{update}(\{A,E\}) = \{a_4,a_6,e_1\}$, $U_{old}(\{A,E\}) + U_{nc}(\{A,E\}) \geq U_{updated}(\{A,E\})$.

Lemma 3. *If $S_{old}(c) \cap S_{nc}(c)$ is empty, then $U_{updated}(c) = U_{old}(c) + U_{nc}(c)$.*

Proof: $S_{old}(c) \cap S_{nc}(c) = \varnothing$, which means that $S_{old}(c)$ and $S_{nc}(c)$ do not contain any same instances, all of the instances in $S_{old}(c)$ and $S_{nc}(c)$ must be calculated. So $U_{update}(c) = U_{old}(c) + U_{nc}(c)$.

For co-location $\{C,E\}$ in Fig. 3(a) and (c), $S_{old}(\{C,E\}) = \{c_3,e_1,e_2\}$, $S_{nc}(\{C,E\}) = \{c_4,e_4\}$. $U_{old}(\{C,E\}) + U_{nc}(\{C,E\}) = 140$, $S_{updated}(\{C,E\}) = \{c_3,c_4,e_1,e_2,e_4\}$, obviously, $U_{updated}(\{C,E\}) = U_{old}(\{C,E\}) + U_{nc}(\{C,E\})$.

We can obtain the following corollary from Lemma 3.

Corollary 1. *If a co-location c is high utility both in original database and in new-changed set, and $S_{old}(c) \cap S_{nc}(c)$ is empty, then c must be a high utility pattern in updated database.*

Lemma 4. *If $S_{sold}(c) \cap S_{nc}(c) \neq \varnothing$, then $U_{updated}(c) = U_{old}(c) + U_{nc}(c) - U(S_{sold}(c) \cap S_{nc}(c))$.*

Proof: $S_{sold}(c) \cap S_{nc}(c) \neq \varnothing$, it means that $U_{old}(c) + U_{nc}(c)$ must have been re-calculated some same instances, These instances are all contained in $S_{sold}(c) \cap S_{nc}(c)$, $U_{updated}(c) = U_{old}(c) + U_{nc}(c) - U(S_D(c) \cap S_{nc}(c))$.

For example, the co-location $\{A,E\}$ in $S2$ of Fig. 3(b), $S_{old}(\{A,E\}) \cap S_{nc}(\{A,E\}) = \{e_1\}$, $U_{updated}(\{A,E\}) = U_{old}(\{A,E\}) + U_{nc}(\{A,E\}) - U(\{e_1\})$.

Lemma 5. *If $U_{old}(c) + U_{nc}(c) \leq Minutility(S_{updated})$, then the co-location c must not be high utility.*

Proof: From Lemma 2, we can know $U_{old}(c) + U_{nc}(c) \geq U_{updated}(c)$, so when we know $U_{old}(c) + U_{nc}(c) \leq Minutility(S_{updated})$ there must be $U_{updated}(c) \leq Minutility(S_{updated})$, so c must be not a high utility pattern.

Like co-location $\{A,B\}$ in Fig. 3(b), $U_{old}(\{A,B\}) + U_{nc}(\{A,B\}) \leq Minutility(S_{updated})$. $\{A,B\}$ is not a high utility pattern in update database.

4.3.2   The Examples
Example 1: From Table 1, we first get the high utility patterns of $S1$: $\{A,E\}$ and $\{C,E\}$. Then added the fresh data $\{a_6,b_5,c_5,d_3,e_3,e_4\}$ into $S1$ shown in Fig. 3(b). At first, we calculate the total utility of $S2$, and the utility threshold: $Minutility(S2) = 70$. After scanning the neighbor relationships of the fresh data, we get the new neighbors: $\{a_3, b_5\},\{a_3,c_5\},\{a_6,e_4\},\{b_5,c_2\},\{b_4,c_5\},\{c_3,d_3\},\{c_4,e_4\},\{d_1,e_3\},\{d_2,e_3\},\{d_3,e_2\}$, then the new-changed set on $S2$ is $\{a_3,a_6,b_4,b_5,c_2,c_3,c_4,c_5, d_1,d_2,d_3,e_2,e_3,e_4\}$, so we can calculate the $Minutility(new$-$changed\ set\ of\ S2) = 49.75$. The patterns $\{A,B\}$, $\{A,E\}$ and $\{C,E\}$ are contained in the changed patterns. $\{A,E\}$ and $\{C,E\}$ belong to the case 1: they are high utility patterns both in $S1$ and new-changed set $S2$. $S_{S1}(\{A,E\}) \cap S_{nc1}(\{A,E\})$ is

Table 1. The examples

The original database S1	Minutility $(S1) = (5 \times 4 + 4 \times 5 + 4 \times 10 + 2\times7 + 2 \times 40) \times 0.25 = 43.5$			
	patterns	Instance set	$U_{old}(c)$	High Utility Patterns:{A,E}{C,E} Non-high Utility Set:{A,B}{A,C} {B,D}{B,C}
	{A,B}	$a_1,a_2,a_3,$ $b_1,b_2,$ b_4	27	
	{A,C}	$a_1,a_2,a_3,$ $c_1,c_2,$ c_3	32	
	{A,E}	a_4,e_1	44	
	{B,D}	$b_1,b_3,d_1,$ d_2	24	
	{B,C}	$b_1,b_3,b_4,$ c_1,c_2	35	
	{C,E}	c_3,e_1,e_2	90	
New-changed set of S2	Minutility(New-Changed Set of S2) = $(2 \times 4+2 \times 5+4 \times 10 + 3\times7 + 3\times40) \times 0.25 = 49.75$ Minutility(S2) = 70			
	patterns	Instance set	$U_{nc}(c)$	High utility patterns:{D,E}{C,E}{A,E} Non-high utility set:{A,B}{A,C}{B, C}{C,D}
	{A,B}	a_3,b_5	9	
	{A,C}	a_3,c_5	14	
	{A,E}	a_6,e_4	44	
	{B,C}	$b_4,b_5,c_2,$ c_5	30	
	{C,D}	c_3,d_3	17	
	{C,E}	c_4,e_4	50	
	{D,E}	$d_1,d_2,d_3,$ e_2,e_3	101	
New-changed set of S3	Minutility(New-Changed Set of S3) = $(2 \times 4+4 \times 5+5 \times 10 + 3\times7 + 3\times40) \times 0.25 = 54.75$ Minutility(S3) = 65			
	patterns	Instance set	$U_{nc}(c)$	High utility patterns:{D,E}{B,C} Non-high utility set:{A,B}{A,C}{A, E}{B,C}{C,D}
	{A,B}	a_3,b_5	9	
	{A,C}	a_3,c_5	14	
	{A,E}	a_6,e_1	44	
	{B,C}	$b_1,b_3,b_4,$ $b_5,c_2,$ $c_5,c_6,$ c_7	60	
	{C,D}	c_3,d_3	17	
	{D,E}	$d_1,d_2,d_3,$ e_2,e_3	101	

empty, so {A,E} is a high utility pattern in updated spatial database. The pattern {A,B} belongs to the case 4, because $U_{old}(\{A,B\}) + U_{nc1}(\{A,B\}) \leq Minutility(S2)$.

Example 2: From Table 1, co-location {A,E} is high utility in $S1$, {B,C} is not a high utility pattern; Then we added the fresh data $\{a_6,b_5,c_5,c_6,c_7,d_3\}$, which is shown in Fig. 3(c). $Minutility(S3) = 65$, and the new-changed set is $\{a_3,a_6,b_1,b_3,b_4,b_5,c_2,c_3,c_5,c_6, c_7,d_1,d_2,d_3,e_1,e_2,e_3\}$, the $Minutility(new$-$changed$ set of $S3) = 54.75$. The changed 2-co-location patterns are {A,B}, {A,C}, {A,E}, {B,C}, {C,D} and {D,E}. The co-location {A,E} belongs to case 2, $S_{S2}(\{A,E\}) \cap S_{nc2}(\{A,E\}) = \{e_1\}$. $U_{old}(\{A, E\}) + U_{nc2}(\{A,E\}) - U(\{e_1\}) \leq Minutility(S3)$. And {B,C} belongs to case 3, $S_{S2}(\{B, C\}) \cap S_{nc}(\{B,C\}) = \{b_1,b_3,b_4,c_2\}$, $U_{update}(\{B,C\}) = U_{old}(\{B,C\}) + U_{nc2}(\{B, C\}) - U(\{b_1,b_3,b_4,c_2\}) \geq Minutility(S3)$.

4.4 Unchanged Patterns

We just calculate the patterns which appeared in new-changed set. Because under the condition of the original high utility patterns we have known, the changed patterns must relate with the fresh data and many patterns are not changed. If we re-scan the whole update spatial database and re-calculate all patterns, we will waste a lot of computational time on the unchanged patterns. In order to not repeat calculation, our algorithm just computes the changed patterns, which can reduce the computational time and improve execution efficiency.

The algorithm just focuses on the changed patterns which appeared in new-changed set. We need to get the new neighbor relationships, and generate the changed candidates. After calculating all changed patterns, we should deal with the original and unchanged patterns to get the complete high utility set. Co-location c is unchanged and c is a high utility co-location in S_{old}. If $U_{old}(c) \geq Minutility(S_{updated})$, c is a high utility pattern in $S_{updated}$. If the unchanged co-location c is a non-high utility pattern in S_{old}, it must be not high utility, because $U_{old}(c)$ must less than $Minutility(S_{updated})$.

5 Experimental Evaluation

We evaluate the PUCP algorithm with the basic algorithm and EPA algorithm [21] using synthetic data sets and two real data sets. We conducted the following experiments: First, we examine the effects of our algorithm design decision. Using the same neighbor degree [4], we compared the computational costs of generating 2-co-location high utility patterns. We also compared the costs of generating neighborhood relationships using different neighbor degree. Second, we examine the stability of the PUCP algorithm according to the number of points, the neighbor distance thresholds, and the utility threshold ratios. All algorithms are implemented in C# of Visual Studio 2013. And all experiments are conducted on a desktop with Inter Core 3 CPU, 2 GB memory and Windows 7.

5.1   Extended Pruning Algorithm (EPA)

EPA proposed in [21] is based on join-less algorithm. EPA used a pruning method to decrease the patterns that need to calculate. They defined a pruning threshold called extend pattern utility ratio EPUR, which extended the real utility of co-locations and made the utility of patterns to satisfy the downward closure property, then made the extended utility ratio the total utility of spatial databases. EPUR and the downward closure property can prune many low utility patterns in early stage and improve the efficiency of execution.

5.2   Effects of Algorithm Design Decision

The computation costs of generating 2-co-location high utility patterns. Neighbor degree is the average number of neighbor points per each point. We used an original database, whose neighbor degree is 1.42, and added different number of fresh data, i.e., the size of original database is 10500, and the sizes of fresh data are 500, 1000, 2000, 3000, 4000. In Fig. 6(a), the execution time of three algorithms is increased with the increased number of points. However, the execution time of PUCP shows much less than the other two algorithms, because we just mainly focus on the new data points, which decreased the number of data points that have to be dealt with.

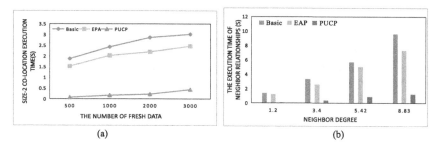

(a)   (b)

Fig. 6. Effects of algorithm design decision ((a) The computation costs of generating 2-co-location high utility patterns; (b) The cost of generating neighborhood relationships with different degree) (Color figure online)

The cost of generating neighborhood relationships with different degree. Neighbor relationships are the foundation of getting the high utility co-locations, so we used different neighbor degrees to evaluate the three algorithms. As we can see from Fig. 6 (b), the cost of the PUCP is well below than the basic and EPA, because it just calculate the neighbor relationships in new-changed set. With the increased neighbor degree, the execution time of three algorithms is increased, because the increased neighbor degree causes the more neighbors for a data point.

5.3 Stability of PUCP Algorithm

Effect of the number of points. We compared the execution time of three algorithms with different number of data points. We used two different spatial frame sizes [4] which can determine the density of the spatial database, 986×986 and 15000×15000, and used the *increment ratio* $(0 \sim 1)$ to limit the number of fresh data points, i.e., the number of original data points is 10000 and *increment ratio* is *20 %, the number of fresh data is* $10000 \times 20 \% = 2000$. At first, the increment ratio is 20 %, and the original data increased from $10000 \sim 50000$. Second, the number of original data is 10000, and the increment ratio is increased from 10 % $\sim$ 90 %. From Fig. 7(a), the performance of PUCP is better than basic algorithm and EPA in spatial database with increased density, by increasing the total number of spatial data points. In any frame size, PUCP shows low execution time. The basic algorithm and EPA in frame size 15000×15000 also shows less cost, but in 986×986, their execution time is dramatically increased due to the increased density. From Fig. 7(b), the computation cost of PUCP is also lower than the basic algorithm and EAP in any frame sizes with increased the *increment ratio*. As shown in Fig. 7, PUCP shows stability to large dense data sets.

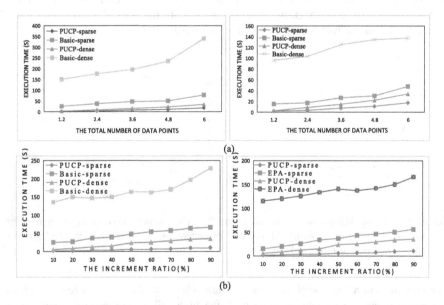

Fig. 7. Effect of the number of points ((a) Different number of original data points; (b) Different increment ratio)

Effect of the utility threshold ratio. In Fig. 8(a), the original data points set 25000, the increment ratio is 20 %, $\xi = 0.25$, $d = 15$. With increasing of ξ from 20 % $\sim$ 40 %, PUCP and EPA algorithms' execution time is less than the basic algorithm. However, PUCP just need deal with new-changed sets, so the performance of the PUCP is better than EPA and basic algorithm. With the utility threshold ratio is increased, EAP and

PUCP performance is better because in that more and more patterns' utilities are less than the utility thresholds of $S_{updated}$, so more patterns are pruned.

Effect of the distance threshold d. In Fig. 8(b), the original data points is 20000, the increment ratio is 20 %, and the utility threshold ratio is 0.25. Similarly, total running time is increased with the increase of d, because a larger value of d means more instances could form cliques which bring more join operations. The performance of PUCP is still better than the basic algorithm and EAP. As shown in Fig. 8, PUCP shows the stability while changing the utility threshold ratio and the distance threshold.

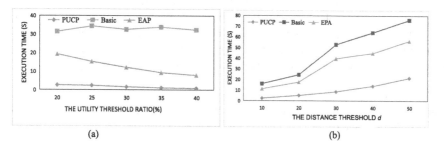

(a)   (b)

Fig. 8. Effect of the utility threshold ratio and the distance threshold d ((a) Different utility threshold ratio; (b) Different distance threshold d

5.4   Evaluation with Real Data Sets

We used two different real plants data sets (two vegetation distribution data sets of the "Three Parallel Rivers of Yunnan Protected Areas") to evaluate PUCP algorithm. At first, the increment ratio is selected 20 %, and the total number of original data points is 300. We changed the distance threshold from 10 to 50. In Fig. 9(a), the PUCP method shows much better performance at different distance thresholds. Second, we used another data set which the total number of old points is 10500. Then the data sets added different number of fresh points. From Fig. 9(b), we can see the execution time of PUCP increase more slowly than the other algorithms with the increase of fresh data points. According to the two results of experiments, our PUCP algorithm shows much better performance.

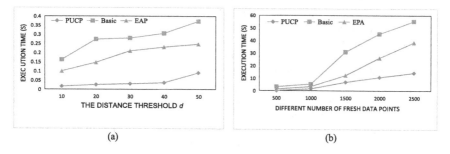

(a)   (b)

Fig. 9. Real data sets ((a) Different distance threshold d; (b) Different number of fresh data points)

6 Conclusion and the Future Work

Focusing on the problem of mining the high utility co-location patterns in the updated database which increased fresh data points into the original database, we define a series of new concepts, divide the changed co-location patterns into four cases and use different criterion to solve the changed patterns which belong to different cases. At the same time, using some methods to identify the unchanged patterns and avoid to testing of many co-locations which are not high utility patterns. So the proposed PUCP algorithm is complete and efficient in finding and updating the high utility patterns. In the experiment evaluation using synthetic and real data sets, we examined the effects of our algorithm design decision and the stability of the PUCP algorithm. The PUCP algorithm efficiently performs in experiments and shows stability to dense and sparse data. In future work, we plan to explore methods to mine high utility spatial co-location patterns, which taking into account the increase and decrease of data points.

Acknowledgment. This work was supported in part by grants (No. 61472346, No. 61262069) from the National Natural Science Foundation of China and in part by a grant (No. 2015FB149, No. 2015FB114) from the Science Foundation of Yunnan Province.

References

1. Koperski, k, Han, J.: A two-phase algorithm for fast discovery of high utility itemsets. In: Egenhofer, M., Herring, J.R. (eds.) SSD 1995. LNCS, vol. 951, pp. 47–66. Springer, Heidelberg (1995)
2. Morimoto, Y.: Mining frequent neighboring class sets in spatial databases. In: 7th ACM SIGKDD International Conference on Knowledge Discovery and Data Mining, pp. 353–358. ACM, New York (2001)
3. Huang, Y., Shekhar, S., Xiong, H.: Discovering co-location patterns from spatial data sets: a general approach. IEEE Trans. Know. Data Eng. (TKDE) **16**(12), 1472–1485 (2004)
4. Yoo, J.S., Shekhar, S.: A join-less approach for co-location pattern mining: a summary of results. IEEE Tran. Knowl. Data Eng. (TKDE) **18**(10), 1323–1337 (2006)
5. Yao, H., Hamilton, H.J., Butz, C.J.: A foundational approach to mining item set utilities from database. In: 4th SIAM International Conference on Data Mining, pp. 215–221 (2004)
6. Yao. H, Hamilton. H.J, and Geng. L.: A unified framework for utility-based measure for mining item sets. In: ACM SIGKDD 2nd Workshop on Utility-Based Data mining, pp. 28–27. ACM, New York (2006)
7. Liu, Y., Liao, W.-k., Choudhary, A.K.: A two-phase algorithm for fast discovery of high utility itemsets. In: Ho, T.-B., Cheung, D., Liu, H. (eds.) PAKDD 2005. LNCS (LNAI), vol. 3518, pp. 689–695. Springer, Heidelberg (2005)
8. Yoo, J.S., Vasudevan, H.: Effectively updating co-location patterns in evolving spatial databases. In: 6th International Conferences on Pervasive Patterns and Applications (PATTERNS 2014), pp. 96–99 (2014)
9. Yoo, J.S, Boulware, D.: Incremental and parallel spatial association mining. In: IEEE International Conference on Big Data (IEEE BigData 2014), pp. 75–76. IEEE Computer Society, New York (2014)

10. Ahmed, C.F., Tanbeer, S.K., Jeong, B.S.: A novel approach for mining high-utility sequential patterns in sequence databases. Electron. Telecommun. Res. Inst. **181**, 4878–4894 (2011)
11. Yin, J., Zheng, Z., Gao, L.: Uspan: an efficient algorithm for mining high utility sequential patterns. In: 18th ACM SIGKDD international conference on Knowledge discovery and data mining, pp. 660–668. ACM, New York (2012)
12. Ahmed, C.F., Tanbeer, S.K., Jeong, B.S., Choi, H.J.: A framework for mining interesting high utility patterns with strong frequency affinity. Inf. Sci. **181**, 4878–4894 (2011). Elsevier Inc.
13. Tseng, V.S., Wu, C.W., Shie, B.E.: UP-growth: an efficient algorithm for high utility itemset mining. In: 16th ACM SIGKDD International Conference on Knowledge Discovery and Data Mining, pp. 253–262. ACM, New York (2010)
14. Lin, C.W., Lan, G.C., Hong, T.P.: An incremental mining algorithm for high utility item sets. Expert Syst. Appl. **39**(8), 7173–7180 (2012)
15. Ahmed, C.F., Tanbeer, S.K., Jeong, B.S.: BS efficient tree structures for high utility pattern mining in incremental databases. IEEE Trans. Knowl. Data Eng. (TKDE) **21**(12), 1708–1721 (2009)
16. Cheung, D., Lee, S.D., Kao, D.: A general incremental technique for maintaining discovered association rules. In: 5th International Conference on Databases Systems for Advanced Applications, pp. 185–194. ACM, New York (1997)
17. Thomas, S., Bodagala, S., Alsabti, K., Ranka, S.: An efficient algorithm for the incremental updation of association rules in large databases. In: ACM International Conference on knowledge Discovery and Data Mining, pp. 263–266. ACM, NewYork (1997)
18. Thomas, S., Chakravarthy, S.: Incremental mining of constrained associations. In: Prasanna, V.K., Vajapeyam, S., Valero, M. (eds.) HiPC 2000. LNCS, vol. 1970, pp. 547–558. Springer, Heidelberg (2000)
19. Cheung, D., Han, J., Ng, V., Wong, C.Y.: Maintenance of discovered association rules in large databases: an incremental updating technique. In: IEEE International Conference on Data Engineering, pp. 106–114. IEEE Computer Society, New York (1996)
20. Ayan, N., Tansel, A., Arkyn, E.: An efficiently algorithm to update large itemsets with early pruning. In: International Conference on Knowledge Discovery and Data Mining, pp. 287–291. ACM, New York (1999)
21. Yang, S., Wang, L.Z., Bao, X.G., Lu, J.L.: A framework for mining spatial high utility co-location patterns. In: 12th International Conference on Fuzzy Systems and Knowledge Discovery (FSKD 2015), pp. 631–637 (2015)
22. Lu, J.L., Wang, L.Z., Xiao, Q., Shang, Y.: Incremental mining of co-locations from spatial database. In: 12th International Conference on Fuzzy Systems and Knowledge Discovery (FSKD 2015), pp. 612–617 (2015)

Mining Top-k Distinguishing Sequential Patterns with Flexible Gap Constraints

Chao Gao[1], Lei Duan[1,2(✉)], Guozhu Dong[3], Haiqing Zhang[4], Hao Yang[1], and Changjie Tang[1]

[1] School of Computer Science, Sichuan University, Chengdu, China
{gaochaochn,hyang.cn}@outlook.com,
{leiduan,cjtang}@scu.edu.cn
[2] West China School of Public Health, Sichuan University, Chengdu, China
[3] Department of Computer Science and Engineering,
Wright State University, Dayton, USA
guozhu.dong@wright.edu
[4] School of Software Engineering, Chengdu University of Information Technology,
Chengdu, China
haiqing_zhang_zhq@163.com

Abstract. Distinguishing sequential pattern (DSP) mining has been widely employed in many applications, such as building classifiers and comparing/analyzing protein families. However, in previous studies on DSP mining, the gap constraints are very rigid – they are identical for all discovered patterns and at all positions in the discovered patterns, in addition to being predetermined. This paper considers a more flexible way to handle gap constraint, allowing the gap constraints between different pairs of adjacent elements in a pattern to be different and allowing different patterns to use different gap constraints. The associated DSPs will be called DSPs with flexible gap constraints. After discussing the importance of specifying/determining gap constraints *flexibly* in DSP mining, we present GepDSP, a heuristic mining method based on Gene Expression Programming, for mining DSPs with flexible gap constraints. Our empirical study on real-world data sets demonstrates that GepDSP is effective and efficient, and DSPs with flexible gap constraints are more effective in capturing discriminating sequential patterns.

Keywords: Contrast mining · Sequential pattern · Gene expression programming

1 Introduction

Distinguishing sequential pattern (DSP) mining is an interesting data mining problem concerned with discovering sequential patterns that discriminate different classes of sequences. Specifically, DSP mining aims at extracting sequential patterns that occur frequently in sequences of one class but infrequently in

This work was supported in part by NSFC 61572332, and China Postdoctoral Science Foundation 2014M552371.

B. Cui et al. (Eds.): WAIM 2016, Part I, LNCS 9658, pp. 82–94, 2016.
DOI: 10.1007/978-3-319-39937-9_7

sequences of another class. DSP mining has been proven to be useful in sequential data analysis, such as building accurate classifiers, performing anomaly detection, and analyzing protein/DNA data sets. In general, DSP mining is a specific type of contrast data mining [1].

By studying the existing algorithms on DSP mining, such as those proposed in [2–4], we found that *setting proper gap constraints is critical in DSP mining*. Gap constraint has been widely used in sequential pattern mining, such as [5–7]. Specifically, a gap constraint specifies the minimum and maximum numbers of wildcards allowed when matching two adjacent elements in a sequential pattern with a sequence. However, it is difficult for users, who often lack sufficient priori knowledge, to set proper gap constraints.

Moreover, in previous studies on DSP mining, the way of setting gap constraints is *rigid*, since (i) for all discovered DSPs, the gap constraints are identical, and (ii) all gap constraints between every two adjacent elements in a pattern are the same. The above two weaknesses may imply that existing algorithms cannot find discriminating patterns in given datasets when they exist.

Example 1 shows that, for a specified pattern, setting different gap constraints may lead to different matches.

Example 1. Consider the set of sequences listed in Table 1. Let P be the pattern <a, g, t>. We consider three situations:

 (i) the gap constraints between a and g, and between g and t, are both $[0, 2]$;
 (ii) the gap constraints between a and g, and between g and t, are both $[1, 3]$;
 (iii) the gap constraint between a and g is $[0, 2]$, and the gap constraint between g and t is $[1, 3]$.

We can see that in case (i), P only matches $S1$ (<a̲, g̲, c, a, t̲, a, c>); in case (ii), P only matches $S2$ (<a̲, t, g̲, a, c, c, t̲>); and in case (iii), P matches $S1$, $S2$ and $S3$ (<a̲, g̲, c, a, t̲, a, c>; <a̲, t, g̲, a, c, c, t̲>; <a̲, g̲, a, c, a, t̲>).

Table 1. A sequential data set

Id	Sequence
$S1$	<a, g, c, a, t, a, c>
$S2$	<a, t, g, a, c, c, t>
$S3$	<a, g, a, c, a, t>
$S4$	<a, c, a, a, t, g, t, a, c>

Obviously, we will suffer *the combinatorial explosion problem* if we enumerate all possible gap constrains when generating candidate patterns. Indeed, the number of candidate patterns grows exponentially when the size of alphabet and the maximal sequence length increase (see Sect. 2.2).

We note that, for many applications, finding a set of optimal DSPs is much better than finding none or very few less-than-optimal DSPs, and is often enough in helping us solve the underlying data analysis problem. To overcome the limitations discussed above, we propose an evolutionary computing method, named GepDSP, to discover DSPs with flexible gap constraints. GepDSP is designed based on the framework of Gene Expression Programming (GEP), which seeks the optimal solution through evolving the set of candidate solutions. Our experiments show that GepDSP is effective to find optimal DSPs without any pre-fixed gap constraints in reasonable running time.

This paper makes the following main contributions on mining DSPs: (1) introducing a novel data mining problem of mining top-k DSPs with flexible gap constraints; (2) introducing evolutionary computing technique into DSP mining, and designing a heuristic method for discovering top-k DSPs with flexible gap constraints; (3) conducting extensive experiments on three real-world data sets involving thousands of sequences, to evaluate our proposed mining algorithm, and to demonstrate that the proposed algorithm can find DSPs with flexible gap constraints, and it is effective and efficient.

In the rest of the paper, we formulate the problem definition in Sect. 2, and review related works in Sect. 3. In Sect. 4, we present the critical techniques of our heuristic methods. We report a systematic empirical study using real-world data in Sect. 5, and conclude the paper in Sect. 6.

2   Problem Formulation and Analysis

In this section, we first give a formal definition for distinguishing sequential patterns with flexible gap constraints. To that end, we will also need to define some concepts concerning sequence matching with respect to "gap constraint sequence". Then, we investigate the complexity of the problem.

2.1   Problem Definition

We start with some preliminaries. Let Σ be an *alphabet*, which is a finite set of distinct *items*. A *sequence* S is an ordered list of items with the form of $S =< e_1, e_2, \ldots, e_n >$, where $e_i \in \Sigma$ for all $1 \leq i \leq n$. The *length* of S, denoted by $||S||$, is the number of items in S. We denote by $S[i]$ the i-th item in S ($1 \leq i \leq ||S||$). The *gap* between $S[i]$ and $S[j]$ ($1 \leq i < j \leq ||S||$) is the number of items between $S[i]$ and $S[j]$.

A *gap constraint* γ is an interval specified by two nonnegative integers with the form of $\gamma = [\gamma.min, \gamma.max]$ such that $\mathcal{G}_{min} \leq \gamma.min < \gamma.max \leq \mathcal{G}_{max}$, where $\mathcal{G}_{min}$ and $\mathcal{G}_{max}$ are user-specified bounds[1]. In previous studies one fixed gap constraint such as γ will be enforced on all pairs of consecutive elements in all sequential patterns. In this paper it will be used on one pair of consecutive elements in one sequential pattern.

---

[1] We can set the values of $\mathcal{G}_{min}$ and $\mathcal{G}_{max}$ to 0 and the maximal sequence length (of the dataset under consideration), respectively, as default.

We now introduce a concept to allow us to use different gap constraints for different pairs of consecutive elements in sequence patterns. A *gap constraint sequence* Γ is an ordered list of gap constraints with the form of $\Gamma =< \gamma_1, \gamma_2, ..., \gamma_m >$. We denote by $||\Gamma||$ the number of gap constraints in Γ, and denote by $\Gamma[i]$ the i-th gap constraint in Γ ($1 \leq i \leq ||\Gamma||$).

Given a sequence S', we say sequence S *matches* S' with gap constraint sequence Γ ($||\Gamma|| = ||S|| - 1$), denoted by $S \sqsubseteq_\Gamma S'$, if there exist integers $1 \leq k_1 < k_2 < ... < k_{||S||} \leq ||S'||$, such that

- (i) $S'[k_i] = S[i]$, for all $1 \leq i \leq ||S||$, and
- (ii) $k_{j+1} - k_j - 1 \in \Gamma[j]$, for all $1 \leq j \leq ||S|| - 1$.

Given a set of sequences D and a gap constraint sequence Γ, the *support* of a sequence S in D satisfying Γ, denoted by $Sup((S, \Gamma), D)$, is defined by

$$Sup((S, \Gamma), D) = \frac{|\{S' \in D \mid S \sqsubseteq_\Gamma S'\}|}{|D|} \tag{1}$$

where $|D|$ is the number of sequences in D.

Given two sets of sequences, D_+ and D_-, the *contrast score* of (S, Γ) from D_- to D_+, denoted by $cScore((S, \Gamma))$ is defined by

$$cScore((S, \Gamma), D_+, D_-) = Sup((S, \Gamma), D_+) - Sup((S, \Gamma), D_-) \tag{2}$$

Definition 1 (Top-k Distinguishing Sequential Pattern with Flexible Gap Constraints). *Given two sets of sequences D_+ and D_-, a sequence S, a gap constraint sequence Γ, and an integer k, the tuple $P = (S, \Gamma)$ is a top-k distinguishing sequential pattern (kDSP) with flexible gap constraints, if there does not exist a set $\mathcal{P} = \{P_i = (S_i, \Gamma_i) \mid cScore(P_i, D_+, D_-) > cScore(P, D_+, D_-), P_i \neq P\}$ satisfying $|\mathcal{P}| > k$.*

Given D_+, D_-, k, $\mathcal{G}_{min}$, and $\mathcal{G}_{max}$, the problem of mining top-k distinguishing sequential patterns with flexible gap constraints is to find all tuples in the form of (S, Γ) with k highest contrast scores from D_- to D_+.

2.2 Complexity Analysis

To find all kDSPs, a straightforward way is enumerating all candidate patterns and computing their contrast scores (Eq. 2). From Definition 1, we know that a candidate pattern P consists of a sequence S and a gap constraint sequence Γ satisfying $||S|| = ||\Gamma|| + 1$. Moreover, by combining S and Γ, pattern P can be represented in the following form

$$P = (S, \Gamma) =< S[1], \Gamma[1], S[2], \Gamma[2], ..., \Gamma[||S|| - 1], S[||S||] >$$

where, $S[i] \in \Sigma$ ($1 \leq i \leq ||S||$), and $0 \leq \Gamma[j].min < \Gamma[j].max \leq \mathcal{L}$ ($1 \leq j \leq ||S|| - 1$, $\mathcal{L}$ is the maximal sequence length in the given data set). Thus, the complexity to enumerate all candidate patterns with the constraints of $||S|| \leq l$ from classes of D_- to D_+ is $O((|D_+ + D_-|)|\Sigma|^l \mathcal{L}^l)$.

Obviously, the discovery of all kDSPs is time consuming, especially when the number of sequences, the size of alphabet, and the maximal sequence length are large. Our empirical study (Sect. 5) on the real world data sets show that it is even impractical to find all kDSPs with fixed and predetermined gap constraints within reasonable time. To break this limitation, we design a GEP-based method to discover all kDSPs. We will present the details in Sect. 4.

3   Related Work

3.1   Distinguishing Sequential Pattern Mining

As distinguishing sequential patterns can describe contrast information between different classes of sequences, the discovery of distinguishing sequential patterns has many interesting applications. For example, She et al. [8] predicted outer membrane by looking for subsequences that discriminate outer membrane proteins (OMPs) from non-OMPs. Shah et al. [9] built a classifier using contrast patterns as features for peptide folding prediction.

We now discuss previous studies on the discovery of distinguishing sequential patterns satisfying gap constraints. Ji et al. [2] designed an approach to find the minimal distinguishing subsequences with gap constraints, which occur frequently in sequences of one class but infrequently in sequences of another class. Wang et al. [3] brought the density concept into distinguishing sequential pattern mining, and proposed a method to mine contrast patterns satisfying both frequency and density thresholds. Yang et al. [4] studied the problem of mining contrast sequential patterns satisfying gap constraints from sequences in which elements are itemsets. Yang et al. [10] proposed an approach to find k distinguishing sequential patterns with largest contrast measure values.

To the best of our knowledge, there are no previous methods tackling exactly the same problem as kDSP mining with flexible gap constraints. The most related previous work is [10], which finds the top-k DSPs. However, it is considerably different from our work since their method is limited on using predetermined and fixed gap constraints (for a discovered pattern, all gap constraints between every pair of elements are predetermined and identical).

3.2   Preliminary Concepts of GEP (Gene Expression Programming)

The GEP framework aims to seek the optimal solution for some given measure; it is very similar to those of Genetic Algorithms (GA) and Genetic Programming (GP). However, GEP is a genotype/phenotype system that is quite different from GA and GP. There are two main parts in GEP, called chromosomes and expression trees, respectively. For an individual in GEP, the genotype is represented by a linear string of fixed length (chromosome), and the phenotype is represented by a hierarchical structure (expression tree) that contains semantic information of the individual. Such representations allow GEP to be flexible to encode candidate solutions as individuals and to perform genetic operations.

Based on the natural selection principle, i.e. "the fittest survives", GEP finds an optimal solution by iteratively evolving a population of individuals (which encode candidate solutions), through genetic operations, such as selection, crossover, and mutation. The details of GEP can be found in [11]. GEP has been widely used in data mining research fields, such as symbolic regression [12], classification [13], and time series prediction [14].

4   Design of GepDSP

In this section, we present our GEP-based algorithm, namely GepDSP, to discover top-k distinguishing sequential patterns with flexible gap constraints.

4.1   Candidate Pattern Expression

The most critical issue of using GEP in kDSP mining is how to express each candidate pattern as an individual in GepDSP. In GEP, the basic unit of an individual is a *gene* which is composed by a *head* part and a *tail*. The head part contains terminals and functions, while the tail part contains only terminals. In the case of an individual is composed by more than one gene, a linking operator is used to connect two genes. GEP uses decoding/encoding rules to guarantee the validity of individuals [11]. Limited by space, we omit the details here.

Please recall that a kDSP P consists of a sequence S and a gap constraint sequence Γ, and can be represented as $< S[1], \Gamma[1], S[2], \Gamma[2], ..., \Gamma[||S|| - 1], S[||S||] >$. Naturally, we treat $S[i]$ as a terminal, and treat $\Gamma[i]$ as a function that connect two sequence elements (i.e. $S[i]$ and $S[i+1]$). In other words, we take elements from Σ as terminals and take gap constraints as functions.

Figure 1 illustrates the components of a two-gene individual, the expression tree (for the individual) and subtrees (for the two genes), and the corresponding pattern. The two genes in the individual have the same head length (3) and total length (7), but their expression trees are different, and so are their coding regions. For *gene*1, the coding region is the whole string. For *gene*2, the coding region is the first 5 elements (thus, the expression tree does not contain the cc at the end). The expression trees of *gene*1 and *gene*2 are connected by a linking gap ($[0, 3]$) to compose the expression tree of the individual.

4.2   Pattern Mining

GepDSP starts with a random generation of certain number of individuals (candidate patterns) to make up the initial population. For each individual, GepDSP evaluates its fitness according to the contrast score (Eq. 2).

On the basis of the principle of natural selection and survival for the fittest, GepDSP selects the individuals according to their fitness, and reproduces new individuals by genetic modifications. Note that, during the process of evolution, we always keep the k individuals with top highest contrast scores.

The individuals of each new generation undergo the same evolution process including evaluation, selection, and reproduction with modification. The evolution process repeats until some stop condition (e.g. number of generations, contrast scores of solutions, etc.) is satisfied.

Algorithm 1 gives the pseudo-code of GepDSP.

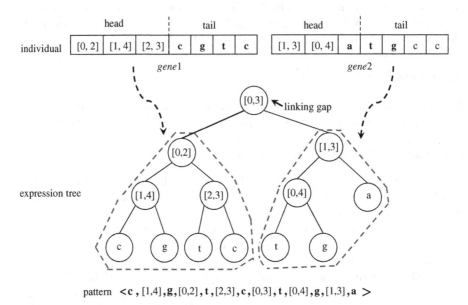

pattern < **c** , [1,4], **g**, [0,2], **t**, [2,3], **c**, [0,3], **t**, [0,4], **g**, [1,3], **a** >

Fig. 1. How to parse/understand the expression (tree) of an individual in GepDSP. (The coding region of a gene, which starts from the first element, is determined by the level-based transversal of the expression tree that produces a valid sequential pattern.)

Algorithm 1. GepDSP

Input: D_+ and D_-: two classes of sequences, k: an integer, $\mathcal{G}_{min}$ and $\mathcal{G}_{max}$: two integers satisfying $\mathcal{G}_{min} < \mathcal{G}_{max}$
Output: R: the set of kDSPs
 1: $R \leftarrow \emptyset$; // the set of discovered kDSPs
 2: $Pop \leftarrow CreateInitialPoplution$; // generate a set of candidate kDSPs as the initial population
 3: $Fit \leftarrow FitnessEvaluation(Pop)$; // evaluate the fitness of each candidate kDSP
 4: $R \leftarrow FetchTopKPatterns(Pop, Fit, k)$; // get k patterns with top contrast scores
 5: **if** the stop condition is true **then**
 6:     **return** R;
 7: **end if**
 8: $Pop' \leftarrow Selection(Pop)$; // select $|Pop| - |R|$ individuals from Pop;
 9: $Pop \leftarrow R \cup Pop'$;
10: $GeneticOperation(Pop)$; // perform genetic operations on each candidate kDSP
11: goto Step 3;

---

5   Empirical Evaluation

In this section, we report a systematic empirical study on real data sets to verify the effectiveness and efficiency of GepDSP. Specifically, we first verify that

Table 2. Characteristics of sequence data sets

Data sets	Di-heme cytochrome		ABC-2		Concanavalin	
	+	−	+	−	+	−
Num. of sequences	2014	2421	13058	7250	3528	3098
Size of Σ	20	21	21	21	21	21
Avg. sequence length	180.42	204.20	204.66	340.59	157.39	131.10
Min. sequence length	46	41	34	77	57	43
Max. sequence length	257	437	302	892	222	182

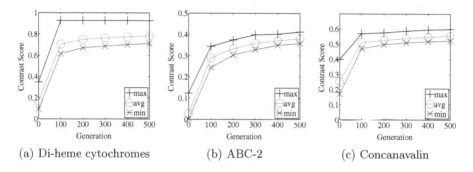

(a) Di-heme cytochromes   (b) ABC-2   (c) Concanavalin

Fig. 2. Convergence of GepDSP in terms of computed contrast score ($k = 10$)

GepDSP can discover kDSPs with flexible gap constraints, that the mined DSPs have higher contrast scores compared with kDSPs with fixed gap constraints. Second, we verify that in the case of the gap constraints are fixed (set when the algorithm starts), GepDSP can discover similar kDSPs within shorter time compared with kDSP-Miner [10]. For the sake of brevity, we denote by GepDSP-fix the version of GepDSP using fixed gap constraints.

We collect three data sets from Pfam [15], which is a protein families database. Di-heme cytochrome is a superfamily which includes a variety of different heme binding cytochromes. ABC-2 is a clan containing 12 families. Concanavalin is a superfamily including a diverse range of carbohydrate binding domains and glycosyl hydrolase enzymes. Table 2 lists the characteristic of each data set, the symbol "+" represents the positive class and "−" represents the negative class.

All algorithms were implemented in Java and compiled by JDK 8. All experiments were conducted on a computer with an Intel Core i7-3770 3.40 GHz CPU, and 16 GB main memory, running Windows 10 operating system.

5.1 Effectiveness on Mining kDSPs with Flexible Gap Constraints

We run GepDSP on three data sets, respectively, to find kDSPs with flexible gap constraints. In the experiments we set $\mathcal{G}_{min} = 0$ and $\mathcal{G}_{max} = 20$, although we can run GepDSP with any gap constraints; this setting will allow competing methods to finish. We set $k = 10$ and parameters in [11] for GEP evolution as default. All results of GepDSP/GepDSP-fix are the average value of 10 independent runs.

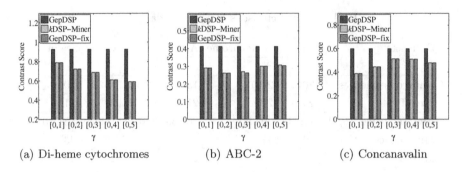

(a) Di-heme cytochromes   (b) ABC-2   (c) Concanavalin

Fig. 3. Comparison on maximum contrast scores (Color figure online)

Table 3. Top-10 kDSPs with contrast scores by GepDSP (500 generations) and kDSP-Miner ($\gamma = [0,1]$) and GepDSP-fix (500 generations, $\gamma = [0,1]$) on Di-heme cytochromes

GepDSP	kDSP-Miner	GepDSP-fix
<H,[13, 14],H> : 0.927	<H,[0, 1],W> : 0.789	<H,[0, 1],W> : 0.789
<H,[13, 14],H,[2, 16],A> : 0.841	<L,[0, 1],H,[0, 1],W> : 0.456	<L,[0, 1],H,[0, 1],W> : 0.456
<A,[0, 16],H,[13, 14],H> : 0.821	<H,[0, 1],F> : 0.455	<H,[0, 1],F> : 0.455
<H,[13, 14],H,[2, 16],L> : 0.813	<H,[0, 1],W,[0, 1],L> : 0.415	<H,[0, 1],W,[0, 1],L> : 0.415
<L,[4, 16],H, [13, 14],H> : 0.796	<R,[0, 1],L,[0, 1],H> : 0.413	<H,[0, 1],Y> : 0.394
<P,[9, 19],H,[13, 14],H> : 0.757	<H,[0, 1],Y> : 0.394	<V,[0, 1],H> : 0.390
<H,[13, 14],H,[0, 18],R> : 0.755	<V,[0, 1],H> : 0.390	<H,[0, 1],I> : 0.385
<H,[13, 14],H,[0, 18],G> : 0.740	<H,[0, 1],I> : 0.385	<H,[0, 1],N> : 0.321
<H,[3, 12],A,[0, 7],H> : 0.728	<H,[0, 1],N> : 0.321	<H,[0, 1],H> : 0.316
<H,[3, 12],V,[0, 6],H> : 0.726	<H,[0, 1],H> : 0.316	<W,[0, 1],H> : 0.314

For each data set, we record the maximal, average, and minimal contrast scores of the discovered top-10 kDSPs by GepDSP in different generations. As shown in Fig. 2, GepDSP can find a stable result quickly (around 200 generations).

For comparison, we run kDSP-Miner and GepDSP-fix using different gap constraints, respectively. Figure 3 compares the maximum contrast scores obtained by the three algorithms. We can see that most of the patterns with higher contrast scores were discovered by GepDSP.

Table 3 lists the top-10 patterns obtained on the Di-heme cytochromes data set by the three algorithms. We can see that GepDSP can find patterns with flexible gap constraints and higher contrast scores.

5.2 Effectiveness on Mining kDSPs with Fixed Gap Constraints

Structure Similarity of Mined Patterns: We compare the patterns discovered by GepDSP-fix with those by kDSP-Miner. First, we define the structure

Table 4. Similarities between top-10 kDSPs discovered by GepDSP-fix and kDSP-Miner on Di-heme cytochromes

Gap constraint	Number of generations					
	25	100	200	300	400	500
[0, 1]	0.825	0.875	0.922	0.927	0.930	0.933
[0, 2]	0.675	0.748	0.837	0.847	0.853	0.860
[0, 3]	0.698	0.818	0.852	0.882	0.899	0.899
[0, 4]	0.495	0.724	0.789	0.820	0.840	0.863
[0, 5]	0.473	0.700	0.751	0.804	0.854	0.912

Table 5. Similarities between top-10 kDSPs discovered by GepDSP-fix and kDSP-Miner on ABC-2

Gap constraint	Number of generations					
	25	100	200	300	400	500
[0, 1]	0.683	0.740	0.793	0.800	0.807	0.813
[0, 2]	0.602	0.614	0.636	0.640	0.664	0.693
[0, 3]	0.568	0.617	0.718	0.705	0.744	0.837
[0, 4]	0.587	0.668	0.675	0.708	0.725	0.736
[0, 5]	0.662	0.724	0.754	0.777	0.787	0.801

similarity between two kDSPs. Given two patterns P_1 and P_2, the structure similarity between them is defined by

$$sim(P_1, P_2) = 1 - \frac{ED(P_1, P_2)}{max\{|P_1|, |P_2|\}}$$

where $ED(P_1, P_2)$ is the *Edit Distance* [16].

Let $\mathbb{R}^k$ be the set of all permutations of the set $\{i \mid 1 \leq i \leq k\}$. Given two lists of top-k patterns ℓ_1 and ℓ_2, we measure the similarity between ℓ_1 and ℓ_2, denoted by $AvgSim(\ell_1, \ell_2)$, according to Eq. 3:

$$AvgSim(\ell_1, \ell_2) = \frac{1}{k}max\{\sum_{i=1}^{k} sim(\ell_1[i], \ell_2[R[i]]) \mid R \in \mathbb{R}^k\} \qquad (3)$$

where $\ell_1[i]$ is the i-th pattern in ℓ_1, and $R[i]$ is the i-th element in R.

Please note that $AvgSim(\ell_1, \ell_2) \in [0.0, 1.0]$; the larger the value of $AvgSim(\ell_1, \ell_2)$ is, the more similar ℓ_1 and ℓ_2 are.

Tables 4, 5, 6 list the similarities between top-10 kDSPs discovered by GepDSP-fix and those by kDSP-Miner on the three data sets. We can see that the results of GepDSP-fix are similar to those of kDSP-Miner, especially, when the number of generations is large.

Table 6. Similarities between top-10 kDSPs discovered by GepDSP-fix and kDSP-Miner on Concanavalin

Gap constraint	Number of generations					
	25	100	200	300	400	500
$[0,1]$	0.842	0.895	0.927	0.943	0.950	0.957
$[0,2]$	0.715	0.762	0.782	0.783	0.785	0.792
$[0,3]$	0.885	0.910	0.927	0.937	0.953	0.953
$[0,4]$	0.755	0.862	0.890	0.907	0.910	0.910
$[0,5]$	0.812	0.943	0.980	0.990	1.0	1.0

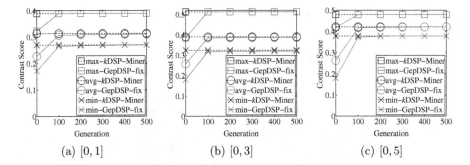

(a) $[0,1]$ (b) $[0,3]$ (c) $[0,5]$

Fig. 4. Comparison on contrast scores vs different gap constraints on Concanavalin ($k = 10$)

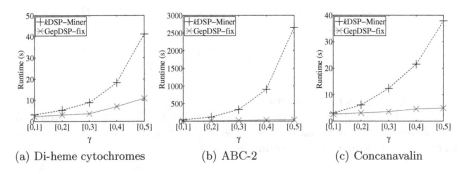

(a) Di-heme cytochromes   (b) ABC-2   (c) Concanavalin

Fig. 5. Running time w.r.t. gap constraint (k=10)

Contrast Score Comparison: Figure 4 illustrates the maximal, average, and minimal contrast scores of top-10 kDSPs discovered by GepDSP-fix and those of kDSP-Miner. We can see that GepDSP-fix can find patterns whose contrast scores are close to kDSPS discovered by kDSP-Miner around 100 generations.

Efficiency Comparison: Figure 5 shows the running time of GepDSP-fix and kDSP-Miner w.r.t. gap constraints. We can see that GepDSP-fix (500 generations) is more efficient than kDSP-Miner.

6   Conclusions

In this paper, we proposed and studied a new problem of mining distinguishing sequential patterns with flexible gap constraints. For the sake of efficiency, we proposed a heuristic method, called GepDSP, based on Gene Expression Programming, to find optimal DSPs with flexible gap constraints. Our experiments verify the effectiveness and efficiency of GepDSP.

There are many research problems that are worthy of future study. For example, how to evaluate the quality of discovered DSPs with flexible gap constraints, how to design operators to improve the evolution efficiency, and how to evaluate the goodness of the results of GepDSP when the ground truth is not available. It is also interesting to explore strategies for incorporating domain knowledge into mining DSPs with flexible gap constraints.

References

1. Dong, G., Bailey, J.: Contrast Data Mining: Concepts, Algorithms, and Applications. CRC Press, Boca Raton (2012)
2. Ji, X., Bailey, J., Dong, G.: Mining minimal distinguishing subsequence patterns with gap constraints. Knowl. Inf. Syst. **11**(3), 259–286 (2007)
3. Wang, X., Duan, L., Dong, G., Yu, Z., Tang, C.: Efficient mining of density-aware distinguishing sequential patterns with gap constraints. In: Bhowmick, S.S., Dyreson, C.E., Jensen, C.S., Lee, M.L., Muliantara, A., Thalheim, B. (eds.) DASFAA 2014, Part I. LNCS, vol. 8421, pp. 372–387. Springer, Heidelberg (2014)
4. Yang, H., Duan, L., Dong, G., Nummenmaa, J., Tang, C., Li, X.: Mining itemset-based distinguishing sequential patterns with gap constraint. In: Proceedings of the 20th International Conference on Database Systems for Advanced Applications, pp. 39–54 (2015)
5. Li, C., Yang, Q., Wang, J., Li, M.: Efficient mining of gap-constrained subsequences and its various applications. ACM Trans. Knowl. Discov. Data (TKDD) **6**(1), 2:1–2:39 (2012)
6. Zhang, M., Kao, B., Cheung, D.W., Yip, K.Y.: Mining periodic patterns with gap requirement from sequences. ACM Trans. Knowl. Discov. Data (TKDD) **1**(2), 7 (2007)
7. Xie, F., Wu, X., Hu, X., Gao, J., Guo, D., Fei, Y., Hua, E.: MAIL: mining sequential patterns with wildcards. Int. J. Data Min. Bioinf. **8**(1), 1–23 (2013)
8. She, R., Chen, F., Wang, K., Ester, M., Gardy, J.L., Brinkman, F.S.L.: Frequent-subsequence-based prediction of outer membrane proteins. In: Proceedings of the 9th ACM SIGKDD International Conference on Knowledge Discovery and Data Mining, pp. 436–445 (2003)
9. Shah, C.C., Zhu, X., Khoshgoftaar, T.M., Beyer, J.: Contrast pattern mining with gap constraints for peptide folding prediction. In: Proceedings of the 21st International Florida Artificial Intelligence Research Society Conference, pp. 95–100 (2008)
10. Yang, H., Duan, L., Hu, B., Deng, S., Wang, W., Qin, P.: Mining top-k distinguishing sequential patterns with gap constraint. J. Softw. **26**(11), 2994–3009 (2015)
11. Ferreira, C.: Gene Expression Programming: Mathematical Modeling by an Artificial Intelligence. SCI, vol. 21. Springer, Heidelberg (2006)

12. Peng, Y., Yuan, C., Qin, X., Huang, J., Shi, Y.: An improved gene expression programming approach for symbolic regression problems. Neurocomputing **137**, 293–301 (2014)
13. Zhou, C., Xiao, W., Tirpak, T.M., Nelson, P.C.: Evolving accurate and compact classification rules with gene expression programming. IEEE Trans. Evol. Comput. **7**(6), 519–531 (2003)
14. Duan, L., Tang, C., Li, X., Dong, G., Wang, X., Zuo, J., Jiang, M., Li, Z., Zhang, Y.: Mining effective multi-segment sliding window for pathogen incidence rate prediction. Data Knowl. Eng. **87**, 425–444 (2013)
15. Finn, R.D., Bateman, A., Clements, J., Coggill, P., Eberhardt, R.Y., Eddy, S.R., Heger, A., Hetherington, K., Holm, L., Mistry, J., et al.: Pfam: the protein families database. Nucleic Acids Res. **42**(D1), 222–230 (2014)
16. Navarro, G.: A guided tour to approximate string matching. ACM Comput. Surv. **33**(1), 31–88 (2001)

A Novel Chinese Text Mining Method
for E-Commerce Review Spam Detection

Xiu Li$^{(\boxtimes)}$ and Xinwei Yan

Graduate School at Shenzhen, Tsinghua University,
Shenzhen 518000, China
li.xiu@sz.tsinghua.edu.cn,
yanxw13@mails.tsinghua.edu.cn

Abstract. Review spam is increasingly rampant in China, which seriously hampers the development of the vigorous e-commerce market. In this paper, we propose a novel Chinese text mining method to detect review spam automatically and efficiently. We correctly extract keywords in complicated review text and conduct fine-grained analysis to recognize the semantic orientation. We study the spammers' behavior patterns and come up with four effective features to describe untruthful comments. We train classifier to classify reviews into spam or non-spam. Experiments are conducted to demonstrate the excellent performance of our algorithm.

Keywords: E-Commerce · Review spam · Chinese text mining · Subject logic · Logistic regression

1 Introduction

After the crazy development in these years, China's e-commerce market gradually exposes some problems, among which selling fake product is the biggest issue. In early 2015, the government reported a sample survey showing that 62.5 % of the goods in *Taobao*, China's largest retail e-commerce platform, are not authentic. This is not consistent with the credit status *Taobao* claimed and customers have lost confidence on the credit ratings given by the e-commerce platform. As a result, viewing reviews generated by other customers has become an essential step in online shopping. In fact, a large proportion of positive comments attract potential customers and arouse their desire to buy [1]. So reviews have become important resources for online stores to chase financial gains, which leads to the spread of review spam. Review spam refers to reviews have been deliberately written to sound authentic and deceive the reader. Merchants usually reshape their businesses by accumulating a large amount of fake favorable reviews, which is called *hyper spam*. They can also use spammed reviews to defame their competitors, which is called *defaming spam*. It is reported that there are more than one thousand active companies that create fraudulent reviews in China. Hence detection and blocking of review spam is significant to the development of China's e-commerce.

There is extensive research work in the detection of review spam and current methods can be divided into two categories: content based or non-content based.

© Springer International Publishing Switzerland 2016
B. Cui et al. (Eds.): WAIM 2016, Part I, LNCS 9658, pp. 95–106, 2016.
DOI: 10.1007/978-3-319-39937-9_8

Non-content based methods focus on spammers' behavior patterns such as ID duplication, rating singleton, time interval between purchase and review and geographic information [2]. These features are not very reliable and the false detection rate is relatively high. Content based methods mainly extract features through text mining and build machine learning models to classify reviews into spam or non-spam. Some approaches have achieved good results on the English data sets. But Chinese text is more complex than English and it requires more work to recognize the meaning of reviews. In terms of the present stage, the accuracy is limited by lack of effective features.

In this paper, we propose a novel and effective content based method and make several contributions towards Chinese text review spam detection. (1) We analyze the grammatical features of Chinese reviews and come up with an adjective-oriented keywords extraction method. We use Latent Dirichlet Allocation as a supplement to make improvement on the recall rate. (2) We discover spammers' behavior patterns in reviews and put forward four original effective features: Universality degree, Outlier degree of sentiment, Uncertainty degree and Similarity degree. We train classifier using these features and classify reviews into spam or non-spam. (3) We conduct experiments to compare our method with others and evaluate its reliability. The achievements of this paper have been applied to practice.

The rest of this paper is organized as follows. Section 2 briefly reviews related work. Section 3 introduces our method in detail. We show empirical evaluation with discussion in Sect. 4. In Sect. 5, we present the conclusion and future work.

2 Related Work

In this section, we introduce some well performed content based text mining methods for review spam. Basically they can be divided into three categories: genre identification, detecting duplication and text categorization.

2.1 Genre Identification

Ott et al. prove that the genre of text has an influence on the distribution of parts-of-speech (POS) of reviews [3]. They test differences in POS distribution between spammed reviews and truthful reviews. Ravi et al. propose a framework for spam detection based on a set of frequently occurring sequential patterns [4]. They introduce a variant called "min-closed" to capture the patterns and extend the PRISM algorithm to detect review spam. However, spammers now tend to copy the format of truthful reviews, which makes the identification rate greatly drop.

2.2 Finding Contend Duplication

Liu and Jindal first propose a review spam detection method based on duplication studies [5]. They find that spammers usually use templates to create reviews and only change the product name when spam a single product. Atefeh et al. also point out that creating unique fake reviews is very time-consuming and spammers overwhelmingly tend to copy the existing reviews [6]. But the similarity between two reviews is difficult to calculate unless the semantics are fully understood.

2.3   Text Categorization

Machine learning based classifiers are learned for the two classes – spam or non-spam. Ahn et al. propose a method for spam prevention called CAPTCHAs [7]. They make some contributions but their techniques impair the consumer experience. Biro et al. use topic mode Latent Dirichlet Allocation to learn spam topics [8]. Sahami et al. use naive-bayes to do the classification [9]. Chen et al. come up with a method that can identify spam content and spammers simultaneously [10]. In the field of Chinese text, Hu and Liu apply Apriori algorithm to feature extraction and identify semantic orientation of an opinion word by the set of seed adjectives [11]. Zhao and Zhou use templates to extract product features and the corresponding opinion words [12]. The accuracies of these approaches are limited by the effectiveness of the features. Obviously, semantic recognition has been the major challenge in the Chinese text mining.

3   Methodology

In this section, we present our text mining approach for Chinese e-commerce review spam detection in detail. Firstly, we introduce how to extract keywords from reviews. Then we come up with a new variable "Universality degree" to distinguish words. Next, we conduct sentiment analysis at the attribute level and calculate the uncertainty of sentence. At last, we select several effective features and use machine learning classifier to detect review spam.

3.1   Keywords Extraction

As the basic step of text mining, keywords extraction is related to how to automatically discover what attributes are evaluated in reviews and how sentiments for attributes are expressed. For every review sentence, the keywords include attribute words, related sentiment words, negative adverbs and degree adverbs. For example, in the sentence "手机很漂亮,但电池不耐用"("The cellphone is very beautiful, but the battery is not durable"), "手机"("cellphone") and "电池" ("battery") are attribute words. "漂亮"("beautiful") and "耐用" ("durable") are related sentiment words. "不"("not") is a negative adverb and "很"("very") is a degree adverb. Generally speaking, the adverbs are around the sentiment words and it's easy to match them. The biggest challenge is correctly discovering pairs of {attribute, sentiment} which we call senti-attribute. If we mismatch a senti-attribute, like durable-cellphone or beautiful-battery, the meaning of the whole sentence is misunderstood.

The traditional language models always identify attributes first and then find the sentiment words belong to them. But in Chinese reviews, sentiment words are generally more than attributes. These models often cause some sentiment words cannot match with the attributes, which makes part of the sentence's meaning lost. Some researchers extends the Latent Dirichlet Allocation method to find all the senti-attributes [13]. However, it damages the words' location information which is very valuable in Chinese text. In view of this, our keywords extraction is sentiment words oriented and we use Latent Dirichlet Allocation as a supplement.

Our assumption is that every sentiment word is related to a unique attribute and every attribute may have one or more sentiment words corresponding to it in a single review. Through our researches, we find that 95 % of Chinese reviews have such a structure that the words are arranged in the order of nouns, adverbs and adjectives. Hence, after parts-of-speech (POS) tagging through word segmentation tools, we first target adjectives in the sentence. They are almost all sentiment words. Then we look for nouns in front of the sentiment words. The search range is five words and we set the nearest noun as an attribute. We can identify most of senti-attributes through this way and we use Latent Dirichlet Allocation to find the related attributes of sentiment words that fail to match. For a certain product, assume that we find n attributes $ATTRI = \{attri_1, attri_2, \ldots attri_n\}$ in all m reviews. In review j ($j \leq m$), there is a sentiment word $senti$ which is related to a certain attribute but fail to match. $senti$ obeys a multinomial distribution on $ATTRI$. Assume the parameter is θ_n and we get:

$$senti \sim Multinomial(\theta_n)$$

The prior distribution of a multinomial distribution is Dirichlet distribution, hence:

$$\theta_n \sim Dirichlet(\alpha_n)$$

where α_n is the parameter of the Dirichlet distribution. When α_n is known, θ_n can be estimated by Gibbs sampling [14]. The probability that $senti$ is related to $attri_i$ ($i \leq n$) can be calculate:

$$\theta_i = \frac{\alpha_i + C_i^j(senti)}{1 + \sum_{i'=1}^{n} C_{i'}^j(senti)} \tag{1}$$

where $C_i^j(senti)$ is the number of senti-attributes that contain sentiment word $senti$ and attribute $attri_i$ in review j. $\sum_{i'=1}^{n} C_{i'}^j(senti)$ is the total number of senti-attributes that contain sentiment word $senti$ in review j. Obviously, when $senti$ only appears once in review j, the multinomial distribution parameter θ_i equals the prior distribution parameter α_i. We can calculate α_i with word-frequency statistics of all senti-attributes in all m reviews of the product:

$$\alpha_i = \frac{\sum_{j=1}^{m} C_i^j(senti)}{\sum_{i=1}^{n} \sum_{j=1}^{m} C_i^j(senti)} \tag{2}$$

where $\sum_{j=1}^{m} C_i^j(senti)$ is the number of senti-attributes that contain sentiment word *senti* and attribute *attri*$_i$ in all *m* reviews. $\sum_{i=1}^{n}\sum_{j=1}^{m} C_i^j(senti)$ is the number of senti-attributes that contain sentiment word *senti* in all *m* reviews. We match *senti* with attribute *attri*$_i$ when. So far we have found all pairs of {attribute, sentiment} in reviews.

3.2   Universality of Word

As mentioned above, it is very time-consuming to create unique fake reviews for spammers. When they use templates, the words are relatively fixed. We assume that spammers may not use very unique words to express praise and disparage, like using "兼容"("compatible") to describe a cell phone system. Instead, they will just use "好" or "棒" ("good" or "great") in hyper spam in order to save time. We come up with a new variable "Universality degree" to distinguish words of senti-attributes. If a sentiment word can describe many attributes, it has a high universality degree. Likewise, spam reviews may not contain unique attributes of product, like "待机时间"("standby time") of cell phone. Spammers tend to comment on some obvious attributes such as "外观", "质量" or "价格"("outlook", "quality" or "price").

For a sentiment word *senti*, we assume that the set of senti-attributes that contains sentiment word *senti* is *Senti_Attri* = {*senti_attri*$_1$, *senti_attri*$_2$, ... *senti_attri*$_n$}. In a certain review, *senti* may related to *Attri*$_j$($j \leq n$). We define the "Universality degree" of sentiment word *senti* in this review as follows:

$$Uni\_senti = \frac{Num(senti\_attri_j)}{\sum_{i=1}^{n} Num(senti\_attri_i)} \qquad (3)$$

where $Num(senti\_attri_j)$ is the number of times that *senti_attri*$_j$ appears in all reviews. $\sum_{j=1}^{n} Num(senti\_attri_j)$ is the number of times that senti-attributes contains *senti* appears in all reviews. Obviously, *Uni_senti* ranges from *0* to *1*.

Similarly, assume that a certain attribute *attri* appears in reviews of *m* kinds of product. We present the review set as $R = \{R_{p1}, R_{p2}, ... R_{pm}\}$. In a certain review, *attri* may be an attribute of product $i(i \leq m)$. We define the "Universality degree" of attribute *attri* in this review as follows:

$$Uni\_attri = \frac{Num(R_{pi})}{\sum_{j=1}^{m} Num(R_{pj})} \qquad (4)$$

where $Num(R_{pi})$ is the number of times that *attri* appears in reviews of product i. $\sum_{j=1}^{m} Num(R_{pj})$ is the number of times that *attri* appears in reviews of all kinds of products. *Uni_attri* ranges from *0* to *1*, too.

Assume review r contains p attributes and q sentiment words. The "Universality degree" of review r is the average number of the words:

$$Uni\_r = \frac{1}{2}\left(\frac{\sum_{i=1}^{p} Uni\_attri_i}{p} + \frac{\sum_{j=1}^{q} Uni\_senti_j}{q}\right) \qquad (5)$$

The "Universality degree" of review r also ranges from *0* to *1*. We take it as **Feature 1** in classification for review spam detection.

3.3   Sentiment Analysis

In this paragraph, we aim at recognizing sentimental polarity of sentiment words automatically. Current techniques basically use existing dictionaries and thus the sentimental polarity of a word is static. However, some words can be positive or negative in different contexts. For example, the adjective "大"("big or large") is positive when it describes a cellphone screen. But it is negative in the review "手机耗电量很大"("the power consumption of the cellphone is very large"). So the polarity of a sentiment word may depend on the attribute it describes. Benefiting from universality degrees of sentiment words we proposed, we find an important regular pattern that the sentiment word with a low universality degree has definite sentimental polarity. We take advantage of this observation to automatically build seeds set for each attribute so that the algorithm become self-adaptive to different products.

For attribute *attri*, there is a set of sentiment words $S = \{senti_1, senti_2, \ldots senti_n\}$ may describe it. We select some of the words to build the seeds set:

$$senti_i \in Seed \ if \ Uni\_senti_i < threshold \qquad (6)$$

Then we divide *Seed* into a positive set *Seed_p* and a negative set *Seed_n* by comparing the distances from a seed to "好"("good") and "坏"("bad") in HowNet dictionary:

$$seed \in \begin{cases} Seed\_p & ifDistance(seed, "good") < Distance(seed, "bad") \\ Seed\_n & ifDistance(seed, "good") > Distance(seed, "bad") \end{cases} \qquad (7)$$

The polarity of $senti_i$ can be determined by:

$$P\_senti_i = \begin{cases} 1 & if \ \frac{1}{n_1}\sum\limits_{j=1}^{n_1} distance\left(senti_i, Seed\_p_j\right) < \frac{1}{n_2}\sum\limits_{j=1}^{n_2} distance\left(senti_i, Seed\_n_j\right) \\ -1 & if \ \frac{1}{n_1}\sum\limits_{j=1}^{n_1} distance\left(senti_i, Seed\_p_j\right) > \frac{1}{n_2}\sum\limits_{j=1}^{n_2} distance\left(senti_i, Seed\_n_j\right) \end{cases}$$
$$(8)$$

Therefore, we obtain the sentimental polarity of every sentiment word. Assume there are m attributes in k reviews of a product. We can get the attitude on every attribute in a review by calculating the average polarity of related sentiment words.

$$P\_attri_j = \begin{cases} 1 & if \ \frac{1}{n}\sum\limits_{i=1}^{n} P\_senti_i > 0 \\ 0 & if \ \frac{1}{n}\sum\limits_{i=1}^{n} P\_senti_i = 0 \\ -1 & if \ \frac{1}{n}\sum\limits_{i=1}^{n} P\_senti_i < 0 \end{cases} \qquad (9)$$

P_attri_j can be -1, 0 or 1. The sentimental vector of a review r can be expressed as:

$$\mathbf{Senti\_r} = [P\_attri_1, P\_attri_2, \ldots P\_attri_m]$$

Respectively, the average sentimental vector of k reviews is:

$$\mathbf{Senti\_av} = [P\_attri\_av_1, P\_attri\_av_2, \ldots P\_attri\_av_m]$$

Since the attitudes of spam reviews are often inconsistent with most of the truthful reviews, we define "Outlier degree of sentiment" of a review as follows and take it as *Feature 2*.

$$Out\_r = \left\| \frac{1}{2}(\mathbf{Senti\_r} - \mathbf{Senti\_av}) \right\|_2 = \frac{1}{2}\sqrt{\sum\limits_{i=1}^{m}(P\_attri_i - P\_attri\_av_i)^2} \qquad (10)$$

3.4   Uncertainty of Review

When consumers make comments on products, they always express both positive and negative opinions. It is a basic feature of human thinking that we do not have a complete view of a proposition. Spammers are weak in this kind of behavior because they have to make their viewpoints clear to interfere with the judgment of buyers. A. Jøsang. proposed the concept of Subject Logic and made uncertainty a parameter of opinion [15]. Subject Logic uses a four variable tuple $\omega(x) = <b, d, u, a>$ to model the sentimental attitude. The variables are subject to the constraints that $b + d + u = 1$ and $b, d, u, a \in [0, 1]$. Their specific meanings are shown in Table 1.

Table 1. Meanings of Subject Logic parameters

Alphabetic symbol	Variable	Meaning
b	Belief	The belief mass in support of the proposition being true
d	Disbelief	The belief mass in support of the proposition being false
u	Uncertainty	The amount of uncommitted belief mass
a	Base rate	The priori probability of belief

We apply Subject Logic to product reviews to obtain the uncertainty degree. We first define a new operation $Count(Set)$ to count the number of elements in a set. For example, given that $A = \{a_1, a_2, \ldots a_n\}$, we get $Count(A) = n$. Suppose that the sentimental vector of a review r is **Senti_r** $= [P\_attri_1, P\_attri_2, \ldots P\_attri_m]$, then we define:

$$P\_attri \in \begin{cases} P\_p & if\ P\_attri > 0 \\ P\_n & if\ P\_attri < 0 \end{cases} \qquad (11)$$

$$b = \frac{Count(P\_p)}{Count(P\_p) + Count(P\_n)} \left| \frac{Count(P\_p) - Count(P\_n)}{Count(P\_p) + Count(P\_n)} \right| \qquad (12)$$

$$d = \frac{Count(P\_n)}{Count(P\_p) + Count(P\_n)} \left| \frac{Count(P\_p) - Count(P\_n)}{Count(P\_p) + Count(P\_n)} \right| \qquad (13)$$

$$u = 1 - \left| \frac{Count(P\_p) - Count(P\_n)}{Count(P\_p) + Count(P\_n)} \right| \qquad (14)$$

We take u, the "Uncertainty degree" of a review, as **Feature 3** in classification.

3.5   Similarity of Reviews

Some researchers point out that spammers usually allocate a specific time interval to post spam reviews [16]. Spam attacks have to burst in reviews so as not to be overwhelmed by truthful comments. Since spammers always express the same opinions and even use templates, fraudulent reviews should have high similarities in a certain period of time. We measure the similarity of two reviews by counting the number of attributes they both contain. Assume there are m attributes in k reviews of a product. Review r_1 contains n_1 attributes and r_2 contains n_2 attributes. The number of attributes they both contain is $n_3 (n_3 \leq \min\{n_1, n_2\})$. We define the similarity of r_1 and r_2 as follows:

$$Sim(r_1, r_2) = \frac{n_3}{n_1 + n_2 - n_3} \qquad (15)$$

Assume there are m reviews in a certain period of time, the similarity degree of a review can be calculated by:

$$Sim\_r = \frac{1}{m} \sum_{i=1}^{m} Sim(r, r_i) \tag{16}$$

The time interval is a changeable variable which depends on the daily volume of business. We take the "Similarity degree" of a review as **Feature 4** in classification for review spam detection.

3.6   Classification

In this paragraph, we classify reviews into spam or non-spam using the four features: **University degree, Outlier degree, Uncertainty degree** and **Similarity degree**. We choose Logistic Regression as the binary classifier because it is simple and effective.

We introduce Logistic Regression briefly. Suppose that $\mathbf{\Theta} = [\theta_1, \theta_2, \ldots \theta_n]$ is the parameter vector we get by training, the classifier function will be:

$$h(\mathbf{X}) = \frac{1}{1 + e^{-\Theta \mathbf{X}}} \tag{17}$$

The test data is positive with a probability of $h(\mathbf{X})$. Respectively, it is negative with a probability of $(1 - h(\mathbf{X}))$.

The biggest problem is that there is no public Chinese spam review data set. Strictly speaking, no one can prove an online review is totally fraudulent. But Christopher, G. H. has proved that human has the ability to distinguish spam reviews from truthful reviews [17]. We follow his method to build the training set by human assessors. Since all four features range from *0* to *1*, we do not have to normalize them. Finally, we use the trained classifier to divide the review set into spam or non-spam.

4   Experiment

In this section, we present experiments we have conducted to evaluate the performance of our method by comparing with other methods.

4.1   Data Set

The data we use is real review data on *Taobao* platform. Since the limit quantity of online reviews that can be seen by customers for a single product is two thousand, we build our set that contains twenty thousand reviews covering ten kinds of products including cellphone, kingdle, skirt and so on.

4.2   Text Mining Performance

We choose **Precision, Recall, F1-means** and **Accuracy of sentiment recognition** to evaluate the performance of our Chinese text mining technique:

$$Precision = \frac{Number\ of\ correct\ identified\ senti - attributes}{Number\ of\ all\ identified\ senti - attributes} \quad (18)$$

$$Recall = \frac{Number\ of\ correct\ identified\ senti - attributes}{Number\ of\ all\ correct\ senti - attributes} \quad (19)$$

$$F1 - means = \frac{2 \times Precision \times Recall}{Precision + Recall} \quad (20)$$

We randomly choose one thousand reviews to conduct the experiments. All these reviews are read by three human assessors. They vote to determine whether the result is correct. We compare our keywords extraction results with some famous methods mentioned earlier [11–13] (Table 2).

Table 2. Recognition result of pairs of {sentiment words, attribute}

Methods	Precision	Recall	F1-means
Templates	0.64	0.39	0.48
Apriori	0.69	0.54	0.61
LDA	0.76	0.71	0.73
Our method	**0.86**	**0.83**	**0.84**

As we can see, LDA greatly improves the recall rate because it can identify a lot more keywords. But it does not improve the precision because it destroys the structure of Chinese sentences. Our method solve this problem and achieve good performance. Then we compare the accuracy of sentiment recognition with three wide-used methods: PMI [18], SMSO [19] and LE [20] (Table 3).

Table 3. Accuracy of sentiment recognition

Methods	Accuracy
PMI	0.73
SM + SO	0.76
LE	0.82
Our method	**0.85**

Obviously, our method has a high recognition rate in terms of the existing Chinese dictionaries.

4.3   Spam Detection Performance

Our reviews are labeled by three professors from Shenzhen E-commerce Security Center. Similarly, we calculate the **Precision, Recall**, and **F1-means** to evaluate the performance (Table 4).

Table 4. Accuracy of review spam detection

Methods	Precision	Recall	F1-means
Naïve Bayes	0.52	0.57	0.54
Duplication detection	0.69	0.64	0.66
Frequency pattern	0.73	0.66	0.69
Our method	**0.88**	**0.71**	**0.78**

The recall rate of review spam is respectively low because it is very hard to detect all the fraudulent reviews. The good thing is the precision of our method is pretty high.

5 Conclusion and Future Work

In this paper, we incorporate Chinese text mining into the review spam detection. Firstly, we combine grammatical method with Latent Dirichlet Allocation in keywords extraction. Then we analyze behavior patterns of spammers and propose four effective features. Next, we train the binary classifier to detect the spam reviews. Finally, the experiment demonstrate the efficiency of the proposed method.

Although we have made several improvements in the review spam detection, researching is endless. Our main future work is to discovery more features of spam. Besides, we plan to build a Chinese dictionary of e-commerce related words in order to improve the text mining performance. Furthermore, we may combine content based detection methods with non-content based methods to discover spam reviews and spammers at the same time.

References

1. Ho-Dac, N.N., Carson, S.J., Moore, W.L.: The effects of positive and negative online customer reviews: do brand strength and category maturity matter? J. Mark. **77**(6), 37–53 (2013)
2. Xie, S., Wang, G., Lin, S., Yu, P.: Review spam detection via temporal pattern discovery. In: Proceedings of the 18th ACM SIGKDD International Conference on Knowledge Discovery and Data Mining. ACM (2012)
3. Ott, M., Cardie, C., Hancock, J.: Estimating the prevalence of deception in online review communities. In: Proceedings of the 21st International Conference on World Wide Web. ACM (2012)
4. Ravi, K., Srinivasan, H.S., Krishnan, K.: Comment spam detection by sequence mining. In: WSDM 2012. ACM (2012)
5. Nitin, J., Bing, L.: Opinion spam and analysis. In: WSDM 2008. ACM (2008)
6. Atefeh, H., Mohammad, A.T., Naomie, S.: Detection of review spam: a survey. Expert Syst. Appl. **42**, 3634–3642 (2015)
7. von Ahn, L., Blum, M., Hopper, N., Langford, J.: CAPTCHA: using hard AI problems for security. In: TREC (2005)
8. Biro, I., Szabo, J., Benczur, A.A.: Latent Dirichlet allocation in web spam filtering. In: AIRWeb (2008)

9. Sahami, M., Dumais, S., Heckerman, D., Horvitz, E.: A Bayesian approach to filtering junk e-mails. In: AAAI Workshop on Learning for Text Categorization (1998)
10. Chen, F., Tan, P.N., Jain, A.K.: A co-classification framework for detecting web spam and spammers in social media Web sites. In: CIKM (2009)
11. Hu, M., Liu, B.: Mining and summarizing customer reviews. In: Proceedings of the ACM SIGKDD International Conference on Knowledge Discovery and Data Mining, pp. 168–177 (2004)
12. Zhao, W., Zhou, Y.: A template-based approach to extract product features and sentiment words. In: Proceedings of the International Conference on Natural Language Processing and Knowledge Engineering, pp. 1–5 (2009)
13. Yohan, J., Alice, O.: Aspect and sentiment unification model for online review analyses. In: WSDM 2011. ACM (2011)
14. Blei, D.M., Ng, A.Y., Jordan, M.I., Lafferty, J.: Latent Dirichlet allocation. J. Mach. Learn. Res. 3, 993–1022 (2003)
15. Audun, J.: A logic for uncertain probabilities. Int. J. Uncertainty Fuzziness Knowl. Based Syst. 9(3), 1–31 (2001)
16. Xie, S., Wang, G., Yu, P.: Review spam detection via temporal pattern discovery. In: Proceedings of the 18th ACM SIGKDD International Conference on Knowledge Discovery and Data Mining. ACM (2012)
17. Christopher, G.H.: Detecting deceptive opinion spam using human computation. In: Association for the Advancement of Artificial Intelligence. AAAI (2012)
18. Turney, P., Littman, M.: Measuring praise and criticism: inference of semantic orientation from association. ACM Trans. Inf. Syst. 21(4), 315–346 (2003)
19. Gamon, M., Aue. A.: Automatic identification of sentiment vocabulary exploiting low association with known sentiment terms. In: Proceedings of ACL (2005)
20. Kanayama, H., Nasukawa, T.: Fully automatic lexicon expansion for domain-oriented sentiment analysis. In: Proceedings of EMNLP (2006)

Spatial and Temporal Databases

Retrieving Routes of Interest Over Road Networks

Wengen Li[1,2](✉), Jiannong Cao[1], Jihong Guan[2],
Man Lung Yiu[1], and Shuigeng Zhou[3]

[1] The Hong Kong Polytechnic University, Hong Kong, China
{cswgli,csjcao,csmlyiu}@comp.polyu.edu.hk
[2] Tongji University, Shanghai, China
{8lwengen,jhguan}@tongji.edu.cn
[3] Fudan University, Shanghai, China
sgzhou@fudan.edu.cn

Abstract. In this paper, we propose route of interest (ROI) query which allows users to specify their interests with query keywords and returns a route such that (i) its distance is less than a distance threshold and (ii) its relevance to the query keywords is maximized. ROI query is particularly helpful for tourists and city explorers. For example, a tourist may wish to find a route from a scenic spot to her hotel to cover many artware shops. It is challenging to efficiently answer ROI query due to its NP-hard complexity. Novelly, we propose an adaptive route sampling framework that adaptively computes a route according to a given response time, and gradually improve the quality of the route with time. Moreover, we design a suite of route sampling techniques under this framework. Experiments on real data suggest that our proposed solution can return high quality routes within a short response time.

Keywords: Route of interest · Route sampling · Randomized algorithm

1 Introduction

The widespread use of mobile devices and mobile applications enables users to share information and experiences about points of interest (POIs) on the Web. These POIs can be restaurants, pubs and scenic places. For example, customers can comment on restaurants in recommendation websites and share their experiences in social networks; restaurant owners may post advertisements on their features in online business directories. This large amount of textual data makes it possible for users to query on POIs by using both locations and textual information.

In the past decade, many queries have been proposed to retrieve POIs by combining the spatial proximity (e.g., distances of POIs from the user) and the textual relevance (between the textual descriptions of POIs and user's query keywords). These works can be roughly classified into the following four categories.

© Springer International Publishing Switzerland 2016
B. Cui et al. (Eds.): WAIM 2016, Part I, LNCS 9658, pp. 109–123, 2016.
DOI: 10.1007/978-3-319-39937-9_9

- each POI that independently matches with query keywords [8,17,22],
- a group of POIs that collectively cover all query keywords [3,4,12],
- a set of POIs enclosed in a region [5,9,15,19], e.g., rectangles and sub-graphs,
- an optimal route that covers required POIs [1,2,13,18].

In this paper, we consider the scenarios where users have start and end locations, and wish to explore some relevant POIs along their journeys. Existing route queries aim to find the shortest route that covers POIs of specific categories [13,18] or POIs containing the query keywords [1,2]. An example is *"find the shortest route that passes through a restaurant and a pub"*. Such queries are suitable when users need to visit each category of POI once. In contrast, we focus on a different case where the user wants to explore as many relevant POIs as possible. For example, a tourist may ask: *"find a route that covers as many restaurants and pubs as possible"*. This would offer more choices for the tourist (i.e., user) to choose from, and thus improve her satisfaction.

To deal with the above issues, we propose *route of interest (ROI) query*. Given the user's interests (i.e., query keywords) and travel requirements (i.e., start s and end e locations, route distance threshold), the ROI query returns the route from s to e such that: (i) the length of the route is bounded by the given distance threshold, and (ii) the total number of relevant POIs on the route is maximized. The length is restricted because users may not want to travel a very long route. Essentially, the goal of ROI query is to find a route that is most relevant to query keywords while satisfying the given distance threshold.

We will prove that ROI query is NP-hard (cf. Sect. 3.2). For such a computationally expensive problem, the typical solutions are either to develop an exact solution (e.g., branch-and-bound) that guarantees the optimal result but incurs significant time cost, or to develop a heuristics solution (e.g., nearest neighbor heuristics) that offers quick results but of lower quality. We visualize the result quality and the response time of these two solutions in Fig. 1. In this paper, we advocate on an adaptive solution that provides flexible trade-offs between result quality and response time. We take the response time limit into account and propose an adaptive route sampling framework which computes a route according to the given response time. In summary, our contributions are presented as follows.

- We propose a novel route query termed route of interest (ROI) query, and prove that it is NP-hard (Sect. 3).
- We design an adaptive route sampling framework that can generate routes of high quality within a given response time limit. In addition, we propose a suite of route sampling techniques to improve the processing performance of the proposed framework, including uniform route sampling, prioritized route sampling, and dynamic prioritized route sampling (Sect. 4).
- We conduct extensive experiments on three real datasets to evaluate the performance of our proposal (Sect. 5).

In addition, the related work is reviewed in Sect. 2 and conclusion is presented in Sect. 6.

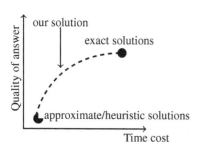

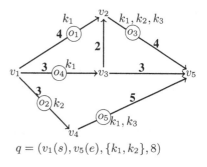

$$q = (v_1(s), v_5(e), \{k_1, k_2\}, 8)$$

Fig. 1. Quality of results vs. response time for exact and approximate/heuristic solutions.

Fig. 2. An example of road network and a ROI query on the road network

2 Related Work

Those works conducted to retrieve POIs can be roughly classified into four categories as elaborated below.

- **Retrieving single POIs** [6,10,14,17,21,22]: In this category, each returned POI matches with the query keywords independently. Some works [6,21] use query keywords to filter POIs before sorting them according to their distances to the query location. In some other works [14], the spatial proximity (between the query and the locations of POIs) and the textual relevance (between query keywords and the textual description of POIs) are combined by a linear function to rank POIs.
- **Retrieving a set of POIs** [3,4,12,16]: In these studies, a set of POIs collectively satisfies the query keywords. Concretely, the goal of these queries is to find a set of POIs collectively covering the query keywords while the internal distance of the set of POIs is minimized.
- **Retrieving a region of POIs** [5,9,15,19]: In these works, a region, e.g., rectangle, circle and polygon, is predefined. The objective is to retrieve a region which has the largest number of POIs relevant to the given query keywords.
- **Retrieving a route to cover POIs** [1,2,7,13,18,20]: Roughly, we can further put these studies into two sub-categories, i.e., *route searching with categorical constraints* [7,13,18], and *route searching with keyword constraints* [1,2,20]. In the first sub-category, the whole set of POIs are divided into different categories and the goal is to find a route that cover POIs from some required categories once. In route searching with keyword constraints, each query retrieves a route that cover the given query keywords [1,2,20].

All these queries described above are different from our ROI query which retrieves the optimal route that cover as many as possible relevant POIs under the distance constraint.

3   Problem Statement and Hardness

3.1   Problem Definition

Before defining ROI query formally, we first give the definitions of road networks and points of interest (POI) as below.

Definition 1 (Road Network). *A road network is modeled as a weighted directed graph $G = (V, E)$, where each vertex $v_i \in V$ represents an intersection of roads or a road terminal; and each edge $e_{i,j} \in E$ represents the directed road segment from v_i to v_j and its length is $|e_{i,j}|$.* □

Definition 2 (Point of Interest, POI). *Each POI, o_i, is denoted by a triplet $o_i = (l, e, \mathcal{D})$, where l is the spatial location consisting of longitude and latitude; e is the edge on which o_i resides; and $\mathcal{D}$ is the textual description of o_i, e.g., the category, service contents, and users' comments. All POIs on a road network are denoted by $\mathcal{O}$.* □

Figure 2 illustrates an example of road network with five vertices, seven edges and five POIs. The length of each edge is marked besides the corresponding edge. A simple route R on G consists of a sequence of vertices, i.e., $R = (v_1, \ldots, v_\gamma)$, and $v_i \neq v_j$ if $i \neq j$ $(1 \leq i, j \leq \gamma)$. The length of R, $dist(R)$, is the sum of the lengths of all edges on R, i.e., $dist(R) = \sum_{i=1}^{\gamma-1} |e_{i,i+1}|$. The shortest route from v_i to v_j is represented as $SR(v_i, v_j)$. In addition, if $o.e$ is an edge of R, we say $o \in R$.

With the preparation above, we have the formal definition of *route of interest (ROI) query* as below.

Definition 3 (Route of Interest (ROI) Query). *Given a road network G and the POIs $\mathcal{O}$, a route of interest query is denoted by $q = (s, e, K, L)$, where s and e are the start and end locations, respectively, K is a set of query keywords to describe uers' interests, and L is a distance threshold. The objective of q is to find a route R^* between s and e such that:*

$$R^* = arg \max_{R \in \mathcal{R}_L(s,e)} S(R) \tag{1}$$

where

- *$\mathcal{R}_L(s, e)$ is the set of feasible routes between s and e whose lengths are less than L, i.e., $\mathcal{R}_L(s, e) = \{R | R \in \mathcal{R}(s, e) \land dist(R) \leq L\}$ and $\mathcal{R}(s, e)$ represents all the simple routes between s and e;*
- *$S(R)$ is the score of R with respect to the query keywords of ROI query q and computed by $S(R) = \Sigma_{o_i \in R} |o_i.\mathcal{D} \cap K|$, where $|o_i.\mathcal{D} \cap K|$ computes the number of query keywords contained by $o_i.\mathcal{D}$ and is denoted by $S(o_i)$.* □

Meanwhile, we have the score of an edge $S(e) = \Sigma_{o_i \in e} S(o_i)$, thus $S(R) = \Sigma_{e \in R} S(e)$.

Table 1. Routes for the ROI query in Fig. 1

Routes	Length	POIs on route	Score
$R_1 = (v_1, v_3, v_5)$	6	$\{o_4\}$	1
$R_2 = (v_1, v_2, v_5)$	8	$\{o_1, o_3\}$	3
$R_3 = (v_1, v_3, v_2, v_5)$	9	$\{o_3, o_4\}$	3
$R_4 = (v_1, v_4, v_5)$	8	$\{o_2, o_5\}$	2

Example 1. Figure 2 illustrates a ROI query, where $s = v_1$ and $e = v_5$ are the start and end locations, respectively, $K = \{k_1, k_2\}$ specifies two query keywords, and the distance threshold is $L = 8$. We have four possible routes as presented in Table 1. R_3 should be pruned because it is not a feasible route ($dist(R_3) = 9 > 8$). Other three routes are feasible routes because all of them satisfy the distance threshold. The final query result is R_2 since its score is larger than that of routes R_1 and R_4. □

3.2   Problem Complexity

We prove that ROI query problem is NP-hard by a reduction from the Hamiltonian Path (Ham-Path) problem which is a well-known NP-hard problem. Given a graph $G(V, E)$, a Ham-Path decides whether there exists a simple path in G that visits every vertex in V once. First, we define the decision version of ROI query problem, i.e., *Decision-ROI*, as follows.

Definition 4 (Decision-ROI). *Given a road network G, Decision-ROI decides whether there exists a path from s to e such that its length is at most L and its score is at least X, where X is a float number and $X \geq 0$.* □

Then, we have the following theorem.

Theorem 1. *Decision-ROI problem is NP-hard.* □

Proof. The theorem can be proved by a reduction from Ham-Path problem to Decision-ROI problem. Assume that graph $G(V, E)$ is an instance of Ham-Path problem and that graph $G'(V', E')$ is an instance of Decision-ROI problem. First, we convert G to G'. Initially, we let G' have the same vertices of G, i.e., $V = V'$. If $e_{i,j} \in E$, we set the length and score of each edge $e_{i,j}$ in G' as $|e_{i,j}| = 1$ and $S(e_{i,j}) = 1$. Otherwise, we have $|e_{i,j}| = +\infty$ and $S(e_{i,j}) = 0$. Two additional vertices $v_s = s$ and $v_e = e$ are added to G' such that, for any vertex $v_i \in V'$, $|e_{s,i}| = |e_{e,i}| = 1$ and $S(e_{s,i}) = S(e_{e,i}) = 0$. Now we have $V' = V \cup \{v_s, v_e\}$.

We can prove that a Ham-Path problem on graph G is equivalent to a Decision-ROI problem on graph G'. Concretely, G has a Ham-Path if and only if there exists a path in G' from v_s to v_e such that the length of the path is at most $n + 1$ and the total score of the path is at least $n - 1$. This can be proved from two aspects as below.

- If there is a Ham-Path P in G, the number of vertices in P should be n. By adding v_s and v_e to the two ends of P, respectively, we can generate a new path P' whose length and score are $n + 1$ and $n - 1$, respectively.
- If there is a path P' in G' such that $dist(P') \leq n + 1$ and $S(P') \geq n - 1$, the number of edges in P' (except for the edges connected to v_s and v_e) should be at least $n - 1$ because each edge has a score of 1 at most. Therefore, the total number of edges in P' is at least $n + 1$ if the two edges connected to v_s and v_e are counted. Considering that the length of P' is at most $n + 1$ and each edge has a length of 1, we can conclude that the number of edges in P' is $n + 1$ and the number of vertices in P' is $n + 2$. Thus, a Ham-Path on G is obtained by removing v_s and v_e from P'.

The proof is completed.   □

4   An Adaptive Route Sampling Framework

4.1   Overview of the Framework

As ROI query problem is NP-hard, it is impossible to design algorithms with polynomial time complexity as long as P$\neq$NP. Therefore, it is reasonable to design approximate solutions for ROI queries. Meanwhile, to overcome the low efficiency of exact solutions and the low quality of approximate/heuristic solutions, we propose a route sampling framework, as illustrated in Fig. 3, which adaptively generates query results of different qualities according to the given response time. Taking as input the road network and a ROI query, the sample route module generates a feasible route with a certain route sampling method each time. The current optimal route is updated if the new feasible route has a larger score. Then, the response time limit is checked. If there is still spare time, more loops are executed repeatedly to generate new feasible routes. Otherwise, the current optimal route is returned as the query result.

Actually, this route sampling framework involves two key issues, i.e., sampling a feasible (i.e., the feasibility issue) and good (i.e., the quality issue) route each time. We will elaborate these two issues in the following sub-sections.

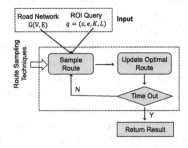

Fig. 3. The adaptive route sampling framework for ROI queries

4.2 Sampling a Feasible Route

We sample a feasible route R by starting from s and adding edges sequentially until the end location e is reached. The challenging issue is how to ensure that R is feasible. Assume that the current expanding vertex is v_i with adjacent vertices $adj(v_i)$, i.e., $adj(v_i) = \{v_j | e_{i,j} \in E\}$. After removing those visited vertices, we get the unvisited adjacent vertices of v_i, i.e., $adj_u(v_i)$. Then, a vertex v_j is selected from $adj_u(v)$ according to selection probability $p_{i,j}$ and concatenated to R to generate a longer partial route which covers more relevant POIs, where $0 \leq p_{i,j} \leq 1$ and $\sum_{v_j \in adj_u(v_i)} p_{i,j} = 1$. The details of $p_{i,j}$ will be elaborated in Sect. 4.3. However, some vertices from $adj_u(v)$ may violate the distance threshold. In order to guarantee a feasible route, the vertex v_j selected from $adj_u(v)$ must survive from the distance pruning as below.

Lemma 1 (Distance pruning). *For any vertex $v_j \in adj_u(v_i)$, v_j can be selected for route expansion if and only if the following condition holds:*

$$dist(R) + |e_{i,j}| + dist(SR(v_j, e)) \leq L \qquad (2)$$

where R is the partial route from s to v_i, $|e_{i,j}|$ is the length of the edge from v_i to v_j, and $dist(SR(v_j, e))$ is the shortest distance between v_j and the end location e. □

Proof. The proof is obvious because any vertex $v_j \in adj_u(v_i)$ satisfying Eq. (2) can generate a feasible route. □

We record those vertices in $adj_u(v_i)$ surviving from distance pruning of Lemma 1 as $adj'_u(v_i)$. Therefore, the selection of vertices to expand the route should be constrained by the distance threshold L and we call it *selection constrained* (SC) expansion. Then, we have the following theorem.

Theorem 2. *SC expansion can always generate a feasible route.* □

Proof. Assume the current partial route and its newly added vertex are R and v_i, respectively. Initially, $R = \langle s \rangle$ and $v_i = s$. In each expansion, an adjacent node v_j satisfying Eq. (2) is selected. We can always generate a feasible route $R' = R \oplus \{v_j\} \oplus SP(v_j, e)$ because it satisfies the distance threshold L. Here $x \oplus y$ represents a concatenation operation between x and y. □

Example 2. Take the ROI query in Fig. 2 for example. Initially, we have $R = \langle v_1 \rangle$ and $adj'_u(v_1) = \{v_2, v_3, v_4\}$. Assume that v_3 is selected and $R = \langle v_1, v_3 \rangle$. Meanwhile, we have $adj_u(v_3) = \{v_2, v_5\}$. As $dist(R) + |e_{3,2}| + dist(SR(v_2, e)) = 3 + 2 + 4 = 9 > L = 8$, we have $adj'_u(v_3) = \{v_5\}$. Finally, a feasible route $R = \langle v_1, v_3, v_5 \rangle$ is sampled. □

In SC expansion, $dist(SR(v_j, e))$ is computed online. To accelerate this computation, we first conduct a backward expansion from the end location e on the reverse graph of G (generated by changing the direction of each edge in G) to compute the shortest distances from e to all vertices that are less than L away from e. These distances are stored and can be checked quickly during the SC expansion, thus avoiding the computation of the network distance between v_j and e for each partial route.

The process of feasible route sampling is outlined in Algorithm 1, which receives as input an ROI query q and returns as output a feasible route R.

Algorithm 1. Sample a feasible route

Input: a ROI query $q = (s, e, K, L)$
Output: a feasible route R

1 Initialize route $R \leftarrow \langle s \rangle$ and $v_i \leftarrow s$
2 **while** *not reach e* **do**
3 $\quad adj'_u(v_i) \leftarrow \emptyset$
4 $\quad$ **foreach** $v_j \in adj(v_i)$ **do**
5 $\quad\quad R' \leftarrow R \oplus \{v_j\} \oplus SP(v_j, e)$
6 $\quad\quad$ **if** $v_j \notin R \wedge dist(R') \leq L$ **then**
7 $\quad\quad\quad adj'_u(v_i).\text{add}(v_j)$
8 $\quad$ let $p_{i,j}$ be the probability to select vertex v_j
9 $\quad$ pick v_j from $adj'_u(v_i)$ according to $p_{i,j}$
10 $\quad R.\text{add}(v_j)$ and $v_i \leftarrow v_j$
11 **return** R

4.3   Route Sampling Techniques

Though SC expansion guarantees a feasible route, how to select a vertex from $adj'_u(v_i)$ is not yet tackled because selection probability $p_{i,j}$ is still unknown. The quality issue depends on the techniques used to compute $p_{i,j}$. In the following subsections, we will elaborate different techniques to compute $p_{i,j}$. Note that, other route sampling techniques can be also added to the framework without much effort.

A. Uniform Route Sampling. Straightforwardly, during the expansion, a vertex can be selected from $adj'_u(v_i)$ uniformly, i.e., $p_{i,j} = \frac{1}{|adj'_u(v_i)|}$. By doing this repeatedly until the end location e is reached, a feasible route is generated.

Example 3. Take the ROI query in Fig. 2 for example. We start from v_1 with the partial route $R = \langle v_1 \rangle$. Initially, by Lemma 1, we have $adj'_u(v_1) = \{v_2, v_3, v_4\}$. Then, a vertex is selected from $adj'_u(v_1)$ randomly with equal probability. We assume v_2 is selected and $R = \langle v_1, v_2 \rangle$. Next, we have $adj_u(v_2) = \{v_3, v_5\}$. However, v_3 is removed because $dist(R) + |e_{3,2}| + dist(SR(v_2, e)) = 3 + 2 + 4 = 9 > L = 8$. Therefore, only one vertex is left, i.e., $adj'_u(v_2) = \{v_5\}$. Now, the final vertex v_5 is reached and we get a feasible route $R = \langle v_1, v_2, v_5 \rangle$ whose score is $S(R) = S(e_{1,2}) + S(e_{2,5}) = 3$. □

B. Prioritized Route Sampling. In uniform route sampling, all vertices in $adj'_u(v_i)$ are selected equally. Therefore, in most cases, we cannot obtain a good route unless enough number of feasible routes are sampled. Intuitively, a good route sampling method should give higher priority to those routes with large scores. However, it is non-trivial to achieve this because we do not have the whole route before the route sampling is finished. In other words, we only have a partial route when selecting a vertex from $adj'_u(v_i)$. Therefore, selecting v_j from $adj'_u(v_i)$ can be done in a greedy fashion. To this end, we define the desirability of selecting a vertex $v_j \in adj'_u(v_i)$ as below.

$$\eta_{i,j} = 1 + \frac{S(e_{i,j})}{|e_{i,j}| \cdot S^+_{avg}(G)} \tag{3}$$

where $S^+_{avg}(G)$ is the maximum average score over the whole road network, i.e.,

$$S^+_{avg}(G) = \max_{e \in E} S_{avg}(e) = \max_{e \in E} \frac{S(e)}{|e|} \tag{4}$$

Then, we compute the selection probability of v_j by $p_{i,j} = \frac{\eta_{i,j}}{\sum_{v_h \in adj'_u(v_i)} \eta_{i,h}}$. Thus, the scores of adjacent edges are taken into consideration while selecting the next vertex to expand. Though this is just a local greedy strategy, it improves the quality of returned routes compared with uniform route sampling, which is validated in the experiments.

Example 4. Take the ROI query in Fig. 2 for example. We have $S(e_{1,2}) = S(e_{1,3}) = S(e_{1,4}) = S(e_{4,5}) = 1$, $S(e_{2,3}) = S(e_{3,5}) = 0$, and $S(e_{2,5}) = 2$. By Eq. (4), we have $S^+_{avg}(G) = 0.5$. Initially, we start from v_1 with route $P = \langle v_1 \rangle$ and have $adj'_u(v_1) = \{v_2, v_3, v_4\}$ by Lemma 1. Particularly, for vertex v_2, we have

$$\eta_{1,2} = 1 + \frac{S(e_{1,2})}{|e_{1,2}| \cdot S^+_{avg}(G)} = 1 + \frac{1}{4 * 0.5} = 1.5$$

In such a way, we have $\eta_{1,3} = \eta_{1,4} = 1.67$. Further, we have

$$p_{1,2} = \frac{\eta_{1,2}}{\sum_{v_h \in adj'_u(v_1)} \eta_{i,h}} = \frac{1.5}{1.5 + 1.67 + 1.67} = 0.31.$$

Similarly, we have $p_{1,3} = p_{1,4} = 0.345$. Then a vertex is selected from $adj'_u(v_1)$ according to the computed probability $p_{i,j}$. Without loss of generality, we assume v_2 is selected and $R = \langle v_1, v_2 \rangle$. Next, we have $adj_u(v_2) = \{v_3, v_5\}$. However, v_3 is removed due to the distance violation as discussed previously. Therefore, only one vertex is left, i.e., $adj'_u(v_2) = \{v_5\}$ and we have $p_{2,5} = 1$. Now, the final vertex v_5 is reached and we have a feasible route $R = \langle v_1, v_2, v_5 \rangle$ whose score is $S(P) = S(e_{1,2}) + S(e_{2,5}) = 3$. □

C. Dynamic Prioritized Route Sampling. To take advantage of previously sampled routes, we further propose dynamic prioritized route sampling technique, which adjusts the selection probability $p_{i,j}$ dynamically according to the knowledge of sampled routes. Assuming that a set of routes have been sampled by using the prioritized route sampling, different strategies to update $p_{i,j}$ are elaborated below.

DPri-Sample 1. Intuitively, as more and more routes are sampled, the probability to obtain better routes increases. Therefore, for each edge $e_{i,j} \in R$, we update its desirability by $\eta'_{i,j} = \frac{\eta_{i,j}}{1+\mathcal{R}(e_{i,j})}$, where $\mathcal{R}(e_{i,j})$ is the number of sampled routes that pass edge $e_{i,j}$. Accordingly, $p_{i,j}$ is adjusted as below.

$$p_{i,j} = \frac{\eta_{i,j}/(1+\mathcal{R}(e_{i,j}))}{\Sigma_{v_h \in adj'_u(v_i)}\{\eta_{i,h}/(1+\mathcal{R}(e_{i,h}))\}} \tag{5}$$

DPri-Sample 2. Intuitively, we tend to travel these edges covered by routes of high scores. Therefore, we increase the desirability of an edge $e_{i,j}$ with respect to the scores of those routes that travel through $e_{i,j}$, i.e.,

$$p_{i,j} = \frac{\eta_{i,j} \cdot (1 + \Sigma_{k=1}^m \Delta\tau_{i,j}^k)}{\Sigma_{v_h \in adj'_u(v_i)}\eta_{i,h} \cdot (1 + \Sigma_{k=1}^m \Delta\tau_{i,h}^k)} \tag{6}$$

where $\Delta\tau_{i,j}^k$ is the increase ratio contributed by the k-th route R_k and computed with the follow function.

$$\Delta\tau_{i,j}^k = \begin{cases} \frac{S(R_k)}{dist(R_k) \cdot S_{avg}^+(G)} & \text{if route } R_k \text{ travels on edge } e_{i,j} \\ 0 & \text{otherwise} \end{cases} \tag{7}$$

where $S(R_k)$ is the score of route R_k and $S_{avg}^+(G)$ is the maximum average score over the whole road network computed by Eq. (4).

DPri-Sample 3. Instead of updating $p_{i,j}$ once a feasible route is sampled, we update $p_{i,j}$ when a set of m' routes have been sampled. Thus, $p_{i,j}$ is computed as below.

$$p_{i,j} = \frac{\eta_{i,j} \cdot \tau_{i,j}}{\Sigma_{v_h \in adj'_u(v_i)}(\eta_{i,h} \cdot \tau_{i,h})} \tag{8}$$

where the increase ratio $\tau_{i,j}$ is adjusted by $\tau'_{i,j} = (1-\rho) \cdot \tau_{i,j} + \Sigma_{k=1}^{m'}\Delta\tau_{i,j}^k$ and $\rho \in [0,1]$. Initially, we have $\tau_{i,j} = 1$ for all edges $e_{i,j}$.

D. Quality Analysis. Assuming the maximum degree of vertices in road network G is d, in the worst case, the number of feasible routes is $|\mathcal{R}_L(s,e)| = d^{\frac{L}{L_{min}}}$, where L_{min} is the length of the shortest edge in G. For simplicity, we use ζ to denote $d^{\frac{L}{L_{min}}}$. In uniform sampling, each feasible route has the equal probability to be sampled. Therefore, in each iteration, the probability that the optimal path is selected is $Pr\{R^*\} = \frac{1}{\zeta}$. After I iterations, the probability that we can obtain the optimal route is $1 - (1 - \frac{1}{\zeta})^I$. To guarantee that the probability that we obtain the optimal result is larger than $(1 - \varepsilon)$, $\varepsilon \in (0,1)$, the number of iterations should be $I \geq \frac{log\varepsilon}{log(\zeta-1)-log\zeta}$. The analysis of other sampling methods are similar, thus being omitted for saving space.

5   Experiments

5.1   Data Sets and Setup

We use three real datasets, i.e., the road networks and POIs of Dublin, Beijing and London. All these datasets are extracted from OpenStreetMap (OSM)[1]. The details of three datasets are presented in Table 2.

Table 2. Statistics of three data sets.

Data sets	#vertices	#edges	#PoIs
Dublin (DL)	62,975	82,730	5,297
Beijing (BJ)	46,029	62,778	21,192
London (LN)	209,045	282,267	34,341

In our experiments, the performance of three route sampling schemes i.e., uniform route sampling (Uni), prioritized route sampling (Pri), and dynamic prioritized route sampling (DPri) are evaluated. We use DPri1, DPri2 and DPri3 to denote the three different update strategies in DPri. In addition, we revise one heuristic algorithm termed Nearest Neighbor Heuristic (NNH) in [11] to process ROI query and use it as a baseline. NNH expands from the start location s, and selects the nearest POI containing query keywords to expand until the distance threshold L is used out.

We generate 50 ROI queries for each data set. The two query locations of each query are randomly selected from vertices V, and query keywords are randomly selected from the corresponding vocabulary built based on the textual description of POIs. The average score of returned routes for 50 ROI queries is used as the metric to evaluate the performance of solutions. Note that the response time is implicitly evaluated since it is an adjustable parameter in our solution.

The settings of used parameters are summarized in Table 3 with their default values in bold font. We use London dataset as the default dataset if no specific statement. $|V(s,e)|$ denotes the number of vertices whose sum distance to two query locations, i.e., s and e, is less than the distance threshold L. $|V(s,e)|$ is computed with two expansions from s and e, respectively. In addition, the distance threshold of each ROI query is specified with a certain deviation from $dist(SR(s,e))$.

5.2   Experimental Results

Varying parameters in DPri3: Figure 4 illustrates the average scores of 50 ROI queries over London dataset with $T = 1$ second while varying the parameters m' and ρ. As suggested by Fig. 4(a), the optimal value of m' is $0.03*|V(s,e)|$.

---

[1] https://www.openstreetmap.org/.

Table 3. Parameters and their default values in experiments.

Parameters	Meaning	Values
m'	number of sampled routes to update $p_{i,j}$	{0.005, 0.01, 0.02, **0.03**, 0.05, 0.1, 0.3, 0.5}* $|V(s, e)|$
ρ	the decrease ratio in DPri3	0.01, 0.05, 0.1 **0.2**, 0.3, 0.4, 0.5
L	distance threshold	{1.2, **1.3**, 1.4, 1.5, 1.6}*$dist(SR(s, e))$
T	response time limit	**1**, 3, 5, 7, 9 (s)

A smaller m' will degrade DPri3 to the Pri sampling while a larger m' will slow the update of $p_{i,j}$, thus reducing the efficiency of route sampling. Another parameter used in DPri3 is the decrease ratio ρ which should be set to 0.2, as suggested by Fig. 4(b). A smaller value than 0.2 results in a quick increase of $p_{i,j}$ for frequently visited edges while reducing the chance to explore unvisited edges. On the contrary, a larger value than 0.2 will weaken the impacts of previously sampled routes.

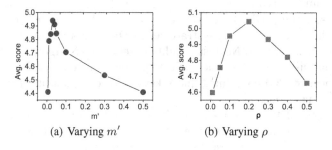

(a) Varying m' (b) Varying ρ

Fig. 4. Varying parameters in DPri3.

Comparing Different Route Sampling Methods: Figure 5 illustrates the performance of different route sampling methods on three datasets. Pri outperforms Uni because it considers the scores of edges rather than selecting edges uniformly. DPri1 does not achieve a better performance than Pri because it greatly reduces the chances to visit those edges that have been visited many times before. However, these edges may still lead to some optimal routes. Differently, both DPri2 and DPri3 achieve a better performance than other route sampling methods. Particularly, the scores of query results returned by DPri3 are much better than that of other route sampling methods. Therefore, to simplify the presentation, we only report the results of DPri3 for dynamic prioritized route sampling in the following experiments.

Varying Response Time: Figure 6(a) illustrates the results of ROI query while varying the response time limit. For Uni, Pri, and DPri3, the average scores of query results gradually increases with the increase of response time limit. Actually, given a response time large enough, the route sampling framework can

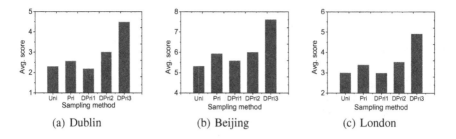

(a) Dublin   (b) Beijing   (c) London

Fig. 5. Comparing different route sampling methods on three different data sets.

find results with very high quality. The query results obtained by NNH keep the same while varying the response time limit because it only selects one route greedily as the query result.

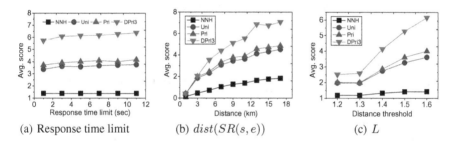

(a) Response time limit   (b) $dist(SR(s,e))$ (c) L

Fig. 6. Varying response time limit, distance $dist(SR(s,e))$, and distance threshold L.

Varying Distance $dist(SR(s,e))$ and Distance Threshold L: We generate 20 distance intervals from 0 Km to 20 km with a distance span of 1 km. For each distance interval, we randomly generate 50 ROI queries such that $dist(SR(s,e))$ is within the corresponding distance interval. According to Fig. 6(b), the average score increases with the increase of $dist(SR(s,e))$ and DPri3 achieves a much better performance than Uni, Pri, and NNH. Figure 6(c) illustrates the query results while varying the distance threshold L. As suggested by Fig. 6(c), DPri3 still has the best performance.

6   Conclusion

In this paper, we study the ROI query which retrieves a route according to users' keywords described interest. The returned route should cover the most relevant POIs while has a length less than a distance specified. To efficiently solve ROI query problem, we proposed an adaptive route sampling framework with a suite of route sampling techniques to generate query results according to the given response time. Though ROI query in this work focuses on road networks, it can also find applications in other scenarios, e.g., route searching in shopping malls and theme parks.

Acknowledgments. This work is partially supported by the National Natural Science Foundation of China under Grant No.61373036 and the NSFC/RGC Joint Research Scheme No. N_PolyU519/12.

References

1. Cao, X., Chen, L., Cong, G., Guan, J., Phan, N.-T., Xiao, X.: KORS: keyword-aware optimal route search system. In: ICDE, pp. 1340–1343 (2013)
2. Cao, X., Chen, L., Cong, G., Xiao, X.: Keyword-aware optimal route search. PVLDB **5**(11), 1136–1147 (2012)
3. Cao, X., Cong, G., Guo, T., Jensen, C.S., Ooi, B.C.: Efficient processing of spatial group keyword queries. ACM Trans. Database Syst. **40**(2), 13 (2015)
4. Cao, X., Cong, G., Jensen, C.S., Ooi, B.C.: Collective spatial keyword querying. In: SIGMOD, pp. 373–384 (2011)
5. Cao, X., Cong, G., Jensen, C.S., Yiu, M.L.: Retrieving regions of interest for user exploration. PVLDB **7**(9), 733–744 (2014)
6. Cary, A., Wolfson, O., Rishe, N.: Efficient and scalable method for processing top-k spatial boolean queries. In: Gertz, M., Ludäscher, B. (eds.) SSDBM 2010. LNCS, vol. 6187, pp. 87–95. Springer, Heidelberg (2010)
7. Chen, H., Ku, W.-S., Sun, M.-T., Zimmermann, R.: The multi-rule partial sequenced route query. In: ACM-GIS, p. 10 (2008)
8. Chen, L., Cong, G., Jensen, C.S., Dingming, W.: Spatial keyword query processing: an experimental evaluation. PVLDB **6**(3), 217–228 (2013)
9. Choi, D.-W., Chung, C.-W., Tao, Y.: A scalable algorithm for maximizing range sum in spatial databases. PVLDB **5**(11), 1088–1099 (2012)
10. Christoforaki, M., He, J., Dimopoulos, C., Markowetz, A., Suel, T.: Text vs. space: efficient geo-search query processing. In: CIKM, pp. 423–432 (2011)
11. Deng, D., Shahabi, C., Demiryurek, U.: Maximizing the number of worker's self-selected tasks in spatial crowdsourcing. In: SIGSPATIAL, pp. 314–323 (2013)
12. Guo, T., Cao, X., Cong, G.: Efficient algorithms for answering the m-closest keywords query. In: SIGMOD, pp. 405–418 (2015)
13. Li, F., Cheng, D., Hadjieleftheriou, M., Kollios, G., Teng, S.-H.: On trip planning queries in spatial databases. In: Medeiros, C.B., Egenhofer, M., Bertino, E. (eds.) SSTD 2005. LNCS, vol. 3633, pp. 273–290. Springer, Heidelberg (2005)
14. Li, Z., Lee, K.C.K., Zang, B., Lee, W.-C., Lee, D.L., Wang, X.: Ir-tree: an efficient index for geographic document search. IEEE Trans. Knowl. Data Eng. **23**(4), 585–599 (2011)
15. Liu, J., Yu, G., Sun, H.: Subject-oriented top-k hot region queries in spatial dataset. In: CIKM, pp. 2409–2412 (2011)
16. Long, C., Wong, R.C.-W., Wang, K., Fu, A.W.-C.: Collective spatial keyword queries: a distance owner-driven approach. In: SIGMOD, pp. 689–700 (2013)
17. Rocha-Junior, J.B., Nørvåg, K.: Top-k spatial keyword queries on road networks. In: EDBT, pp. 168–179 (2012)
18. Sharifzadeh, M., Kolahdouzan, M.R., Shahabi, C.: The optimal sequenced route query. VLDB J. **17**(4), 765–787 (2008)

19. Tao, Y., Xiaocheng, H., Choi, D.-W., Chung, C.-W.: Approximate maxrs in spatial databases. PVLDB **6**(13), 1546–1557 (2013)
20. Zeng, Y., Chen, X., Cao, X., Qin, S., Cavazza, M., Xiang, Y.: Optimal route search with the coverage of users' preferences. In: IJCAI, pp. 2118–2124 (2015)
21. Zhang, C., Zhang, Y., Zhang, W., Lin, X.: Inverted linear quadtree: efficient top k spatial keyword search. In: ICDE, pp. 901–912 (2013)
22. Zheng, K., Han, S., Zheng, B., Shang, S., Jiajie, X., Liu, J., Zhou, X.: Interactive top-k spatial keyword queries. In: ICDE, pp. 423–434 (2015)

Semantic-Aware Trajectory Compression
with Urban Road Network

Na Ta$^{(\boxtimes)}$, Guoliang Li, Bole Chen, and Jianhua Feng

Department of Computer Science, Tsinghua University, Beijing 100084, China
dqn13@mails.tsinghua.edu.cn, cbl_tae@163.com,
{liguoliang,fengjh}@tsinghua.edu.cn

Abstract. Vehicles are generating more and more trajectories, which
are expensive to store and mange. Thus it calls for vehicular trajectory compression techniques. We propose a semantic-aware compression
framework that includes two steps: Map Matching(MM) and Semantic Trajectory Compression(STC). On the one hand, due to measurement errors, trajectories cannot be precisely mapped to real roads. In
the MM step, we utilize multidimensional information(including distance
and direction) to find the most matching roads and generate aligned trajectories. On the other hand, some unnecessary points in trajectories can
be reduced based on the roads. In the STC step, we extract a set of *crucial points* from the aligned trajectory, which capture the major driving
semantics of the trajectory. Meanwhile, our decompression method is
fairly lightweight and can efficiently support various applications. Empirical study shows that MM achieves high matching quality, STC achieves
more than 8 times compression ratio, and decompression is efficient on
real datasets.

1 Introduction

The advancement of vehicle positioning techniques have produced high-volume
trajectory data. A vehicular trajectory is a sequence of discrete locations at
sampling time, reporting instant position, velocity, direction, etc., of the vehicle.
A growing number of applications have been proposed to utilize trajectory data,
e.g., frequent path finder [11], taxi pick-up recommending system [13], data
management for traffic analysis [3] and adaptive trajectory storage [4].

However, raw GPS trajectories are not instantly usable because: (1) The
measurement error of GPS devices calls for **Map Matching**, i.e., fitting raw
trajectories into the underlying road networks; (2) The *sampling error* due to
transmission failure causes unstable sampling rate, making it hard to tell the
exact route; and (3) The *huge volume* of trajectories makes the management
and processing expensive, calling for **Compression** algorithms.

A number of map matching [1,2,10,17] and trajectory compression [5,8,9,
15] algorithms have been proposed separately. However, current map matching
techniques do not care about compression while existing compression algorithms
rarely take into consideration urban road networks constraining running vehicles.

B. Cui et al. (Eds.): WAIM 2016, Part I, LNCS 9658, pp. 124–136, 2016.
DOI: 10.1007/978-3-319-39937-9_10

To alleviate these problems, we combine the two processes to effectively compress massive vehicular trajectories, producing standardized(fit into the road network) and lightweight(less storage) trajectories.

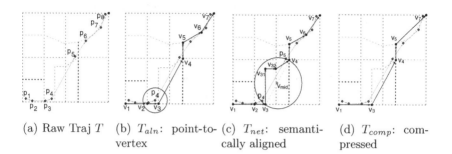

(a) Raw Traj T (b) T_{aln}: point-to-vertex   (c) T_{net}: semantically aligned   (d) T_{comp}: compressed

Fig. 1. A running example

Our map matching(MM) process has two steps: point-to-vertex matching and semantic alignment. In the first step, we exploit self-semantic of a single sample point p to find p's matching road vertex, i.e., distances and direction angle differences of p to potential matching road edges. Figure 1(a) presents a fragment of road network where roads are denoted by dotted lines. The solid line from the lower left corner to the upper right corner represents a raw trajectory T, and dots along T are sample points($|T| = 8$). Each sample point of T is mapped to a vertex(the black dots) in the road network to form an aligned trajectory T_{aln} ($|T_{aln}| = 7$) in Fig. 1(b). In the second step, as T_{aln} may not be well fit to the road network(e.g., not fully connected), we fit T_{aln} to the underlying road network using correlative-semantic of adjacent points in T_{aln} and construct the semantically aligned trajectory T_{net} ($|T_{net}| = 9$) in Fig. 1(c).

In the compression(STC) process, we propose to maintain the core semantics of trajectories for storage reduction, where the compressed trajectories are represented as a list of *crucial points*, such as street intersections, road turning points, etc., plus the beginning and end of the trajectory. After STC, T is finally compressed to T_{comp} ($|T_{comp}| = 5$) in Fig. 1(d). On average, STC can eliminate more than 80 % of raw points(Sect. 6).

Many real applications, e.g., location-based services, require to efficiently decompress the compressed trajectories for instant feedback. Under our semantic settings, the recovery of trajectories is efficient as we can easily retrieve the travelled roads to restore the trajectory within the road network.

In summary, we make the following contributions:

- We propose a method to fit raw trajectories into urban road networks using single and correlative semantics, effectively revealing the actual traces and eliminating the measurement and sampling errors by GPS devices.
- We devise an efficient semantic trajectory compression algorithm, achieving high compression ratio even with average quality digital maps; meanwhile,

the decompression recovers the original trajectory to the minimum loss ratio, facilitating applications such as online query answering.
- We combine a set of practical metrics to form a benchmark of evaluating trajectory compression algorithms both physically and semantically. Experimental results on real-world datasets show that our methods achieve high quality and good performance, and scale well.

The rest of the paper is organized as follows. Section 2 formally defines the problems and reviews related work. Sections 3, 4, 5 present our solution and algorithms in details. Section 6 experimentally evaluates the performance of our algorithms on real datasets. Finally, Sect. 7 concludes the paper.

2 Preliminaries

2.1 Problem Formulation

Before we formally define the problems of map-matching and semantic trajectory compression/decompression, we first introduce some concepts.

Definition 1 Road Network. A road network is a directed graph $G(V, E)$, where $V = \{v_i(x_i, y_i)\}$ is the set of vertices, a vertex v_i is represented by a pair of latitude x_i and longitude y_i; and $E = \{e_j(v_k, v_m)\}$ is the set of edges which are road fragments directly connecting vertices in V.

Definition 2 Trajectory T. A trajectory is a sequence of historical locations of a vehicle, $T = \{p_i(l_i, t_i, spd_i, head_i) \mid 1 \leq i \leq |T|\}$, where $l_i = (x_i, y_i)$ represents the latitude x_i and longitude y_i of the vehicle at time t_i, spd_i is the instantaneous speed, $head_i$ is the angle of driving direction to north clockwise, and the size of the trajectory, $|T|$, is the number of sample points in T.

Road edges are generally rather short in a digital map(within 50 m on average in our settings) and the vertices are rather dense. Thus we can map each trajectory point to a uniquely aligned road vertex and get an aligned trajectory, then construct a semantically aligned trajectory by connecting vertices by edges. We take into consideration both the instant driving direction of certain sample point and its distance to vertices when choosing the aligned vertex.

Definition 3 Aligned Vertex v_{aln}. A road is a **potential matching road** of a sample point p if p's driving direction and the direction of the tangent line to p's closest point on the road is within an angle threshold α. The potential matching road with the minimum distance to p is called the **optimal matching road**, on which the closest vertex to p is called the aligned vertex of p.

Definition 4 Aligned Trajectory T_{aln}. $T_{aln} = \{(v_i, t_i, spd_i, head_i) \mid 1 \leq i \leq |T_{aln}|\}$ is the aligned trajectory of T, where v_i is the aligned vertex of p_i.

The aligned vertices in T_{aln} of two neighboring points in T may not be directly connected by certain edges in the road network, so we need to find the connecting path. To this end, we define semantically aligned trajectory.

Definition 5 *Semantically Aligned Trajectory* T_{net}. T_{net} is a sequence of points fully connecting points in T_{aln} with road edges. $T_{net} = \{(v_i, t_i, spd_i, head_i) \mid 1 \leq i \leq |T_{net}|\}$, where $v_i(x_i, y_i)$ is (1) an aligned vertex of some point in T; or (2) a vertex along the shortest path between two adjacent aligned vertices in T_{aln}.[1]

In the semantically aligned trajectory T_{net}, some points can be omitted. For example, in Fig. 1(c), v_6 on the light blue road can be omitted, because if we know that the car has gone through the light blue road, it must pass by v_6. In contrast, v_5 and v_7 are crucial and should be reserved, otherwise we can not induce which road to go from v_4. Thus the STC tries to preserve only *crucial points* in a semantically aligned trajectory T_{net}.

Definition 6 *Crucial Point*. A point p_{cr} in a semantically aligned trajectory T_{net} is a crucial point if: (1) p_{cr} and its predecessor point p'_{cr} or its successive point p''_{cr} are on different roads in the road network G; or (2) the angle difference between lines $\overline{p'_{cr}p_{cr}}$ and $\overline{p_{cr}p''_{cr}}$ is less than a predefined bound β.

Definition 7 *Semantically Compressed Trajectory* T_{comp}. T_{comp} is a subset of a semantically aligned trajectory T_{net}, where each point is a crucial point.

Now we formalize the problems we tackle in this paper.

Map Matching(MM). Given an original trajectory T, find the semantically aligned trajectory T_{net} in the road network G that reveals the real route of T.

Semantic Trajectory Compression(STC). Given a semantically aligned trajectory T_{net}, compress the trajectory to only contain semantically crucial points, aiming to achieve the minimal storage requirement.

Semantic Trajectory Decompression(STD). Given a semantically compressed trajectory T_{comp}, restore the trajectory within the constraint of the road network, aiming to maximize both physical and semantical similarities to T_{net}.

2.2   Related Work

We review related work on map matching, trajectory compression and exploration of semantics of trajectories.

Map Matching. In [1] the Fréchet distance is used to measure the similarity of a line segment pattern(the trajectory) to certain part of a geometric graph as a bigger pattern(a digital map). The time complexity is $\mathcal{O}(mn \log^2 mn)$, where m and n are the number of edges and vertices of the map. Brakatsoulas et al. [2]

---

[1] If an element e_i in T_{net} is an aligned vertex to a sample point p, then the timestamp, speed and heading of e_i are directly copied from p; otherwise, e_i is along the Dijkstra path between the vertices of two samples p_i and p_{i+1}, these information are interpolated accordingly. If more than one sample points are aligned to e_i, we use the mean values these points to assign the timestamp, speed, etc. to e_i.

propose incremental and global matching strategies. The incremental method uses a "look ahead"strategy, and it runs in $\mathcal{O}(na^{l+1})$ time, where n is the number of sample points, a is the adjacent road edges to each point and l is the number of edges in the look-ahead strategy. The global algorithm extends [1] with average and weak Fréchet distance to handle outliers. The time complexity is $\mathcal{O}(mn \log mn)$. The algorithms of [1] and [2] haven't been tested on real datasets. Yin et al. [17] use the Dijkstra's shortest path to measure the distance between a trajectory and a list of road edges in the network. The path of the smallest distance then constitutes the matching route. Lou et al. [10] try to complement the missing route for low-sampling-rate trajectories. Candidate edges with small distance and high relative distance ratio are chosen to the map-matched trajectory. The time complexity is $\mathcal{O}(nk^2m \log m + nk^2)$, where k is the maximum number of candidate for a sampling point. Our MM algorithm is agile because it examines self-semantic of a single trajectory point to ensure the simplicity; meanwhile, the correlative-semantic of point pairs are utilized to guarantee that the inter-relationship between adjacent sample points facilitate the accuracy when revealing the actual trace.

Trajectory Compression Compression by Line Simplification. Initial works adopt the line simplification methodology [5, 7, 12]. Points are preserved if their removal will significantly influence the physical form of trajectories. Quality of compression can be measured by the Euclidian Distance(ED) between original and compressed trajectories; or Synchronized Euclidian Distance(SED) if the temporal factor is considered. The Douglas-Peucker(DP) algorithm [5] repeatedly removes points if the maximum ED between original and compressed trajectories is within a given threshold. DP could not handle the temporal feature. Using SED instead, the Top-Down Time Ratio [12] overcomes the limitation. The Opening Window algorithm [7] uses a sliding window to discard the maximum number of points that would not incur a spatial error within a threshold.

Compression using Speed and Direction. Dead Reckoning [16] keeps both coordination and velocity of a point p, and removes any subsequent points if their locations can be calculated from p, with a given SED. The first point p' after p in T that can not be estimated in this way is kept, until the end of T. Direction-Preserving Trajectory Simplification [8] uses the direction information of a trajectory to form more meaningful compressions to assist Map Matching and other applications. In [9] an improved solution is proposed to minimize the spatial error while achieving a pre-defined simplification size.

Advanced Compression Algorithms. MMTC [6] proposes to solve trajectory compression problem combined with map matching, on-line and off-line. The MM step uses *Network distance* to measure the similarity between trajectory fragments and roads. The TC step is a cost-optimization problem adopting the Minimum Description Length principle. PRESS [15] separately compresses the spatial and temporal dimensions of trajectories. The Huffman encoding of a shortest path travelled between two nodes in the road network is used to reduce the storage. The temporal dimension is denoted as the travelled distance till time t_i, albeit "this representation is storage consuming"[15]. Time Synchronized Network Distance and Network Synchronized Time Difference are used to bound the inaccuracy introduced by compression. In contrast, our STC algorithm compresses the temporal and spatial information together and does not need to separate these two dimensions. Experimental results(Sect. 6) show that our algorithm achieves higher compression ratio while maintaining nice data quality.

Semantics of Trajectories. The algorithm in [14] aims to reduce the storage of human movement traces constrained in urban transportation networks. It maintains *reference points* to represent the semantics of a short trip by a person. Their decompression process requires an iterative probing step to find the actual road taken. The applicability is different from ours in that this algorithm deals with relatively high sampling rate, low speed and short (both the duration and distance travelled) trajectories generated by human, while our algorithm tackles massive long-lasting vehicle trajectories with uncertain sampling rates.

3   The Two-Step Map Matching (MM)

We use two steps, namely the point-to-vertex matching(Sect. 3.1) and semantic alignment(Sect. 3.2) to fit raw trajectories into road networks and get the connected path travelled.

3.1   Point-to-Vertex Matching

The point-to-vertex matching aims to produce the aligned trajectory T_{aln} for a raw trajectory T by finding the aligned vertex for each sample point of T. For each trajectory point p_i, the algorithm uses a circle c of radius r centered at p_i to construct the candidate set S_i of potential matching edges to p_i (note initial radius value does not impact the matching result). For each road edge e_j within or intersecting c, if its angle difference to p_i's direction is within a threshold α, e_j is added to S_i. If S_i is empty, we enlarge the searching circle by doubling its radius util some edge satisfying both the angle and distance conditions are added into S_i. Then the edge in S_i with the minimal distance to p_i is selected as p_i's matching edge e_{opt}, and e_{opt}'s closer vertex to p_i is the aligned vertex of p_i.

Algorithm 1 shows the pseudo code. In line 6, procedure GetCandidates finds all edges with angle difference no more than α to point p_i's direction in the searching circle c. Once a non-empty set S_i is constructed, procedure MinDist in line 10 returns the edge in S_i with minimal distance to p_i.

Figure 1(b) illustrates how the point-to-vertex matching works. For sample point p_4, its searching circle encloses three roads(pink, green and red), but the horizontal red road is not added to S_4(candidate set of p_4) as its angle difference to p_4 exceeds given threshold. So S_4 contains two vertical roads, the green one to the right is closer to p_4, thus its closer vertex v_3 is p_4's aligned vertex.

Algorithm 1. Point-to-Vertex Match	**Algorithm 2.** Semantic Alignment
Input: Raw trajectory T, road network $G(V, E)$, angle diff. threshold α **Output:** Aligned trajectory T_{aln} of T	**Input:** Point-to-vertex matched trajectory T_{aln}, road network $G(V, E)$ **Output:** Semantically Aligned trajectory T_{net} of T

```
1  begin
2  │   T_aln = {};
3  │   for each raw sample point p_i do
4  │   │   S_i = {}, construct circle c with r_0;
5  │   │   while (S_i is empty) do
6  │   │   │   S_i = GetCandidates(p_i, c, α);
7  │   │   │   if (S_i is empty) then
8  │   │   │   │   enlarge c: r = 2 * r;
9  │   │   │   else
10 │   │   │   │   e_opt = MinDist(S_i, p_i);
11 │   │   │   │   break;
12 │   │   v_i = e_opt's closer vertex to p_i;
13 │   │   T_aln = T_aln ∪ {v_i};
14 │   return T_aln;
```

```
1  begin
2  │   T_net = {};
3  │   for unconnected points v_i, v_{i+1} in T_aln
   │   do
4  │   │   v_mid = midpoint of v_i and v_{i+1};
5  │   │   d = length(v_i v_{i+1});
6  │   │   while (sp(v_i v_{i+1}) is incomplete) do
7  │   │   │   construct circle c with d, v_mid;
8  │   │   │   sp(v_i v_{i+1}) =
   │   │   │   ShortestPath(c, v_i, v_{i+1});
9  │   │   │   d = 2 * d;
10 │   │   L_i = vertex list of sp(v_i v_{i+1});
11 │   │   T_net = T_net ∪ L_i;
12 │   return T_net;
```

3.2 Semantic Alignment

Adjacent vertices in the aligned trajectory T_{aln} may not be directly connected on the road network, thus we need to construct the whole traveling trace. Assuming the shortest path is taken between unconnected aligned vertices of successive raw sample points, we propose an efficient *localized* Dijkstra algorithm to identify the fully-connected semantically aligned trajectory T_{net}. Our *localized* algorithm searches over a much smaller area than the whole road network G checked by the global Dijkstra algorithm.

For two unconnected adjacent vertices v_i and v_{i+1} in T_{aln}, we find the shortest path $sp(v_i, v_{i+1})$ as follows: we construct the searching area as a circle centered at the midpoint v_{mid} of line $\overline{v_i v_{i+1}}$, with the diameter set to the length of $\overline{v_i v_{i+1}}$. Then we use the Dijkstra algorithm to find $sp(v_i, v_{i+1})$ using road edges covered by c. If $sp(v_i, v_{i+1})$ can not be found, we double the diameter of c repeatedly util $sp(v_i, v_{i+1})$ is established. Connecting all neighboring vertices in T_{aln} in such manner, we get the semantically aligned trajectory T_{net}. Algorithm 2 shows the pseudo code.

Lemma 1 establishes the validity of our localized Dijkstra algorithm.

Lemma 1. *A Dijkstra shortest path between two adjacent points in T_{aln} always exists. (Proof is omitted due to space limits.)*

In Fig. 1(c), to find $sp(v_3, v_4)$ where v_3 and v_4 are aligned vertices of p_4 and p_5, we use the midpoint v_{mid} of $\overline{v_3 v_4}$ as the center and the length of $\overline{v_3 v_4}$ as the diameter to circle the searching area c. As $sp(v_3, v_4) = (v_3, v_{31}, v_{32}, v_4)$ exists in c, the construction is done. The rest of T_{aln} are processed similarly.

4 The Semantic Trajectory Compression (STC)

STC utilizes two types of semantics: geometry and street semantics. In *geometry* compression, it uses road segments to tell if certain points in T_{net} can be omitted. A road segment is a sequence of continuous road edges of similar directions(e.g. $\pm 5°$ to due north), it can be part of a real road or a whole road. Geometry compression only preserves end points of a road segment and generates the intermediately compressed trajectory T_{tmp}. If a road name to road edge mapping relationship M is available, in the *street* compression, we use M to remove points in T_{tmp} that are not start/end of certain road or intersections, and output the compressed trajectory T_{comp}.

It is noteworthy that the compression performance is influenced by the quality of the underlying digital road map. Consider two copies of the road network of Beijing, G_s with 58,624 vertices and 130,714 edges is a sparse network used by [10], G_d with 1,285,215 vertices and 2,690,296 edges is ours. The edge number of G_d is 20.6 times that of G_s. First, the sizes of map-matched trajectories are of different orders of magnitude even under the same MM algorithm. E.g., a sample trajectory T has 97 points in G_s and 3,105 points in G_d after MM. Second, the more edges a road network has for the same street, the easier to compress trajectories. This is because the aligned trajectory has much more edges in the denser network than in the sparser network, and a greater portion of each aligned trajectory can be discarded to generate the compressed trajectory. Using geometry and street compression, our STC algorithm effectively overcomes this network quality issue, since only core semantics are reserved(Sect. 6.2). Algorithm 3 gives the pseudo code.

Algorithm 3. STC	**Algorithm 4. STD**
Input: Road network $G(V,E)$, road name to road edge mapping M, aligned traj. $T_{net} = \{v_1, v_2, \ldots, v_{|T_{net}|}\}$, angle difference bound β	**Input**: Road network $G(V,E)$, $T_{comp} = \{v_1, v_2, \ldots, v_{|T_{comp}|}\}$
Output: Semantically compressed T_{comp}	**Output**: Recovered traj. T_{rec} in G
1 **begin**	1 **begin**
2 $T_{tmp} = \{v_1\}$;	2 $T_{rec} = \{\}$, i=1;
3 **for** *(each $v_{i(2\leq i<|T_{net}|)}$ in T_{net})* **do**	3 **for** *(each $v_{i(1\leq i<|T_{comp}|)}$ in T_{comp})* **do**
4 **if** (Angle$(\overline{v_{i-1}v_i}, \overline{v_i v_{i+1}}) < \beta$) **then**	4 **if** *(v_i and v_{i+1} on same road)* **then**
5 $T_{tmp} = T_{tmp} \cup \{v_i\}$;	5 $S_{rec} = $ FetchRoad(v_i, v_{i+1});
6 $T_{tmp} = T_{tmp} \cup \{v_{|T_{net}|}\}$;	6 **else**
7 $T_{comp} = \{v_1\}$;	7 $S_{rec} = $ RecoverPath(G, v_i, v_{i+1});
8 **for** *(each $v_{j(2\leq j<|T_{tmp}|)}$ in T_{tmp})* **do**	8 $T_{rec} = T_{rec} \cup S_{rec}$;
9 **if** v_{j-1} and v_j on diff. roads **then**	9 **return** T_{rec};
10 $T_{comp} = T_{comp} \cup \{v_j\}$;	
11 $T_{comp} = T_{comp} \cup \{v_{|T_{tmp}|}\}$;	
12 **return** T_{comp};	

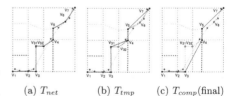

(a) T_{net} (b) T_{tmp} (c) T_{comp}(final)

Fig. 2. The semantic trajectory compression

In the **geometry compression** (Algorithm 3 lines 2–6), STC uses an angle bound β(e.g., 160°) to test if a sequence of continuous vertices in T_{net} belong to the same road segment. Procedure Angle in line 4 returns the angle formed by lines $\overline{v_{i-1}v_i}$ and $\overline{v_iv_{i+1}}$. In Fig. 2(a), the map-matched trajectory $|T_{net}| = 9$, the part from v_1 to v_3 is on the red road, v_3 to v_4 on the green road, v_4 to v_5 on the black road, and v_5 to v_7 on the blue road. For v_2, as

$\mathtt{Angle}(\overline{v_1 v_2}, \overline{v_2 v_3}) > \beta$, v_2 is discarded. For v_3, $\mathtt{Angle}(\overline{v_2 v_3}, \overline{v_3 v_{31}}) < \beta$, we take that the trajectory may have taken a turn at v_3 on the road network, thus v_3 is reserved in T_{tmp}. Similarly, v_{32} and v_6 are discarded. After geometry compression, $T_{tmp} = \{v_1, v_3, v_{31}, v_4, v_5, v_7\}$ with $|T_{tmp}| = 6$ in Fig. 2(b).

In the **street compression**(Algorithm 3 lines 7–12), we explore the possibility to extremely squeeze T_{tmp} into containing only the start/end vertices of each real-world street it traverses. For vertices that are not on the same road to its predecessor, it is added to T_{comp} (lines 8–10), otherwise it is discarded. In Fig. 2(c), since v_{31} is on the same road to v_3, v_{31} is removed from T_{tmp}. The final compression result $T_{comp} = \{v_1, v_3, v_4, v_5, v_7\}$ with $|T_{comp}| = 5$.

5   The Semantic Decompression (STD)

Before STC, MM places the shortest paths between unconnected vertices in T_{aln} to produce T_{net}; reversely, STD finds the *local* Dijkstra shortest paths for neighboring points in T_{comp}. Since we have the heading direction for each point in T_{comp}, we do not need to probe which direction to take if there are more than one outgoing edges from the aligned vertex. In contrast, the method in [14] stores a *description* of a reference point in the compressed trajectory, it has to recursively test which edge to go until the next reference point is reached. In addition, if adjacent points in T_{comp} are in the same road, STD simply fetches the part of the road between these two points to recover the trajectory. For example, in Fig. 2(c), v_3 and v_4 are on the same green road, so we fetch all edges from v_3 to v_4 to recover the aligned trajectory fragment $(v_3, v_{31}, v_{32}, v_4)$. Algorithm 4 shows the pseudo code of STD.

Moving direction of the recovered points are assigned from the direction of the corresponding road edge. The speed and timestamp are interpolated using the average speed between two successive points $p_i = (v_i, t_i, spd_i, head_i)$ and $p_{i+1} = (v_{i+1}, t_{i+1}, spd_{i+1}, head_{i+1})$ in T_{comp} assuming constant velocity. For a piece of recovered trajectory $T_{rec}(p_i, p_{i+1}) = (p_i, p_{i1}, p_{i2}, \ldots, p_{il}, p_{i+1})$, the speed for each recovered node $p_{ik}(1 \leq k \leq l)$ is calculated as: $spd_{ik} = len(p_i, p_{i+1})/(t_{i+1} - t_i)$, where $len(p_i, p_{i+1})$ is the total length of $T_{rec}(p_i, p_{i+1})$. The timestamp t_{ik} for p_{ik} is calculated as: $t_{ik} = t_i + (t_{i+1} - t_i)/len(p_i, p_{i+1}) * len(p_i, p_{ik})$, where $len(p_i, p_{ik})$ is the sub total length from p_i to p_{ik} in the recovered trajectory.

6   Experiments

In this section, we evaluate our proposed methods and report the experiment results on (1)the quality(i.e., map matching accuracy, compression ratio and decompression precision); and (2)the efficiency(the running time).

Experimental Setting. All algorithms were implemented in C++. All the experiments were conducted in a machine with 1.3 GHz Intel Core i5 CPU, 4 GB RAM, running Mac OS X Yosemite 10.10.1.

Dataset. We used a **real road network** G_d of Beijing with 1,285,215 vertices(36 MB) and 2,690,296 edges(83 MB). And a **real trajectory dataset** of 8041 taxies in Beijing, May, 2009, which contained about 12 million trajectories (7.4 GB). Positional update rate ranged from 10 s to about 5 min.

6.1 Evaluating Map Matching

Evaluation Metrics. (1) *Accuracy*: the ratio of correctly matched road edges to total number of road edges in a map-matched trajectory. (2) *Similarity*: the maximum point-to-point distance from the original trajectory to its map-matched trajectory. (3) *Runtime efficiency*.

Comparison algorithm. We implemented the state-of-the-art map matching algorithm in [6] and compared it with our MM algorithm.

Experiment results.

(1) *Accuracy*. We compared the matching accuracy at different sampling rates. As the sample rate is unevenly distributed in our real trajectory data, we selected 1,000 trajectories with 1-min sample rate, and removed certain sample points to generate three sets of trajectories with sample rate of 2-min/3-min/ 4-min. We labelled the map match for these four sets as the ground truth. In Fig. 3(a) we can see that MM outperformed the matching algorithm of [6], especially when the sample rate became lower. We also tested the accuracy without specifying a fixed sample rate. Our MM algorithm achieved an accuracy of 81.46 % while the method in [6] had 64.82 % accuracy.

(2) *Similarity*. In Fig. 3(b), we compared the distribution in percentage of the maximum deviation from map-matched trajectories to original trajectories. Under our MM algorithm, 68.94 % of deviation fell into the 40 to 60 m interval, while more than half(55.22 %) of the trajectories suffered a maximum matching deviation over 60 m under the algorithm of [6].

(3) *Efficiency*. In Fig. 3(c), we compared the average execution time of matching trajectories of different sizes onto the road network. While both algorithms exhibited near-linear growth rate of running time over trajectory sizes, MM ran faster, because of the efficient matching edge selection technique.

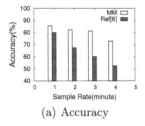
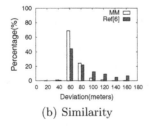
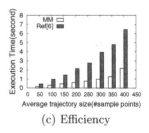

(a) Accuracy   (b) Similarity   (c) Efficiency

Fig. 3. Evaluating map matching

r_{comp}	STC	PRESS
Max.	9.2	8.2
Min.	7.8	3.1
Avg.	8.4	5.9

Network	#Vertices	#Edges	Avg. r_{comp}
G_d	1,285,215	2,690,296	8.412
G_s	426,196	946,434	5.381
G_s/G_d	0.332	0.352	0.640

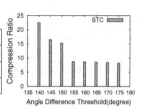

Fig. 4. Evaluating semantic compression

6.2   Evaluating Semantic Trajectory Compression

Evaluation metrics. (1) *Compression ratio*: $r_{comp} = \frac{|T_{net}|}{|T_{comp}|}$. (2) *Compression adaptivity* to different qualities of road networks. (3) *Runtime efficiency*.

Comparison algorithm. We compared our STC method with the state-of-the-art trajectory compression method PRESS [15].

Experiment results.

(1) *Compression Ratio.* We compared the maximum, minimum and average compression ratio of [15] in the left table of Fig. 4. While [15] can achieve a maximum compression ratio of 8.2, the average compression ratio was around 5.9 times, about 70 % of our STC, which has introduced semantics of trajectories to guarantee high compression ratio.

Next in the right figure of Fig. 4 we compared the impact of the angle difference bound β to the average compression ratio in the geometry compression step. β varied from 140° to 175° with a 5° step. Since a smaller β value grouped more nodes in a trajectory into one road segment, we can see that the compression rate decreased as β grew. For $\beta \geq 155°$, the STC algorithm achieved good compression quality(similarity within 200 m) and compression ratio exhibited minor improvement when α increased; for $\beta < 155°$, the compression ratio was greatly raised; however, compressed trajectories differed too much from original trajectories in physical shape, so it was not acceptable. A proper threshold should be larger than 155°.

(2) *Compression Adaptivity.* We tested the correlation of the compression quality and road network precision levels to verify the effectiveness and stability of STC. We compared a sparse road network G_s with a dense version G_d (the table in the middle of Fig. 4). As numbers of vertices and edges reduced, the aligned trajectory of the same raw trajectory contained less vertices, making it harder to achieve high compression ratio. The network density shrank about 70 % from G_d to G_s, but STC can still achieve an average compression ratio around 64 % with G_s to that with G_d, meaning that STC was able to smooth the impact to the compression ratio caused by the precision of road networks.

(3) *Compression Efficiency.* The compressing time of STC per single trajectory was always *less than 1 ms*, with the number of points in a map-matched trajectory varying from 1500 to 15000, exhibiting good efficiency. The time consuming step in an MM+STC framework is the map matching process, which took more than 99 % of the execution time.

6.3 Evaluating Semantic Trajectory Decompression

Evaluation metrics. (1) *Accuracy*: the ratio of correctly recovered edges to total number of edges in the ground truth. (2) *Similarity*: the maximum point-to-point distance from original trajectory to the recovered trajectory T_{rec}. (3) *Runtime efficiency*.

Experiment results. As there are little work quantitatively evaluating the performance of a decompression algorithm, we only present the result of our own. We did not group different sample rates since our STD algorithm was able to handle varying sample rates.

(1) *Accuracy*. STD achieved satisfying performance with maximum of 83.77 %, minimum of 67.20 %, and 77.12 % average accuracy.

(2) *Similarity*. In Fig. 5(a), we evaluated the distribution in percentage of the maximum deviation from the recovered trajectories to the original ones. Positional deviation caused by decompression is within acceptable error level, meaning that the decompression can produce practical result.

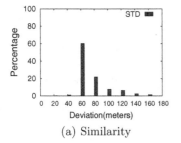

(a) Similarity

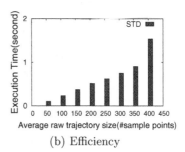

(b) Efficiency

Fig. 5. Evaluating decompression

(3) Efficiency. The average execution time of decompression increased near linearly to the size of trajectories. T_{comp} may contain more nodes than T as T was sometimes sparsely sampled, thus the step to find the Dijkstra shortest path in two adjacent nodes in T_{comp} took less time as the searching area was smaller than that for T under the MM algorithm. Figure 5(b) shows the details.

7 Conclusion

In this paper, we have studied the map matching and semantic trajectory compression problems. We proposed to effectively combine the matching of trajectories into road networks and the semantic compression to reduce storage requirement. We also devised an efficient semantic decompression method. Experimental study on real datasets verified the validity and efficiency of our algorithms.

Acknowledgement. This research is supported in part by the 973 Program of China (2015CB358700), the NSF of China (61272090, 61373024), Tencent, Huawei, Shenzhou, FDCT/116/2013/A3, MYRG105(Y1-L3)-FST13-GZ, and the Chinese Special Project of Science and Technology (2013zx01039-002-002).

References

1. Alt, H., Efrat, A., Rote, G., Wenk, C.: Matching planar maps. J. Algorithms **49**(2), 262–283 (2003). SODA 2003 Proceedings of the 14th Annual ACM-SIAM Symposium on Discrete Algorithms
2. Brakatsoulas, S., Pfoser, D., Salas, R., Wenk, C.: On map-matching vehicle tracking data. In: VLDB 2005, pp. 853–864. VLDB Endowment (2005)
3. Brakatsoulas, S., Pfoser, D., Tryfona, N.: Practical data management techniques for vehicle tracking data. In: ICDE 2005, pp. 324–325. IEEE (2005)
4. Cudre-Mauroux, P., Wu, E., Madden, S.: Trajstore: an adaptive storage system for very large trajectory data sets. In: ICDE 2010, pp. 109–120. IEEE (2010)
5. Douglas, D.H., Peucker, T.K.: Algorithms for the reduction of the number of points required to represent a digitized line or its caricature. Cartographica **10**(2), 112–122 (1973)
6. Kellaris, G., Pelekis, N., Theodoridis, Y.: Map-matched trajectory compression. J. Syst. Soft. **86**(6), 1566–1579 (2013)
7. Keogh, E., Chu, S., Hart, D., Pazzani, M.: An online algorithm for segmenting time series. In: ICDM 2001, pp. 289–296. IEEE (2001)
8. Long, C., Wong, R.C.-W., Jagadish, H.: Direction-preserving trajectory simplification. Proc. VLDB Endow. **6**(10), 949–960 (2013). VLDB 2013
9. Long, C., Wong, R.C.-W., Jagadish, H.: Trajectory simplification: on minimizing the direction-based error. Proc. VLDB Endow. **8**(1), 49–60 (2014). VLDB 2014
10. Lou, Y., Zhang, C., Zheng, Y., Xie, X., Wang, W., Huang, Y.: Map-matching for low-sampling-rate gps trajectories. In: SIGSPATIAL, pp. 352–361. ACM (2009)
11. Luo, W., Tan, H., Chen, L., Ni, L.M.: Finding time period-based most frequent path in big trajectory data. In: SIGMOD 2013, pp. 713–724. ACM (2013)
12. Meratnia, N., Park, Y.-Y.: Spatiotemporal compression techniques for moving point objects. In: Bertino, E., Christodoulakis, S., Plexousakis, D., Christophides, V., Koubarakis, M., Böhm, K. (eds.) EDBT 2004. LNCS, vol. 2992, pp. 765–782. Springer, Heidelberg (2004)
13. Qu, M., Zhu, H., Liu, J., Liu, G., Xiong, H.: A cost-effective recommender system for taxi drivers. In: Proceedings of the 20th ACM SIGKDD International Conference on Knowledge Discovery and Data Mining, pp. 45–54. ACM (2014)
14. Richter, K.-F., Schmid, F., Laube, P.: Semantic trajectory compression: representing urban movement in a nutshell. J. Spat. Inf. Sci. **4**, 3–30 (2015)
15. Song, R., Sun, W., Zheng, B., Zheng, Y.: Press: a novel framework of trajectory compression in road networks. arXiv preprint 2014. arXiv:1402.1546
16. Trajcevski, G., Cao, H., Scheuermanny, P., Wolfsonz, O., Vaccaro, D.: On-line data reduction and the quality of history in moving objects databases. In: 5th International workshop on Data Engineering for wireless and Mobile Access, pp. 19–26. ACM (2006)
17. Yin, H., Wolfson, O.: A weight-based map matching method in moving objects databases. In: Proceedings of the 16th International Conference on Scientific and Statistical Database Management, 2004, pp. 437–438. IEEE (2004)

Discovering Underground Roads
from Trajectories Without Road Network

Qiuge Song, Jiali Mao, and Cheqing Jin[✉]

School of Computer Science and Software Engineering,
Institute for Data Science and Engineering,
East China Normal University, Shanghai, China
sugar_song@sina.com, maojl1231@163.com, cqjin@sei.ecnu.edu.cn

Abstract. With the wide application of GPS-enabled electronic devices, huge amounts of positional information data have been accumulated, so that it's critical to discover inherent knowledge from such massive data. In this paper, we address this topic by proposing two issues, including how to discover the underpasses for pedestrians to cross the roads, and how to discover the tunnels providing passageways for vehicles. Subsequently, we propose a three-step framework to deal with the issues, including an incremental clustering phase, a sub-trajectory detecting phase and a cluster filtering phase. Experiments upon real-life data sets demonstrate the effectiveness and efficiency of the proposed framework.

1 Introduction

With the rapid development of the location-based technology and the popular use of GPS-enabled electronic devices, people's lifestyle has been changed profoundly. Researchers have extracted intelligence and communal behaviours from such large amounts of positional data to obtain a better understanding of the underlying dynamics of an individual, a community, or a city. It enables many innovative applications in traffic management, city planning, disease containment, etc.

Today's commercially available digital maps have achieved an accuracy in the range of a few to a few ten meters as well as a coverage for the major highway road network and urban regions that enables useful automotive applications, e.g., navigation and route guidance systems. However, the digital map may be incomplete due to the following reasons. First, after the map being publication, there may be some existing roads being blocked for maintenance or some new roads just coming into service which results in missing some latest information on the map. Second, the digital map in use may only contain specific information for the current application, and it may not contain some other important information. Instead of the expense of a complete road survey, GPS trajectory data can be utilized to generate entirely new sections of the road map at a fraction of the cost. Researches in [2–10] have found some road information from trajectories effectively, e.g., the road centerlines, the road intersections and so on.

© Springer International Publishing Switzerland 2016
B. Cui et al. (Eds.): WAIM 2016, Part I, LNCS 9658, pp. 137–150, 2016.
DOI: 10.1007/978-3-319-39937-9_11

In general, there are two main types of underground roads: underpasses and tunnels. The underpass is an important kind of pedestrian crossing facility built near the major urban roads that facilitates easing traffic congestion and protecting pedestrians' safety. The tunnel is a passage for vehicles, which is usually built under the ground, through a hill, under the sea or underneath a railway or another road in metropolitan areas. In underground roads, environment and illumination have a great impact on the legibility of the traffic signs. A digital map with underground road information can remind drivers to slow down and turn on the headlights before entering tunnels, and it can also inform drivers which tunnel exit they should choose. However, due to some reasons, newly-constructed underground roads are not updated timely in the digital map. What's more, the road network of a city can't be easily gained. Hence, how to discover the underground roads without road network in a city remains a big challenge.

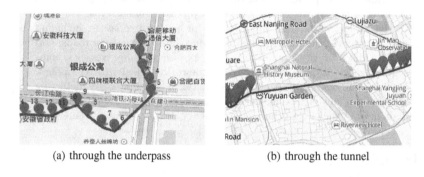

(a) through the underpass   (b) through the tunnel

Fig. 1. Two trajectories through the underground roads

Fortunately, the difference of users' trajectories inside and outside the underground provides us a way to discover the underground roads in a city. Figure 1(a) shows the walking trajectory of a person going through an underpass (between points 5 and 6). Every red dot represents a GPS point in the trajectory. Obviously, for the points from 1 to 5, and from 6 to 14, the gap between each pair of consecutive points is similar and is significantly shorter than the gap between points 5 and 6 comparably. The same situation is shown in Fig. 1(b). The reason behind is that the GPS-enabled electronic devices cannot receive signals when the user is going through the underground roads. As shown in Fig. 1, the gaps between every pair of consecutive points before entering or after exiting an underground road are similar, because the user was moving on the ground and the GPS-enabled devices can receive signals normally.

It doesn't seem difficult to discover the underground roads when the GPS devices always work well and all users move at the same speed all the time. However, such hypothesis cannot hold in real applications due to the following two reasons. First, the signals in urban canyons are weak and imprecise. Second, each user may change the speed very frequently, so that the gaps between two consecutive points vary a lot. Additionally, without the road network, we can't apply map-matching algorithm to the trajectories and even we can't judge whether a person was crossing a road or not. Hence, discovering the underground roads without road network is indeed non-trivial.

Till now, there's abundant work on mining trajectory data. However, to the best of our knowledge, only the work [1] discovers underpasses from trajectories. In this paper, we propose a three-step framework to deal with this issue. It includes an incremental clustering phase, where a number of line segments related to the aforementioned pattern are detected and incrementally maintained, a sub-trajectory detecting phase, where the number of the sub-trajectories passing a certain area is counted and stored, and a cluster filtering phase, where all the clusters are filtered and the underground road locations are obtained. The main contributions of this paper are listed below.

- We formally address the problem of discovering the underground roads in a city by using trajectories without any road network.
- We propose a three-step framework to deal with this issue. With this framework, we present two methods (*UNClu* and *TUClu*) to discover the underpasses from walking trajectories and the tunnels from vehicular trajectories respectively.
- We conduct a series of experiments to evaluate the performance of two proposed methods upon two real-life data sets respectively.

The rest of the paper is organized as follows. In Sect. 2, we review some work related to this topic. The problem is defined formally in Sect. 3. Subsequently, we describe our framework in detail in Sect. 4. Experimental results are given in Sect. 5. We conclude our work briefly in the last section.

2   Related Work

Previous studies have elaborated on trajectory mining. A good survey [2] classified the existing work into three main categories, including social dynamics which refers to the study of the collective behaviour of a city's population, traffic dynamics which studies the flow of the movement through the road network, and operational dynamics which refers to the study and analysis of taxi driver's modus operandi. Early in 1999, Rogers et al. firstly attempted to make a road map from GPS traces in [3]. It used an initial map to refine the centerline of the road by a hierarchical agglomerative clustering algorithm. Guo et al. [4] presented initial simulation work with a similar goal of finding the centerline of the road through the statistical analysis. This paper also uses an incremental clustering algorithm, but we don't need any road network. Without using any prior road map, Worrall and Nebot [5] implemented the approach to construct roads using clustering technique, Cao and Krumm [6] clarified and then merged GPS traces to create a routable road map. Chen and Krumm [7] found traffic lanes by modeling the spread of GPS traces across multiple lanes as a mixture of Gaussians. Different from above, we discover the underground roads rather than the whole road network, which need extracting specific features. Fathi and

Krumm [8] firstly used GPS traces to pinpoint road intersections by a supervised learning algorithm. However, it's unsuited to our issue because we don't need any training data. Biagioni and Eriksson [9] presented an extensible map inference pipeline, designed to mitigate GPS error and admit less-frequently traveled roads, while our work makes good use of the line segments of missing sampling points. Biagioni and Eriksson [10] made a survey and comparative evaluation on inferring road maps from GPS traces including above work. The datasets used in aforementioned methods are all vehicular GPS trajectories, but we take the walking trajectories into account besides the vehicular trajectories in our research. Compared with vehicular trajectories, the walking trajectories have low sampling frequency, low accuracy and changeable directions. The above mentioned methods can't be directly applied to the walking trajectory dataset.

There are also many researches working on mining walking trajectories. Li et al. [11] proposed a framework referred to as hierarchical-graph-based similarity measurement to consistently model each individual's location history and effectively measure the similarity among users. It took into account both the sequence property of people's movement behaviors and the hierarchy property of geographic spaces. In [12], based on multiple users GPS trajectories, Zheng et al. aimed to mine interesting locations and classical travel sequences in a given geospatial region. Nevertheless, the aforementioned methods for mining walking trajectories can't be used to detect underground roads.

To the best of our knowledge, only the work [1] intends to discover underpasses from trajectories. However, it can only find underpasses with the aid of the road network. In this paper, we propose a general framework to detect all kinds of underground roads including underpasses for pedestrians and tunnels for vehicles. What's more, we don't need the road network at all. Although Stockx et al. [13] could detect subway stations when a user was taking a subway, it detected them by using the accelerometer and the gyroscope embedded in a smart-phone instead of analyzing trajectory data.

3   Problem Definition

Definition 1 (Trajectory). *Given the trajectory database D, a trajectory $T_k \in D$ is a sub-sequence of GPS points affiliated to an object o_k, denoted as $T_k = \{(p_1, t_1), (p_2, t_2), ..., (p_i, t_i), ...\}$, where p_i is the location of the object o_k at the time stamp t_i in 2-D space (i.e., $p_i = (x_i, y_i)$). Such records arrive in chronological order, i.e., $\forall i < j$, $t_i < t_j$. A line segment L_i refers to a line connecting two adjacent points, i.e., $L_i = (s_i, e_i) = (p_i, p_{i+1})$. Correspondingly, a trajectory is also denoted as $\{L_1, L_2, ...\}$.*

After analyzing the trajectories, we find lots of line segments of missing sampling points like the pattern shown in Fig. 1. Our major idea is to cluster this kind of line segments as defined below. In addition, we compute the total number of sub-trajectories within the range of a cluster. Our goal is to detect the clusters where underground roads exist.

Definition 2 (Line Segment of Missing Sampling Points, LM). *An LM $L_i = (p_i, p_{i+1})$ is a line segment in a trajectory missing sampling points between t_i and t_{i+1}. Correspondingly, the time interval of the LM is denoted as $\Delta t_i = t_{i+1} - t_i$.*

Definition 3 (Sub-trajectory Passing a certain area, SP). *Assuming a certain area has an underground road, an SP for the area is a sub-trajectory that a user goes along the underground road, no matter whether he/she goes through the underground road or on the ground, denoted as $SP = \{p_{j-n}, ..., p_{j-1}, p_j\}$ where p_j is the jth point in one trajectory.*

Definition 4 (Cluster Feature, CF). *CF for a cluster of LMs $\{L_1, L_2, ..., L_n\}$, is of the form $(L_{cen}, \Delta t_{max}, N_{LM}, N_{SP})$.*

- *L_{cen} : the representative line segment of the cluster, denoted as $L_{cen} = (s_{cen}, e_{cen})$ where s_{cen} and e_{cen} are the starting point and the ending point of L_{cen};*
- *Δt_{max} : the maximal time interval of the set of LMs in the cluster;*
- *N_{LM} : the number of LMs in the cluster, i.e., $N_{LM} = n$;*
- *N_{SP} : the number of SPs for the area of the cluster.*

It's different to calculate L_{cen} in *UNClu* and *TUClu*, since the planforms of underpasses and tunnels are different. The planform of an underpass can be expressed as a rectangle and there is no directional limit when going through the underpass. Hence, given a CF, we can obtain the minimal bounding rectangle (MBR) of all LMs contained in the cluster. And s_{cen} and e_{cen} of L_{cen} in *UNClu* are just the bottom left vertex and the top right vertex of the MBR. The tunnels are generally long and the planform of a tunnel can be expressed as a line segment in a direction. s_{cen} and e_{cen} of L_{cen} in *TUClu* can be calculated with $s_{cen} = \frac{\sum_{i=1}^{n} L_i.s_i}{n}$ and $e_{cen} = \frac{\sum_{i=1}^{n} L_i.e_i}{n}$. With N_{LM} and N_{SP} of a CF, we can calculate the proportion of LMs among all SPs, i.e., $R_{LS} = \frac{N_{LM}}{N_{SP}}$. Note that an LM inserted into a cluster is also an SP for the area of the cluster.

When clustering LMs, we use the distance function between a line segment and a cluster, which is in fact defined between the line segment and the representative line segment of the cluster. It's adapted from similarity measures in the area of pattern recognition [14]. We assign the longer line segment to L_i and the other shorter one to L_j without losing generality.

Definition 5. *The distance function is the sum of the perpendicular distance, the parallel distance and the angle distance which is defined as*

$$dist(L_i, L_j) = \frac{d_{\perp 1}^2 + d_{\perp 2}^2}{d_{\perp 1} + d_{\perp 2}} + \frac{d_{\|1} + d_{\|2}}{2} + \begin{cases} \| L_j \| \times sin(\theta), & 0^o \leq \theta < 90^o \\ \| L_j \|, & 90^o \leq \theta < 180^o \end{cases} \quad (1)$$

where $d_{\perp 1}$ is the Euclidean distance between s_j and p_s, $d_{\perp 2}$ is that between e_j and p_e, $d_{\|1}$ is the Euclidean distances of p_s to s_i and $d_{\|2}$ is that of p_e to e_i,

*supposing that p_s and p_e are the projection points of the points s_j and e_j onto L_i
respectively. $\| L_j \|$ denotes the length of L_j, θ denotes the smaller intersecting
angle between L_i and L_j.*

Different from the definition in [14], we compute the parallel distance with
$\frac{d_{\|1}+d_{\|2}}{2}$ instead of $min(d_{\|1}, d_{\|2})$ which can help us distinguish the directions of
the two line segments.

4 Our Framework

In this section, we describe our framework in detail. At first, we introduce the
framework in Sect. 4.1. Then, each of the next three subsections describes a key
module in this framework. Finally, we analyze the performance of our framework
in Sect. 4.5.

4.1 The Overall Framework

Algorithm 1 describes the overall framework. The input is the trajectory database D. This framework processes each trajectory T_k in D one by one. For each
line segment L_i in T_k, at first, it tries to judge whether L_i is an LM or not. If it is
an LM, we try to insert L_i into the clusters (see Sect. 4.2). Then, it tries to judge
whether the sub-trajectory $\{L_{i-n}, ..., L_i\}$ is an SP for each cluster C_j or not, and
if it is, then the variable N_{SP} of C_j is updated accordingly (see Sect. 4.3). Finally,
we filter all clusters and derive the clusters where an underground road exists
apiece (see Sect. 4.4).

Algorithm 1. FindingUndergroundRoads

Input: D: the trajectory database
Output: Z: the set of CFs for generated clusters
Initialize: Z is empty
 1: **for** each trajectory $T_k \in D$ **do**
 2:     **for** each line segment $L_i \in T_k$ **do**
 3:         **if** L_i is an LM **then**   //see Sect. 4.2
 4:             $Z \leftarrow Z.InsertToClusters(L_i)$;   //see Sect. 4.2
 5:         **end if**
 6:         **for** each CF $C_j \in Z$ **do**
 7:             **if** $\{L_{i-n}, ..., L_i\}$ is an SP for C_j **then**   //see Sect. 4.3
 8:                 $C_j.N_{SP} \leftarrow C_j.N_{SP} + 1$;
 9:             **end if**
10:         **end for**
11:     **end for**
12: **end for**
13: $Z \leftarrow FilteringClusters(Z)$;   //see Sect. 4.4
14: output Z;

4.2   LM Clustering

This module aims at detecting whether a line segment L_i is an LM or not. If it is, we insert L_i into the existing clusters. The most critical task in detecting an LM is to check whether the sampling rate changes suddenly or not. As shown in Fig. 2, no matter which type the situation is, the time interval of the LM is longer than other that of normal line segments because of missing sampling points. Assuming that the speed is constant in a short period of time, the length of the LM is also longer.

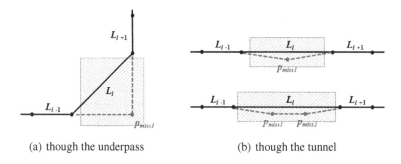

(a) though the underpass   (b) though the tunnel

Fig. 2. LMs though the underground roads

Therefore, a line segment L_i is an LM if it satisfies two constraints: the time constraint and the distance constraint. The time constraint is given as

$$\Delta t_i > \alpha_1 \times (\Delta t_{i-1} + \Delta t_{i+1}) \tag{2}$$

where Δt_i is the time interval of the L_i. The distance constraint is given as

$$S_i > \alpha_2 \times (S_{i-1} + S_{i+1}) \tag{3}$$

where S_i is the length of the L_i. And α_1 and α_2 are tuning parameters for two constraints which vary with different types of data sets. For instance, α_2 for the underpass data set as in Fig. 2(a) (approximately $\frac{\sqrt{2}}{2}$) is smaller than α_2 for the tunnel data set as in Fig. 2(b) (approximately 1).

After finding an LM L_i, we insert L_i into the existing clusters. The goal of Algorithm 2 is generating and incrementally maintaining clusters by applying hierarchical clustering method. It proceeds as follows. For an LM L_i, we find the closest cluster C_j. If the distance between L_i and L_{cen} of C_j is shorter than a distance threshold β, L_i will be inserted into C_j. Otherwise, a new cluster C_{new} will be created for L_i. When the number of clusters exceeds m, we need to merge two closest clusters to make room for the new created cluster and ensure the efficiency of the algorithm. Note that β is related to the length of L_{cen} for C_j, which is computed as $min(length(Cj.Lcen)/2, \beta)$.

Algorithm 2. InsertToClusters

Input: L_i: an LM extracted from trajectories
          Z: the set of CFs for generated clusters
Parameter: β: the maximum distance limitation
          m: the maximum number of the clusters
Output: Z: the set of CFs for generated clusters after inserting the LM
1: Find the closest $C_j \in Z$ to L_i;
2: $\beta \leftarrow min(length(C_j.L_{cen})/2, \beta)$;
3: **if** $distance(L_i, C_j.L_{cen}) \leq \beta$ **then**
4:    $C_j \leftarrow C_j \cup \{L_i\}$;
5: **else**
6:    **if** $|Z| = m$ **then**
7:       Merge two closest clusters;
8:    **end if**
9:    $C_{new} \leftarrow \{L_i\}$;
10:    $Z \leftarrow Z \cup C_{newly}$;
11: **end if**
12: output Z;

4.3   SP Detection

This module aims at detecting whether the sub-trajectory $\{L_{i-n}, ..., L_i\}$ is an SP for cluster C_j or not, and if it is, the variable N_{SP} of C_j will be updated accordingly. The sub-trajectory across an underpass is usually distinct from that along a tunnel, which necessitates the different methods to detect SPs in *UNClu* and *TUClu* respectively.

UNClu. Because the underpasses are usually located at the intersections of two roads, we can simply regard an underpass as a rectangle then the diagonal of it becomes L_{cen} of a cluster. The pedestrians can walk across the road in various directions, So there's no directional limit on the SP. The sub-trajectory $\{L_{i-n}, ..., L_i\}$ is treated as an SP for the cluster C_j if it meets the following two conditions. Firstly, L_{i-n} and L_i intersect with two different edges of the rectangle. Secondly, the time interval from L_{i-n} to L_i is shorter than Δt_{max} of C_j. As shown in Fig. 3(a), ST_1 and ST_2 are both SPs.

TUClu. There are usually two tunnels in either opposite direction at the same location. We can limit the direction of the sub-trajectory. Besides that the time interval from L_{i-n} to L_i is shorter than Δt_{max} of C_j, the starting point and the ending point must be within the circles which are centered at s_{cen} and e_{cen} with a radius of β respectively, then we can say we find an SP for C_j. Note that β is computed as $MIN(len/2, \beta)$, where len is the length of L_{cen} for C_j.

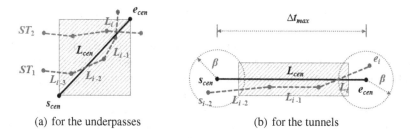

(a) for the underpasses   (b) for the tunnels

Fig. 3. Two SP detection methods in UNClu and TUClu

4.4 Filtering

This module aims at filtering all clusters to find the locations where underground roads exist. In general, the area of a cluster containing an underground road may have a greater value of N_{LM} since there are a large number of people going through the underground roads. Once they are under the ground, it is prone to miss sampling points in their trajectories. However, the value of N_{LM} is associated with the traffic flow in the area of the cluster. Hence, the constraint of N_{LM} for C_i is given as

$$C_i.N_{LM} > \gamma \times \frac{\sum_j C_j.N_{LM}}{n} \qquad (4)$$

where $C_j \in C_\epsilon$, the distance between C_i and C_j is shorter than the parameter ϵ, and n is the number of clusters in C_ϵ. So $\frac{\sum_j C_j.N_{LM}}{n}$ is just the average number of LMs in that area which reflects the density of trajectories. γ is a tuning parameter to adjust the threshold of N_{LM} of C_i.

However, some other locations like urban canyons also have plenty of LMs, which is maybe even larger than that in the location of underground roads. Because there are a larger flow of people there and more LMs are produced for other causes. Therefore, judging whether there is an underground road in the area of one cluster only according to N_{LM} is not comprehensive. We also consider the constraint of R_{LS} of C_i which is given as

$$C_i.R_{LS} \geq \frac{\delta l_i}{l_{max}} \qquad (5)$$

where l_i is the length of L_{cen} for C_i and l_{max} is the maximal length of L_{cen} for all clusters. Since it is more prone to miss sampling points and there are fewer SPs in longer underground roads. δ is a tuning parameter to adjust the threshold of R_{LS} of C_i.

4.5 Performance Analysis

The space complexity of our framework is $O(m)$, where m is the number of clusters and can be set according to the memory of the computer. Concerning the time complexity, the cost of traversing all line segments of trajectories is $O(n)$, where n is the number of line segments of all trajectories. For each LM L_i, the cost of incorporating it into the nearest cluster is $O(m)$. If we record the nearest cluster for each cluster, we can merge two closest clusters in $O(m)$. And SP detection stage costs $O(m)$. The cost of cluster filtering stage is also $O(m)$. So the total processing cost of our framework is $O(nm)$. When the data set is large, process can be hastened if the clusters are indexed by an R-tree [15]. In our framework, the SPs for a cluster before the generation of clusters will be missed, but it's negligible when the data set is large enough. Consequently, our framework scales to large data sets very well.

5   Experiment

In this section, we conduct extensive experiments upon real-life data sets. All code was written in Java, and run on a computer with 8GB memory and Intel i5 CPU.

5.1   Dataset Description

We use four real-life datasets, including *HefeiTraj*, *ShanghaiTraj*, *HefeiUnderpasses* and *ShanghaiTunnels*.

- *HefeiTraj:* this dataset describes the walking records gathered by volunteers in early 2014 in the city center of Hefei. After de-duplication, it has 90,037 trajectories, each containing a sequence of coordinates.
- *ShanghaiTraj:* this dataset describes the trajectories of the taxis in Shanghai, China. It contains the GPS logs of about 30,000 taxis during three months (October, November, December) in 2013, covering about 93 % of main road network of Shanghai.
- *HefeiUnderpasses:* this dataset describes the locations of all seven underpasses in the central area of Hefei by using street view maps.
- *ShanghaiTunnels:* this dataset contains the locations of all tunnels in Shanghai, China by using street view maps.

5.2   Settings

This section introduces (1) how we evaluate the methods, and (2) which baseline method was compared with.

Criteria. We have several parameters used in our framework, including α_1, α_2, β, ϵ, γ and δ. Among them, β is related to the length of L_{cen}. As for ϵ, the larger the value is, the better the effectiveness is, while the lower the efficiency is. Hence, we only test the parameters α_1, α_2 used in LM clustering stage, and γ, δ used in cluster filtering stage.

- *Density Ratio*: we use *Density Ratio* defined as the ratio between the local LM density and the global LM density to measure the performance of LM detection. The local LM density is the average number of LMs in the areas where underground roads exist, and the global LM density is that in the whole area. Hence, as for the density ratio, the larger, the better.
- *F-measure*: we evaluate the effectiveness of the algorithms by using *F1-measure* which is interpreted as a weighted average of *Precision* and *Recall*.

Baseline. We compare our methods with the naive approach that clusters the line segments which are longer than d_{max}. The baseline methods upon *HefeiTraj* and *ShanghaiTraj* data sets are shortened as "UBase" and "TBase" respectively. Proved by experiments, when $d_{max} = 40$ in UBase and $d_{max} = 500$ in TBase, we obtain the best performance respectively. Although *Recall* is high here, *Precision* is low. Hence, their values of *F-measure* are still too low (0.182 in UBase and 0.464 in TBase).

5.3   Effectiveness

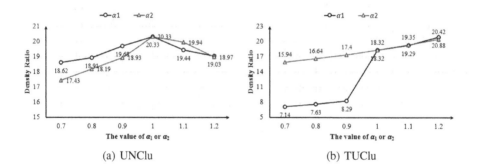

(a) UNClu   (b) TUClu

Fig. 4. Density Ratio in UNClu and TUClu

UNClu. Figure 4(a) shows the change of *Density Ratio* under different values of α_1 and α_2 in UNClu. When α_2 is fixed, the value of *Density Ratio* firstly grows, then decreases with the increment of α_1. When $\alpha_1 = 1$, it reaches the peak. When α_1 is fixed, the value of *Density Ratio* also reaches the peak when $\alpha_2 = 1$. Since the lengths of the underpasses are not long, and it usually misses only one sampling point when a pedestrian walks through an underpass. We can also observe the effect of γ and δ after fixing α_1 and α_2, as shown in Table 1. When δ is fixed, the precision firstly increases, then decreases while the recall decreases with the increment of γ. When γ is fixed, the precision grows and the recall decreases with the increment of δ. According to *F-measure*, when we

Table 1. Performance of different parameters (γ, δ) in UNClu

γ	0.5	0.6	0.7	0.8	0.9	δ	0.07	0.08	0.09	0.10	0.11
Precision	0.667	**0.750**	**0.750**	0.714	0.667	**Precision**	0.583	0.667	**0.750**	0.800	1.000
Recall	0.857	**0.857**	**0.857**	0.714	0.571	**Recall**	1.000	0.857	**0.857**	0.571	0.429
F-measure	0.750	**0.800**	**0.800**	0.714	0.615	**F-measure**	0.737	0.750	**0.800**	0.666	0.600

Table 2. Performance of different parameters (γ, δ) in TUClu

γ	0.8	1.0	1.2	1.4	1.6	δ	1.0	2.0	3.0	4.0	5.0
Precision	0.838	**0.862**	0.859	0.857	0.855	**Precision**	0.862	0.875	**0.930**	0.925	0.930
Recall	0.950	**0.933**	0.917	0.900	0.883	**Recall**	0.933	0.933	**0.883**	0.817	0.667
F-measure	0.890	**0.896**	0.887	0.878	0.869	**F-measure**	0.896	0.903	**0.906**	0.868	0.777

set $\gamma = 0.6$ or 0.7, $\delta = 0.09$, we obtain good effectiveness which is significantly better than that of UBase. We can't find the underpasses which are small-scale and short.

TUClu. Figure 4(b) shows the change of *Density Ratio* under different values of α_1 and α_2 in TUClu. When α_2 is fixed, the value of *Density Ratio* grows with the increment of α_1. When $\alpha_1 = 1$, it grows most significantly. When α_1 is fixed, the value of *Density Ratio* grows with the increment of α_2 very slowly. Since the speed of the vehicles doesn't change very frequently compared with that of walkers, and the time constraint is consistent with the distance constraint. We can also observe the same effect of γ and δ as that in UNClu after fixing α_1 and α_2, as shown in Table 2. According to *F-measure*, when we set $\gamma = 1$, $\delta = 3$, we obtain good effectiveness which is nearly two times better than that of TBase. δ in TUClu is larger than δ in UNClu, since tunnels are usually longer than underpasses. We can discover most tunnels in Shanghai. The tunnels that we can't find are shorter than 400 meters in length or in the areas that have fewer or even none trajectories. The areas that we find but no tunnels exist in fact are usually underground parking lots.

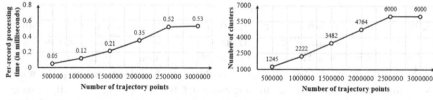

(a) Execution time versus number of points   (b) Number of clusters versus number of points

Fig. 5. Efficiency of TUClu

5.4   Efficiency

Figure 5(a) shows the average execution time of TUClu as the number of trajectory points grows. We can observe that the per-record processing time increases with the increment of the number of trajectory points. Mainly because the number of clusters increases as shown in Fig. 5(b). However, the efficiency will be steady when the number of clusters reaches the given threshold m. The efficiency of UNClu is similar to that of TUClu.

6   Conclusion

We propose a novel framework to detect underground roads from raw trajectories without road network. To the best of our knowledge, we are the first to address the issue that detects both the underpasses and the tunnels. We find underground roads mainly in three steps. Initially, we detect whether the line segments that belong to different trajectories are LMs or not, and group the identified LMs into clusters. Subsequently, we judge whether or not the present sub-trajectory is an SP for a cluster, meanwhile, we update two critical variables N_{LM} and N_{SP} for the cluster accordingly. Finally, we filter clusters and obtain the clusters satisfying two constraints. The experimental results show that our approach can find the underground roads effectively and efficiently.

Acknowledgment. Our research is supported by the 973 program of China (No. 2012CB316203), NSFC (61370101, U1501252), Shanghai Knowledge Service Platform Project (No. ZF1213), Innovation Program of Shanghai Municipal Education Commission (14ZZ045).

References

1. Song, Q., Jin, C., Wang, X., Gao, M., Zhou, A.: Discovering underpasses from walking trajectories. In: 31st IEEE International Conference on Data Engineering Workshops, pp. 129–133 (2015)
2. Castro, P., Zhang, D., Chen, C., Li, S., Pan, G.: From taxi GPS traces to social and community dynamics: A survey. ACM Comput. Surv. **46**, 17 (2013)
3. Rogers, S., Langley, P., Wilson, C.: Mining GPS data to augment road models. In: International Conference on Knowledge Discovery and Data Mining, pp. 104–113 (1999)
4. Guo, T., Iwamura, K., Koga, M.: Towards high accuracy road maps generation from massive GPS traces data. In: IEEE International Geoscience and Remote Sensing Symposium, pp. 667–670 (2007)
5. Worrall, S., Nebot, E.: Automated process for generating digitised maps through GPS data compression. In: Australasian Conference on Robotics and Automation (2007)
6. Cao, L., Krumm, J.: From GPS traces to a routable road map. In: 17th ACM SIGSPATIAL International Conference on Advances in Geographic Information Systems, pp. 3–12 (2009)

7. Chen, Y., Krumm, J.: Probabilistic modeling of traffic lanes from GPS traces. In: Proceedings of GIS, pp. 81–88 (2010)
8. Fathi, A., Krumm, J.: Detecting road intersections from GPS traces. In: Fabrikant, S.I., Reichenbacher, T., van Kreveld, M., Schlieder, C. (eds.) GIScience 2010. LNCS, vol. 6292, pp. 56–69. Springer, Heidelberg (2010)
9. Biagioni, J., Eriksson, J.: Map inference in the face of noise and disparity. In: Proceedings of the 20th International Conference on Advances in Geographic Information Systems, pp. 79–88. ACM (2012)
10. Biagioni, J., Eriksson, J.: Inferring road maps from global positioning system traces: Survey and comparative evaluation. Transp. Res. Rec. J. Transp. Res. Board **2291**, 61–71 (2012)
11. Li, Q., Zheng, Y., Xie, X., Chen, Y., Liu, W., Ma, W.: Mining user similarity based on location history. In: Proceedings of the 16th ACM SIGSPATIAL International Conference on Advances in Geographic Information Systems (2008)
12. Zheng, Y., Zhang, L., Xie, X., Ma, W.: Mining interesting locations and travel sequences from GPS trajectories. In: Proceedings of the 18th International Conference on World Wide Web, pp. 791–800. ACM (2009)
13. Stockx, T., Hecht, B., Schoning, J.: SubwayPS: towards smartphone positioning in underground public transportation systems. In: Proceedings of the 22nd ACM SIGSPATIAL International Conference on Advances in Geographic Information Systems, pp. 93–102 (2014)
14. Chen, J., Leung, M., Gao, Y.: Noisy logo recognition using line segment Hausdorff distance. Pattern Recogn. **36**(4), 943–955 (2003)
15. Guttman, A.: R-trees: a dynamic index structure for spatial searching. In: Proceedings of SIGMOD, pp. 47–57 (1984)

Ridesharing Recommendation: Whether and Where Should I Wait?

Chengcheng Dai[(✉)]

Department of Computer Science, City University of Hong Kong,
Kowloon Tong, Hong Kong
chengcdai2-c@my.cityu.edu.hk

Abstract. Ridesharing brings significant social and environmental benefits, e.g., saving energy consumption and satisfying people's commute demand. In this paper, we propose a recommendation framework to predict and recommend whether and ˙where should the users wait to rideshare. In the framework, we utilize a large-scale GPS data set generated by over 7,000 taxis in a period of one month in Nanjing, China to model the arrival patterns of occupied taxis from different sources. The underlying road network is first grouped into a number of road clusters. GPS data are categorized to different clusters according to where their sources are located. Then we use a kernel density estimation approach to personalize the arrival pattern of taxis departing from each cluster rather than a universal distribution for all clusters. Given a query, we compute the potential of ridesharing and where should the user wait by investigating the probabilities of possible destinations based on ridesharing requirements. Users are recommended to take a taxi directly if the potential to rideshare with others is not high enough. Experimental results show that the accuracy about whether ridesharing or not and the ridesharing successful ratio are respectively about 3 times and at most 40 % better than the naive "stay-as-where-you-are" strategy. This shows that about 500 users can save 4–8 min with our recommendation. Given 9 RMB as the starting taxi fare and suppose users can save half of the total fare by ridesharing, users can save 10.828-44.062 RMB.

1 Introduction

Due to the emergency of saving energy consumption and assuaging traffic congestion while satisfying people's needs in commute and willings to save money in ride, ridesharing enabled by low cost geo-location devices, smartphones, social networks and wireless networks has recently received a lot of attention [1–4].

Taxis are considered as a major means of transportation in modern cities. In many big cities, taxis are equipped with GPS sensors to report their locations, speed, direction and occupation periodically. In Fig. 1 we gather statistics of the pick-up actions from a large-scale GPS data set generated by over 7,000 taxis in a period of one month in Nanjing, China. There are roughly 3 peaks per day where the pick-up number is above 6,000. We partition Nanjing metropolitan area into

© Springer International Publishing Switzerland 2016
B. Cui et al. (Eds.): WAIM 2016, Part I, LNCS 9658, pp. 151–163, 2016.
DOI: 10.1007/978-3-319-39937-9_12

30×30 grids with equal intervals. We count all the pick-up and drop-off events in 9:30am–9:35am on June 25th in each region and analysis the hotspots. As shown in Fig. 2a and b, both pick-up and drop-off have hotspots, indicating that many queries are likely to get ridesharing. Ridesharing gives the potential to solve congestion, pollution and environmental problems as well as saves money for users.

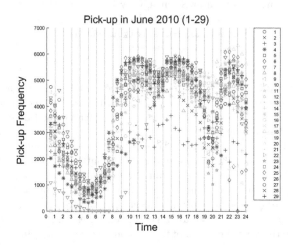

Fig. 1. Pick-up frequency in 30 min of Nanjing.

Ridesharing can be classified into carpooling [5], real-time taxi ridesharing [1,2,6], slugging [3] and Dial-a-Ride [7]. In slugging, passengers change their sources and destinations to join the trips of drives while drivers of others change their routes to pick up and drop off passengers. In real-time taxi ridesharing, ridesharing becomes an optimization problem to allocate a query to proper taxis considering extra cost in terms of both distance and waiting time. In Carpooling or recurring ridesharing, the driver is associated with her own trip. In Dial-a-Ride vehicles need to return to the same location (depot) after the trip.

Previous works on ridesharing mostly focus on the driver side by considering which taxi should the coming request be assigned to for the minimum extra travel time or travel distances. On the other hand, we focus on the user side of ridesharing to help the users decide whether they can rideshare and where should they wait if they are likely to get ridesharing. We use slugging as the ridesharing type since slugging is shown to be effective in reducing vehicle travel distance as a form of ridesharing [3].

Consider the scenario of slugging, Alice raises a ridesharing query $Q = (id, timestamp, l_s, l_d, t_s, t_e, t_w)$ where id is user id, $timestamp$ is when the query is submitted, l_s and l_d are respectively the source and the destination of the user, t_s is the maximum walking time to a new place, and t_e is the maximum walking time after she left the taxi to her own destination. t_w is the maximum waiting time at the new place for ridesharing. At present there is no taxi to rideshare

Alice in the system. Alice may either take a taxi directly or try again after a short time period. With the observation that Alice may also walk to some place nearby during waiting in order to increase her chance to rideshare, we propose a framework to predict the probability for Alice to rideshare and recommend whether and where should she wait. The key challenges are (i) how to embed query satisfaction requirements with recommendation, which makes our problem more complicated than finding passengers or taxis [8–10]; (ii) how to develop effective machine learning algorithms for ridesharing recommendation.

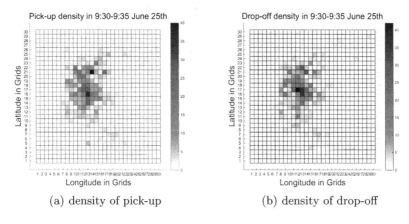

(a) density of pick-up   (b) density of drop-off

Fig. 2. Hotspots for pick-ups and drop-offs.

In the framework, ridesharing recommendation is based on the probabilities to have other passengers departing from somewhere within time t_s from source l_s towards somewhere within time t_e from destination l_e. Since taxi appearance is too dynamic for a single road, the underlying road network is grouped into road clusters. For each road cluster that are within time t_s from source l_s, we investigate the probability for taxis to depart from somewhere in it and have somewhere that is within time t_e to destination l_d as their destination by kernel density estimation [11]. Only road clusters with the probabilities greater than a threshold are recommended to users. In case many road clusters satisfies the condition, we return top-k clusters according the probabilities. If no cluster is returned for Q, the user will be suggested not to wait for ridesharing and take a taxi directly. Thus users can either save time or save money.

Rather than calculating a common distribution for all road clusters [12,13] to describe possible taxi destinations, we derive unique distributions for each road cluster. Given any new location (lon, lat), we can obtain the probability to have a taxi that departs from a certain cluster towards location (lon, lat). In reality, the probabilistic distributions should be various since clusters have different features such as Point-of-Interests (POIs). For instance, at noon taxi passengers from clusters with POIs of companies are likely to have restaurants as destinations while passengers from clusters with POIs of residencies may have companies as destinations.

To the best of our knowledge, this is the first work to study ridesharing from the aspect of predicting whether and where should a user wait. The main contributions of this paper can be summarized as follows:

- We design a recommendation framework to help users decide whether and where to wait for ridesharing.
- We explore the arrival patterns of road clusters based on kernel density estimation for ridesharing recommendation.
- Experimental results based on real GPS data set show that the accuracy about whether ridesharing or not and the ridesharing successful ratio are respectively about 3 times and at most 40 % better than the naive "stay-as-where-you-are" strategy, i.e., users wait at their sources for ridesharing. About 500 users can save 4–8 min with our recommendation. Given 9 RMB as the starting taxi fare and suppose users can save half of the total fare by ridesharing, users can save 10.828-44.062 RMB.

The rest of this paper is organized as follows. Section 2 highlights related works. Section 3 delineates the proposed recommendation framework. Section 4 analyzes the performance. Finally Sect. 5 concludes this paper.

2   Related Works

In this section, we highlight related works in both ridesharing and recommender systems based on taxi GPS data set.

2.1   Ridesharing

Ridesharing is transformed to an optimization problem about matching one driver and multiple ridesharing queries considering extra cost in terms of both distance and waiting time. We summarize the related works as dynamic matching and other issues like fair payment mechanism.

Dynamic Ridesharing Matching. Current dynamic ridesharing can be classified into four types, namely slugging, taxi ridesharing, carpooling and Dial-a-Ride. In slugging passengers change their sources and destinations to join the trips of drivers. From the passenger's point of view, this requires the source and destination of driver to be close to those of the passenger. In this paper we adapted slugging as our ridesharing type. The closeness is controlled by t_s and t_e in the query Q. With the walking speed, we can easily get the maximum walking distances of each query. In the other three types drivers change their routes to pick up and drop off passengers. Carpooling [5] is ridesharing based on private cars where the driver is associated with her own trip. Carpooling considers computing the best route for a given set of requests. In contrast to carpooling, taxi ridesharing [1,2,6] is more challenging as both passengers' queries and taxis' positions are highly dynamic and are real-time in most cases.

Besides, pricing mechanism is required to incite the driver. In Dial-a-Ride [7], vehicles need to return to the same location (depot) after the trip, which can be treated as the carpooling problem with additional restrictions about depot. In this paper, we adapt slugging as the ridesharing model to recommend whether and where should a user wait for ridesharing. Ridesharing recommendation is different from travelling plan problems [14] where the sequence of must-visit locations is known in advance and the target is to decide the optimal visit order. As an online recommendation problem, we need to provide ridesharing suggestions for each request in real time.

Other Issues. Though ridesharing is envisioned as a promising solution to mitigate traffic congestion and air pollution for metropolitan cities, people raise social discomfort and safety concerns about traveling with strangers. Social ridesharing with friends [14] is studied to overcome these barriers. Another interesting issue about ridesharing is fair payment mechanism, including when taking a ride with friends [15] and pricing mechanism to incite the taxi drivers in taxi ridesharing [1]. Trip grouping [16] is to group similar trips where the sources and destinations are close to each other according to certain heuristics. The difference is that there is no waiting time or cost constraints in trip grouping.

2.2 Recommender Systems

GPS records of taxis take down information including ID, time, longitude, latitude, speed, direction and occupation, which reflect the patterns of both passengers and taxi drivers. Applications based on taxi GPS trajectory data including urban planning [17,18], route prediction [19] and recommender systems [8–10,20]. We here focus on discussing recommender systems.

Current recommender systems provide services for either passengers or taxi drivers. A passenger-finding strategy based on L1-norm SVM [10] is proposed to determine whether a taxi should hunting or waiting for passengers. TaxiRec [8] evaluates the passenger-finding potentials of road clusters based on supervised learning and recommends the top-k road clusters for taxi drivers. T-Finder [9,20] recommends some locations instead road clusters by utilizing historical data for both passengers and taxis. There is no training process comparing to supervised learning techniques [8]. Comparing to recommending road clusters or locations for taxis or users [8,9,20], besides finding a taxi, ridesharing recommendation also needs to consider the possible destinations of the coming taxis to predict whether ridesharing can be successful or not. This makes the problem much more challenging.

3 Ridesharing Recommendation

3.1 Preliminaries

Road Network. We model a road network as a direct graph $G(V, E)$, where E and V are sets of road segments and intersections of road segments. The travel cost of each road segment (u, v) may be either time or distance measure. Since they can be converted from one to the other with the moving speed, they are used interchangeably. In addition, a grid index structure is built on the underlying road network. Given a location (lon, lat), we can find out a road segment on which the location is located.

Taxi. There are three possible status for a taxi: occupied $(\mathcal{O})$, cruising $(\mathcal{C})$ and parked $(\mathcal{P})$. A taxi can pick up a passenger $(\mathcal{P} \rightarrow \mathcal{O}$ and $\mathcal{C} \rightarrow \mathcal{O})$ or drop off a passenger $(\mathcal{O} \rightarrow \mathcal{P}$ and $\mathcal{O} \rightarrow \mathcal{C})$. Taxis mentioned in this paper refers to occupied taxis, i.e., taxis with passengers, and are willing to rideshare queries.

Ridesharing Query. A ridesharing query is defined as $Q = (id, timestamp, l_s, l_d, t_s, t_e, t_w)$ where id is the user id, $timestamp$ is when the query is submitted, l_s and l_d are respectively the source and the destination of the user, t_s is the maximum walking time from l_s to a new place; and t_e is the maximum walking time from the destination of taxi to l_d. t_w is the maximum waiting time for ridesharing at the new place for ridesharing.

Query Satisfaction. Q can rideshare with a taxi if and only if (i) the walking time to a new waiting place for the user is no longer than t_s; (ii) the walking time from the destination of the taxi to the user's own destination l_d is no longer than t_e; (iii) the waiting time at the new place for taxi is no longer than t_w.

Problem Definition. Given a ridesharing query $Q = (id, timestamp, l_s, l_d, t_s, t_e, t_w)$, we aim to recommend the user to rideshare or take a taxi directly by herself. For users that are recommended to rideshare with others, we also provide places that are easier to get ridesharing for users.

3.2   Road Segment Clustering

Since a single road segment is not a proper evaluation unit, we adapt the road segment clustering method proposed in [8], where k-means [11] is used to partition road segments[1] $\{r_1, r_2, ..., r_N\}$ into k clusters $\{C_1, C_2, ..., C_k\}$, to minimize the intra-cluster sum of square $\operatorname{argmin} \sum_{i=1}^{k} \sum_{r_j \in C_i} ||r_j - \mu_i||$, where $\mu_i = \frac{1}{n_i} \sum_{r_j \in C_i} r_j$ and n_i is the number of road segments in the i-th cluster. Interesting readers may refer to the original paper [8] for the details of road segment clustering. Figure 3 shows the clustering result in the urban area of Nanjing when the number of clusters is set to 1,000. Since we built a grid index on the road networks, by identifying the road segment where location (lon, lat) located, we can get the cluster that the location belongs to.

3.3   Kernel Density Estimation

Kernel density estimation can be used with arbitrary distributions and does not assume a fixed distribution in advance, which is used to predict the probability for taxis to depart from somewhere in cluster C_i and have somewhere (lon, lat) as

---

[1] The mid-point of a road segment is considered as its representative point.

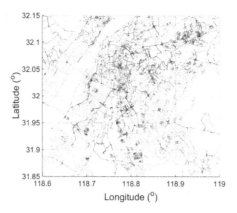

Fig. 3. Clustering on the road segments in urban area of Nanjing.

their destination. The estimation process consists of two steps: sample collection and distribution estimation.

Sample Collection. Recall that taxis departing from different clusters have different distributions of destinations since clusters have various features such as POI distributions. Given the GPS data, in order to personalize the destination distributions of each cluster, the start location decides which cluster will utilize the sample.

Intuitively ridesharing can succeed or not is related to not only locations of source and destination, but also when the query is submitted since traffic directions in modern city depend on time. Consider the traffics between companies and residencies, in the morning most traffics are likely to be from residencies to companies while in the evening traffics take the opposite direction. If only the destinations of GPS data are considered, we may recommend a user to rideshare even the query is from companies to residencies in the morning.

To avoid this, we collect samples as: $s = (lon, lat, TimeLabel)$ where (lon, lat) is the destination of the taxi and $TimeLabel$ is the time label of the sample, indicating when the trip departing from a certain cluster to (lon, lat) happens. We divide a day into 48 time intervals, with the unit of 30 min (i.e., (1) 0:00 to 0:30, (2) 0:30 to 1:00, ..., (48) 23:30 to 0:00) and label them from 1 to 48. The label of the time interval containing the start time of the GPS data decides the $TimeLabel$ of a sample. We discretize time because ridesharing consider taxis appearing in a time period to rideshare a query.

Distribution Estimation. Let $S^c =< s_1, s_2, ..., s_n >$ be the samples for a certain cluster c that follows an unknown density p. Its kernel density estimator over S^c for a new sample s_{n+1} is given by:

$$p(s_{n+1}|S^c) = \frac{1}{n\sigma^3} \sum_{i=1}^{n} K\left(\frac{s_{n+1} - s_i}{\sigma}\right), \tag{1}$$

where each sample $s_i = (lon_i, lat_i, TimeLabel_i)^T$ is a three-dimensional column vector with the longitude (lon_i) and latitude (lat_i) and $TimeLabel$ ($TimeLabel_i$). $K(.)$ is the kernel function, σ is the optimal bandwidth[2] [21]. In this paper we apply the widely used multi-dimensional normal kernel:

$$K(\mathbf{x}) = \frac{1}{(2\pi)^{d/2}|\boldsymbol{\Sigma}|^{1/2}} \exp(-\frac{1}{2}(\mathbf{x} - \boldsymbol{\mu})^T \boldsymbol{\Sigma}^{-1}(\mathbf{x} - \boldsymbol{\mu})), \qquad (2)$$

where $\mathbf{x}$ is a real d-dimensional column vector and $\mathbf{x} \sim \mathcal{N}(\boldsymbol{\mu}, \boldsymbol{\Sigma})$. We consider the time domain as the third dimension due to the nature of ridesharing.

3.4   Recommendation Framework

We now present the ridesharing framework in Algorithm 1. Given a ridesharing query $Q = (id, timestamp, l_s, l_d, t_s, t_e, t_w)$, network expansion technique [22] is used to select all reachable road segments from l_s in t_s time. Clusters that contain any of these road segments are added to the candidate cluster set $L = \{C_i | i = 1, ..., N\}$ (Line 1). Similarly we also get road segments that are reachable within t_e from l_d as $D = \{r_j | j = 1, 2, ..., M\}$ (Line 2). Denote $prob(C_i \rightarrow r_j)$ as the probability for taxis to depart from somewhere in cluster C_i and have somewhere on road r_j as their destination. For each candidate cluster C_i, we compute $prob(C_i \rightarrow r_j)$ for each road segment r_j in D with Eq. 1. The total probability P is used as the ridesharing potential of cluster C_i (Line 3 to 10). If no valid cluster exists for Q, the user will be suggested not to wait for ridesharing and take taxi directly (Line 12). The top-k clusters whose potentials are no less than a certain threshold are recommended for rideharing (Line 14).

4   Experiments

4.1   Experiment Settings

Dataset. The large-scale GPS data set is generated by over 7,000 taxis in a period of one month in Nanjing[3], China. Each GPS record includes ID, time, longitude, latitude, speed, direction and occupation. We only take the occupied trips into consideration to rideshare queries [1,2]. We divide data set into the training set and the testing set in terms of the start time of each record. In practice we can only utilize the past data to predict the future. We take data in June 1st – June 28th as training data. 1,000 records in June 29th are randomly selected as ridesharing queries. The start time, source and destination of trip are treated as $timestamp$, l_s and l_d in the queries.

Comparison of Performances. We compare our ridesharing recommendation with the naive "stay-as-where-you-are" strategy, denoted as RR and SAWYA

---

[2] Optimal bandwidth $\sigma = 0.969n^{-\frac{1}{7}}\sqrt{\frac{1}{3}\sum_i s_{ii}}$ where s_{ii} is the marginal variance.
[3] Road networks are obtained from OpenStreetMap. http://www.openstreetmap.org.

Algorithm 1. Ridesharing recommendation

---

input : $Q = (id, timestamp, l_c, l_d, t_s, t_e, t_w)$

output: Whether and where to wait for ridesharing

1 Get clusters $L = \{C_i | i = 1, ..., N\}$ that are reachable from l_s in t_s time;

2 Get roads $D = \{r_j | j = 1, ..., M\}$ that are reachable from l_d in t_e time;

3 **foreach** *Candidate cluster C_i in L* **do**

4 $\quad$ $P = 0$;

5 $\quad$ **foreach** *Possible destination r_j in D* **do**

6 $\quad\quad$ Take the mid-point (r_j^{lon}, r_j^{lat}) of r_j and the label *TimeLabel* of the time interval containing *timestamp* as the new sample $s_{n+1} = (r_j^{lon}, r_j^{lat}, TimeLabel)$;

7 $\quad\quad$ Compute $prob(C_i \rightarrow r_j)$ with Eq. 1;

8 $\quad\quad$ $P+ = prob(C_i \rightarrow r_j)$;

9 $\quad$ **if** $P \geq threshold$ **then**

10 $\quad\quad$ Add C_i to answer set;

11 **if** *Answer set is empty* **then**

12 $\quad$ Recommend user don't wait for ridesharing and take taxi directly;

13 **else**

14 $\quad$ Sort according to P and recommend top-k clusters for users to rideshare.

---

respectively. In SAWYA, the users wait for ridesharing at where they are, i.e., the cluster that l_s is in. Recall in a ridesharing query we have walking time t_s and waiting time t_w, in SAWYA $t_s = 0$ and we define the new waiting time t'_w as $t_s + t_w$.

When evaluating the performance of SAWYA, during t'_w time, if there is any taxi whose source is in the same cluster as Q and destination is reachable from l_d of Q within t_e, SAWYA is considered to give users an accurate *to-rideshare*. Similarly, for each recommended cluster in our recommendation framework, if any taxi whose source is in them and destination is reachable from l_d of Q within t_e time, our recommendation is considered as an accurate *to-rideshare*. On the other hand, if a user is recommended not to wait for ridesharing and there is no taxi to rideshare, our recommendation is considered as an accurate *not-to-rideshare*.

Parameters. We study three parameters t_s, t_e and t_w about their influence on the performances of RR and SAWYA, as shown in Table 1. The walking speed is set to $1.4 \, \text{m/s}$ [23]. We recommend top-k clusters where k is set to 5.

4.2 Performance Metrics

Ridesharing Successful Ratio. We measure the ratio of successful ridesharing of both RR and SAWYA by *RSRatio*, defined as *RSRatio* = no. of accurate *to-rideshare* / no. of queries.

Table 1. Overview about parameters

Parameter	Default value	Range
t_s	4 min	[2 min, 4 min, 6 min, 8 min, 10 min]
t_e	4 min	[2 min, 4 min, 6 min, 8 min, 10 min]
t_w	4 min	[2 min, 4 min, 6 min, 8 min, 10 min]

Prediction Accuracy. We measure the accuracy of predicting whether the user should wait for ridesharing or not by *accuracy*, defined as *accuracy* = (no. of accurate *not-to-rideshare* + no. of accurate *to-rideshare*) / no. of queries.

Recommendation Accuracy. To evaluate the quality of cluster recommendations, it is important to find out how many clusters that actually have taxis to rideshare a query are discovered by our framework. For this purpose, we employ standard metrics, i.e., *precision* and *recall*:

$$precision = \frac{\text{no. of discovered clusters}}{k}, recall = \frac{\text{no. of discovered clusters}}{\text{no. of positive clusters}}.$$

Positive clusters are clusters with taxis to rideshare Q and discovered clusters are the positive clusters in the recommended clusters. Precision and recall are averages over all queries to obtain the overall performance.

4.3 Experiment Results

Effect of Walking Time t_s. Table 2 depicts the effect of walking time t_s from current locations of users to new places. Since users can walk to farther places, the number of candidate clusters increases for both RR and SAWYA, which relaxes the requirement of ridesharing. As the number of taxis to rideshare queries increases, both RR and SAWYA have better ridesharing successful ratio (*RSRatio*) and prediction accuracy (*Accuracy*). RR outperforms SAWYA in terms of *RSRatio* by at most 40 % since it can efficiently discover new places to rideshare for users. RR outperforms SAWYA in terms of *accuracy* by about 3 times since RR makes prediction for both *not-to-rideshare* and *to-rideshare*. This shows that about 500 users can save 4–8 min with RR. Given 9 RMB as the starting taxi fare and suppose users can save half of the total fare by ridesharing,

Table 2. Effect of walking time t_s (min).

Metrics	RR					SAWYA				
	2	4	6	8	10	2	4	6	8	10
RSRatio	0.143	0.196	0.234	0.297	0.310	0.122	0.141	0.173	0.203	0.215
Accuracy	0.695	0.710	0.726	0.738	0.751	0.122	0.141	0.173	0.203	0.215
Precision	0.316	0.317	0.356	0.395	0.433	-	-	-	-	-
Recall	0.875	0.867	0.746	0.615	0.516	-	-	-	-	-

Table 3. Effect of walking time t_e (min).

Metrics	RR					SAWYA				
	2	4	6	8	10	2	4	6	8	10
RSRatio	0.111	0.196	0.229	0.262	0.288	0.091	0.142	0.186	0.232	0.275
Accuracy	0.667	0.710	0.762	0.770	0.785	0.091	0.142	0.186	0.232	0.275
Precision	0.314	0.317	0.472	0.601	0.636	-	-	-	-	-
Recall	0.857	0.867	0.871	0.902	0.917	-	-	-	-	-

Table 4. Effect of waiting time t_w (min).

Metrics	RR					SAWYA				
	2	4	6	8	10	2	4	6	8	10
RSRatio	0.156	0.196	0.203	0.235	0.254	0.122	0.141	0.173	0.203	0.215
Accuracy	0.681	0.710	0.722	0.749	0.776	0.122	0.141	0.173	0.203	0.215
Precision	0.315	0.317	0.365	0.402	0.507	-	-	-	-	-
Recall	0.860	0.867	0.868	0.881	0.886	-	-	-	-	-

users can save 10.828-44.062 RMB by ridesharing. *Precision* increases as more positive clusters are discovered. *Recall* decreases since a longer t_s leads to more candidate clusters while we only recommend top-k to users.

Effect of Walking Time t_e. Table 3 depicts the effect of walking time t_e from the destinations of taxis to those of users. As t_e increases from 2 min to 10 min, more destinations of taxis become reachable for users, which increases the number of taxis to rideshare queries. Both RR and SAWYA achieve better ridesharing successful ratio (*RSRatio*) and prediction accuracy (*Accuracy*). *Precision* and *recall* both increase because the number of discovered clusters and positive clusters increase with more taxis to rideshare queries.

Effect of Waiting Time t_w. Table 4 depicts the effect of waiting time t_w at the waiting location. As t_w increases, users can wait for taxis for a longer time. Thus more taxis are taken into consideration and increase the probability to rideshare queries. The ridesharing successful ratio (*RSRatio*) and prediction accuracy (*Accuracy*) increase for both RR and SAWYA. As the number of taxis increases with t_w, both the number of discovered clusters and the number of positive clusters increases, leading to the increase in *precision* and *recall*.

5 Conclusion

In this paper, we proposed a recommendation framework based on kernel density estimation to predict whether and where should a user wait for ridesharing. In the framework, we grouped road segments into clusters and modeled the arrival patterns of taxis from different clusters. Given a query, we compute the potential of ridesharing by investigating the probabilities of possible destinations based on ridesharing requirements. Experimental results show that the accuracy about whether ridesharing or not and the ridesharing successful ratio are respectively about 3 times and at most 40 % better than the naive "stay-as-where-you-are" strategy. In the future work, we will study how to incorporate the influence of Point-of-Interests (POIs) into our ridesharing recommendation.

Acknowledgment. The author would like to thank Dr. Yanhua Li and Dr. Jia Zeng for providing their Nanjing taxi GPS dataset. The author also would like to thank anonymous reviewers and Dr. Cong Wang.

References

1. Ma, S., Zheng, Y., Wolfson, O.: Real-time city-scale taxi ridesharing. IEEE Trans. Knowl. Data Eng. **27**(7), 1782–1795 (2015)
2. Huang, Y., Bastani, F., Jin, R., Wang, X.S.: Large scale real-time ridesharing with service guarantee on road networks. PVLDB **7**(14), 2017–2028 (2014)
3. Ma, S., Wolfson, O.: Analysis and evaluation of the slugging form of ridesharing. In: SIGSPATIAL, pp. 64–73 (2013)
4. Kamar, E., Horvitz, E.: Collaboration and shared plans in the open world: studies of ridesharing. In: IJCAI, p. 187 (2009)
5. Calvo, R.W., de Luigi, F., Haastrup, P., Maniezzo, V.: A distributed geographic information system for the daily car pooling problem. Comput. OR **31**(13), 2263–2278 (2004)
6. Cao, B., Alarabi, L., Mokbel, M.F., Basalamah, A.: SHAREK: a scalable dynamic ride sharing system. In: MDM, pp. 4–13 (2015)
7. Attanasio, A., Cordeau, J., Ghiani, G., Laporte, G.: Parallel tabu search heuristics for the dynamic multi-vehicle dial-a-ride problem. Parallel Comput. **30**(3), 377–387 (2004)
8. Wang, R., Chow, C.Y., Lyu, Y., Lee, V.C.S., Kwong, S., Li, Y., Zeng, J.: Taxirec: recommending road clusters to taxi drivers using ranking-based extreme learning machines. In: SIGSPATIAL (2015)
9. Yuan, N.J., Zheng, Y., Zhang, L., Xie, X.: T-finder: A recommender system for finding passengers and vacant taxis. IEEE Trans. Knowl. Data Eng. **25**(10), 2390–2403 (2013)
10. Li, B., Zhang, D., Sun, L., Chen, C., Li, S., Qi, G., Yang, Q.: Hunting or waiting? discovering passenger-finding strategies from a large-scale real-world taxi dataset. In: PerCom., pp. 63–68 (2011)
11. Bishop, C.M.: Pattern Recognition and Machine Learning. Springer, New York (2006)
12. Cheng, C., Yang, H., King, I., Lyu, M.R.: Fused matrix factorization with geographical and social influence in location-based social networks. In: AAAI (2012)
13. Ye, M., Yin, P., Lee, W., Lee, D.L.: Exploiting geographical influence for collaborative point-of-interest recommendation. In: SIGIR, pp. 325–334 (2011)
14. Bistaffa, F., Farinelli, A., Ramchurn, S.D.: Sharing rides with friends: a coalition formation algorithm for ridesharing. In: AAAI, pp. 608–614 (2015)
15. Bistaffa, F., Filippo, A., Chalkiadakis, G., Ramchurn, S.D.: Recommending fair payments for large-scale social ridesharing. In: RecSys., pp. 139–146 (2015)

16. Gidófalvi, G., Pedersen, T.B., Risch, T., Zeitler, E.: Highly scalable trip grouping for large-scale collective transportation systems. In: EDBT, pp. 678–689 (2008)
17. Liu, Y., Kang, C., Gao, S., Xiao, Y., Tian, Y.: Understanding intra-urban trip patterns from taxi trajectory data. J. Geog. Syst. **14**(4), 463–483 (2012)
18. Zheng, Y., Liu, Y., Yuan, J., Xie, X.: Urban computing with taxicabs. In: UbiComp., pp. 89–98 (2011)
19. Yuan, J., Zheng, Y., Xie, X., Sun, G.: T-drive: Enhancing driving directions with taxi drivers' intelligence. IEEE Trans. Knowl. Data Eng. **25**(1), 220–232 (2013)
20. Yuan, J., Zheng, Y., Zhang, L., Xie, X., Sun, G.: Where to find my next passenger. In: UbiComp., pp. 109–118 (2011)
21. Silverman, B.W.: Density Estimation for Statistics and Data Analysis, vol. 26. CRC Press, Boca Raton (1986)
22. Papadias, D., Zhang, J., Mamoulis, N., Tao, Y.: Query processing in spatial network databases. In: VLDB, pp. 802–813 (2003)
23. Browning, R.C., Baker, E.A., Herron, J.A., Kram, R.: Effects of obesity and sex on the energetic cost and preferred speed of walking. J. Appl. Physiol. **100**(2), 390–398 (2006)

Keyword-aware Optimal Location Query in Road Network

Jinling Bao[1,2(✉)], Xingshan Liu[1], Rui Zhou[3], and Bin Wang[1]

[1] College of Information Science and Engineering, Northeastern University,
Shenyang, China
baojinling@research.neu.edu.cn, binwang@mail.neu.edu.cn
[2] Department of Computer Science, Baicheng Normal College, Baicheng, China
[3] Centre for Applied Informatics, College of Engineering and Science,
Victoria University, Melbourne, Australia
rui.zhou@vu.edu.au

Abstract. In this paper, we study a very useful type of optimal location query, motivated by the following real application: for property renting or purchasing, a client often wants to find a residence such that the sum of the distances between this residence and its nearest facilities is minimal, and meanwhile the residence should be on one of the client-selected road segments (representing where the client prefers to live). The facilities are categorized with keywords, eg., school, hospital and supermarket, and in this problem one facility for each category is required. To the best of our knowledge, this type of query has not been studied before. To tackle this problem, we propose a basic algorithm based on dividing roads (edges) into sub-intervals and find the optimal locations by only inspecting the endpoints of the sub-intervals. We also propose an improved algorithm with keyword filtering and edge pruning strategies. Finally, we demonstrate the efficiency of our algorithms with extensive experiments on large-scale real datasets.

Keywords: Optimal location query · Keyword-aware · In Road network

1 Introduction

Optimal location (OL) queries are a type of spatial queries particularly useful for strategic planning of resources in many applications, such as location-based service, profile-based marketing and location planning. An optimal location query usually considers a set of facilities, a set of clients, and a set of candidate locations. The objective of this type of query is to identify a location at which a planned activity can optimize a certain cost metric between the surrounding facilities and the potential clients [1–8]. For example, a client often wants to find a residence such that the sum of the distances between this residence and its nearest school, hospital and supermarket is minimal, and meanwhile the residence should be on one of the client-selected road segments.

B. Cui et al. (Eds.): WAIM 2016, Part I, LNCS 9658, pp. 164–177, 2016.
DOI: 10.1007/978-3-319-39937-9_13

Motivated by the example, this paper studies a novel type of OL query, *keyword-aware optimal location* query, denoted as KOL query, in road network. The problem can be formalized as follows: given a road network $G = (V, E)$, a user-selected edge set $E_u \subseteq E$, a query keyword set ψ_u (e.g., hospital, school, and market), and an object set O where each object covers a query keyword, the objective is to find a location p on any user-selected edge $e \in E_u$, such that there exists a set of objects covering all query keywords, satisfying the sum of the distances between p and each of the objects is minimal. This kind of optimal location query is particularly useful for location planning.

To the best of our knowledge, despite its importance and practicality, no existing work has studied the above KOL query in road network. The most relevant works [1,12] cannot be easily adapted to tackle this problem either. The work [1] introduced MinSum location query asking for a candidate location on which a new facility can be built to minimize the total cost of the clients. Here, all the facilities and clients contribute to the cost metric, while, in our setting, only one candidate client location and a group of facilities covering query keywords need to contribute, and therefore pruning strategies which utilizing total cost upper-bounds in partitions of the network cannot be adapted to our problem. On the other hand, with regard to keyword-aware, the recent work [12] studied m-closest keywords query with the objective to minimize the largest distance between keyword facilities. However, Euclidean distance was used as the distance function, which does not suit road networks and the cost metric is also different.

Based on the unique characteristics of KOL problem, we design a basic algorithm, named MinSum-KOL. The main idea is that we find the local optimal location on each user-selected edge, then find the global optimal location in user-selected edges set. In order to find the optimal location efficiently, we design an improved algorithm, named KOL-KFEP. There are two improved strategies. One strategy is keyword filtering. The other strategy is edge pruning.

In summary, we make the following three major contributions:

(1) We conduct a novel study on keyword-aware optimal location query in road network. Given a set of road segments and a set of query keywords, the optimal location query is to find a location p with smallest sum of distances from the location p to object covering each query keyword.
(2) We propose an algorithm, named MinSum-KOL, to answer the KOL query. In order to find the optimal location more efficiently, we propose two improved strategies. They are keyword filtering and edge pruning, respectively. Based on them, we design an improved algorithm, named KOL-KFEP.
(3) We demonstrate the efficiency of our algorithms with extensive experiments on large-scale real datasets.

The rest of this paper is organized as follows: Sect. 2 reviews the related work and Sect. 3 gives the problem formulation. In Sect. 4, we introduce our basic method, which is further improved in Sect. 5. Our experimental study is given in Sect. 6, and we finally conclude this paper in Sect. 7.

2   Related Work

Our problem is closely related to aggregate nearest neighbor (NN) search, collective spatial keyword queries and MinSum optimal location queries. We classify the related work into two types work, namely optimal location queries with non-road network setting and road network setting.

2.1   Optimal Location Queries with Non-road Network Setting

Papadias et al. [9] studied a group nearest neighbor (GNN) query. Given two sets of points P and Q, a GNN query retrieves the point(s) of P with the smallest sum of distances to all points in Q. They utilize R-trees and Euclidean distance bounds to converge to the result, by minimizing the I/O and computational cost. Cao et al. [10] and Long et al. [11] studied the problem of retrieving a group of spatial web objects such that the group's keywords cover the query's keywords and such that objects are nearest to the query location and have the lowest inter-object distances. A recent work [12] studies the m-closest keywords(mCK) query, which finds a group of objects such that they cover all query keywords and have the smallest diameter, which is defined as the largest distance between any pair of objects in the group.

Above work studies the sum of Euclidean distance. In this paper, we study the problem considering the network distance. The aggregate network distance requires more expensive network traversal than the aggregate Euclidean distance. None of the solutions developed therein is applicable when the keyword-aware optimal location queries reside in a road network.

2.2   Optimal Location Queries with Road Network Setting

Aggregate nearest neighbor (ANN) queries [13] return the object that minimizes an aggregate distance function with respect to a set of query points in road networks. The work considers sum functions and techniques that utilize Euclidean distance bounds and R-tree. In KOL problem, we take keywords into account in location. During finding the optimal location, the nearest objects covering query keywords varies at different locations. It performs incremental NN queries for each location, and combines the results. In our problem, the candidate location is the user-selected edges set, not fixed points, if we also use the Euclidean distance bounds and R-tree index, it incurs more incremental computation. So the method [13] cannot hold our problem efficiently.

Recently, Xiao et al. [1] studied the MinSum location query problem in road network. In work [1], given a road network, a set of clients and a set of facilities, it asks a location where a new facility can be built to minimize the total cost of the clients. The main idea is to augment the road network by creating a vertex for each client and each facility and partition the augmented road network into subnetworks and compute the total cost upper-bound which is the sum of cost from each client to its nearest facility of each sub-network. If the current total cost is smaller than the upper-bounds of all unvisited sub-networks, search terminate;

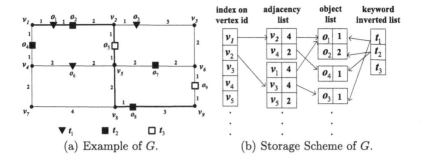

(a) Example of G.　　　　　(b) Storage Scheme of G.

Fig. 1. Running example.

otherwise, move on to the next subgraph. Here, all the facilities and clients contribute to the cost metric, while, in our setting, only one candidate client location and a group of facilities covering query keywords need to contribute, and therefore pruning strategies which utilizing total cost upper-bounds in partitions of the network cannot be adapted to our problem.

3　Problem Formulation

Let $G = (V, E)$ be a road network, which consists of a set of vertices V and a set of edges $E \subseteq V \times V$. Each vertex $v \in V$ represents a road junction; each edge $e \in E$ represents a road segment between two road junctions. For any edge $e = (v_l, v_r)$, v_l (v_r) is the left(right) vertex of e. User-selected set of edges is denoted by $E_u \subseteq E$. O is an object set, ψ is a keyword set, and $\psi = \{t_1, ...t_k\}$, where t_k denotes a keyword and O_{t_k} denotes the object set covering t_k. Query keywords set is denoted as $\psi_u \subseteq \psi$.

Figure 1(a) shows an example of a graph G, which is a road network consisting 9 vertices, 9 objects, and 12 road segments. The squares and triangles represent the objects covering different keywords. The highlighted edges are user-selected edges set $E_u = \{(v_1, v_2), (v_2, v_5), (v_5, v_8), (v_8, v_9)\}$. Figure 1(b) shows the storage scheme for the road network. There is a keyword inverted list. Each keyword points to the object list covering it.

We denote the distance between two locations p_1 and p_2 as $d(p_1, p_2)$, representing the distance of shortest path from p_1 to p_2. We denote location p's nearest object covering keyword t_k as $NNO_{t_k}(p)$. Then the distance between the location p and $NNO_{t_k}(p)$ is denoted by $d(p, NNO_{t_k}(p))$. For ease of expression, the distance is denoted as $d(p, t_k)$. The sum of distances from p to its nearest object covering each query keyword is denoted as $d(p, \psi_u)$, shown as following equation:

$$d(p, \psi_u) = \sum_{t_k \in \psi_u} d(p, t_k) \qquad (1)$$

The KOL query is to find an optimal location which minimizes the sum of distances between this location and object covering each query keyword. Formally, we define the KOL query as follows:

Definition 1. *Keyword-aware Optimal Location Query. Given an undirected connected graph $G = (V, E)$, where $V(E)$ denotes the set of vertices(edges) in G, a user-selected edges set E_u, a query keywords set ψ_u, such that an optimal location $p \in E_u$ with minimum sum of distances to the set ψ_u.*

Where E_u can be any subset of E, ψ_u can be any subset of ψ, i.e., we can have $E_u = E$ and $\psi_u = \psi$. We consider any point $p \in E_u$ as a candidate location. In Fig. 1(a), if query keyword set is $\psi_u = \{t_1, t_2\}$, then $O_{t_1} = \{o_1, o_3, o_6\}$ and $O_{t_2} = \{o_2, o_4, o_7, o_8\}$ are object sets covering query keyword t_1 and t_2, respectively. We can compute the sum of distances from v_1 to nearest objects covering keyword t_1 and t_2: $d(v_1, \psi_u) = d(v_1, o_2) + d(v_1, o_1) = 3$.

Example 1. Given a query $q = (E_u, \psi_u) = (\{(v_1, v_2), (v_2, v_5), (v_5, v_8), (v_8, v_9)\}, \{t_1, t_2\})$, KOL query asks for a location $p \in E_u$, whose sum of distances to its nearest objects covering keyword t_1 and t_2 is minimized. It is not difficult to see the minimum sum of distances is 1 and the optimal point p can be location between node o_1 and node o_2, i.e., the whole interval $[o_1, o_2]$.

4   Basic Algorithm for KOL Queries

In this section, we propose a basic algorithm, named MinSum-KOL, to find the optimal location. The main idea is that we find the local optimal location on each user-selected edge, then combine the results to find the global optimal location.

4.1   Lower Bound on a Sub-interval

Each edge $e = (v_l, v_r)$ may contain multiple objects, where o_i is the i-th object. Thus, we split the whole edge e into a number of sub-intervals with the multiple objects: $[v_l, o_1]; [o_2, o_3], \cdots, [o_k, v_r]$. We represent vertex and object as unified node n_i. Consider a sub-interval $[n_l, n_r]$ of edge e and let p be an arbitrary point on $[n_l, n_r]$. When p moves along edge e, $d(p, t_k)$ changes linearly. We know that $d(p, t_k)$ can be expressed as follows.

   $d(p, t_k) = \min\{d(n_l, t_k) + d(n_l, p), d(n_r, t_k) + d(p, n_r)\}$

   An endpoint is a local optimal location on a sub-interval, if and only if it has the minimum sum of distances among all points on sub-interval. The following theorem shows that one endpoint of sub-interval is always minimal.

Theorem 1. *Given a sub-interval $[n_l, n_r]$, query keywords set ψ_u, and any point $p \in [n_l, n_r]$, if $d(p, \psi_u)$ is smaller than the sum of distances of one endpoint, then $d(p, \psi_u)$ must be larger than the sum of distances of the other endpoint.*

Proof. We assume that $\psi_p^l(\psi_p^r)$ is the subset of keyword in ψ_u, such that the shortest path from p to its nearest object passes through $n_l(n_r)$. First of all, $d(n_l, t_k) \leq d(n_r, t_k) + d(n_l, n_r)$, and similarly, $d(n_r, t_k) \leq d(n_l, t_k) + d(n_l, n_r)$. Next,

$d(p, \psi_u) = \sum_{t_k \in \psi_p^l} d(p, t_k) + \sum_{t_k \in \psi_p^r} d(p, t_k)$
$\qquad = \sum_{t_k \in \psi_p^l} d((n_l, t_k) + d(n_l, p)) + \sum_{t_k \in \psi_p^r} (d(n_r, t_k) + d(p, n_r))$
$\qquad = \sum_{t_k \in \psi_p^l} d(n_l, t_k) + |\psi_p^l| d(n_l, p) + \sum_{t_k \in \psi_p^r} d(n_r, t_k) + |\psi_p^r| d(p, n_r),$

The number of n_l's nearest object covering query keywords, whose shortest path passes through n_l, is lager than $|\psi_p^l|$, and similarly, the number of n_l's nearest object, whose shortest path passes through n_r, is smaller than $|\psi_p^r|$, then

$d(n_l, \psi_u) = \sum_{t_k \in \psi_u} d(n_l, t_k)$
$\qquad \leq \sum_{t_k \in \psi_p^l} d(n_l, t_k) + \sum_{t_k \in \psi_p^r} d(n_l, t_k)$
$\qquad \leq \sum_{t_k \in \psi_p^l} d(n_l, t_k) + \sum_{t_k \in \psi_p^r} d(n_r, t_k) + |\psi_p^r| d(n_l, n_r).$

Assume that $d(p, \psi_u) \leq d(n_l, \psi_u)$, we have,

$\sum_{t_k \in \psi_p^l} d(n_l, t_k) + |\psi_p^l| d(n_l, p) + \sum_{t_k \in \psi_p^r} d(n_r, t_k) + |\psi_p^r| d(p, n_l)$
$\leq \sum_{t_k \in \psi_p^l} d(n_l, t_k) + \sum_{t_k \in \psi_p^r} d(n_r, t_k) + |\psi_p^r| d(n_l, n_r),$

which leads to $|\psi_p^l| d(n_l, p) + |\psi_p^r| d(p, n_r) \leq |\psi_p^r| d(n_l, n_r).$

Since $d(n_l, n_r) = d(n_l, p) + d(p, n_r)$, we have $|\psi_p^l| d(n_l, p) \leq |\psi_p^r| d(n_l, p)$, which means that $|\psi_p^l| \leq |\psi_p^r|$. Similarly,

$d(n_r, \psi_u) = \sum_{c_k \in \psi_u} d(n_r, t_k)$
$\qquad \leq \sum_{t_k \in \psi_p^l} d(n_r, t_k) + \sum_{t_k \in \psi_p^r} d(n_r, t_k)$
$\qquad \leq \sum_{t_k \in \psi_p^l} d(n_l, t_k) + |\psi_p^l| d(n_l, n_r) + \sum_{t_k \in \psi_p^r} d(n_r, t_k),$

because by $|\psi_p^l| \leq |\psi_p^r|$, $d(p, \psi_u) \geq d(n_r, \psi_u).$

Based on Theorem 1, for any point p in a certain sub-interval, $d(p, \psi_u)$ can be lower-bounded by the $\min\{d(n_l, \psi_u), d(n_r, \psi_u)\}$.

By Theorem 1, if the endpoints of a sub-interval have different merits, then the endpoint with the smaller sum of distances should be the only local optimal location. If the sum of distances of one endpoint is equal to that of the other endpoint, either all points on sub-interval have the same sum of distances, or two endpoints have smaller sum of distances than any other points.

Lemma 1. *Given any two adjacent nodes n_l, n_r, $\psi_l (\psi_r)$ is the set of node n_l's $(n_r$'s) nearest objects passing through n_l (n_r), and any location $p \in [n_l, n_r]$, if $NNO_{t_k}(n_l) = NNO_{t_k}(n_r)$ and $|\psi_l| = |\psi_r|$, then $d(p, \psi_u) = d(n_l, \psi_u) = d(n_r, \psi_u).$*

Proof. From $NNO_{t_k}(n_l) = NNO_{t_k}(n_r)$, and $|\psi_l| = |\psi_r|$, we know that
$d(n_l, t_k) = d(n_r, t_k) + d(n_l, n_r), t_k \in \psi_r$, and
$d(n_r, t_k) = d(n_l, t_k) + d(n_l, n_r), t_k \in \psi_l$. Therefore,
$d(n_l, \psi_u) = \sum_{t_k \in \psi_l} d(n_l, t_k) + \sum_{t_k \in \psi_r} d(n_r, t_k) + |\psi_r| d(n_l, n_r).$
$d(n_r, \psi_u) = \sum_{t_k \in \psi_l} d(n_r, t_k) + |\psi_l| d(n_l, n_r) + \sum_{t_k \in \psi_r} d(n_r, t_k).$
$d(p, \psi_u) = \sum_{t_k \in \psi_l} d(n_r, t_k) + |\psi_l| d(n_l, p) + \sum_{t_k \in \psi_r} d(n_r, t_k) + |\psi_r| d(p, n_r)$
$\qquad = \sum_{t_k \in \psi_l} d(n_r, t_k) + \sum_{t_k \in \psi_r} d(n_r, t_k) + |\psi_r| d(n_l, n_r).$
We can conclude that $d(p, \psi_u) = d(n_l, \psi_u) = d(n_r, \psi_u).$

We can determine the result is the interval between two adjacent nodes if they satisfy the condition of Lemma 1.

4.2   Finding Shortest Distances Between Vertices and Objects

In order to compute the shortest distance between a vertex and its nearest object efficiently, we adopt an algorithm which runs Dijkstra's algorithm once only. Firstly, we find the shortest distance from a vertex v_i to its nearest object covering keyword t_k along one of the edges adjacent to v_i if there is an object on one of these edges, otherwise the distance is ∞. Next, We create a new node called the *virtual node*, denoted by VN_{t_k} in the network. For each vertex v_i, we create an edge (VN_{t_k}, v_i) and set the length of this edge to be $d(v_i, t_k)$. It is easy to verify that the time of computing $d(v_i, t_k)$ for all vertices is $\mathcal{O}(|V|)$. Finally, we execute the Dijkstra's algorithm starting from this virtual node VN_{t_k} once only. We guarantee that the distance between each vertex v_i and its nearest object is exactly equal to the shortest distance between VN and v. The idea of using the concept of *virtual node* was studied in [14]. Since the Dijkstra's algorithm takes $\mathcal{O}(|V| \log |V|)$ time [15], finding the shortest distances between each node and its nearest object covering each keyword takes $\mathcal{O}(|\psi||V| \log |V|)$ time.

In order to accelerate the algorithm, we pre-compute each vertex v_i's nearest object covering t_k and the corresponding distance, storing in an adjacency list, respectively.

4.3   Finding Shortest Distances Between Objects

Consider an object o_i covering keyword t_k on an edge $e = (v_l, v_r)$, its nearest object covering keyword t_k is itself. If there is no object covering other keyword t_m on e, the shortest path from o_i to its nearest object covering keyword t_m passes through v_l or v_r. Otherwise if there are objects covering keyword t_m on e, the shortest path from o_i to its nearest object covering keyword t_m passes through v_l, v_r or the nearest object covering keyword t_m is on e.

Lemma 2. *Consider an object covering keyword t_k on an edge $e = (v_l, v_r)$. If there is no object covering other keyword t_m on e, then $d(t_k, t_m) = \min\{d(t_k, v_l) + d(v_l, t_m), d(t_k, v_r) + d(v_r, t_m)\}$. Otherwise, $d(t_k, t_m) = \min\{d(t_k, t_m), d(t_k, vl) + d(v_l, t_m), d(t_k, v_r) + d(v_r, t_m)\}$.*

It is easy to verify the correctness of the above lemma. Let l_o be the greatest number of objects along an edge. The running time of finding distance for one object covering one keyword takes $\mathcal{O}(l_o)$ time, and the overall running time for all objects takes $\mathcal{O}(l_o|\psi||O|)$ time.

4.4   MinSum-KOL Algorithm

The pseudo code of MinSum-KOL is presented as Algorithm 1. Initialization optimal location set $R \leftarrow \emptyset$; $bestDist \leftarrow \infty$, storing the current minimum sum of distances (line 1). We first compute the distance $d(n_i, t_k)$ between each node n_i and its nearest object covering query keyword t_k, then compute the sum of distances of node n_i (lines 2–6). Next, we find the nodes with the minimum sum of distances and update $bestDist$ and the result set R (lines 7–8). For two adjacent nodes of set R in same edge, if they satisfy conditions of Lemma 1, then we replace the nodes with their interval in result set R (lines 9–11). Lastly, we return the result set R (line 12).

Algorithm 1. MinSum-KOL

Input: A graph G and a query $q = (E_u, \psi_u)$;
Output: The optimal location set R
1 Initialization optimal location set $R \leftarrow \emptyset$; minimum sum of distances
   $bestDist \leftarrow \infty$;
2 **for** *each edge* $e_j \in E_u$ **do**
3   **for** *each node* n_i **do**
4   **for** *each keyword* $t_k \in \psi_u$ **do**
5   compute the distance $d(n_i, t_k)$;
6              $d(n_i, \psi_u) \leftarrow d(n_i, \psi_u) + d(n_i, t_k)$;
7   **if** $d(n_i, \psi_u) \leq bestDist$ **then**
8   update $bestDist$ and R;

9 **for** *any two nodes* n_l *and* n_r *of set* R *in same edge* **do**
10   **if** $NNO_{\psi_u}(n_l) = NNO_{\psi_u}(n_r)$ & $\&|\psi_l| = |\psi_r|$ **then**
11   replace nodes n_l and n_r with $[n_l, n_r]$ in set R;

12 **return** R;

The optimal locations on E_u can be found by computing the sum of distance of each node to its nearest object covering each keyword, which takes $\mathcal{O}(|V + O||\psi|)$ time given the distance from each vertex to its nearest object covering each keyword.

5 Improved Algorithm for KOL Queries

When the dataset is large, MinSum-KOL incurs a significant overhead. In order to find the minimum sum of distances efficiently, we propose two improved strategies in this section. One is keyword filtering strategy, the other is edge pruning strategy. An improved algorithm KOL-KFEP is then proposed based on _keyword filtering_ and _edge pruning_ strategies.

In the algorithm Minsum-KOL, given a query keywords set ψ_u, we have to compute the distance of each node to its nearest object covering query keyword. An object does not cover query keyword, it has no connection with the result. We can filter the object nodes not covering query keyword.

5.1 Edge Pruning Strategy

Denote the edge $e's$ nearest object covering the keyword t_k as $NNO_{t_k}(e)$, and the distance from an edge e to its nearest object covering the keyword t_k as $d(e, t_k)$.

Lemma 3. *Given an edge e, and a keyword t_k, if there is an object covering the keyword t_k in edge e, $d(e, t_k) = 0$; else $d(e, t_k) = min\{d(v_l, t_k), d(v_r, t_k)\}$.*

It is easy to verify the correctness of the above lemma. For example, in Fig. 1(a), assume edge $e = (v_1, v_2)$, then $NNO_{t_1}(e)$ is o_1, $NNO_{t_3}(e)$ is o_5, $d(e, t_1) = 0$, $d(e, t_3) = min\{d(v_1, t_3), d(v_2, t_3)\} = 1$.

Theorem 2. *Given a keywords set ψ_u, the sum of distances from any edge e to object covering each query keyword and the sum of distances from any point $p \in e$ to object covering each query keyword, $d(e, \psi_u) \le d(p, \psi_u), p \in e$.*

Proof. Since $d(e, \psi_u) = \sum_{t_k \in \psi_u} d(e, t_k)$, and the point p is arbitrary location on e, $d(p, \psi_u) = \sum_{t_k \in \psi_u} d(p, t_k)$. Assume that edge e includes the object covering keyword t_k, then $d(e, t_k) = 0 \le d(p, t_k)$ by Lemma 3. Assume that edge e does not include the object covering keyword t_k, $d(e, t_k) = min\{d(v_l, t_k), d(v_r, t_k)\}$, while $d(p, t_k) = min\{d(v_l, t_k) + d(v_l, p), d(v_r, t_k) + d(p, v_r)\}$, therefore, $d(e, t_k) \le d(p, t_k)$. Hence $d(e, \psi_u) = \sum_{t_k \in \psi_u} d(e, t_k) \le \sum_{t_k \in \psi_u} d(p, t_k) = d(p, \psi_u)$.

The sum of edge e is lower bound of point $p \in e$. If $d(e_2, \psi_u) > d(p, \psi_u), p \in e_1$, then $d(p', \psi_u), p' \in e_2 > d(p, \psi_u), p \in e_1$, the location $p' \in e_2$ can not be the location with minimum sum of distances. So edge e_2 can be pruned.

5.2 KOL-KFEP Algorithm

The pseudo code is presented in Algorithm 2. Initialization optimal location set $R \leftarrow \emptyset$, the variable $bestDist \leftarrow \infty$, storing the current minimum sum of distances, and minimum priority queue $Q \leftarrow \emptyset$ (line 1). We get the distance from each edge $e_j \in E_u$ to nearest object covering each query keyword. If there is an object covering a query keyword on e_j, we compute $d(e_j, \psi_u)$ and compare $d(e_j, \psi_u)$ and $bestDist$, if $d(e_j, \psi_u) \le bestDist$, push e_j into priority queue Q, else we prune edge e_j. If there is no object covering query keyword on e_j, we compute the sum of distances of v_l and v_r; if $min\{d(v_l, \psi_u), d(v_l, \psi_u)\} \le bestDist$, we update the result set R and $bestDist$ with this node and corresponding sum of distances(lines 2–9). Next, we dequeue each edge e_j from priority queue Q, if $d(e_j, \psi_u) > bestDist$, and the processing terminates early. Otherwise, we compute the sum of distances of each node $n_i \in e_j$, and find the nodes with minimum sum of distances (lines 10–17). Then we determine whether the result is interval (line 18). Lastly, return the result set R (line 19).

Take KOL query as Example 1, $q = (\{(v_1, v_2), (v_2, v_5), (v_5, v_8), (v_8, v_9)\}, \{t_1, t_2\})$. Firstly, we compute the sum of distances of each edge to object covering query keyword t_1 and t_2. Then we get $d((v_1, v_2), \psi_u) = 0$. In edges (v_2, v_5) and (v_5, v_8), there is no object covering keyword t_1 and t_2, and we thus only compute the sum of distances of endpoint in these edges. Since $d(v_2, \psi_u) = 3$ is smallest, we update $bestDist = 3$ and result set $R = \{v_2\}$. Next, we get the sum of distances edge $d((v_8, v_9), \psi_u) = 4$. It is lager than current $bestDist$, so we prune the edge (v_8, v_9) and compute the sum of distances of each node in edge (v_1, v_2). We get the minimum sum of distances is 1, and the result is $R = \{o_1, o_2\}$. Lastly, we get the result $R = \{[o_1, o_2]\}$ based on Lemma 1.

Algorithm 2. KOL-KFEP Algorithm

Input: A graph G and a query $q = (E_u, \psi_u)$;
Output: The optimal location set R

1 Initialization optimal location set $R \leftarrow \emptyset$; minimum sum of distances
   $bestDist \leftarrow \infty$; minimum priority queue $Q \leftarrow \emptyset$;

2 **for** *each edge $e_j \in E_u$* **do**

3   **if** *object covering no keyword in set ψ_u* **then**

4   **if** $\min\{d(v_l, \psi_u), d(v_r, \psi_u)\} \leq bestDist$ **then**

5   update $bestDist$ R;

6   **else**

7   compute the sum of distances $d(e_j, \psi_u)$;

8   **if** $d(e_j, \psi_u) \leq bestDist$ **then**

9               $Q.push(e_j)$;

10 **for** $!Q.empty()$ **do**

11      $e_j \leftarrow Q.pop()$;

12   **if** $d(e_j, \psi_u) > bestdist$ **then**

13   break;

14   **else**

15   **for** *each node $n_i \in e_j$* **do**

16   compute the sum of distances $d(n_i, \psi_u)$;

17              $\cdots$ `// see lines 4--6 in Algorithm 1`

18 $\cdots$ `// see lines 9--11 in Algorithm 1`

19 return R;

---

6 Experiments

This section evaluates our proposed MinSum-KOL and KOL-KFEP algorithms
in terms of query performance.

6.1 Data and Queries

Real-life datasets are tested: (1) CA[1] comprises of the road network of the
California, which contains $21,047$ vertices, $21,692$ edges and $80,930$ objects covering 62 keywords. (2) SF (See footnote 1) comprises of the road network of San
Francisco, which contains $174,955$ nodes, $223,000$ edges, and $250,000$ crawled
objects (with Google Place API) covering 100 keywords. We mapped each object
to on an edge in the road network. We generated 5 query sets for each dataset.
The number of query keywords are $2, 4, 6, 8,$ and 10. Each set has 50 queries.
Other default settings are as follows: The user-selected edges set is $E_u = E$ and
the number of query keywords is 6. All algorithms were implemented in C++
Linux, and run on an Intel(TM) i5-2400 CPU @3.10 GHz with 4 GB RAM.

---

[1] http://www.cs.utah.edu/~lifeifei/SpatialDataset.htm.

6.2   Experimental Results

We evaluated MinSum-KOL and KOL-KFEP algorithms in terms of query performance varying five parameters: (1) the number of vertices($|V|$), (2) the number of objects $|O|$, (3) the number of keywords $|\psi|$, (4) the number of query keywords $|\psi_u|$, and (5) the percentage $\tau = |E_u|/|E|$ of edges, in CA and SF.

(1) Effect of $|V|$. Our first set of experiments focus on effect of vertices. Figure 2 illustrates the running time of our solutions. By extracting sub-network of CA (SF) road network, we obtain 5 sub-datasets containing 4,000 (30,000), 8,000 (60,000), 12,000 (90,000), 16,000 (120,000) and 20,000 (150,000) vertices. The number of objects and keywords vary between 5,000 (40,000) and 80,000(200,000), 10 (15) and 62 (100), respectively. The running time of two algorithms increases when $|V|$ increases, since a larger $|V|$ leads to more edges in E_u. KOL-KFEP performs much better than MinSum-KOL when $|V|$ increases, because the keyword filtering and edge pruning strategies reduce the search space. The running time of two algorithms in SF is larger than in CA, for the size of dataset SF is lager than CA.

(2) Effect of $|O|$. This set of experiments investigate the scalability of our solutions by varying the number of objects $|O|$. We varied the number of objects, with the number of keywords fixed, i.e., the number of objects covering a particular keyword increases when the number of objects increases. Figure 3 illustrates the running time when the number of objects changes from 5,000 (40,000) to

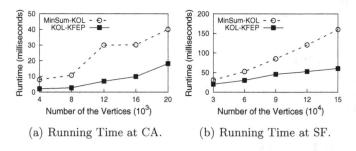

(a) Running Time at CA.   (b) Running Time at SF.

Fig. 2. Running time vary the number of vertices ($|V|$).

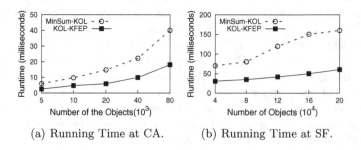

(a) Running Time at CA.   (b) Running Time at SF.

Fig. 3. Running time vary the number of objects ($|O|$).

80,000 (200,000) in CA (SF). Since a larger $|O|$ covering a particular keyword leads to more search space and search time, the running time of two algorithms increases when the number of objects increases and KOL-KFEP is faster than MinSum-KOL.

(3) Effect of $|\psi|$. We varied the number of keywords, with the number of objects is fixed, i.e., the number of objects covering a particular keyword decreases when the number of keywords increases. Figure 4 illustrates the result. The running time of two algorithms increases in CA and SF, and KOL-KFEP is always faster than MinSum-KOL when the number of keywords changes from 10 to 60 in CA. The pattern is similar to that on the SF dataset shown in Fig. 4(b).

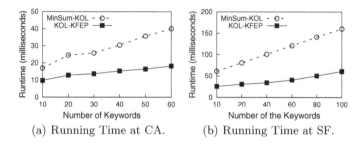

(a) Running Time at CA.   (b) Running Time at SF.

Fig. 4. Running time vary the number of keywords ($|\psi|$).

(4) Effect of $|\psi_u|$. As shown in Fig. 5, the running time of two algorithms increases when the number of query keywords changes from 2 to 10 in CA and SF. When $|\psi_u|$ increases, the number of search object also increases. Again, KOL-KFEP is faster than MinSum-KOL in CA and SF.

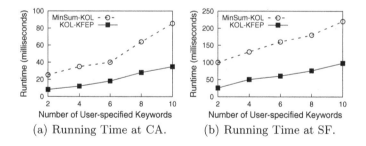

(a) Running Time at CA.   (b) Running Time at SF.

Fig. 5. Running time vary the number of query keywords ($|\psi_u|$).

(5) Effect of the Percentage τ. Figure 6 illustrates the running time of our solutions when τ changes from 20 % to 100 %. The running time of two algorithms increases when τ increases, since a larger τ leads to more edges in E_u. The running time of MinSum-KOL is proportional with τ, but the running time of KOL-KFEP is not, since the effect of its improved strategies.

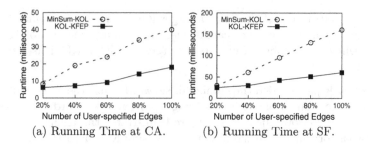

(a) Running Time at CA.   (b) Running Time at SF.

Fig. 6. Running time vary the percentage τ.

In summary, algorithm KOL-KFEP is more efficient than MinSum-KOL, because the keyword filtering and edge pruning strategies can reduce the search space. In CA and SF, the ratios $|O|/|E|$ are about 4 and 1, respectively. Obviously, the average number of objects in each edge is 4 in CA and 1 in SF. The pattern of query performance is similar in dataset CA and SF, since performance of MinSum-KOL is related to the size of dataset and performance of KOL-KFEP has benefited from the improved strategies. From the result of experiments, we concluded that edge pruning strategy is more efficient in dataset CA, while the keyword filtering strategy is more efficient in dataset SF.

7   Conclusions

In this paper, we have studied a novel type of query, keyword-aware optimal location query in road network. We have proposed a basic algorithm based on dividing road segments into sub-intervals to find the optimal locations and an improved algorithm with keyword filtering and edge pruning strategies. We demonstrate the efficiency of our algorithms with extensive experiments on large-scale real datasets. In future work, we will study the monitoring of optimal locations when objects and keywords have updates.

Acknowledgments. The work is partially supported by the NSF of China (Nos. 61272178, 61572122), the NSF of China for Outstanding Young Scholars (No. 61322208), and the NSF of China for Key Program (No. 61532021).

References

1. Xiao, X., Yao, B., Li, F.: Optimal location queries in road network databases. In: ICDE, pp. 804–815. IEEE Press, Hannover (2011)
2. Zhang, D., Du, Y., Xia, T., Tao, Y.: Progressive computation of the min-dist optimal location query. In: VLDB, pp. 643–654. VLDB Endowment Inc., Seoul (2006)
3. Cabello, S., Díaz-Báñez, J.M., Langerman, S., Seara, C., Ventura, I.: Reverse facility location problems. In: CCCG, pp. 68–71. Windsor (2005)

4. Chen, Z., Liu, Y., Wong, R., Xiong, J., Mai, G., Long, C.: Efficient algorithms for optimal location queries in road networks. In: Sigmod, pp. 123–134. ACM Press, Snowbird (2014)
5. Cardinal, J., Langerman, S.: Min-max-min geometric facility location problems. In: EWCG, Delphi, pp. 149–152 (2006)
6. Choi, D.W., Chung, C.W., Tao, Y.: A scalable algorithm for maximizing range sum in spatial databases. In: VLDB, pp. 1088–1099. VLDB Endowment Inc., Istanbul (2012)
7. Yan, D., Wong, R.C.W., Ng, W.: Efficient methods for finding influential locations with adaptive grids. In: CIKM, pp. 1475–1484. ACM Press, Glasgow (2011)
8. Qi, J., Zhang, R., Kulik, L., Lin, D., Xue, Y.: The min-dist location selection query. In: ICDE, pp. 366–377. IEEE Press, Washington (2012)
9. Papadias, D., Shen, Q., Tao, Y.: Group nearest neighbor queries. In: ICDE, pp. 301–312. IEEE Press (2004)
10. Cao, X., Cong, G., Jensen, C.: Collective spatial keyword querying. In: Sigmod, pp. 373–384. ACM Press, Athens (2011)
11. Long, C., Wong, R., Wang, K.: Collective spatial keyword queries: a distance owner-driven approach. In: Sigmod. ACM Press, New York (2013)
12. Guo, T., Cao, X., Cong, G.: Efficient algorithms for answering the m-closest keywords query. In: Sigmod. ACM Press, Melbourne (2015)
13. Yiu, M., Mamoulis, N., Papadias, D.: Aggregate nearest neighbor queries in road networks. TKDE 17(6), 820–833 (2005). IEEE Press
14. Erwig, M.: The graph voronoi diagram with applications. Networks 36(3), 156–163 (2000)
15. Dijkstra, E.W.: A note on two problems in connexion with graphs. Numer. Math. 1(1), 269–271 (1959)

Point-of-Interest Recommendations by Unifying Multiple Correlations

Ce Cheng[1]([✉]), Jiajin Huang[1], and Ning Zhong[1,2]

[1] International WIC Institute, Beijing University of Technology,
Beijing 100024, China
chengcesmile@163.com, hjj@emails.bjut.edu.cn
[2] Deptartment of Life Science and Informatics, Maebashi Institute of Technology,
Maebashi 371-0816, Japan
zhong@maebashi-it.ac.jp

Abstract. In recent years, we have witnessed the development of location-based services which benefit users and businesses. This paper aims to provide a unified framework for location-aware recommender systems with the consideration of social influence, categorical influence and geographical influence for users' preference. In the framework, we model the three types of information as functions following a power-law distribution, respectively. And then we unify different information in a framework and learn the exact function by using gradient descent methods. The experimental results on real-world data sets show that our recommendations are more effective than baseline methods.

Keywords: Location-aware recommender systems · Power-law distribution · Gradient descent methods

1 Introduction

With the development of smart mobile devices, location-based social network services (LBSNs for short), such as Gowalla and Foursquare, have attracted more and more attentions of users. In LBSNs, we call locations points-of-interest (POIs for short). An LBSN allows users to establish social links, check in POIs (e.g., restaurants), and rate their visiting POIs. These historical data bridge the gap between users' preferences and POIs, which provides us a chance to recommend preferred POIs for users. How to make high quality recommendations in LBSNs is becoming a very valuable problem.

Most existing POI recommendation methods (e.g., memory based and model based collaborative filtering) employ historical check-in data of users to compute a score between a user and a POI, and recommend unvisited POIs with higher scores for the user. As data used in these methods only show whether a user checked in a POI, recommendation quality will be low for users who have only visited a very small number of POIs.

In order to improve location recommendation quality to users, social influence, geographical influence, and categorical information have been studied in

© Springer International Publishing Switzerland 2016
B. Cui et al. (Eds.): WAIM 2016, Part I, LNCS 9658, pp. 178–190, 2016.
DOI: 10.1007/978-3-319-39937-9_14

POI recommendations. For social influence, friends may have more similar behaviors since they might have lots of common tastes and interests [12]. For geographical influence, geographically closer POIs may have similar characteristics than the POIs that are far from each other [9]. Users may prefer POIs closer to a POI that they liked [12]. For categorical information, they reflect users' preferences and the nature characteristics of POIs [1, 14].

In real-world data sets, Zhang et al. [14] modeled the social check-in frequency or rating and the categorical popularity as a power-law distribution to make recommendations. Ye et al. [12] modeled the willingness of a user moving from one place to another as a function of their distance following a power-law distribution. Though their works used social influence, geographical influence, and categorical information, they didn't model these three types of information as a power-law distribution in a unified framework. In this paper, we model the three types of information as a function following a power-law distribution, respectively. And then we unify them in a framework and learn the unified function by using gradient descent methods. At last, the learned function is used to compute a score between a user and a POI.

The rest of the paper is organized as follows. Section 2 gives an overview of some related work. Section 3 presents models of our system. Section 4 provides extensive experiments and results based on the real data set. Finally, Sect. 5 draws the conclusions and gives a further view.

2   Related Work

Many recent studies have tried to improve the POI recommendation quality by exploiting the social, categorical and geographical influence.

Social Information for POI Recommendations. To improve the quality of location-based recommender systems, social links between users have been widely used. Many studies employed the similarities between users extracted from social links into the traditional collaborative filtering methods [11, 12]. Also many recent studies fused the social information with the matrix factorization model. Jamali et al. [5] proposed the SocialMF model and Ma et al. [10] presented SoRec model, both of which incorporate social relationships as weights of latent factor, and Cheng et al. [3] included the social information into the PMF model as regularization term. Zhang et al. [14] utilized the social correlations between users to deduce the relevance score of users on the unvisited POIs through aggregating the visit frequency of their friends.

Categorical Information for POI Recommendations. The category of a POI implies the property of the POI and what kind of activities attract the user through their history check-in. Thus with the help of the category information of POIs, we can extract the specific preference of a user. Ying et al. used the category tag of the POI to consider the Preference-triggered Intentions of a user on a POI [13]. Bao et al. proposed a weighted category hierarchy to model users' preferences and infers the expertise of each user through the category information of locations. Zhang et al. combined the category bias of a user and the popularity of a POI to model the category correlations for POI recommendations [14].

Geographical Information for POI Recommendations. As the check-in behaviors of users on POIs would be affected by geographical proximity significantly, the geographical information can be exploited for POI recommendations. Bao et al. [1] and Wang et al. [11] payed attention to the influence of POIs that are close to the current location of the user. Lian et al. [8] applied the geographical latent factor model to derive latent features of POIs. Li et al. [7] proposed a ranking based geographical factorization method. Cheng et al. [3] proposed a Multi-center Gaussian Model to capture the geographical influence. Ye et al. [12] used a power law distribution to model the check-in probability to the distance between two POIs visited by the same user. Zhang et al. [14] developed an adaptive kernel estimation method to model the geographical correlations.

Each mainly focuses on one or two type of information following power distributions in location-based recommender systems. As the power distribution is a widely used distribution, we have been motivated to in-depth break limitations of related works by combining social, categorical and geographical influence in a unified framework.

3   Models

Let U be the user set with m users, P be the POI set with n POIs and L be the category set with k categories. We identify some necessary matrixes just as follows:

User-POI matrix: $T \in \mathbb{R}^{m \times n}$. For each entry $t_{i,j}$ in T represents the visit times of u_i in U to p_j in P obtained from an LBSN which contains users' historical check-in data, and m and n are the user number and POI number in the LBSN, respectively. Note that users always visit a very small part of POIs, most entries in T are zero.

User-User matrix: $S \in \mathbb{R}^{m \times m}$. For each entry $s_{i,i'}$ in S indicates the social relation between u_i and $u_{i'}$ in U. $s_{i,i'} = 1$ denotes that there exists relationship between two users, and $s_{i,i'} = 0$ means none relation.

User-Category matrix: $C \in \mathbb{R}^{m \times k}$. For each entry $c_{i,r}$ in C represents the check-in times of u_i to POIs that belong to l_r in L. Each POI may belong to multiple categories and k is the total number of category.

Category-POI matrix: $D \in \mathbb{R}^{k \times n}$. For each entry $d_{r,j}$ in D represents the correlation between l_r in L and p_j in P, if p_j belongs to l_r in L, $d_{r,j} = 1$; otherwise $d_{r,j} = 0$. Note that most entries of D are zero, because a POI is only attached to a few categories.

POI-POI matrix: $O \in \mathbb{R}^{n \times n}$. For each entry $o_{j,j'}$ in O indicates the correlation between p_j and $p_{j'}$ in P. Each POI is connected with a pair of geographical latitude and longitude coordinates. We use the reciprocal of the distance between p_j and $p_{j'}$ to present $o_{j,j'}$. Thus if two POIs are closer, the more they are relevant.

3.1   Social Correlations

In LBSNs, friends have social links with each other, and they may have similar interests. Users are easy to be affected by their friends' check-in behavior when they choose a POI. Here, we can deduce the preference of u_i to an unvisited p_j by considering the social correlations between u_i and her friends who have visited the POI. Specially, the social check-in times $x_{i,j}$ can be computed as follows [14].

$$x_{i,j} = \sum_{u'_i \in U} s_{i,i'} \cdot t_{i',j} \qquad (1)$$

Figure 1(a) describes the distribution of the social check-in times through analyzing on a real-world data set Foursquare. From Fig. 1(a), we can see that the social check-in times follows a power law distribution. Note that since the values of x we evaluate are no more than 30, the dots in Fig. 1(a) is sparse. So we can assume the probability density function of the social check-in times random variable x by a power law distribution as follows [14].

$$f_{So}(x_{i,j}) = a \cdot x_{i,j}^{-\alpha}, a > 0, x_{i,j} \geq 0, \alpha > 0 \qquad (2)$$

The function f_{So} in Eq. (2) decreases monotonically as the social check-in times x increases. However, friends share more similar interests with each other and thus the social relevance score should be increasing as the social check-in times increases. So we use the reciprocal of f_{So} to describe the social relevance score of $x_{i,j}$.

$$F_{So}(x_{i,j}) = a \cdot x_{i,j}^{\alpha}, a > 0, x_{i,j} \geq 0, \alpha > 0 \qquad (3)$$

where F_{So} is the social relevance score whose curve passes through the point $(0, 0)$. And Eq. (3) describes the positive correlation between the social check-in times random variable x and the social relevance score F_{So} in a reasonable way.

3.2   Category Correlations

The category of a POI implies the property of the POI and thus can reflect interests of users who have checked the POI. For instance, a user visiting a

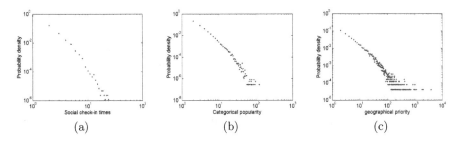

(a)   (b)   (c)

Fig. 1. Distribution on Foursquare through social, category, geographical aspects

Chinese restaurant indicates that he likes Chinese food. So it is more reasonable to recommend a Chinese restaurant than a pizza parlor to him. Actually, a user usually checks a certain subset of the categories with a stable tastes. Here, we utilize category correlations (i.e., user-category correlation and category-POI correlation) to predict the categorical popularity $y_{i,j}$ of u_i to an unvisited p_j. Specially, $y_{i,j}$ can be computed as follows [14].

$$y_{i,j} = \sum_{l_r \in L} c_{i,r} \cdot d_{r,j} \tag{4}$$

We also conduct analysis on Foursquare to observe the real distribution of categorical popularity y. From Fig. 1(b), we can see that y follows a power law distribution very well. So we can introduce a monotonically decreasing power law distribution function to depict the probability density of y, the details can be seen as follows.

$$f_{C_a}(y_{i,j}) = b \cdot y_{i,j}^{-\beta}, b > 0, y_{i,j} \geq 0, \beta > 0 \tag{5}$$

But the categorical relevance score is increasing monotonely with the categorical popularity because people always be interested in some certain categories and prefer POIs belong to those categories. Thus we represent positive correlations between the categorical popularity and categorical relevance score by the reciprocal of probability density of y. Specially, it can be seen as follows.

$$F_{C_a}(y_{i,j}) = b \cdot y_{i,j}^{\beta}, b > 0, y_{i,j} \geq 0, \beta > 0 \tag{6}$$

3.3   Geographical Correlations

In LBSNs, geographical information of POIs have a great influence on users' check-in behaviors. Users usually prefer to travel a limited distance and Ye et al. [12] proposed that the POIs visited by the same user tend to be clustered geographically. Thus, we argue that the geographical priority of u_i on an unvisited p_j have a close relationship with the distance between p_j and POIs that u_i has checked.

$$z_{i,j} = \sum_{l'_j \in P} o_{j',j} t_{i,j'} \tag{7}$$

where $o_{j',j}$ means the correlations between $p_{j'}$ and p_j which is inversely associated with the distance between them. $t_{i,j'}$ represents the preference weight of u_i on p_j. Figure 1(c) describes the real density probability distribution of geographical priority z based on our analysis on Foursquare. We can observe that geographical priority follows a power law distribution very well. By the same way, the probability density function can be described by Eq. (8).

$$f_{Geo}(z_{i,j}) = c \cdot z_{i,j}^{-\gamma}, c > 0, z_{i,j} \geq 0, \gamma > 0 \tag{8}$$

Note that geographical relevance scores increase monotonically with the geographical priority since users prefer to visit nearby interested locations. Thus we apply the similar process to represent the positive correlations by the corresponding reciprocal form, i.e., Eq. (9).

$$F_{Geo}(z_{i,j}) = c \cdot z_{i,j}^{\gamma}, c > 0, \gamma > 0 \tag{9}$$

3.4   POI Recommendations

We fuse the social, categorical and geographical relevance score, i.e., Eqs. (3), (6) and (9), into a unified preference score $g'(i, j)$ based on the product rule:

$$g'_{i,j} = F_{S_o}(x_{i,j}) \cdot F_{C_a}(y_{i,j}) \cdot F_{Geo}(z_{i,j}) \tag{10}$$

Equation (10) can be transformed into the logarithm form to fit a linear model just as follows.

$$\ln g'_{i,j} = \ln abc + \alpha \cdot \ln x_{i,j} + \beta \cdot \ln y_{i,j} + \gamma \cdot \ln z_{i,j} \tag{11}$$

More specifically, let $g_{i,j} = \ln g'_{i,j}$, $q = \ln abc$, $w_1 = \alpha$, $w_2 = \beta$, $w_3 = \gamma$ and $h_{i,j}^1 = \ln x_{i,j}$, $h_{i,j}^2 = \ln y_{i,j}$, $h_{i,j}^3 = \ln z_{i,j}$. We obtain the corresponding equation as follows:

$$g_{i,j} = q + \sum_{d=1}^{3} w_d \cdot h_{i,j}^d \tag{12}$$

where w_d is the weight coefficient. We approach w_d by gradient descent method and the objective function is just as follows:

$$E = \frac{1}{2} \sum_{i,j} (g_{i,j} - t_{i,j})^2 + \frac{\lambda}{2} \sum_{d=1}^{3} \|w_d\|^2 \tag{13}$$

where the first term of the objective function represents the prediction error value on T, and the second term is the regularization term which can avoid overfitting.

Linear Iteration. We use the gradient descent to get the weight variable w_d. The gradient descent algorithm has been widely used for many machine learning tasks [2,4,5]. The main process involves scanning the training set and iteratively updating parameters. Specially, the gradients of the objective function E with respect to w_d are computed as Eq. (14). In each iteration the variable w_d is updated in the opposite direction of the gradient and the details can be seen in Eqs. (15) and (16).

$$\frac{\partial E}{\partial w_d} = \sum_{i,j} (g_{i,j} - t_{i,j}) h_{i,j}^d + \lambda w_d \tag{14}$$

$$\Delta w_d = -\eta \frac{\partial E}{\partial w_d} \tag{15}$$

$$w_d \leftarrow w_d + \Delta w_d \tag{16}$$

After obtaining the converged w_d, we can get the missing values in User-POI matrix T.

Nonlinear Iteration. Here we use $EG^{\pm}$ algorithm to get the weight variable w_d. $EG^{\pm}$ is short for the exponentiated gradient algorithm which have positive and negative weights. $EG^{\pm}$ is a widely used nonlinear gradient descent method. It uses the components of the gradient in the exponents of factors that are used in updating the weight vector multiplicatively [6]. Kivinen et al. presented the specific and detailed statement of $EG^{\pm}$ [6]. We first initialize the weight vectors $w^+ = (1/6, 1/6, 1/6)$, $w^- = (1/6, 1/6, 1/6)$ and $w = (1/3, 1/3, 1/3)$. Then the weight vectors can be updated according to the following rules.

$$w_d^+ \leftarrow \frac{w_d^+ \cdot e^{\Delta w_d}}{\sum_{d=1}^{3}(w_d^+ e^{\Delta w_d} + w_d^- e^{-\Delta w_d})} \tag{17}$$

$$w_d^- \leftarrow \frac{w_d^- \cdot e^{-\Delta w_d}}{\sum_{d=1}^{3}(w_d^+ e^{\Delta w_d} + w_d^- e^{-\Delta w_d})} \tag{18}$$

$$w_d = w_d^+ - w_d^- \tag{19}$$

where Δw_d can be got through Eqs. (14) and (15).

4 Experiments and Results

We first describe the data set and evaluation metrics in Sect. 4.1. Then, we present the results of our experiments containing different method comparison, studying on three components, and analyzing on the influence of parameters of our experiments in Sect. 4.2.

4.1 Data Set and Evaluation Metrics

Data Set. Foursquare [1] is a real-world data set consisting of check-in records on POIs obtained from Foursquare user histories. Users can leave a tip to share experience with their friends by their mobile phones when they check a POI, each tip contains the category information and ID number of the POI, meanwhile the latitude and longitude information of the POI can also be recorded by their mobile phones. In this data set, we select locations which have been visited more than five times, users whose check-in number more than twelve times, the final data set includes of 493 users, 3292 locations and 11399 check-in records.

Evaluation Metrics. We evaluate our models in terms of ranking by presenting each user with a top-K list of POIs sorted by the predicted values. More specifically, the predicted rating of u_i on p_j is obtained from Eq. (12), in which we use the finally updated vector w.

The evaluation metrics are precision and recall. In detail, they can be computed as follows:

$$Precision = \frac{1}{m} \sum_{u_i \in U} \frac{|R(u_i) \bigcap T(u_i)|}{|R(u_i)|} \tag{20}$$

$$Recall = \frac{1}{m} \sum_{u_i \in U} \frac{|R(u_i) \bigcap T(u_i)|}{|T(u_i)|} \tag{21}$$

where $R(u_i)$ is a top-K list of a targeted user based on descending order of the prediction values, $T(u_i)$ is the set of POIs a user has visited in the test data. Note that the precision and recall for the whole recommender system are calculated by averaging all the precision and recall values of all the users, respectively.

4.2   Results

First, we compare our SCG with the state-of-the-art POI recommendation technique. Second, we study the recommendation quality of the three components in SCG. Last, we investigate the impact of sensitive parameters.

Comparison. Our methods integrate social information, categorical information and geographical information into a framework by the product rule to make POI recommendations. The weights corresponding to each information are obtained dynamically through machine learning, and specifically, we propose linear iteration and nonlinear iteration. Here we call our models as L-SCG and NL-SCG with respect to linear iteration and nonlinear iteration, respectively. Meanwhile, we introduce two state-of-the-art POI recommendation techniques as follows:

- OSCG [12,14]: This method also incorporates social correlations, categorical correlations and geographical correlations, which we call original SCG method (OSCG for short). OSCG model the social factor and categorical factor by a power-law distribution, respectively [14], while instead of computing adaptive kernel estimation in [14], the geographical factor is modeled by a power-law distribution [12]. Finally the three factors are fused by linear combination to make POI recommendation.
- USG [12]: This method involves user preference, social influence and geographical influence, but ignores category impact. It fuses these three factors into a framework by linear combination. What's more the weight vector is set in a static way, i.e., the weighting parameters are manually tuned.

Our experiment is performed on Foursquare. We set $\eta = 2 \times 10^{-5}$ for L-SCG and $\eta = 2 \times 10^{-4}$ for NL-SCG, $\lambda = 0.5$, and iteration times as 50 for NL-SCG and 70 for L-SCG, the details can be discussed in the following sections. To demonstrate the effectiveness of our models, we compare our methods with the state-of-the-art OSCG and USG with different amounts of data used for training. In addition, the Foursquare data is randomly divided by different proportions of

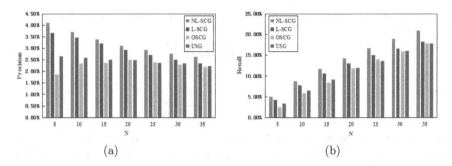

(a)   (b)

Fig. 2. Recommendation accuracy of different methods based on 4/5 data for training
(Color figure online)

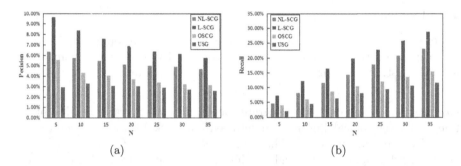

(a)   (b)

Fig. 3. Recommendation accuracy of different methods based on 2/3 data for training
(Color figure online)

training data and testing data as 2:1 and 4:1, respectively. Figures 2 and 3 show
the corresponding results.

From Figs. 2 and 3, we can see that our methods both NL-SCG and
L-SCG (SCG for short) outperform OSCG and USG, which indicates that (1)
it is an more accurate way to reach the weights dynamically through machine
learning instead of static manual adjustment; (2) it is more reasonable than linear combination to fuse the influence of different factors in a framework through
the product role. It must be noted that NL-SCG improves L-SCG when the proportion of train set and test set is 4:1, while L-SCG has a better performance
when the proportion is 2:1, this indicates that gradient descent is more sensitive
to the size of training set.

Study on Three Components in SCG. In this section, we study the three
components in our models, namely, SocCo corresponding to social correlations,
CateCo corresponding to category correlations and GeoCo corresponding to geographical correlations. The basic idea behind SocCo is that friends have similar
tastes in LBSNs and we only consider social influence to make POI recommendations. CateCo only contains the category correlations considering people are

always like some certain kinds of items. GeoCo only incorporates geographical correlations with regard to most people prefer to visit nearby POIs. The details corresponding to the three components can be seen as Eqs. (3), (6) and (9). We use linear gradient descent to get their exponential parameters.

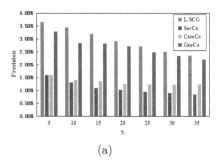 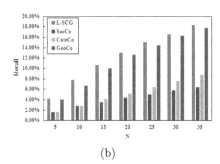

(a)   (b)

Fig. 4. Recommendation accuracy of the three components of SCG (Color figure online)

Figure 4 shows the recommendation accuracy of L-SCG along with its three component factors, namely, social correlations, categorical correlations and geographical correlations. From Fig. 4, we can conclude that: (1) each of the three correlations plays an important role in SCG for POI recommendations; (2) the performance of GeoCo is much higher than SocCo and CateCo which indicates geographical correlations provide comparable results against both social and categorical correlations, since in LBSNs geographical influence matters a lot; (3) it is obvious to see that no matter in precision or recall, our method L-SCG always show the best performance. This indicates that the more correlations are unified, the more helpful it is to enhance the recommendation quality. In other words, people can be influenced by the social, categorical and geographical correlations in different degrees. It is reasonable to unify these correlations together to obtain a comprehensive consideration.

Study on Parameter in SCG. In this section, we study the impact of parameter on SCG based on 2/3 data for training. The learning rate η controls the step size for searching weight vector w in each iteration, the regularization parameter λ can avoid over-fitting and iteration times determine the best state of our experiments during the iterations.

Figure 5 presents the effect of different learning rate η on SCG. From Fig. 5(a), we have the following observations: (1) the optimal learning rate in L-SCG is $\eta = 2 \times 10^{-5}$ which go towards the right direction, since with the increasing of iterations, the performance is better; (2) a bigger η can results in the divergency of the model, e.g., the model is no longer convergent after the second iteration when $\eta = 2 \times 10^{-4}$; (3) a smaller η can lead the model convergence rate too slow.

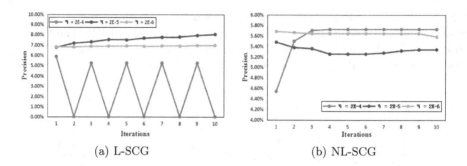

(a) L-SCG   (b) NL-SCG

Fig. 5. Effect of different learning rate η

When η is 2×10^{-6}, it is hard to reach a good performance. From Fig. 5(b), we can easily see that the optimal learning rate in NL-SCG is $\eta = 2 \times 10^{-4}$ which can lead to the best state of the performance.

We analyze the effect of regularization parameter λ on the final result. We set λ as 0.05, 0.5 and 5, respectively, but the corresponding results are almost the same and thus we argue that regularization parameter have little influence on the our model.

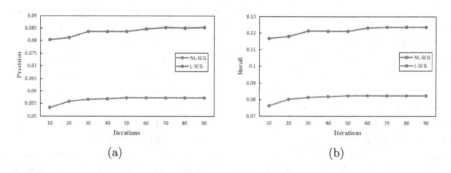

(a)   (b)

Fig. 6. Effects of iteration times

Figure 6 shows the effects of iteration times on both NL-SCG and L-SCG method based on 2/3 data for training, we can see that NL-SCG can get a stable state in a faster speed than L-SCG. Actually NL-SCG can reach its best state after 50 iteration times while L-SCG iterate 70 times.

5   Conclusion and Future Work

This paper presents a new POI recommendation approach unifying multi correlations, namely, social correlations, category correlations and geographical correlations. The experimental results on the Foursquare data set show that our methods significantly improve the recommendation accuracy of the baseline methods.

However, since our data sets have a high sparsity rate (99.29 %), the whole precision and recall of our experiments are low. In the future, on one hand, we are going to focus on data sparsity problems to make recommendation more accurate; on the other hand, we will consider to incorporate time and sentiment features extracted from semantic tips in the prediction function.

References

1. Bao, J., Zheng, Y., Mokbel, M.F.: Location-based and preference-aware recommendation using sparse geo-social networking data. In: Proceedings of the 2012 SIGSPATIAL International Conference on Advances in Geographic Information Systems (GIS 2012), pp. 199–208 (2012)
2. Bottou, L.: Large-scale machine learning with stochastic gradient descent. In: Proceedings of the 2010 International Conference on Computational Statistics (COMPSTAT 2010), pp. 177–187 (2010)
3. Cheng, C., Yang, H.Q., King, I., Lyu, M.R.: Fused matrix factorization with geographical and social influence in location-based social networks. In: Proceedings of the 2012 AAAI Conference on Artificial Intelligence (AAAI 2012), pp. 17–23 (2012)
4. Cheung, K.W., Tian, L.F.: Learning user similarity and rating style for collarborative recommendation. Inf. Retrieval 7(3–4), 395–410 (2004)
5. Jamali, M., Ester, M.: A transitivity aware matrix factorization model for recommendation in social networks. In: Proceedings of the 2011 International Joint Conference on Aritificial Intelligence (IJCAI 2011), pp. 2644–2649 (2011)
6. Kivinen, J., Warmuth, M.K.: Exponentiated gradient versus gradent descent for linear predictors. Inf. Comput. 132(l), 1–64 (1997)
7. Li, X.T., Cong, G., Li, X.L., Pham, T.N., Krishnaswamy, S.: Rank-GeoFM: a ranking based geographical factorization method for point of interest recommendation. In: Proceedings of the 2015 ACM International Conference on Research and Development in Information Retrieval (SIGIR 2015), pp. 433–442 (2015)
8. Lian, D., Zhao, C., Xie, X., Sun, G., Chen, E., Rui, Y.: GeoMF: joint geographical modeling and matrix factorization for point-of-interest recommendation. In: Proceedings of the 2014 ACM SIGKDD International Conference on Knowledge Discovery and Data Mining (SIGKDD 2014), pp. 831–840 (2014)
9. Liu, B., Xiong, H., Papadimitriou, S., Fu, Y., Yao, Z.: A general geographical probabilistic factor model for point of interest recommendation. IEEE Trans. Knowl. Data Eng. 27(5), 1167–1179 (2015)
10. Ma, H., Yang, H.X., Lyu, M.R., King, I.R.: SoRec: social recommendation using probabilistic matrix factorization. In: Proceedings of the 2008 ACM Conference on Information and Knowledge Management (CIKM 2008), pp. 931–940 (2008)
11. Wang, H., Terrovitis, M., Mamoulis, N.: Location recommendation in location-based social networks using user check-in data. In: Proceedings of the 2013 ACM SIGSPATIAL International Conference on Advances in Geographic Information Systems (SIGSPATIAL 2013), pp. 374–383 (2013)
12. Ye, M., Yin, P., Lee, W.C., Lee, D.L.: Exploiting geographical influence for collaborative point-of-interest recommendations. In: Proceedings of the 2011 ACM International Conference on Research and Development in Information Retrieval (SIGIR 2011), pp. 325–334 (2011)

13. Ying, J.J.C., Lee, W.C., Tseng, V.S.: Mining geographic-temporal-semantic patterns in trajectories for location prediction. ACM Trans. Intell. Syst. Technol. 5(1), 1–33 (2014)
14. Zhang, J.D., Chow, C.Y.: GeoSoCa: exploiting geographical, social and categorical correlations for point-of-interest recommendations. In: Proceedings of the 2015 ACM SIGIR International Conference on Research and Development in Information Retrieval (SIGIR 2015), pp. 443–452 (2015)

Top-k Team Recommendation in Spatial Crowdsourcing

Dawei Gao[1], Yongxin Tong[1(✉)], Jieying She[2],
Tianshu Song[1], Lei Chen[2], and Ke Xu[1]

[1] SKLSDE Lab, IRC, Beihang University, Beijing, China
{david_gao,yxtong,songts,kexu}@buaa.edu.cn
[2] The Hong Kong University of Science and Technology, Hong Kong SAR, China
{jshe,leichen}@cse.ust.hk

Abstract. With the rapid development of Mobile Internet and Online
To Offline (O2O) marketing model, various spatial crowdsourcing
platforms, such as Gigwalk and Gmission, are getting popular. Most
existing studies assume that spatial crowdsourced tasks are simple
and trivial. However, many real crowdsourced tasks are complex and
need to be collaboratively finished by a team of crowd workers with
different skills. Therefore, an important issue of spatial crowdsourcing platforms is to recommend some suitable teams of crowd workers to satisfy the requirements of skills in a task. In this paper, to
address the issue, we first propose a more practical problem, called
Top-k Team Recommendation in spatial crowdsourcing (TopkTR)
problem. We prove that the TopkTR problem is NP-hard and design a
two-level-based framework, which includes an approximation algorithm
with provable approximation ratio and an exact algorithm with pruning
techniques to address it. Finally, we verify the effectiveness and efficiency
of the proposed methods through extensive experiments on real and synthetic datasets.

1 Introduction

Recently, thanks to the development and wide use of smartphones and mobile
Internet, the studies of crowdsourcing are switching from traditional crowdsourcing problems [15,16] to the issues in spatial crowdsourcing markets, such as
Gigwalk, Waze, Gmission, etc., where crowd workers (workers for short in this
paper) are paid to perform spatial crowsourced tasks (tasks for short in this
paper) that are requested on a mobile crowdsourcing platform [17].

Most existing studies on spatial crowdsourcing mainly focus on the problems
of task assignment [6,7,13,14,17], which are to assign tasks to suitable workers,
and assume that tasks are all simple and trivial. However, in real applications,
there are many complex spatial crowdsourced tasks, which often need to be collaboratively completed by a team of crowd workers with different skills. Imagine
the following scenario. David is a social enthusiast and usually organizes different types of parties on weekends. On the coming Saturday, he intends to hold

B. Cui et al. (Eds.): WAIM 2016, Part I, LNCS 9658, pp. 191–204, 2016.
DOI: 10.1007/978-3-319-39937-9_15

Table 1. The skill, payoff and capacity information of crowd workers

	w_1	w_2	w_3	w_4	w_5
Skills	$\{e_1, e_2\}$	$\{e_1\}$	$\{e_2, e_3\}$	$\{e_2\}$	$\{e_1, e_2, e_3\}$
Price	2	1	3	1	2
Capacity	1	1	2	1	1

a dance party and needs to recruit some sound engineers, guitarists, cooks and dancers. However, David faces a dilemma that his limited budget cannot afford to recruit all the aforementioned workers. He has to recruit fewer cheap crowd workers who have multiple skills and can take up several responsibilities, e.g. a worker can play the guitar and also manage the sound systems. Therefore, David posts his tasks on a spatial crowdsourcing platform, Gigwalk, and wants to find cheap crowd workers to satisfy his requirements. In fact, many task requestors have the same appeal: *can spatial crowdsourcing platforms recommend several cheaper candidate teams of crowd workers who can satisfy the multiple skills requirement of the tasks?* To further illustrate this motivation, we go through a toy example as follows.

Example 1. Suppose we have five crowd workers $w_1 - w_5$ on a spatial crowd-sourcing platform, whose locations are shown in a 2D space (X, Y) in Fig. 1. Each worker owns different skills, which are shown in the second row in Table 1. Furthermore, each worker has a price for each task and a capacity, which is the maximum number of skills that can be used in a task that he/she performs, which are presented in the third and forth rows in Table 1. Moreover, a team-oriented spatial crowdsourced task and its locality range (the dotted circle) are shown in Fig. 1. Particularly, the task requires that the recruited crowd workers must cover three skills, $\{e_1, e_2, e_3\}$. To help the task requestor save cost, the spatial crowdsourcing platform usually recommends top-k cheapest teams of

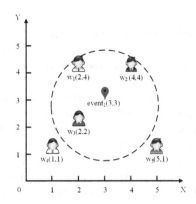

Fig. 1. Locations of the task and the five crowd workers

crowd workers, who can satisfy the requirement of skills. Furthermore, the recommended teams should not have free riders. In other words, each recommended team cannot satisfy the required skills if any worker in the team leaves. Therefore, in this example, the top-2 cheapest teams without free riders are $\{w_2, w_3\}$ and $\{w_1, w_3\}$, respectively, if the parameter $k = 2$.

As discussed above, we propose a novel team recommendation problem in spatial crowdsourcing, called the *top-k team recommendation in spatial crowdsourcing* (Top$k$TR) problem. As the example above indicates, the Top$k$TR problem not only recommends k cheapest teams but also satisfies the constraints of spatial range and skill requirement of tasks, capacity of workers, and no free rider in teams. Notice that the Top-1TR problem can be reduced to the classical team formation problem if the constraints on the capacity of workers and free riders are removed. More importantly, the Top$k$TR problem needs to return k teams instead of the cheapest team, which is its main challenge. We make the following contributions.

- We identify a new type of team-oriented spatial crowdsourcing applications and formally define it as the top-k team recommendation in spatial crowdsourcing (Top$k$TR) problem.
- We prove that the Top$k$TR problem is NP-hard and design a two-level-based framework, which not only includes an exact algorithm to provide the exact solution but also can seamlessly integrate an approximation algorithm to guarantee $\ln |E_t|$ theoretical approximation ratio, where $|E_t|$ is the number of required skills of the task.
- We verify the effectiveness and efficiency of the proposed methods through extensive experiments on real and synthetic datasets.

The rest of the paper is organized as follows. In Sect. 2, we formally define our problem and prove its NP-hardness. In Sect. 3, we present an two-level-based framework and its exact and approximation solutions. Extensive experiments on both synthetic and real datasets are presented in Sect. 4. We review related works and conclude this paper in Sects. 5 and 6, respectively.

2   Problem Statement

We formally define the *Top-k Team Recommendation in spatial crowdsourcing* (Top$k$TR) problem and prove that this problem is NP-hard. For convenience of discussion, we assume $E = <e_1, \cdots, e_m>$ to be a universe of m skills.

Definition 1 (Team-oriented Spatial Crowdsourced Task). *A team-oriented spatial crowdsourced task ("task" for short), denoted by $t = <l_t, E_t, r_t>$, at location l_t in a 2D space is posted to the crowd workers, who are located in the circular range with the radius r_t around l_t, on the platform. Furthermore, $E_t \subseteq E$ is the set of the required skills of the task t for the recruited team of crowd workers.*

Definition 2 (Crowd Worker). *A crowd worker ("worker" for short) is denoted by* $w = <l_w, E_w, p_w, c_w>$, *where* l_w *is the location of the worker in a 2D space,* $E_w \subseteq E$ *is the set of skills that the worker is good at,* p_w *is the pay-off for the worker to complete a crowdsourced task, and* c_w *is the capacity of the worker, namely the maximum number of skills used by the worker to complete a crowdsourced task.*

Note that the team-oriented spatial crowdsourced tasks studied in this paper, e.g. organizing a party, renovating a room, etc., usually need to be completed in teams. Though a worker may be good at multiple required skills, he/she cannot finish all the works by himself/herself. Therefore, we limit the capacity of each worker to balance the workload of the whole team. To simplify the problem, we assume that each worker receives the same payoff for different tasks since the capacity of the used skills of each user can be restricted. On one hand, these workers often have similar workloads and do not need a team leader to do a task. On the other hand, our model can be also easily extended to address the scenario where workers ask for different rewards for his/her different skills. Finally, we define our problem as follows.

Definition 3 (Top$k$TR Problem). *Given a team-oriented spatial crowd-sourced task* t, *a set of crowd workers* W, *and the number of recommended crowdsourced teams* k, *the Top$k$TR problem is to find* k *crowdsourced teams,* $\{g_1, \cdots, g_k\}$ ($\forall g_i \subseteq W, 1 \leq i \leq k$) *with* k *minimum* $Cost(g_i) = \sum_{w \in g_i} p_w$ *such that the following constraints are satisfied:*

- *Skill constraint: each required skill is covered by the skills of at least one worker.*
- *Range constraint: each worker* $w \in g_i$ *must locate in the restricted range of the task* t.
- *Capacity constraint: the number of skills used by each worker* $w \in g_i$ *cannot exceed* w's *capacity* c_w.
- *Free-rider constraint: no team still satisfies the skill constraint if any worker in the team leaves.*

Theorem 1. *The Top$k$TR Problem is NP-hard.*

Proof. When $k = 1$ and the capacity constraint is ignored, such special case of the Top$k$TR problem is equivalent to the team formation problem [8], which has been proven to be NP-hard. Therefore, the Top$k$TR problem is also an NP-hard problem.

3   A Two-Level-Based Framework

To solve the problem effectively, we present a two-level-based algorithm framework. The first level aims to find the current top-1 feasible team with the minimum price, and the second level utilizes the function in the first level to iteratively maintain the top-k best teams. Particularly, the two-level-based framework has a nice property that the whole algorithm can keep the same approximation guarantee of the algorithm as in the first level.

Algorithm 1. Two-Level-based Framework

---

 input : $W = \{w_1, \cdots, w_{|W|}\}, t, k$, and top-1 function top-1$(.,.)$
 output: Top-k teams $G = \{g_1, \cdots, g_k\}$.
1 $Queue \leftarrow \varnothing;\; G \leftarrow \varnothing;$
2 Insert the team generated by the function top-1(W,t) into $Queue$;
3 **while** $Queue \neq \varnothing$ **do**
4   $res \leftarrow$ top of $Queue$;
5   $G \leftarrow G \bigcup \{res\};$
6   **if** $|G| = k$ **then**
7    $\lfloor$ **return** G;
8   Remove top of $Queue$;
9   **foreach** $w \in res$ **do**
10    $\lfloor$ Insert the team generated by the function top-1$(W_{res} - \{w\}, t)$ into $Queue$;

3.1 Overview of the Framework

The main idea of the two-level framework is that the top-2 best team can be discovered if and only if the top-1 best team is found first. In other words, after excluding the top-1 best team from the solution space, not only the size of the solution space is shrunken, but also the global top-2 best team must be the local top-1 best team in the shrunken solution space. The function of finding the local top-1 best team is denoted as the top-1 function in the first level, which will be described in details as the approximation algorithm and the exact algorithm in Sects. 3.1 and 3.2, respectively.

The framework is shown in Algorithm 1. We first initialize an empty priority queue of teams $Queue$, which sorts the elements in non-increasing prices of the teams, and the top-k teams G in lines 1–2. In line 3, we use a given algorithm, which can be exact or approximate, to get the exact or approximate top-1 team and insert it into $Queue$. In lines 4–11, if $Queue$ is not empty, we get the top element res of $Queue$ and insert res into G. For each w in res, we reduce the solution space of res to $W_{res} - \{w\}$, find a solution in it, and insert the solution into $Queue$. We repeat this procedure until we get k teams.

As introduced above, the framework has a nice property that the whole algorithm can keep the same approximation guarantee of the algorithm (top-1 function) in the first level.

Theorem 2. *If the top-1 function top-1$(.,.)$ in the framework is an approximation algorithm with approximate ratio of r, the approximate cost of the i-th team in the approximation top-k teams by the framework keeps the same approximate ratio compared to the cost of the corresponding i-th exact team.*

Proof. We represent the approximation top-k teams generated by the framework as $\{g_1^a, \cdots, g_k^a\}$, and the exact top-k teams is denoted as $\{g_1^{ex}, \cdots, g_k^{ex}\}$. Because the top-1 function top-1$(.,.)$ has approximate ratio of r, $\text{Cost}(g_1^a) \leq r \times \text{Cost}(g_1^{ex})$. When the framework excludes g_1^a from the solution space and utilizes the top-1 function to obtain the other local top-1 team, it has the following two cases: (1) if $g_1^a = g_1^{ex}$, we have $g_2^a \leq r \times g_2^{ex}$; (2) $g_1^a \neq g_1^{ex}, g_2^a \leq r \times g_1^{ex}$.

Algorithm 2. Top-1 Greedy Approximation Algorithm

---

    input : $W = \{w_1, \cdots, w_{|W|}\}, t$
    output: Team g.
1 $g \leftarrow \varnothing$;
2 **while** *the team g cannot satisfy the requirement of E_t* **do**
3      $w \leftarrow argmax_{w \in W}(\frac{MAXITEM(g \bigcup \{w\}) - MAXITEM(g)}{P_w})$;
4      $g \leftarrow g \bigcup \{w\}$;
5 **return** $Refine(g)$

3.2 Top-1 Approximation Algorithm

The main idea of the top-1 approximation algorithm utilizes the greedy strategy to choose the best worker w, who can bring the maximum gain to the current partial team g. Algorithm 2 illustrates the top-1 approximation algorithm. We first initialize a empty team g in line 1. In lines 2–4, when g cannot satisfy the requirement of skills of the task t, denoted by E_t, the algorithm selects a worker w with the maximum gain and the least price for the current team. The function $MAXITEM(.)$ is used to calculate the number of skills in E_t that can be covered by a specific team. In line 5, since g may contain free-rider workers, we have to refine the team.

Example 2. Back to our running example in Example 1. The running process of the top-1 approximation algorithm is shown in Table 2. In the first round, we choose w_2 with the biggest benefit 1. Since $\{w_2\}$ can not handle the task, we proceed to choose w_3 with the biggest benefit of $\frac{2}{3}$. Now, we can handle the task with $\{w_2, w_3\}$ and the price is 4.

Approximation Ratio. The approximation ratio of the algorithm is $O(\ln |E_t|)$. Inspired by [10], it is easy to get the approximation ratio of Algorithm 2. Due to the limited space, the details of the approximation ratio proof are omitted in this paper.

Table 2. The running process of Top-1 approximation algorithm

Round	w_1	w_2	w_3
1	1/2	**1**	2/3
2	1/2		**2/3**

Complexity Analysis. The time consumed by $MAXITEM$ is $O(|E_t|^2 \log(|E_t|))$. Line 3 will be executed at most $|E_t|$ times. The Refine step takes $O(2^{|E_t|})$ time. Thus, the total time complexity is $O(|W||E_t|^3 \log(|E_t|) + 2^{|E_t|})$. Since $|E_t|$ is usually very small in real applications, the algorithm is still efficient.

Finally, the following example illustrates the whole process of the complete approximation algorithm based on the two-level-based framework.

Algorithm 3. Top-1 Exact Algorithm

---

    input : $W = \{w_1, \cdots, w_{|W|}\}, t$
    output: Team g.
1 $C_g \leftarrow$ Price of Top-1 Greedy Approximation Algorithm(W, t);
2 $state \leftarrow \varnothing$;
3   **foreach** $w \in W$ **do**
4   **if** $p_w \leqslant C_g$ **then**
5   **foreach** *cover condition of w as c* **do**
6   **foreach** $s \in state$ **do**
7   **if** $s.P + c.P \leqslant C_g$ **then**
8   Insert new cover condition of $s + c$ into *temp_state*;
9   Insert c into *temp_state*;
10   update *state* using *temp_state* and clear *temp_state*;

11 $T \leftarrow$ cover condition of skills E_t;
12 **return** T;

Example 3. Back to our running example in Example 1. Suppose $k = 2$ and the required skills of the task $t = \{1, 2, 3\}$. We first use the Top-1 greedy approximation algorithm to get a team of $\{w_1, w_3\}$ in the first level of the framework. Then we continue to adopt the Top-1 greedy approximation algorithm to find the local top-1 teams from $W - \{w_1\}$ and $W - \{w_3\}$. The returned teams are $\{w_1, w_3\}$ and $\varnothing$ respectively. Thus, the final top-2 teams generated by the whole framework are $\{w_2, w_3\}$ and $\{w_1, w_3\}$.

3.3 Top-1 Exact Algorithm

Since the number of skills required by a task is often not large, the main idea of the Top-1 exact algorithm is to enumerate the cover state of every proper subset of the intersection of the skills between a worker and a task. For each proper subset, we maintain a cover state of the covered skills and the total price of workers. We update the global cover state when processing each worker. When we have processed all the workers, the cover states of all the required skills of the task are the exact solution.

The exact algorithm is shown in Algorithm 3. We first get a approximate solution using a greedy algorithm and store the price of the solution in C_g in line 1. We then initialize *state* to store the currently best cover state. In lines 4–10, we successively process each worker in W. For worker w, if w_p is not larger than C_g, we enumerate all the cover states of w_p. For each cover state c, we combine it with cover state *state*. If the combined price is not larger than C_g, we store the current cover state in *temp_state*. We finally store c in *temp_state* and use it to update *state*. After we have processed all the workers in W, we check the cover state of the required skills of task t and its associated team is the best team. In line 4 and line 7, we adopt two pruning strategies. In line 4, we use C_g to prune a single worker whose price is too high. In line 7, we use C_g to prune a new cover state whose price is too high.

Example 4. Back to our running example in Example 1. We first use the top-1 approximation algorithm shown in Algorithm 2 to get an approximate solution $T = \{w_2, w_3\}$ with total price of 4, which is used as the current lower bound. Then we maintain the cover state using a triple structure, which contains the covered skills, the workers and the total price of the current optimal team for each possible combination of the required skills. w_1 can cover skill 1 or 2 with price 2, which is less than the lower bound of 4, so the cover state of w_1 can be $\{<\{1\}, \{w_1\}, 2>, <\{2\}, \{w_1\}, 2>\}$. As w_1 is the first worker we process, we just update the current best cover state as $\{<\{1\}, \{w_1\}, 2>, <2, \{w_1\}, 2>\}$. We then proceed to process w_2. We combine the only cover state, $<\{1\}, \{w_2\}, 1>$ with the cover states in *state*, and then we get a new cover state of $<\{1, 2\}, \{w_1, w_2\}, 2>$. After processing w_2, the current best cover state is $\{<\{1\}, \{w_2\}, 1>, <\{2\}, \{w1\}, 2>, <\{1, 2\}, \{w_1, w_2\}, 2>\}$. We can process w_3 similarly and the final cover state is $\{<\{1\}, \{w_2\}, 1>, <\{2\}, \{w_1\}, 2>, <\{1, 2\}, \{w_1, w_2\}, 2>, <\{1, 2, 3\}, \{w_2, w_3\}, 4>\}$ and the best team is $\{w_2, w_3\}$.

Complexity Analysis. Line 3 runs $|W|$ times, line 5 runs $C(|E_t|, |E_t|/2)$ times, and line 8 runs $2^{|E_t|}$ times. Therefore, the total time complexity is $O(|W|(2^{|E_t|}))$. When $|E_t|$ is not too large, the exact algorithm can be used.

4 Experimental Study

4.1 Experimental Setup

We use a real dataset collected from gMission [5], which is a research-based general spatial crowdsourcing platform. In the gMission dataset, every task has a task description, a location, a radius of the restricted range, and the required skills. Each worker is also associated with a location, a set of his/her owning skills, a price, and a capacity of skills that s/he completes a task. Currently, users often recruit crowd workers to organize all kinds of activities on the gMission platform. In this paper, our real dataset includes the information of 11205 crowd workers, where the average number of skills and the average capacity owned by the workers are 5.46 and 4.18, respectively. We also use synthetic dataset for evaluation. In the synthetic dataset, the capacity and the number of skills owned by a worker follow uniform distribution in the range of 1 to 20, respectively. Statistics of the synthetic dataset are shown in Table 3, where we mark our default settings in bold font.

Based on the two-level-based framework, we evaluate an approximation algorithm (Algorithms 1 and 2), called TTR-Greedy, and two exact algorithms (Algorithms 1 and 3), called TTR-Exact (which does not use the proposed pruning rules) and TTR-ExactPrune, and a baseline algorithm in terms of total utility score, running time and memory cost, and study the effect of varying parameters on the performance of the algorithms. The baseline algorithm uses a simple random greedy strategy, which first finds the best team, then randomly removes

Table 3. Synthetic Dataset

Factor	Setting
$|W|$	1000, 3000, **5000**, 7000, 9000
k	4, **8**, 12, 16, 20
$|E_t|$	4, **8**, 12, 16, 20
$\mu_{|E_w|}$	2, 4 **6**, 8, 10
$\sigma_{|E_w|}$	8, 10, **12**, 14, 16
Scalability ($|W|$)	10K, 30K, 50K, 70K, 90K

a worker from the best team from the set of workers, and iteratively finds the other $k - 1$ best teams following the two steps above. The algorithms are implemented in Visual C++ 2010, and the experiments were performed on a machine with Intel(R) Core(TM) i5 2.40 GHz CPU and 4GB main memory.

4.2 Experiment Results

In this subsection, we test the performance of our proposed algorithms through varying different parameters.

Effect of Cardinality of W. The results of varying $|W|$ are presented in Fig. 2a to c. Since TTR-Exact and TTR-ExactPrune return the same utility results, only utility results of TTR-ExactPrune are plotted. We can first observe that the utility decreases as $|W|$ increases, which is reasonable as more high-quality workers can are available. Also, we can see that TTR-Greedy is nearly as good as the exact algorithms. As for running time, TTR-Exact consumes more time with more workers due to larger search space while the TTR-ExactPrune is quite efficient due to its pruning techniques. The other algorithms do not vary much in running time. For memory, TTR-ExactPrune is the most efficient while TTR-Exact and TTR-Greedy are less efficient.

Effect of Parameter k. The results of varying k are presented in Fig. 2d to f. We can observe that the utility, running time and memory generally increase as k increases, which is reasonable as more teams need to be recommended. Again, we can see that TTR-Greedy is nearly as good as the exact algorithms but runs much faster. Also, we can see that the pruning techniques are quite effective as TTR-ExactPrune is much faster than TTR-Exact. Finally, TTR-Greedy is the most inefficient in terms of memory consumption.

Effect of the Number of Required Skills in Tasks. The results are presented in Fig. 2g to i. We can see that the utility values increase first with increasing number of required skills $|E_t|$ but decrease later when $|E_t|$ further increases. The possible reason is that when $|E_t|$ is not large, the required skills are still quite diverse and thus more workers need to be hired to complete the task as $|E_t|$ increases. However, as $|E_t|$ becomes too large, many workers may

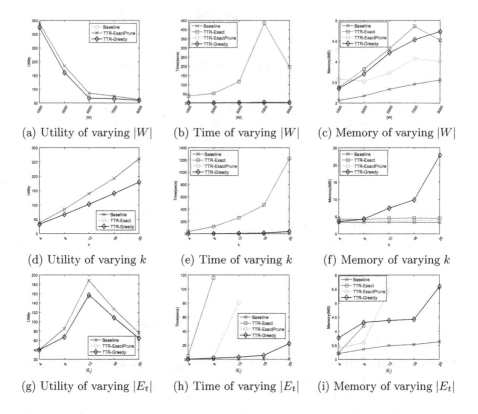

(a) Utility of varying $|W|$ (b) Time of varying $|W|$ (c) Memory of varying $|W|$

(d) Utility of varying k (e) Time of varying k (f) Memory of varying k

(g) Utility of varying $|E_t|$ (h) Time of varying $|E_t|$ (i) Memory of varying $|E_t|$

Fig. 2. Results on varying $|W|$, k, and $|E_t|$.

use their own multiple skills to complete the task and thus less workers may be needed. As for running time and memory, we can observe that the values generally increase. Again, TTR-ExactPrune is highly inefficient compared with the other algorithms. Notice that the exact algorithms run very long time when $|E_t|$ is large, so we do not plot their results when $|E_t|$ is larger than 12.

Effect of the Distribution of the Number of Skills Per Each Worker (μ and σ). The results are presented in Fig. 3a to f. We can first observe that the utility value first increases as μ and σ increase and then drops when μ and σ further increase. The possible reason is that when μ and σ first increase, the skills of workers are more diverse and may not cover the requirements of the tasks and thus more workers are still needed. However, as μ and σ further increase, many workers can utilize their multiple skills and thus less workers are needed. As for running time, TTR-Exact is again very inefficient. Finally, for memory, TTR-ExactPrune is more efficient than TTR-Exact and TTR-Greedy.

Scalability. The results are presented in Fig. 3g to i. Since the exact algorithms are not efficient enough, we only study the scalability of TTR-Greedy. We can

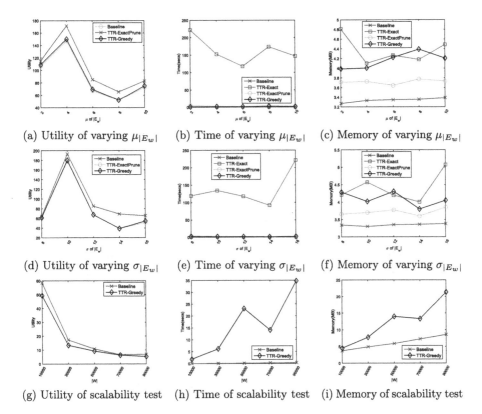

Fig. 3. Results on varying $\mu_{|E_w|}$, $\sigma_{|E_w|}$, and scalability test.

see that the running time and memory consumption TTR-Greedy is still quite small when the scale of data is large.

Real Dataset. The results on real dataset are shown in Fig. 4a to c, where we vary k. We can observe similar patterns as those in Fig. 2d to f. Notice that the exact algorithms are not efficient enough on the dataset, so no result of them when k is larger than 8 is presented.

Conclusion. For utility, TTR-Greedy is nearly as good as the exact algorithms, and TTR-Greedy and the exact algorithms all perform better than the baseline algorithm do. As for running time, TTR-Exact is the most inefficient, while TTR-ExactPrune is much more efficient than TTR-Exact due to its pruning techniques but is still slower than TTR-Greedy.

5   Related Work

In this section, we review related works from two categories, spatial crowdsourcing and team formation.

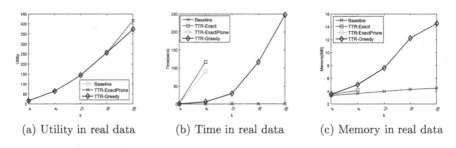

(a) Utility in real data   (b) Time in real data   (c) Memory in real data

Fig. 4. Performance on the real dataset.

5.1 Spatial Crowdsourcing

Most works on spatial crowdsourcing study the task assignment problem. [6,14] aim to maximize the number of tasks that are assigned to workers. Furthermore, the conflict-aware spatial task assignment problems are studied [11,12,18]. Recently, the issue of online task assignment in dynamic spatial crowdsourcing scenarios is proposed [17]. [7] further studies the reliability of crowd workers based on [6]. [13] studies the location privacy protection problem for the workers. [7] studies the route planning problem for a crowd worker and tries to maximize the number of completed tasks. The corresponding online version of [7] is studied in [9]. Although the aforementioned works study the task allocation problem on spatial crowdsourcing, they always assume that spatial crowdsourcing tasks are simple micro-tasks and ignore that some real spatial crowdsourced tasks often need to be collaboratively completed by a team of crowd workers.

5.2 Team Formation Problem

Another closely related topic is the team formation problem [8], which aims to find the minimum cost team of experts according to skills and relationships of users in social networks. [1,2] further studies the workload balance issue in the static and dynamic team formation problem. The capacity constraint of experts is also considered as an variant of the team formation problem in [10]. Moreover, the problems of discovering crowed experts in social media market are also studied [3,4]. The above works only consider to find the minimum cost team, namely top-1 team, instead of top-k teams without free riders. In addition, we address the spatial scenarios rather than the social networks scenarios.

6   Conclusion

In this paper, we study a novel spatial crowdsourcing problem, called the _Top-k Team Recommendation in spatial crowdsourcing_ (TopkTR), which is proven to be NP-hard. To address this problem, we design a two-level-based framework, which not only includes an exact algorithm with pruning techniques to get the exact solution but also seamlessly integrates an approximation algorithm to guarantee theoretical approximation ratio. Finally, we conduct extensive experiments which verify the efficiency and effectiveness of the proposed approaches.

Acknowledgment. This work is supported in part by the National Science Foundation of China (NSFC) under Grant No. 61502021, 61328202, and 61532004, National Grand Fundamental Research 973 Program of China under Grant 2012CB316200, the Hong Kong RGC Project N_HKUST637/13, NSFC Guang Dong Grant No. U1301253, Microsoft Research Asia Gift Grant, Google Faculty Award 2013.

References

1. Anagnostopoulos, A., Becchetti, L., Castillo, C., Gionis, A., Leonardi, S.: Power in unity: forming teams in large-scale community systems. In: CIKM 2010, pp. 599–608 (2010)
2. Anagnostopoulos, A., Becchetti, L., Castillo, C., Gionis, A., Leonardi, S.: Online team formation in social networks. In: WWW 2012, pp. 839–848 (2012)
3. Cao, C.C., She, J., Tong, Y., Chen, L.: Whom to ask?: jury selection for decision making tasks on micro-blog services. Proc. VLDB Endowment **5**(11), 1495–1506 (2012)
4. Cao, C.C., Tong, Y., Chen, L., Jagadish, H.V.: Wisemarket: a new paradigm for managing wisdom of online social users. In: SIGKDD 2013, pp. 455–463 (2013)
5. Chen, Z., Fu, R., Zhao, Z., Liu, Z., Xia, L., Chen, L., Cheng, P., Cao, C.C., Tong, Y., Zhang, C.J.: gMission: a general spatial crowdsourcing platform. Proc. VLDB Endowment **7**(14), 1629–1632 (2014)
6. Kazemi, L., Shahabi, C.: Geocrowd: enabling query answering with spatial crowdsourcing. In: GIS 2012, pp. 189–198 (2012)
7. Kazemi, L., Shahabi, C., Chen, L.: Geotrucrowd: trustworthy query answering with spatial crowdsourcing. In: GIS 2013, pp. 304–313 (2013)
8. Lappas, T., Liu, K., Terzi, E.: Finding a team of experts in social networks. In: SIGKDD 2009, pp. 467–476 (2009)
9. Li, Y., Yiu, M.L., Xu, W.: Oriented online route recommendation for spatial crowdsourcing task workers. In: Claramunt, C., Schneider, M., Wong, R.C.-W., Xiong, L., Loh, W.-K., Shahabi, C., Li, K.-J. (eds.) SSTD 2015. LNCS, vol. 9239, pp. 137–156. Springer, Heidelberg (2015)
10. Majumder, A., Datta, S., Naidu, K.: Capacitated team formation problem on social networks. In: SIGKDD 2012, pp. 1005–1013 (2012)
11. She, J., Tong, Y., Chen, L.: Utility-aware social event-participant planning. In: SIGMOD 2015, pp. 1629–1643 (2015)
12. She, J., Tong, Y., Chen, L., Cao, C.C.: Conflict-aware event-participant arrangement. In: ICDE 2015, pp. 735–746 (2015)
13. To, H., Ghinita, G., Shahabi, C.: A framework for protecting worker location privacy in spatial crowdsourcing. Proc. VLDB Endowment **7**(10), 919–930 (2014)
14. To, H., Shahabi, C., Kazemi, L.: A server-assigned spatial crowdsourcing framework. ACM Trans. Spat. Algorithms Syst. **1**(1), 2 (2015)
15. Tong, Y., Cao, C.C., Chen, L.: TCS: efficient topic discovery over crowd-oriented service data. In: SIGKDD 2014, pp. 861–870 (2014)

16. Tong, Y., Cao, C.C., Zhang, C.J., Li, Y., Chen, L.: Crowdcleaner: Data cleaning for multi-version data on the web via crowdsourcing. In: ICDE 2014, pp. 1182–1185 (2014)
17. Tong, Y., She, J., Ding, B., Wang, L., Chen, L.: Online mobile micro-task allocation in spatial crowdsourcing. In: ICDE 2016 (2016)
18. Tong, Y., She, J., Meng, R.: Bottleneck-aware arrangement over event-based social networks: the max-min approach. World Wide Web J. (to appear). doi:10.1007/s11280-015-0377-6

Explicable Location Prediction Based on Preference Tensor Model

Duoduo Zhang, Ning Yang$^{(\boxtimes)}$, and Yuchi Ma

College of Computer Science, Sichuan University, Chengdu, China
cherry.scu@gmail.com, yangning@scu.edu.cn, scuRichard.Ma@gmail.com

Abstract. Location prediction has been attracting an increasing interest from the data mining community. In real world, however, to provide
more targeted and more personal services, the applications like location-
aware advertising and route recommendation are interested not only in
the predicted location but its explanation as well. In this paper, we investigate the problem of Explicable Location Prediction (ELP) from LBSN
data, which is not easy due to the challenges of the complexity of human
mobility motivation and data sparsity. In this paper, we propose a Preference Tensor Model (PTM) to address the challenges. The core component
of PTM is a preference tensor, each cell of which represents how much
a user prefers to a specific place at a specific time point. The explicable
location prediction can be made via a retrieval of the preference tensor,
and meanwhile a motivation vector is generated as the explanation of the
prediction. To model the complicated motivations of human movement,
we propose two motivation tensors, a social tensor and a personal tensor,
to represent the social cause and the personal cause of human movement.
From the motivation tensors, the motivation vector consisting of a social
ingredient and a personal ingredient can be produced. To deal with data
sparsity, we propose a Social Tensor Decomposition Algorithm (STDA)
and a Personal Tensor Decomposition Algorithm (PTDA), which are
able to fill missing values of a sparse social tensor and a sparse personal
tensor, respectively. Particularly, to achieve a higher accuracy, STDA
fuses an additional social constraint with the decomposition. The experiments conducted on real-world datasets verify the proposed model and
algorithms.

Keywords: Location prediction · Tensor Model · Location Based Social
Network

1 Introduction

Recently, location prediction based on check-in data of Location-Based Social
Networks (LBSNs) has attracted increasing attention from the community of

---

This work was supported by the National Science Foundation of China under Grant
61173099 and the Sci. & Tech. Support Programs of Sichuan Province under Grant
2014JY0220.

© Springer International Publishing Switzerland 2016
B. Cui et al. (Eds.): WAIM 2016, Part I, LNCS 9658, pp. 205–216, 2016.
DOI: 10.1007/978-3-319-39937-9_16

data mining, due to its benefit for location-based applications such as location-aware advertising and marketing [1,10] and epidemic control [5] etc. Although a few techniques are proposed for the prediction of a future check-in location, they often focus on seeking a higher prediction accuracy while little attention is paid to the motivations behind human movements [3,11,14]. However, explicable location prediction is of great value to location-based applications, as it can help the applications provide more targeted services and personal experience to users. In this paper, we investigate the problem of the Explicable Location Prediction (ELP) from check-in data, which is not easy due to the following challenges:

- **Complicated motivations of human movement.** Existing studies [3,4,7,9] show that human mobility is constrained by social relationships and periodic behaviors. For example, shopping can be caused by social reasons, e.g. people want to stay longer with friends, or personal reasons, e.g. people want to buy necessities for daily life, or a mix of both. We need a way to quantitatively represent a combination of multiple motivations, so as to get a better understanding of why a user moves to a location.
- **Data sparsity.** In social networks, users' check-ins are always discontinuous because they only check in at the places they are interested, which results in that for a specific user, check-in data is sparse on both space and time dimensions.

In this paper, we propose a Preference Tensor Model (PTM) for the ELP. The main idea of PTM is inspired by the observation that human movement is not random but driven and explained by a mix of social motivation and personal motivation [12]. The core component of PTM is a preference tensor of a given user, where each cell stores the probability of that user will go to a specific place at a specific time point. An ELP can be made through a retrieval from the preference tensor, as well as a motivation vector as the explanation of the prediction can be generated from two motivation tensors, a social tensor and a personal tensor which are proposed to evaluate the social motivation and personal motivation of a human movement, respectively. To address the challenge of data sparsity, we employ a reconstruction strategy for the filling of missing values. We propose a Social Tensor Decomposition Algorithm (STDA) and a Personal Tensor Decomposition Algorithm (PTDA), which are able to fill missing values of a sparse social tensor and a sparse personal tensor, respectively. Particularly, to achieve a higher accuracy, STDA fuses an additional social constraint with the decomposition.

The main contributions of this paper can be summarized as follows:

(1) We propose a Preference Tensor Model (PTM) for the explicable location prediction. By PTM, a location prediction can be made with a motivation vector as its quantitative explanation.
(2) To model the complicated motivation of a human movement, we propose two motivation tensors, a social tensor and a personal tensor. From the motivation tensors, the mixed motivation vector can be produced as the explanation of a location prediction.

(3) To fill the missing values of a sparse social tensor and a sparse personal tensor, we propose a Social Tensor Decomposition Algorithm (STDA) and and a Personal Tensor Decomposition Algorithm (PTDA). Particularly, to achieve a higher accuracy, STDA fuses an additional social constraint with the decomposition.
(4) Experimental results show that the proposed PTM outperforms the baseline methods.

The rest of the paper is organized as follows. The details of PTM are described in Sect. 2. Motivation tensors are defined in Sect. 3. Data sparsity is addressed in Sect. 4. Experimental results are presented in Sect. 5. We give a brief description of the related work in Sect. 6 and conclude in Sect. 7.

2   Preference Tensor Model

We represent an Explicable Location Prediction (ELP) as a tuple $(r, \boldsymbol{v})$, where r is the predicted location, and $\boldsymbol{v}$ is the motivation vector. The motivation vector plays a role of a quantitative explanation of the location prediction, which is defined as follow:

Definition 1. *Motivation Vector.* *A motivation vector is a vector* $\boldsymbol{v} = (w^s, w^p)$, *where* w^s *and* w^p *represents corresponding weights of social and personal motivations of a check-in.*

Now we describe the details of our PTM, and how an ELP can be made by PTM. At first, for the location prediction, we propose a Preference Tensor, to represent the affinity between a user u, a time slot t, and a location r. Specifically, a preference tensor is a 3-dimensional tensor $\mathcal{T} \in \mathbb{R}^{M \times N \times Q}$, where M, N and Q are the numbers of users, time slots, and locations, respectively. A cell $\mathcal{T}_{u,t,r}$ stores the probability of user u ($u \in \{1, \cdots, M\}$) will go to location r ($r \in \{1, \cdots, N\}$) at time slot t ($t \in \{1, \cdots, Q\}$). We divide $\mathcal{T}_{u,t,r}$ as a sum of a social term and a personal term, i.e.,

$$\mathcal{T}_{u,t,r} = \mathcal{T}^s_{u,t,r} + \mathcal{T}^p_{u,t,r}, \tag{1}$$

where $\mathcal{T}^s \in \mathbb{R}^{M \times N \times Q}$ is a social tensor and $\mathcal{T}^p \in \mathbb{R}^{M \times N \times Q}$ is a personal tensor.

The social tensor and personal tensor are two motivation tensors. We use social tensor $\mathcal{T}^s$ to represent the social ingredient of the motivation of human movement, where a cell $\mathcal{T}^s_{u,t,r}$ measures the preference for location r of user u at time slot t that can be explained under the social context of u . Similarly, the personal tensor $\mathcal{T}^p$ represents the personal ingredient of a motivation, where a cell $\mathcal{T}^p_{u,t,r}$ measures the preference for location r of user u at time slot t that can be explained by personal reasons of u. We will describe the details of how to build the social tensor and personal tensor in next section.

Once we establish the PTM (Eq. (1)), we can predict the location of user u at time t by the following equation:

$$\hat{r} = \underset{r}{\operatorname{argmax}} \, \mathcal{T}_{u,t,r}, \tag{2}$$

where $\hat{r}$ is the predicted location for user u at time slot t. According to the definition of motivation vector, as the explanation of the location prediction, the motivation vector can be computed as $v = (\frac{T^s_{u,t,\hat{r}}}{T_{u,t,\hat{r}}}, \frac{T^p_{u,t,\hat{r}}}{T_{u,t,\hat{r}}})$

3   Construction of Motivation Tensors

In this section, we describe how to construct the two motivation tensors, from a social check-in set and a personal check-in set, respectively. As tensor is a discrete structure, we need to discretize the space and time first. We partition the whole area covered by the total check-in data into grids, and use the cell number to represent the location of the check-in. We also partition one day into 24 time slots with a period of 1 h, and the time of a check-in is instead represented as the time slot number. Now we define the concepts of check-in, social check-in and personal check-in.

Definition 2. Check-in. *A check-in is a triple $Tri(u, t, l)$ representing a user u checks in at location l at time point t.*

A social check-in is a check-in caused by friendship reasons, which is defined as follow:

Definition 3. Social Check-in. *A check-in $Tri(u, t, l)$ is a social check-in, if one of u's friends checked in at the same location at time point t' and $t - t' \leq \delta$, where δ is a threshold given in advance, e.g. one week.*

Different from social check-ins, a personal check-in is caused by personal reasons, which lead to that a user checks in at a place where he/she rarely checks in together with his/her friends.

Definition 4. Personal Check-in. *A check-in $Tri(u, t, l)$ is a personal check-in, if one of u's friends checked in at l at time point t' and $t - t' > \epsilon$, or none of u's friends checked in at l before t, where ϵ is a threshold given in advance.*

Algorithm 1 fulfills the construction of the social tensor and the personal tensor in a straightforward way.

4   Dealing with Sparsity

4.1   Social Tensor Decomposition Algorithm for T^s

To address data sparsity of the social tensor, we propose a Social Tensor Decomposition Algorithm(STDA). Particularly, to achieve a higher accuracy of decomposition, STDA fuses a social constraint represented as a closeness matrix when decomposing T^s. The idea here is that if two users are close enough, they are more likely to share common interest and check in at the same locations. Based on this idea, we define the closeness matrix as follow:

Algorithm 1. Social Tensor and Personal Tensor Construction Algorithm

Input:
    social check-in set C^s, personal check-in set C^p

Output:
    social tensor $\boldsymbol{T}^s \in \mathbb{R}^{M \times N \times Q}$, personal tensor $\boldsymbol{T}^p \in \mathbb{R}^{M \times N \times Q}$

1: Initialize each cell of $\boldsymbol{T}^s$ and $\boldsymbol{T}^p$ with 0;
2: **for** each user i, time slot j, region k **do**
3:    n^s = the number of the occurrences of $Tri(i, j, k)$ in C^s;
4:    n^p = the number of the occurrences of $Tri(i, j, k)$ in C^p;
5:    N_i = the number of the check-ins of user i;
6:    $\boldsymbol{T}^s_{i,j,k} = \frac{n^s}{N_i}$;
7:    $\boldsymbol{T}^p_{i,j,k} = \frac{n^p}{N_i}$;
8: **end for**

Definition 5. *Closeness Matrix. Closeness Matrix, denoted as $\boldsymbol{X}$, where an entry denotes the closeness of user i to user j, and $\boldsymbol{X}_{i,j} = \frac{|C^s_{i,j}|}{|C_i|}$, where $C^s_{i,j}$ is the social check-in set between user i and user j, and C_i is the total check-in set of user i.*

The objective function of STDA is defined as follow:

$$F(\boldsymbol{T}^s, \boldsymbol{\mathcal{I}}, \mathbf{U}, \mathbf{C}, \mathbf{L}) = \frac{1}{2} \|\boldsymbol{T}^s - \boldsymbol{\mathcal{I}} \times_1 \mathbf{U} \times_2 \mathbf{C} \times_3 \mathbf{L}\|^2_F$$

$$+ \frac{\alpha}{2} \|\mathbf{X} - \mathbf{U}\mathbf{V}^T\|^2_F + \frac{\beta}{2} (\|\mathbf{U}\|^2_F + \|\mathbf{C}\|^2_F + \|\mathbf{L}\|^2_F + \|\mathbf{V}\|^2_F), \tag{3}$$

where $\|\cdot\|_F$ represents Frobenius norm; $\mathbf{U} \in \mathbb{R}^{M \times D}$, $\mathbf{C} \in \mathbb{R}^{N \times D}$ and $\mathbf{L} \in \mathbb{R}^{Q \times D}$ are latent factors, and D is the number of latent features; $\mathbf{V} \in \mathbb{R}^{M \times D}$ is a factor matrix of $\mathbf{X}$; $\times_i (1 \leq i \leq 3)$ stands for the tensor multiplication along the i-th mode; α and β are parameters controlling the contributions of different parts. The social constraint represented as $\mathbf{X}$, is imposed via $\mathbf{U}$, as it is shared between $\boldsymbol{T}^s$ and $\mathbf{X}$. Equation (3) involves three terms, where the first and second terms are the errors of the matrix factorization and the tensor decomposition, and the last one is the regularization penalty. As there is no closed-form solution to the minimization of the objective function, we minimize Eq. (3) using gradient descent.

Algorithm 2 gives the procedures of STDA, where the gradients of the objective function can be computed as follows:

$$\partial_{\mathbf{U}_{i*}} F^l = (\widetilde{\boldsymbol{T}}^s_{i,j,k} - \boldsymbol{T}^s_{i,j,k}) \times \boldsymbol{\mathcal{I}} \times_2 \mathbf{C}_{j*} \times_3 \mathbf{L}_{k*} + \alpha(\mathbf{U}_{i*}\mathbf{V}^T - \mathbf{X}_{i*}) + \beta\mathbf{U}_{i*},$$

$$\partial_{\mathbf{C}_{j*}} F^l = (\widetilde{\boldsymbol{T}}^s_{i,j,k} - \boldsymbol{T}^s_{i,j,k}) \times \boldsymbol{\mathcal{I}} \times_1 \mathbf{U}_{i*} \times_3 \mathbf{L}_{k*} + \beta\mathbf{C}_{j*},$$

$$\partial_{\mathbf{L}_{k*}} F^l = (\widetilde{\boldsymbol{T}}^s_{i,j,k} - \boldsymbol{T}^s_{i,j,k}) \times \boldsymbol{\mathcal{I}} \times_1 \mathbf{U}_{i*} \times_2 \mathbf{C}_{j*} + \beta\mathbf{L}_{k*},$$

$$\partial_{\mathbf{V}} F^l = (\mathbf{U}_{i*} \times \mathbf{V}^T - \mathbf{X}) \times \mathbf{U}_{i*}$$

Algorithm 2. Social Tensor Decomposition Algorithm

Input:
   tensor $\mathcal{T}^s$, matrix $\mathbf{X}$, step size θ, parameters α, β,
   the number of latent features D.
Output:
   a filled social tensor $\widetilde{\mathcal{T}}^s$
1: Initialize $\mathbf{U}$, $\mathbf{C}$, $\mathbf{L}$ and $\mathbf{V}$ with random values between 0 and 1;
2: Set $l = 0$;
3: Calculate the initial value of the cost function according to Eq. (3);
4: **while** not converged **do**
5:    **for** each $\mathcal{T}^s_{i,j,k} \neq 0$ **do**
6:       $\mathbf{U}_{i*} = \mathbf{U}_{i*} - \theta \partial_{\mathbf{U}_{i*}} F^l$;
7:       $\mathbf{C}_{j*} = \mathbf{C}_{j*} - \theta \partial_{\mathbf{C}_{j*}} F^l$;
8:       $\mathbf{L}_{k*} = \mathbf{L}_{k*} - \theta \partial_{\mathbf{L}_{k*}} F^l$;
9:       $\mathbf{V} = \mathbf{V} - \theta \partial_{\mathbf{V}} F^l$;
10:      $++l$;
11:      Calculate the l-th iteration of the cost function F^l according to Eq. (3);
12:   **end for**
13: **end while**
14: $\widetilde{\mathcal{T}}^s = \mathcal{I} \times_1 \mathbf{U} \times_2 \mathbf{C} \times_3 \mathbf{L}$;

4.2   Personal Tensor Decomposition Algorithm for $\mathcal{T}^p$

The objective function of PTDA is defined as follow:

$$H(\mathcal{T}^p, \mathcal{I}, \mathbf{U}, \mathbf{C}, \mathbf{L}) = \frac{1}{2} \|\mathcal{T}^p - \mathcal{I} \times_1 \mathbf{U} \times_2 \mathbf{C} \times_3 \mathbf{L}\|_F^2$$
$$+ \frac{\beta}{2}(\|\mathbf{U}\|_F^2 + \|\mathbf{C}\|_F^2 + \|\mathbf{L}\|_F^2), \tag{4}$$

where $\mathbf{U}$ $(\mathbf{U} \in \mathbb{R}^{M \times D})$, $\mathbf{C}$ $(\mathbf{C} \in \mathbb{R}^{N \times D})$ and $\mathbf{L}$ $(\mathbf{L} \in \mathbb{R}^{Q \times D})$ are latent factors, and D is the number of latent features. As shown in Algorithm 3, PTDA also employs the gradient descent to find an optimal solution to Eq. (4).

In Algorithm 3, the personal tensor can be estimated as $\widetilde{\mathcal{T}}^p = \mathcal{I} \times_1 \mathbf{U} \times_2 \mathbf{C} \times_3 \mathbf{L}$, and the gradients of the objective function can be given as follows:

$$\partial_{\mathbf{U}_{i*}} H^l = (\widetilde{\mathcal{T}}^p_{i,j,k} - \mathcal{T}^p_{i,j,k}) \times \mathcal{I} \times_2 \mathbf{C}_{j*} \times_3 \mathbf{L}_{k*} + \beta \mathbf{U}_{i*},$$

$$\partial_{\mathbf{C}_{j*}} H^l = (\widetilde{\mathcal{T}}^p_{i,j,k} - \mathcal{T}^p_{i,j,k}) \times \mathcal{I} \times_1 \mathbf{U}_{i*} \times_3 \mathbf{L}_{k*} + \beta \mathbf{C}_{j*},$$

$$\partial_{\mathbf{L}_{k*}} H^l = (\widetilde{\mathcal{T}}^p_{i,j,k} - \mathcal{T}^p_{i,j,k}) \times \mathcal{I} \times_1 \mathbf{U}_{i*} \times_2 \mathbf{C}_{j*} + \beta \mathbf{L}_{k*}$$

5   Experiment

To verify the performance of PTM, we compare PTM with five baselines on a real-world dataset. All of the experiments are conducted on a PC with Intel Core I5 CPU 3.2 GHZ and 16 GB main memory. The operating system is Windows 7 and all the algorithms are implemented in C#.

Algorithm 3. Personal Tensor Decomposition Algorithm

Input:
　　tensor $\boldsymbol{T}^p$, matrix $\mathbf{X}$, step size θ, parameter β, the number of latent features D.
Output:
　　a filled personal tensor $\widetilde{\boldsymbol{T}}^p$
1: Initialize $\mathbf{U}$, $\mathbf{C}$ and $\mathbf{L}$ with random values between 0 and 1;
2: Set $l = 0$;
3: Calculate the initial value of the cost function according to Eq. (4);
4: **while** not converged **do**
5:　　**for** each $\boldsymbol{T}^p_{i,j,k} \neq 0$ **do**
6:　　　　$\mathbf{U}_{i*} = \mathbf{U}_{i*} - \theta \partial_{\mathbf{U}_{i*}} H^l$;
7:　　　　$\mathbf{C}_{j*} = \mathbf{C}_{j*} - \theta \partial_{\mathbf{C}_{j*}} H^l$;
8:　　　　$\mathbf{L}_{k*} = \mathbf{L}_{k*} - \theta \partial_{\mathbf{L}_{k*}} H^l$;
9:　　　　$++l$;
10:　　　　Calculate the l-th iteration of the cost function $\boldsymbol{H}^l$ according to Eq. (4);
11:　　**end for**
12: **end while**
13: $\widetilde{\boldsymbol{T}}^p = \boldsymbol{\mathcal{I}} \times_1 \mathbf{U} \times_2 \mathbf{C} \times_3 \mathbf{L}$;

5.1　Setting

Dataset. The dataset we use in our experiments is collected from Gowalla[1], a famous LBSN. The dataset contains 6.4 million check-ins and an undirected social network consisting of 19.7 thousand users. We randomly split the dataset into training and testing sets with an 8:2 ratio.

Metrics. We use Root Mean Squared Error (RMSE) and Mean Absolute Error (MAE) to evaluate the prediction accuracy of PTM, which are defined as follows:

$$RMSE = \sqrt{\frac{\sum_{i=1}^{n}(y_i - \hat{y}_i)^2}{n}}, \tag{5}$$

$$MAE = \frac{\sum_{i=1}^{n}|y_i - \hat{y}_i|}{n}, \tag{6}$$

where y_i and $\hat{y}_i$ are the true and estimated values respectively, and n is the number of samples in test set.

　　To evaluate the explanation made by PTM, we propose a new metric, Explicable Accuracy (EAcc), which is defined as

$$EAcc = \frac{\sum I_{u,t,l}}{n}, \tag{7}$$

where $I_{u,t,l} = 1$ if $\widetilde{\boldsymbol{T}}^s_{u,t,l} \geq$ (or $<$) $\widetilde{\boldsymbol{T}}^p_{u,t,l}$ and $\boldsymbol{T}^s_{u,t,l} \geq$ (or $<$) $\boldsymbol{T}^p_{u,t,l}$ on test set, otherwise, 0. Here the idea of EAcc is to use the nonzero cells (i.e., the cells with

---

[1] https://snap.stanford.edu/data/.

indices (u, t, l) such that $\boldsymbol{T}^s_{u,t,l} \neq 0$ and $\boldsymbol{T}^p_{u,t,l} \neq 0$) of the original motivation tensors as the ground truth of which motivation dominates a movement, and check whether the reconstructed values keep the size order between the old values, and if they do, $I_{u,t,l}$ equals 1, otherwise 0.

Baseline Approaches. We compare PTM with the following baselines.

MF (Most-Frequent). [3] MF predicting a location for a user u totally relies on u's history check-ins, which can be calculated as

$$P(u, t, l) = \frac{|Tri|Tri \in \boldsymbol{C_u}, Tri.t = t, Tri.l = l|}{|\boldsymbol{C_u}|}, \tag{8}$$

where $\boldsymbol{C_u}$ is the check-in set of user u.

PMM (Periodic Mobility Model). [3] PMM makes a location prediction for a user based on a time-independent gaussian distribution of the historical locations where the user appeared before.

W3 (Who + Where + When). [14] W3 is a variant of W4 (Who + Where + When), which is a probabilistic model integrating user, location, time and activity information and can be applied to many applications. W3 makes a location prediction by maximize the following posterior probability:

$$P(l|u, s, t) = \frac{\sum_z \sum_r p(u, s, t, r, z, l)}{\sum_z \sum_r \sum_{l'} p(u, s, t, r, z, l')} \tag{9}$$

where u, s, t, r, z, l denote user, workday or holiday, time, region, topic and location, respectively.

RCH (Regularity and Conformity). [11] RCH incorporates regularity and conformity of human mobility in a unified prediction model, and utilizes the interplay between the two factors. It makes a location prediction based on an estimated score for a specific location.

CATD (Context-Aware Tensor Decomposition). [2] CATD makes location predictions adopting a tensor reconstruction strategy. It constructs a sparse tensor using total check-in data and decompose it with additional context constraints collaboratively. The value of each cell in filled tensor can be considered as the probability that a user will check in at a specific region at a specific time point.

5.2 Determination of Parameters

Grid Resolution. Grid resolution in our experiments can affect the performance of PTM significantly. In our experiments, we partition the whole area covered by the total check-ins into grids based on the finding of [8]. Figure 1 shows the RMSE and MAE over different grid sizes. From Fig. 1, we can see that RMSE and MAE both reach the minimal when the grid resolution is $500\,\mathrm{m} \times 1000\,\mathrm{m}$.

The Number of Latent Features D. We tune parameter D from 3 to 9 with respect to RMSE and MAE. The results are shown in Figs. 2(a) and 3(a), which suggest that the best choice of D is 4.

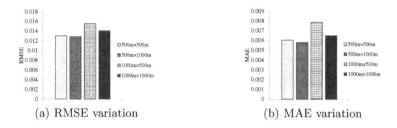

(a) RMSE variation   (b) MAE variation

Fig. 1. RMSE and MAE over different grid resolutions

Contributions of Closeness Matrix and Regularization Penalty. We tune parameters α and β with respect to RMSE and MAE. The results for α are shown in Figs. 2(b) and 3(b), and the results for β shown in Figs. 2(c) and 3(c). The results suggest that $\alpha = 0.3$ and $\beta = 0.05$ is the best choices.

5.3   Prediction Accuracy

Figure 4 shows the RMSEs and MAEs of PTM and the baseline methods. From Fig. 4, we can see that PTM outperforms the baseline methods, which is because PTM (1) considers both the historical movements and the social relationships of users, (2) can handle the sparsity of data, (3) and can distinguish the motivations. For example, MF can not predict new locations for a user where the user never appeared before, since it only considers the history check-ins of that user.

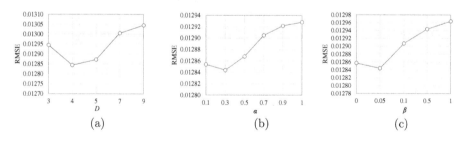

(a)   (b)   (c)

Fig. 2. RMSE over different D, α, β

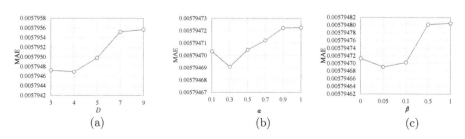

(a)   (b)   (c)

Fig. 3. MAE over different D, α, β

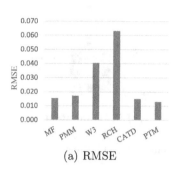 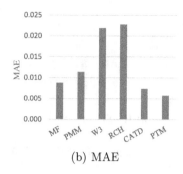

(a) RMSE   (b) MAE

Fig. 4. Accuracy comparison

The baseline methods except for CATD are designed for data that are not sparse, which makes their prediction accuracy lower than the proposed PTM on sparse data. We also note that PTM performs better than CATD, which is because CATD treats the check-ins equivalently without distinguishing social check-ins from personal ones.

5.4 Verification of Motivation Vector

To verify the effectiveness of the predication explanation, represented as motivation vectors, we compare PTM with a naive method, called Naive PTM (NPTM), with respect to EAcc. NPTM is a variant of PTM that fulfills the tensor decomposition without any constraints. Figure 5 shows the EAccs of PTM and NPTM. We can see that the EACC of PTM reaches about 91 %, which is significantly higher than the EAcc (61 %) of NPTM. We argue that this is because we fuse a closeness matrix as the social constraint to the reconstruction of a sparse social tensor.

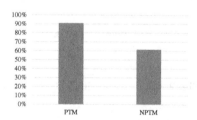

Fig. 5. EAcc comparison

6   Related Work

The existing methods for location prediction roughly fall into two categories, probabilistic model based methods and collaborative model based methods.

Probabilistic Model Based Methods. Yuan et al. [14] propose a probabilistic model W4 (short for who + where + when + what) which reveals the dependencies between user mobility and the spatial, temporal and activity factors. Cho et al. [3] propose a periodic mobility model (PMM) based on the intuition that the human mobility is periodic and often center around home and work office. Yang et al. [12] propose an approach considering both the periodicity and the sociality of human movements. However, these methods often assume that the data is not sparse, and there are sufficient check-ins to support the prediction, which makes them unable to serve our case where the data is sparse.

Collaborative Model Based Methods. Wang et al. [11] propose a hybrid model called RCH by combining both regularity and conformity of human mobility. Lian et al. [6] propose a collaborative exploration and periodically returning (CEPR) model which makes location prediction based on the periodicity of human mobility and the social relationships of users. Ye et al. [13] propose a unified POI recommendation framework which fuses user preferences with social features and geographic features of POIs. Zheng et al. [2] propose a Context-Aware Tensor Decomposition (CATD) model which can utilize additional heterogeneous data as context constraints to optimize the latent feature matrices of the tensor. However, the above methods can not quantitatively explain the motivations of a user movement.

7   Conclusions

In this paper, we propose a Preference Tensor Model (PTM) for the Explicable Location Prediction. PTM makes location prediction via a retrieval from a preference tensor, each cell of which represents how much a user prefers to a specific place at a specific time point. We propose two motivation tensors, a social tensor and a personal tensor, from which a motivation vector consisting of a social ingredient and a personal ingredient can be generated as the explanation of a location prediction. To deal with the data sparsity, we propose a Social Tensor Decomposition Algorithm (STDA) and a Personal Tensor Decomposition Algorithm (PTDA), which are able to reconstruct the sparse motivation tensors by filling the missing values of them. Particularly, to achieve a higher accuracy, STDA fuses an additional social constraint with the decomposition. At last, experimental results on a real-world dataset verify the performance of PTM.

References

1. Agarwal, A., Hosanagar, K., Smith, M.D.: Location, location, location: an analysis of profitability of position in online advertising markets. J. Mark. Res. **48**, 1057–1073 (2011)
2. Bhargava, P., Phan, T., Zhou, J., Lee, J.: Travel time estimation of a path using sparse trajectories. In: Proceedings of the 20th ACM SIGKDD International Conference on Knowledge Discovery and Data Mining, pp. 25–34 (2014)

3. Cho, E., Myers, S.A., Leskovec, J.: Friendship, mobility: user movement in location-based social networks. In: Proceedings of the 17th ACM SIGKDD International Conference on Knowledge Discovery and Data Mining, pp. 1082–1090 (2011)
4. Gonzalez, M.C., Hidalgo, C.A., Barabasi, A.L.: Understanding individual human mobility patterns. Nature **453**, 779–782 (2008)
5. Lan, L., Malbasa, V., Vucetic, S.: Spatial scan for disease mapping on a mobile population. In: Proceedings of the 28th Association for the Advancement of Artificial Intelligence, AAAI 2014, pp. 431–437 (2014)
6. Lian, D., Xie, X., Zheng, V.W., Yuan, N.J., Zhang, F., Chen, E.: CEPR: a collaborative exploration and periodically returning model for location prediction. ACM Trans. Intell. Syst. Technol. (TIST) **6**, 1–27 (2014)
7. Mok, D., Wellman, B., Carrasco, J.: Does distance matter in the age of the internet? Urban Stud. **47**, 2747–2783 (2010)
8. Noulas, A., Scellato, S., Mascolo, C., Pontil, M.: An empirical study of geographic user activity patterns in foursquare. Assoc. Adv. Artif. Intell. **11**, 570–573 (2011)
9. Song, C., Qu, Z., Blumm, N., Barabasi, A.L.: Limits of predictability in human mobility. Science **327**, 1018–1021 (2010)
10. Tan, Z.: Spatial advertisement competition: based on game theory. J. Appl. Math. (2014)
11. Wang, Y., Yuan, N.J., Lian, D., Xu, L., Xie, X., Chen, E., Rui, Y.: Regularity, conformity: Location prediction using heterogeneous mobility data. In: Proceedings of the 21th ACM SIGKDD International Conference on Knowledge Discovery and Data Mining, pp. 1275–1284 (2015)
12. Yang, N., Kong, X., Wang, F., Yu, P.S.: When and where: Predicting human movements based on socail spatial-temporal events. In: Proceedings of 2014 SIAM International Conference on Data Mining (SDM 2014), pp. 515–523 (2014)
13. Ye, M., Yin, P., Lee, W.-C.: Location recommendation for location-based social networks. In: Proceedings of the 18th SIGSPATIAL International Conference on Advances in Geographic Information Systems, pp. 458–461 (2010)
14. Yuan, Q., Cong, G., Ma, Z., Sun, A., Magnenat-Thalmann, N.: Who, where, when, what: discover spatio-temporal topics for twitter users. In: Proceedings of the 19th ACM SIGKDD International Conference on Knowledge Discovery and Data Mining, pp. 605–613 (2013)

Recommender Systems

Random Partition Factorization Machines for Context-Aware Recommendations

Shaoqing Wang[1,2], Cuilan Du[3], Kankan Zhao[1], Cuiping Li[1], Yangxi Li[3(✉)],
Yang Zheng[1], Zheng Wang[1], and Hong Chen[1]

[1] Key Lab of Data Engineering and Knowledge Engineering of MOE,
Renmin University of China, Beijing, China
wsq@ruc.edu.cn
[2] School of Computer Science and Technology, Shandong University of Technology,
Zibo, China
[3] National Computer Network Emergency Response Technical Team/Coordination
Center of China (CNCERT/CC), Beijing, China
liyangxi@outlook.com

Abstract. Fusing hierarchical information implied into contexts can significantly improve predictive accuracy in recommender systems. We propose a Random Partition Factorization Machines (RPFM) by adopting random decision trees to split the contexts hierarchically to better capture the local complex interplay. The intuition here is that local homogeneous contexts tend to generate similar ratings. During prediction, our method goes through from the root to the leaves and borrows from predictions at higher level when there is sparseness at lower level. Other than estimation accuracy of ratings, RPFM also reduces the over-fitting by building an ensemble model on multiple decision trees. We test RPFM on three different benchmark contextual datasets. Experimental results demonstrate that RPFM outperforms state-of-the-art context-aware recommendation methods.

Keywords: Context-aware recommendations · Hierarchical information · Factorization machines · Random decision trees

1 Introduction

With the growth of information, recommender systems have become an important tool to help users to easily find the favorite items. Collaborative Filtering

This work is supported by National Basic Research Program of China (973) (No. 2014CB340403, No. 2012CB316205), National High Technology Research and Development Program of China (863) (No. 2014AA015204) and NSFC under the grant No. 61272137, 61033010, 61202114 and NSSFC (No. 12&ZD220), and the Fundamental Research Funds for the Central Universities, and the Research Funds of Renmin University of China (15XNH113, 15XNLQ06). It was partially done when the authors worked in SA Center for Big Data Research in RUC. This Center is funded by a Chinese National 111 Project Attracting.

© Springer International Publishing Switzerland 2016
B. Cui et al. (Eds.): WAIM 2016, Part I, LNCS 9658, pp. 219–230, 2016.
DOI: 10.1007/978-3-319-39937-9_17

(CF) methods that behind the recommender systems have been developed for many years and is still a hot research topic up to now.

User's decisions (e.g. clicked, purchased, re-tweeted, commented) to the relevant items are made under the certain environments which is often referred to as *context*. The context which includes time, location, mood, companion and so on can be collected easily in real-world applications. Compared to conventional recommendation solely based on user-item interactions, context-aware recommendation (CAR) can significantly improve the recommendation quality,

For this purpose, a great number of context-aware recommendation methods [4,5,8] have been proposed. Among them, Factorization Machines (FM) [8] is currently an influential and popular one. It represents the user-item-context interactions as a linear combination of the latent factors to be inferred from the data and treats the latent factors of user, item, and context equality. Despite its successful application, existing FM model is weak to utilize hierarchical information. In practice, hierarchies can capture broad contextual information at different levels and hence ought to be exploited to improve the recommendation quality. The intuition here is that local homogeneous contexts tend to generate similar ratings. For example, many men who are engaged in IT department like to browse on technology websites in office during the day. However, they enjoy visiting sport websites at home in the evening. Here, users may be arranged in a hierarchy based on gender or occupation, websites may be characterized by contents, and there are natural hierarchies for time and location.

In this paper, we focus on solving the problem of exploiting the hierarchical information to improve the recommendation quality. The main contribution of the paper are summarized as follows:

1. FM model is one of the most successful approaches for context-aware recommendation. However, There is only one set of the model parameters which can be learned from the whole training set. We propose the novel RPFM model which makes use of the intuition that similar ratings can be generated from homogeneous environments.
2. We adopt the k-means cluster method to partition the user-item-context interactions at each node of decision trees. The similarity between the latent factor vectors of FM model can be used to partition the user-item-context interactions. The subset at each node is expected to be more impacted each other.
3. We conduct experiments on three datasets and compare it to five state-of-the-art context-aware recommendations to demonstrate RPFM's performance.

2   Related Works

2.1   Context-Aware Recommendation

In general, there are three types of integration method [2]: (i) contextual pre-filtering method; (ii) contextual post-filtering method; and (iii) contextual modeling method. In contrast to the previous two methods, the contextual modeling method uses all the contextual and user-item information simultaneously to make predictions. More recent works have focused on the third method [4,5,9,10].

2.2 Random Partition on Tree Structure

Fan et al. [3] proposes Random Decision Trees which are applicable for classification and regression to partition the rating matrix and build ensemble. Each time, according to the feature and threshold which were selected randomly, the instances at each intermediate nodes are partitioned into two parts. Zhong et al. [11] proposes Random Partition Matrix Factorization (RPMF), based on a tree structure constructed by using an efficient random partition technique, which explore low-rank approximation to the current sub rating-matrix at each node. RPMF combines the predictions at each node (non-leaf and leaf) on the decision path from root to leaves. Liu et al. [6] handle contextual information by using random decision trees to partition the original user-item-rating matrix such that the ratings with similar contexts are grouped. Matrix factorization is then employed to predict missing ratings of users for items in the partitioned sub-matrix.

3   Preliminaries

Factorization Machines (FM), proposed by Rendle [8], is a general predictor. The model equation for FM of degree $d = 2$ is defined as:

$$\hat{y}(x_i) = \omega_0 + \sum_{j=1}^{p} \omega_j x_{i,j} + \sum_{j=1}^{p} \sum_{j'=j+1}^{p} < \mathbf{v}_j, \mathbf{v}_{j'} > x_{i,j} x_{i,j'}, \tag{1}$$

and

$$< \mathbf{v}_j, \mathbf{v}_{j'} > := \sum_{k=1}^{f} v_{j,k} \cdot v_{j',k}, \tag{2}$$

where the mode parameters Θ that have to be estimated are:

$$\omega_0 \in \mathbb{R}, \quad \mathbf{w} \in \mathbb{R}^p, \quad \mathbf{V} \in \mathbb{R}^{f \times p}. \tag{3}$$

A row vector $\mathbf{v}_i$ of $\mathbf{V}$ represents the i-th variable with f factors. $f \in \mathbb{N}_0^+$ is the dimensionality of the factorization.

The model equation of a factorization machine in Eq. (1) can be computed in linear time $O(f * p)$ because the pairwise interaction can be reformulated:

$$\sum_{j=1}^{p} \sum_{j'=j+1}^{p} < \mathbf{v}_j, \mathbf{v}_{j'} > x_{i,j} x_{i,j'}$$

$$= \frac{1}{2} \sum_{k=1}^{f} ((\sum_{j=1}^{p} v_{j,k} x_{i,j})^2 - \sum_{j=1}^{p} v_{j,k}^2 x_{i,j}^2)$$

Table 1 shows an example of input formation of training set. Here, there are $|U| = 3$ users, $|I| = 4$ items, $|L| = 4$ locations, etc. which are binary indicator variables.

$$U = \{u_1, u_2, u_3\}$$
$$I = \{i_1, i_2, i_3, i_4\}$$
$$L = \{l_1, l_2, l_3, l_4\}$$

For simplicity, we only consider categorical features in the paper. Table 2 shows the model parameters learned from the training set which is shown in Table 1.

Table 1. An example of training set of FM model

	Users			Items				Locations				...	Ratings
x_1	1	0	0	1	0	0	0	1	0	0	0	...	4
x_2	1	0	0	0	1	0	0	0	1	0	0	...	3
x_2	1	0	0	0	1	0	0	0	1	0	0	...	3
x_4	0	1	0	0	0	0	1	0	0	1	0	...	5
x_5	0	0	1	1	0	0	0	0	0	1	0	...	2
x_6	0	0	1	0	1	0	0	0	0	0	1		3

Table 2. An example of parameters' values of FM model

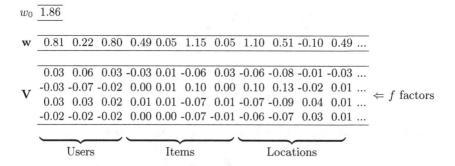

w_0 1.86

w   0.81   0.22   0.80   0.49   0.05   1.15   0.05   1.10   0.51   -0.10   0.49 ...

V
0.03   0.06   0.03   -0.03   0.01   -0.06   0.03   -0.06   -0.08   -0.01   -0.03 ...
-0.03   -0.07   -0.02   0.00   0.01   0.10   0.00   0.10   0.13   -0.02   0.01 ... $\Leftarrow f$ factors
0.03   0.03   0.02   0.01   0.01   -0.07   0.01   -0.07   -0.09   0.04   0.01 ...
-0.02   -0.02   -0.02   0.00   0.00   -0.07   -0.01   -0.06   -0.07   0.03   0.01 ...

Users   Items   Locations

4   Random Partition Factorization Machines

The intuition is that there are similar rating behavior among users under the same or similar contextual environments. Motivated by [11], We describe the proposed Random Partition Factorization Machines (RPFM) for context-aware recommendations. In order to efficiently take advantage of different contextual information, we adopt the idea the random decision trees algorithm.

The rational is to partition the original training set R such that the tuples generated by the similar users, items or contexts are grouped into the same node.

Tuples in the same cluster are expected to be more correlated each other than those in original training set R. The main flow can be found in Algorithm 2 and Fig. 1.

To begin with, there is an input parameter S, the structure of decision trees, which can be generated by Algorithm 1 and determined by cross validation. The parameter S includes contexts for partition at each level, numbers of clusters at each node. The maximial depth of trees can be inferred from the parameter S. For instance, if the value of S is 'C2:4,C3:6,C1:10,C0:5', the meaning is: (i) At the root node of decision trees, the R can be divided into four groups by using k-means method according to the similarity between factor vectors of context C_2. Subsequently, the set at each node of 2nd, 3rd and 4th level of decision trees can be respectively divided into six, ten and five groups according to the similarity between factor vectors of context C_3, C_1 and C_0 using k-means method. (ii) The maximal depth of each tree is five because there are four intermediate levels and one terminal level.

Algorithm 1. GenerateTreesStructure

Input: Depth of decsion trees: h, Numbers of Each Context: $n_0,...,n_{m-1}$
Output: Structure of Decision Trees
1: Initialize contextual information set A;
2: **for** $i=0$ to $m-1$ **do**
3:     $A(i)=C_i$;
4: **end for**
5: $j=0$;
6: **for** $i=0$ to $m-1$ **do**
7:     Select a context C_r from A randomly;
8:     Select $k \in [2, n_r]$ randomly such that the set at each node of $(i+1)$th level of decision trees will be divided into k groups;
9:     Add C_r and k to S;
10:     **if** $j >= h$ **then**
11:         break;
12:     **end if**
13:     $A = A \setminus C_r$;
14:     $j++$;
15: **end for**
16: **return** S;

At each node, we learn the model parameter using FM model.

$$\hat{\omega}_0, \hat{\mathbf{w}}, \hat{\mathbf{V}} = arg \min_{\omega_0, \mathbf{w}, \mathbf{V}} \sum_{i=1}^{|R|} (y(x_i) - \hat{y}(x_i))^2 + \lambda \sum_{j=1}^{p} \|\mathbf{V}_j - \mathbf{V}_j^{pa}\|^2 \qquad (4)$$

where $\| * \|$ is the Frobenius norm and $\mathbf{V}^{pa}$ is the latent factor matrix at parent node. The parameter λ controls the extent of regularization. Equation (4) can be solved using two approaches: (i) stochastic gradient descent (SGD) algorithms,

---

Algorithm 2. RPFM

Input: Training Set: R, Dimensionality of Latent Factor Vectors: f, Number of Trees: N, Structure of Trees: S, Number of Least Support Instances at the Leaf Node: $leastL$, Similarity Function in k-means method: fun, Learning Rate: η, Regularization Values: λ

Output: Tree Ensemble with Model Parameters and Cluster Information at Each Node.

1: Get maximum depth of each tree to h according to S;
2: **for** $i = 1$ to N **do**
3:    Build tree T_i;
4:    **for** $d = 2$ to h **do**
5:       Get number of clusters to k and context for partition to C_r using S and current level d-1;
6:       Learn the parameters Θ using Eq. 1 at each node of the current level d-1;
7:       Partition R into k clusters using context C_r and matrix $\mathbf{V}$ in Θ by taking advantage of k-means method;
8:       **for** $j = 1$ to k **do**
9:          **if** the size of cluster j is less than $leastL$ **then**
10:             Let the current node as leaf node;
11:             break;
12:          **end if**
13:       **end for**
14:       **if** $j > k$ **then**
15:          **for** $j = 1$ to k **do**
16:             Decompose R_{dj} recursively;
17:          **end for**
18:       **end if**
19:    **end for**
20: **end for**
21: **return** $\{T_i\}_{i=1}^N$;

---

which are very popular for optimizing factorization models as they are simple, work well with different loss functions. The SGD algorithm for FM has a linear computational and constant storage complexity [8]. and (ii) alternating least squares (ALS) algorithms, that iteratively solves a least-squares problem per model parameter and updates each model parameter with the optimal solution [9]. Here, $\mathbf{V}$ is $f \times p$ matrix of which f is the dimensionality of factor vectors and $p = n_0 + n_1 + ...n_{m-1}$. n_i is the number of context C_i, m is the number of contextual variables. For simplicity, we denote user set as C_0 and item set as C_1. Each of the $f \times n_i$ sub-matrix is the latent representation of context C_i, as shown in Table 2. The smaller the distance among the factor vectors of context C_i, the greater the similarity.

To partition the training set R, we exact the context and the number of clusters according to the tree structure S and current level. We group the similar latent vectors of context C by making use of the k-means method, In Table 2, suppose we get the context C_1(i.e. Item) and number of clusters $k = 2$ according to input parameter S. Then the initial cluster central points selected randomly are i_1 and i_2. Subsequently, the generated clustering result could be $\{i_1, i_3, i_4\}$ and $\{i_2\}$. Lastly, the training set in the current node can be divided into two groups according to the clustering result of context C_1(i.e. Item) and the value of C_1(i.e. Item) of tuples. In other words, the current node has two child nodes. The subset of one chid node include the tuples whose value of C_1(i.e. Item) $\in \{i_1, i_3, i_4\}$, the remaining tuples are assigned to the other child node.

The partition process stops once one of following conditions is met: (i) the height of a tree exceeds the limitation which can be inferred from the given tree structure parameter S; (ii) the number of tuples at each child node of current node is less than the number of least support tuples $leastL$.

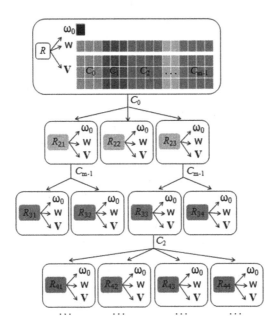

Fig. 1. Random decision trees (one tree)

During training, the function of each non-leaf node is to separate training set by making use of the clustering result of special context, such that the tuples in the subset have more impact each other. However leaf nodes are responsible for prediction.

Note that in different decision trees, the training set is divided differently because that initial k cluster central points are selected randomly at each node of decision trees.

During prediction, for a given case x_i in the test set, we transfer it from root node to leaf node at each tree using the clustering information of each non-leaf node. For instance, the value of S is 'C0:3,C1:2,C2:4' and a test case $x_i = \{u_3, i_1, l_2\}$ corresponding to Table 1. Thus from the root node, the x_i would be transfer to node (e.g. R_{23}) which include i_1 at second level. Then from the node R_{23}, the x_i would be transfer to node (e.g. R_{33}) which include u_3 at third level. Subsequently, from the node R_{33}, the x_i would be transfer to node (e.g. R_{41}) which include l_2 at fourth level. At the target leaf node, the rating can be predicted by taking advantage of Eq. (1) and the parameters learned by the training subset. To the end, the predictions from all trees are combined to obtain the final prediction as shown in Eq. (5)

$$\hat{y}(x_i) = \frac{\sum_{t=1}^{N} \hat{y}_t(x_i)}{N} \tag{5}$$

After partitioning the original training set, the tuples at each leaf node have the more influence on each other. So, the FM model at each leaf node can achieve high quality recommendation. By combining multiple predictions from different decision trees, all subsets in which the tuples are more correlated are comprehensively investigated, personalized and accurate context-aware recommendations can be generated.

5   Experiments

5.1   Datasets

We conduct our experiments on three datasets: the Adom. dataset [1], the Food dataset [7] as well as the Yahoo! Webscope dataset[1].

The **Adom.** dataset [1] contains 1757 ratings by 117 users for 226 movies with many contextual information. The rating scale rang from 1(hate) to 13(absolutely love). However there are missing values in some tuples. After removing the tuples containing missing values, there are 1464 ratings by 84 users for 192 movies in Adom. dataset. We keep 5 contextual information: withwhom, day of the week, if it was on the opening weekend, month, and year seen.

The **Food** dataset [7] contains 6360 ratings (1 to 5 stars) by 212 users for 20 menu items. We select 2 context variables. One context variable captures how hungry the user is: normal, hungry and full. The second one describes if the situation in which the user rates is virtual or real to be hungry.

The **Yahoo!** Webscope dataset contains 221,367 ratings (1 to 5 stars), for 11,915 movies by 7,642 users. There is no contextual information. However the dataset contains user's age and gender features. Just like [9], we also follow [4] and apply their method to generate modified dataset. In other words, we modify the original Yahoo! dataset by replacing the gender feature with a new artificial feature $C \in \{0, 1\}$ that was assigned randomly to the value 1 or 0 for each rating. This feature C represents a contextual condition that can affect the rating. We randomly choose 50 % items from the dataset, and for these items we randomly pick 50 % of the ratings to modify. We increase (or decrease) the rating value by one if $C = 1(C = 0)$ if the rating value was not already 5(1).

5.2   Setup and Metrics

We assess the performance of the models by conducting a 5-fold cross-validation and use the most popular metrics: the Mean Absolute Error (MAE) and Root Mean Square Error (RMSE).

---

[1] R4, http://webscope.sandbox.yahoo.com/catalog.php?datatype=r

5.3   Performance Comparsion

To begin with, we determine the structures of decision trees, i.e., input parameters S, by Algorithm 1. The parameters are 'C2:2,C6:2,C5:2,C3:3,C0:2,C4:5, C1:5', 'C3:3,C2:2,C0:5,C1:4' and 'C3:2,C2:2,C0:2,C1:2' for Adom., Food and Yahoo! dataset respectively. Then, we select 0.01 as the values of learning rate and regularization.

Comparison to Factorization-Based Context-Aware Methods

- **FM** [8] is easily applicable to a wide variety of context by specifying only the input data and achieves fast runtime both in training and prediction.
- **Multiverse** Recommendation [4] is a contextual collaborative filtering model using N dimensional tensor factorization.
- **COT** [5] represents the common semantic effects of contexts as a contextual operating tensor and represents a context as a latent vector.

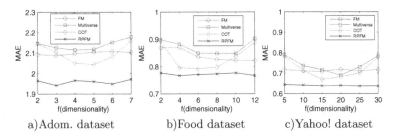

a) Adom. dataset   b) Food dataset   c) Yahoo! dataset

Fig. 2. MAE over three datasets with different dimensionality of latent factor vectors

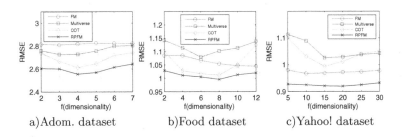

a) Adom. dataset   b) Food dataset   c) Yahoo! dataset

Fig. 3. RMSE over three datasets with different dimensionality of latent factor vectors

Dimensionality of latent factor vectors is one of the important parameters. Though latent factor vectors' dimensionality of various contexts can be different in Multiverse and COT. In order to compare with FM and our proposed RPFM, we just take into account the equal dimensionality of latent factor vectors of various contexts. The scale of three datasets is different, so we run models with

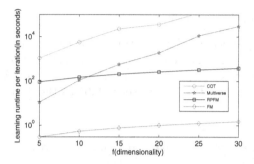

Fig. 4. Learning runtime in seconds for one iteration over the whole training set(in log-y scale) over Yahoo! dataset with different latent dimensions (Color figure online)

$f \in \{2,3,4,5,6,7\}$ over Adom. dataset, $f \in \{2,4,6,8,10,12\}$ over Food dataset and $f \in \{5,10,15,20,25,30\}$ over Yahoo! dataset. Figures 2 and 3 show the result of FM, Multiverse, COT and RPFM over the three real world datasets.

We notice that in all experiment scenarios, dimensionality of latent factor vectors in RPFM is not sensitive and RPFM is more accurate than other recommendation models. These results show that in homogeneous environment which can be obtained by applying random decision trees to partition the original training set, users have similar rating behavior.

High computational complexity for both learning and prediction is one of the main disadvantages of Multiverse and COT. This make them hard to apply for larger dimensionality of latent factor vectors. In contrast to this, the computational complexity of FM and RPFM is linear. In order to compare the runtime of various models, we do experiment on Yahoo! dataset for one full iteration over whole training set. Figure 4 shows that the learning runtime of RPFM is faster than that of Multiverse and COT with increasing the dimensionality, however slower than that of FM which is obvious because RPFM generates an ensemble which reduces the prediction error.

Comparison to Random Partition-Based Context-Aware Methods

- **RPMF** [11] adopted a random partition approach to group similar users and items by taking advantage of decision trees. The tuples at each node of decision trees have more impact each other. Then matrix factorization is applied at each node to predict the missing ratings.
- **SoCo** [6] explicitly handle contextual information which means SoCo partitions the training set based on the values of real contexts. SoCo incorporate social network information to make recommendation. There are not social network information in our selected datasets, So we just consider SoCo without social network information.

Both the number and depth of trees have important impact on the decision tree based prediction methods. Because of space limitations, we just report the experimental result over Food dataset.

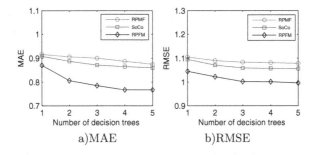

Fig. 5. Impact of number of trees over Food dataset (Color figure online)

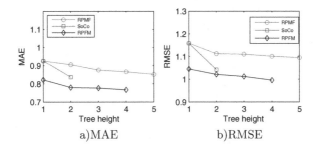

Fig. 6. Impact of depth of trees over Food dataset (Color figure online)

As shown in Fig. 5, we observe that RPFM achieve the best performance compared with RPMF and SoCo. And we notice that MAE/RMSE decreases with increasing number of trees, which means more trees produces higher accuracy. However, when the number of trees increases to around 3, improvements on prediction quality become negligible. We thus conclude that even a small number of trees are sufficient for decision tree based models.

The depth of trees which is one of the input parameters in RPMF can be very large because it can select a latent factor from U, V and a splitting point randomly at each intermediate node during building the decision trees. Here, we define the maximal depth of trees as five in RPMF. In SoCo, the maximal depth of trees equals the number of contextual variables excluding user and item. Specially, the maximal depth of trees over Food dataset is two in SoCo. However, both user and item can be considered as contextual variables in RPFM. Then the maximal depth of trees over Food dataset is four in RPFM. Figure 6 shows that the deeper of trees, the better prediction quality, and RPFM outperforms RPMF and SoCo in terms of MAE and RMSE.

6   Conclusion

In this paper, we propose Random Partition Factorization Machines (RPFM) for context-aware recommendations. RPFM adopts random decision trees to partition the original training set using k-means method. Factorization machines (FM) is then employed to learn the model parameters at each node of the trees and predict missing ratings of users for items under some specific contexts at leaf nodes of the trees. Experimental results demonstrate that RPFM outperforms state-of-the-art context-aware recommendation methods.

References

1. Adomavicius, G., Sankaranarayanan, R., Sen, S., Tuzhilin, A.: Incorporating contextual information in recommender systems using a multidimensional approach. ACM Trans. Inf. Syst. (TOIS) **23**(1), 103–145 (2005)
2. Adomavicius, G., Tuzhilin, A.: Context-aware recommender systems. In: Recommender Systems Handbook, pp. 217–253. Springer, New York (2011)
3. Fan, W., Wang, H., Yu, P.S., Ma, S.: Is random model better? on its accuracy and efficiency. In: ICDM 2003, pp. 51–58. IEEE (2003)
4. Karatzoglou, A., Amatriain, X., Baltrunas, L., Oliver, N.: Multiverse recommendation: n-dimensional tensor factorization for context-aware collaborative filtering. In: Proceedings of the Fourth ACM Conference on Recommender Systems, pp. 79–86. ACM (2010)
5. Liu, Q., Wu, S., Wang, L.: COT: contextual operating tensor for context-aware recommender systems. In: Twenty-Ninth AAAI Conference on Artificial Intelligence (2015)
6. Liu, X., Aberer, K.: SoCo: a social network aided context-aware recommender system. In: Proceedings of the 22nd International Conference on World Wide Web, pp. 781–802. International World Wide Web Conferences Steering Committee (2013)
7. Ono, C., Takishima, Y., Motomura, Y., Asoh, H.: Context-aware preference model based on a study of difference between real and supposed situation data. In: Houben, G.-J., McCalla, G., Pianesi, F., Zancanaro, M. (eds.) UMAP 2009. LNCS, vol. 5535, pp. 102–113. Springer, Heidelberg (2009)
8. Rendle, S.: Factorization machines. In: 2010 IEEE 10th International Conference on Data Mining (ICDM), pp. 995–1000. IEEE (2010)
9. Rendle, S., Gantner, Z., Freudenthaler, C., Schmidt-Thieme, L.: Fast context-aware recommendations with factorization machines. In: SIGIR 2011, pp. 635–644. ACM (2011)
10. Yin, H., Cui, B., Chen, L., Hu, Z., Huang, Z.: A temporal context-aware model for user behavior modeling in social media systems. In: SIGMOD 2014, pp. 1543–1554. ACM (2014)
11. Zhong, E., Fan, W., Yang, Q.: Contextual collaborative filtering via hierarchical matrix factorization. In: SDM, pp. 744–755. SIAM (2012)

A Novel Framework to Process the Quantity and Quality of User Behavior Data in Recommender Systems

Penghua Yu[1], Lanfen Lin[1(✉)], and Yuangang Yao[2]

[1] College of Computer Science and Technology, Zhejiang University,
Hangzhou, China
{yph719,llf}@zju.edu.cn
[2] China Information Technology Security Evaluation Center, Beijing, China
yaoyg@itsec.gov.cn

Abstract. Recommender system has become one of the most popular techniques to cope with the information overload problem. In the past years many algorithms have been proposed to obtain accurate recommendations. Such methods usually put all the collected user data into learning models without a careful consideration of the quantity and quality of individual user feedbacks. Yet in real applications, different types of users tend to represent preferences and opinions in various ways, thus resulting in user data with radically diverse quantity and quality. This characteristic of data influences the performance of recommendations. However, little attention has been devoted to the management of quantity and quality for user data in recommender systems. In this paper, we propose a generic framework to seamlessly exploit different pre-processing and recommendation approaches for ratings of different users. More specifically, we first classify users into groups based on the quantity and quality of their behavior data. In order to handle the user groups diversely, we further propose several data pre-processing strategies. Subsequently, we present a novel transfer latent factor model (TLMF) to transfer learnt models between groups. Finally, we conduct extensive experiments on a large data set and demonstrate the effectiveness of our proposed framework.

Keywords: Recommender system · Quantity and quality · Pre-processing · Transfer learning

1 Introduction

With the rapid growth of online information, techniques that attempt to help people to deal with the information overload issue are becoming indispensable. Recommender system [1] is such a promising technique that produces recommendations for users from a huge collection of items. In general, the degree of a user preference is presented by explicit ratings provided by the given user, or implicit feedbacks inferred from user behaviors. For simplification, we treat all user behavior data as ratings in this work.

Currently, many successful recommendation algorithms [10, 16, 19, 20] have been proposed. Most researchers still focus on introducing and improving more effective and

© Springer International Publishing Switzerland 2016
B. Cui et al. (Eds.): WAIM 2016, Part I, LNCS 9658, pp. 231–243, 2016.
DOI: 10.1007/978-3-319-39937-9_18

efficient recommendation approaches, afterwards the whole data set is utilized for the recommendation performance evaluation [1, 8]. That is they usually put all the collected user ratings into training models, but the quantity and quality of individual data are not well considered. Lately, hybrid methods have been utilized so as to combine advantages of various recommendation approaches [9]. Nevertheless, each component of a given hybrid approach implements a specific recommendation method with the given user data. In a nutshell, recommendation approaches that take all user ratings as input tend to suffer from two major challenges. (1) From the view of *data quantity*, the scalability is a big concern [4] because user data usually increases with a high velocity. Thus the computational cost is extremely expensive if all user ratings are taken as input. (2) From the perspective of *data quality*, traditional methods basically follow the assumption that user ratings correctly reflect their opinions and interests. But prior studies have reported that user ratings are naturally imperfect and noisy [2, 8], which limit the measurable power of a recommender system. It is also known as the magic barrier of recommender system [8], which refers to the upper bound on the rating prediction accuracy.

Several approaches have been introduced to handle the natural noise in recommender systems [3, 5, 11, 13]. O'Mahony et al. [13] classified noise in recommender systems into natural (naturally occurred) and malicious (deliberately inserted) and proposed to remove noisy ratings to improve accuracy. Other researchers proposed to correct noisy ratings in collaborative recommender system by the user re-rating [3] or automatically noise detection methods [11, 14]. Recently, the concept of user coherence [5] has been introduced and shown correlation with the magic barrier. As for the scalable problem, researchers try to speed up the computation by deploying some distributed computing frameworks [18] or by designing parallel strategies [20]. In fact, such methods cannot decrease the essential cost because the data quantity to be used is not changed.

However, to the best of our knowledge, only a few papers have been published to investigate how to manage the quantity and quality of user ratings prior to applying any recommendation approach. In real world recommender systems, different types of users typically represent preferences in various ways, thus resulting in users with radically distinct data quantity and quality. For instance, some users contribute plentiful ratings while others provide only few ratings. At the same time, some users generate consistent ratings while others produce inconsistent ratings. This diversity of data would influence the results of recommendation methods. As a consequence, it would be ideal to process differently for various type of users with diverse data quantity and quality.

In this paper, we attempt to investigate how to process and model the quantity and quality of user ratings to obtain good recommendations with representative and reliable ratings. The contributions of this paper are summarized as follows.

- We propose a generic framework to seamlessly exploit different pre-processing and recommendation approaches for ratings of various users in recommender systems.
- We classify users into groups according to the quantity and quality of their ratings and present several preprocessing strategies for these user groups.
- We finally introduce a novel transfer latent factor model (TLMF) to transfer learning models between distinct user groups.

The remainder of this paper proceeds as follows. In Sect. 2, we discuss previous work on data quantity and quantity in recommender system. In Sect. 3, we describe the proposed framework and its components. The experimental results are reported in Sect. 4. Finally, we make conclusions and discuss the future work in Sect. 5.

2 Related Work

Many successful recommendation algorithms have been proposed in the past years, among which collaborative filtering (CF) [1] is the most popular method that has been widely applied. Neighborhood methods and latent factor models (LFM) are two main directions of the CF. The former approaches focus on computing relationships between items or alternatively between users, e.g., the item based CF predicts a user's rating for an item using ratings of the neighboring items by the same user [17]. The latter LFM approaches attempt to explain observed ratings by characterizing both items and users with a small number of latent factors, which are inferred from item rating patterns [10]. Nowadays, LFM and its variations have become popular because they could combine the good scalability with high predictive accuracy [10, 16, 19].

Recently, inconstancies in user ratings have drawn a lot of attention. Herlocker et al. [8] discussed the noise in user ratings and presented the concept of "*magic barrier*" for recommender systems, which claimed that there is an upper bound on the performance prediction because user ratings are noisy and inconsistent. Hence, researchers realized that the assumption that "collected ratings always reflect user opinions and interest" is not entirely tenable, and started to pay attention to noisy ratings in recommender system rather than blindly modifying current models to enhance recommendations. Several studies have been published to deal with it. O'Mahony et al. [13] classified the noise in recommender systems into natural and malicious and proposed to remove noisy ratings to improve the accuracy. An inter-active method based on re-rating process has been presented [3] that when any noisy rating has been discovered, the system could ask users to rate the corresponding items again by using the test-retest procedure. Li et al. [11] alternatively proposed to auto-matically detect noisy but non-malicious user in social recommender systems based on the assumption that ratings provided by a same user on a closely correlated items should have similar scores. Lately, Bellogín et al. [5] introduced the concept of user coherence and verified that it is correlated with the magic barrier, thus it could be used to discriminate users. This work has also pointed out that different methods should be applied for diverse users. However, it only takes coherence to separate users and further processing and modeling are not discussed.

In terms of the rating quantity, cold start [1] has been a long-standing problem. Eliciting preferences [12, 15] for new users is an effective way to solve the cold start problem. These elicited items are afterwards utilized to ask for ratings by presenting to new users directly. A successful elicitation approach should be able to minimize the user's interaction effort and maximize the information we can obtain from these items so as to increase the recommendation accuracy [13]. Elicitation strategies that based on the value-of-information [13, 15], such as item popularity, entropy and variance, have been firstly considered. Then model based approaches including the greedy algorithm, item clustering, and boosting multiple tree [12] are been employed. In order to boost the elicitations, Chang et al. [6] proposed to ask multiple questions at each trial with groups of items. However, to the best of our knowledge, little literature is available on representative rating sampling for the heavy users with vast ratings. Yet problems, such as whether all ratings for these kinds of users are necessary for recommendations and which ratings are representative, are not clearly answered.

3   The Proposed Framework and Its Components

3.1   A Generic Framework

We introduce the standpoint that users showing different behavioral patterns should be processed and modeled differently in recommender systems. Concretely, we propose a generic framework as illustrated in Fig. 1. Here user behavioral patterns are described by the quantity and quality of their contributed data. We first classify users into groups. Then we design several pre-processing strategies for these subsets. At last, to transfer learning models between user groups we present a new recommendation method.

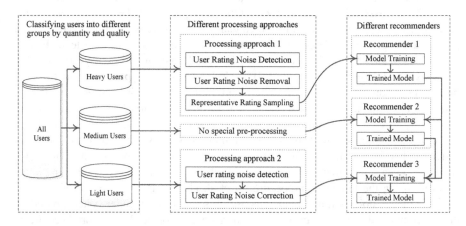

Fig. 1. The processing workflow of the proposed framework.

3.2   Classifying Users into Groups

Data quantity and quality are chosen to differentiate users. The quantity is characterized by the number of ratings a given user has rated. Users are classified into three groups by quantity including heavy, medium, and light. For the data quality, we follow the definition of coherence introduced in [5], and split users into the easy or difficult group. Given a user u, item feature space F. $I(u,f)$ and $I(u,F)$ denote the number of items u that belong to a specific item feature f, and the total of all item feature space. $\overline{r_{uf}}$ is the average value of $I(u,f)$. Thus, we define the coherence of u as follows.

$$c(u) = -\sum_{f\in F}\sqrt{\sum_{i\in I(u,f)}\left(r_{ui} - \overline{r_{uf}}\right)^2}\,\frac{\|I(u,f)\|}{\|I(u,F)\|} \tag{1}$$

By considering both the quantity and quality of user behavior data, we thus classify a given user into one the following six groups.

- Heavy Easy User Group (HEUG), contains users that produce tremendous ratings (heavy) and gain higher coherence (easy means ratings are more consistent);
- Heavy Difficult User Group (HDUG), makes up of users that have plentiful ratings but with lower coherence (difficult as ratings are inconsistent);
- Medium Easy User Group (MEUG), consists of users that produce a medium number of ratings and show high coherence in their ratings;
- Medium Difficult User Group (MDUG), includes users contribute a medium amount of ratings but behave in highly inconsistent manners;
- Light Easy User Group (LEUG), embodies users that generate a few ratings (light) and exhibit high coherence in their ratings;
- Light Difficult User Group (LDUG), takes on users that a few ratings and with lower coherence, which implies users in this groups hardly have reliable data.

3.3   Processing Noisy Ratings

It has been confirmed that the noisy ratings have a negative effect on recommendations. In this section, we present the data quality processing strategies. To begin with, we need to distinguish noisy ratings from these reliable ones by a specific metric.

Noisy Rating Detection. For a rating r_{ui}, we define the rating noisy degree (RND) to detect whether it is noise or not. The item features f_i are utilized to consider the consistent degree of this rating to items have the same features. The RND for r_{ui} is defined as the ratio between features that have high relative deviation, more than the threshold ϑ, to its same feature item set and the total number of features. See in Eq. 2.

$$RND(r_{ui}) = \sum I\left(\sum_{f_i}\frac{|r_{ui} - \overline{r}_{uf}|}{\overline{r}_{uf}} > \vartheta\right)\Big/ \|f_i\| \tag{2}$$

Noisy Rating Processing. Next we propose to deal with noisy ratings diversely for different user groups. We claim that the same processing strategy may harm light users as it would worsen the sparsity problem, which users have relatively few ratings. Thereafter, we present the following different noisy user rating processing procedures:

- Non-processing, which does nothing with the detected noisy ratings.
- Noise Removal (NR), which removes the detected noisy ratings.
- Noise Correction (NC), which corrects those detected noisy ratings. Here we simply correct them with the average rating of items have the same features. It proceeds as,

$$Correction(r_{ui}) = \sum\nolimits_{f_i \in F_i} \bar{r}_{uf_i} \Big/ \|F_i\| \tag{3}$$

It would be more beneficial if we manage noisy ratings variously based on the rating quantity and quality of individual users. We remove noisy ratings for heavy users (both easy and difficult) because those users tend to have abundant data, and perform non-processing for the medium easy user group. As for the light (both easy and difficult) users and the medium difficult users, we correct their noisy data since they provide scare informative behavior data.

3.4 Sampling Representative Ratings

It is necessary to perform sampling processing before training models, especially for heavy users because their ratings probably contain redundant and repeated information. By performing such sampling we want to keep the most information and save the computational cost, thus a good sampling approach should be able to generate as few user ratings as possible meanwhile gain recommendations as accurate as possible.

We adopt a variation of the preference elicitation strategies, namely the **Harmonic mean of Entropy and Logarithm of Frequency (HELF)** [13]. We replaced the entropy with variance because variance is a diversity measure for ordinal data and it is more suitable for measuring the rating diversity. Moreover, we use the logarithm of inversed frequency because we prefer to keep ratings at the long tail instead of these popular ones at the top. Then we take the harmonic mean of these two variables because the it has a good property that it strongly increases or decreases when both of the components increase or decrease. Hence, we define the **Harmonic mean of Logarithm of Inversed Frequency and Variance (HLIFV)** as follows.

$$HLFV(i) = \frac{2 * (1 + LF(i))^{-1} * Var(i)}{(1 + LF(i))^{-1} + Var(i)} \tag{4}$$

$Var(i)$ and $LF(i)$ denote the normalized variance and logarithm frequency of item i, respectively.

3.5 Transferring Models Between User Groups

In its basic form, latent factor model maps both users and items to a joint latent factor space of dimensionality k, such that user-item interactions are modeled as inner products in that space. Then it models directly only the observed ratings by avoiding over-fitting using a regularized model. Take the BiasedMF [10] for example, the objective function is formulated as follows.

$$\min_{b,p,q} \sum\nolimits_{r_{ui} \in R} \left(r_{ui} - (\mu + b_u + b_i + p_u q_i)\right)^2 + \lambda \left(\sum\nolimits_u \|p\|^2 + \sum\nolimits_i \|q\|^2\right) \tag{5}$$

where μ denotes the overall average rating; parameters b_u and b_i indicate observed deviations of user u and item i, respectively, from the average; p_u and q_i denote latent factor of u and i. The stochastic gradient descent solver is typically applied to optimize this objective function, wherein the algorithm loops through all ratings in the training set. Then it modifies parameters (b_u, b_i, p_u, and q_i) by a magnitude proportional to λ in the opposite direction of each gradient.

We have observed that the data quality and quantity for different user groups vary sharply. Hence, we propose to transfer learning item latent factors between user groups. We name it as the transfer latent factor model (TLMF). In the TLMF, a joint latent factor space is used for different user groups. Observed ratings in the source user groups e.g., the easy user group and heavy user group, are used to update user factors and the joint item latent factors. While ratings in the target user groups, e.g., the difficult user group and the light user group, are merely utilized for updating the user factors. The outline of the TLMF is illustrated in Fig. 2.

Note that in TLFM, the training instances of all involved user groups are utilized for latent factor learning. However, not all training examples for an item are adopted to update the latent factor of the item. We stop updating the latent factors of items when they have been well learnt. This means that the latent factors for these items have been well formed or partially learnt from source user groups, and then transferred to assistant the model learning for target user groups. The algorithm proceeds as follows.

Algorithm: The Transfer Latent Factor Model (TLFM) learnt with SGD

Input: The target user groups GT_1, ..., GT_m, the source user group GS_1, ..., GS_m, and the pre-processed ratings for each user group R_{GT1}, ..., R_{GTm}, R_{GS1}, ..., R_{GSm}. The minimum number of ratings for an item $minCnt$,

Method:
1. Initialize user latent factors P, and item latent factors Q randomly
2. Set $C_{Qi} = 0$
3. for rating R_{ui} in R_{GS1}, ..., R_{GSm}:
   Update Pu according to the applied model and update rule
   Update Qi according to the applied model and update rule
   $C_{Qi} = C_{Qi} + 1$
4. for rating R_{ui} in R_{GT1}, ..., R_{GTm}:
   if $C_{Qi} < minCnt$:
   Update Pu according to the applied model and update rule
   Update Qi according to the applied model and update rule
   $C_{Qi} = C_{Qi} + 1$
   else:
   Only update Pu according to the applied model and update rule
5. Repeat 3 and 4 until convergence or reaching maximum interaction count

Output: P, Q

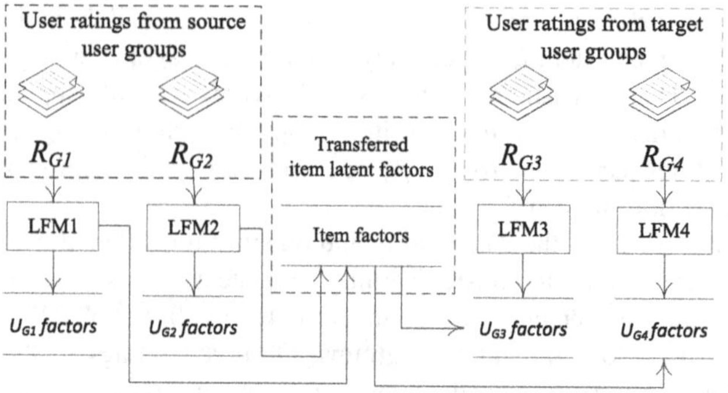

Fig. 2. An illustration of the Transfer Latent Factor Model (TLFM).

4   Experiments

4.1   Data Set

We use a large and real data set – the MovieLens Latest Full (ml-latest)[1], to show effects of proposed framework. It contains 21,000,000 ratings and applied to 30,000 movies by 230,000 users. These data were created between January 09, 1995 and August 06, 2015. Movie genes are adopted as item features, and we remove users with ratings less than 30 and more than 3000.

4.2   Experimental Protocols

We first classify users into six different user groups. For each user groups, we randomly select 20 % ratings of each user as test instances and with the remaining 80 % as training examples. Then we process differently for ratings in each groups and form different sub sets, and we remove noise data in its corresponding test set for evaluations.

We choose the RMSE and Precision@5 in the test data set as metrics (because some light users only have 6 ratings in the test set) to evaluate the performance of different recommendations. For the Precision@5 metric, a threshold of 3.5 is set such that ratings equal to or more than this threshold are seen as relevant. The *minCnt* in TLFM is set to 100 in the following experiments.

4.3   Methods in Comparison

- ItemNN [17]. We first normalize user ratings then apply the cosine as the similarity measure, and choose the neighborhood size to be 30.
- BiasedMF [10], which is formulated in Eq. (5), and we run 100 iterations and use 40 latent factors with learning rate, regularization terms as 0.005 and 0.06.

---

[1] http://grouplens.org/datasets/movielens/.

– Factorization machines (FM) [16]. We run 100 iterations and use 40 latent factors with learning rate, regularization terms as 0.005 and 0.06.

Note that we implement the former two algorithms with the help of Lenskit [7].

4.4 Experimental Results

Analysis on the Data Set and Different User Groups. With 100 and 300 as thresholds to classify light/medium and medium/heavy users, we generate light, medium, and heavy user groups. For each group we keep half users as difficult and the rest half as easy, and form the sub sets created by ratings from each group. The amounts of (users, items, and ratings) for each set are displayed in Table 1. It can be observed that there are considerable number of heavy users in this dataset though light users are of superiority in quantity. We utilize the relative inconsistent degree for noisy rating detection in case of the bias caused by the number of user data, and we set the ϑ as 0.075 for heavy and medium groups and 0.05 for light user groups in the noise detection procedure.

Table 1. Statistical descriptions for sub data sets made up by different user groups.

	Heavy users	Medium users	Light users
Easy users	$\begin{pmatrix} 8253 \\ 18520 \\ 2857282 \end{pmatrix}$	$\begin{pmatrix} 18862 \\ 14423 \\ 2168123 \end{pmatrix}$	$\begin{pmatrix} 30620 \\ 11663 \\ 1153358 \end{pmatrix}$
Difficult users	$\begin{pmatrix} 8253 \\ 25143 \\ 5100663 \end{pmatrix}$	$\begin{pmatrix} 18862 \\ 15428 \\ 2923192 \end{pmatrix}$	$\begin{pmatrix} 30620 \\ 12681 \\ 1583214 \end{pmatrix}$
Sum of above	$\begin{pmatrix} 16506 \\ 26223 \\ 7957945 \end{pmatrix}$	$\begin{pmatrix} 37724 \\ 17123 \\ 5091315 \end{pmatrix}$	$\begin{pmatrix} 61240 \\ 14218 \\ 2736572 \end{pmatrix}$

We then perform the ItemNN, BiasedMF, and FM algorithm on each user group, and present the results of RMSE and Pre@5 in Table 2. It can be seen that the performance of different user groups differ dramatically. From the view of quantity, easy user groups behave preferably better than difficult user groups, while from the view of quantity the more data the better performance we can obtain. Moreover, we also notice that the easy light user group exhibits a better result compared to the difficult heavy user group, e.g., 0.7100 vs. 0.8114 in terms of RMSE for the BiasedMF method. This reveals that both the quantity and quality of user data matter for the recommendation performance.

Comparisons on Different Processing Approaches for Different User Groups. Next in order to determine the effectiveness of the different processing approaches, we implement the BiasedMF methods on distinct user groups that have been processed by different pre-processing approaches. Figure 3 illuminates the results, where "Non" is

Table 2. Results of distinct user groups with various recommendation approaches.

User Groups	Item NN method		BiasedMF Method		FM Method	
	RMSE	Pre@5	RMSE	Pre@5	RMSE	Pre@5
HEUG	0.6911	0.8169	0.6670	0.8462	0.6838	0.9084
HDUG	0.8494	0.8092	0.8114	0.8492	0.8229	0.8926
HUG	0.7957	0.8166	0.7616	0.8507	0.7574	0.9009
MEUG	0.7086	0.8089	0.6920	0.8255	0.6986	0.8721
MDUG	0.9084	0.7980	0.8821	0.8153	0.8717	0.8552
MUG	0.8274	0.8066	0.8006	0.8259	0.7928	0.8668
LEUG	0.7669	0.7100	0.7708	0.7067	0.7669	0.6685
LDUG	0.9429	0.7216	0.9774	0.7096	0.9364	0.6793
LUG	0.8675	0.7184	0.8882	0.7108	0.8669	0.6773

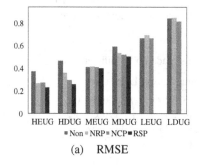

(a)   RMSE

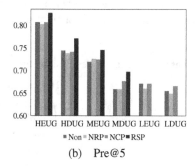

(b)   Pre@5

Fig. 3. Results of the BiasedMF by different processing approaches for different user groups.

no processing, "NRP" means the noise removal processing, "NCP" is noise correction processing, and "RSP" denotes the representative sampling processing. We find that noise removing processing works well for the majority of the groups but not for light groups. Both noise removal and correction show little effect in this case. Besides, it can be seen that the noise processing performs better for difficult user groups.

Performance of the Proposed TLMF Between Different User Groups. We finally conduct a set of experiments to elaborate the performance of the proposed TLMF for the light/medium user groups by transferring latent factors from the heavy and medium/heavy user groups. We report the results on the target MEUG, MDUG, LEUG, and LDUG using different source user groups in Table 3, where "X-nrs" means the X group processed by noise removal and data sampling processing, and the "X-ns" denotes X with noise removal processing. "None" shows no source groups are utilized. We can observe three findings from the results. (1) The performance has been boosted significantly for all user groups, and the LDUG improves about 8 % with transferring knowledge from HEUG-nrs, HDUG-nrs, MEUG-nr; (2) Easy user groups generally gain more improvement compared to difficult groups, and this may be resulted from the fact that the difficult users lack of reliable ratings; (3) As for the quantity of user data,

Table 3. Perforance comparisons of the TLMF between distinct target and source user groups.

Target user group	Source user groups	RMSE	Pre@5
MEUG	None	0.6920	0.8255
	HEUG-nrs	0.6801	0.8344
	HEUG-nrs, HDUG-nrs	0.6696	0.8413
MDUG	None	0.8821	0.8153
	HEUG-nrs	0.8473	0.8259
	HEUG-nrs, HDUG-nrs	0.8025	0.8396
LEUG	None	0.7708	0.7067
	HEUG-nrs	0.7611	0.7099
	HEUG-nrs, HDUG-nrs	0.7533	0.7125
	HEUG-nrs, HDUG-nrs, MEUG-nr	0.7405	0.7186
LDUG	None	0.9774	0.7096
	HEUG-nrs	0.9418	0.7126
	HEUG-nrs, HDUG-nrs	0.9126	0.7166
	HEUG-nrs, HDUG-nrs, MEUG-nr	0.8985	0.7201

light users could obtain more promotion from the TLMF compared to the medium users because light users typically suffer from the cold start problem, which indicates that the system could not gain enough information from their rare ratings.

5   Conclusions

In this paper, we investigated how to manage and model user data of different quantity and quality so as to obtain desirable recommendations with representative and reliable ratings. We suggested to exploit various pre-processing and recommendation methods for ratings of various users, and preliminarily explored a generic framework for this. In this framework, we first classified users into groups according to the quantity (heavy, medium, and light) and quality (easy and difficult) of their contributed ratings. We have verified that recommendation performances of distinct groups differ radically. Next we proposed several pre-processing approaches to cope with the data quality (noisy rating detection and processing) and data quantity (representative rating sampling) issue. In order to further boost the performance of light and difficult groups, we introduced a novel transfer latent factor model (TLMF) to transfer item latent factors learned from ratings of heavy and easy groups. Finally, we demonstrated that our proposed method significantly enhances the recommendation performance.

There remain several interesting and important directions worthy of further research. The first direction is to design some effective and efficient model-based approaches to process heterogeneous user feedbacks within the proposed framework. For example, with the prevalence of smart mobile phones, recommendations on the social networks and LBSNs have become pressing. In such media, users contribute heterogeneous data that could better capture the user data quality. Another promising consideration is to broadly evaluate metrics of recommendation performance except the accuracy metrics, such as the diversity, novelty, and coverage, and it is also necessary to perform some online studies to further prove the effectiveness of different processing approaches. The third direction is to further investigate the corresponding processing and model methods for the dynamic setting. We have confirmed the proposed framework on a static large data set, but real applications always collect user data in a stream-like situation. Our current processing approaches cannot apply to the dynamic recommender systems, but in our ongoing work we intend to employ some steam mining approaches to enable the dynamic recommendations.

Acknowledgments. This work is supported by grants from the Doctoral Program of the Ministry of Education of China (Grant No. 20110101110065), the National Key Technology R&D Program of China (Grant No. 2012BAD35B01-3), and the National Natural Science Foundation of China (Grant No. U1536118). We would also like to thank the GroupLens Research Group for the contribution of the publicly available data set and the open source Lenskit.

References

1. Adomavicius, G., Tuzhilin, A.: Toward the next generation of recommender systems: a survey of the state-of-the-art and possible extensions. IEEE Trans. Knowl. Data Eng. **17**(6), 734–749 (2005)
2. Amatriain, X., Pujol, J.M., Oliver, N.: I like it… I like it not: evaluating user ratings noise in recommender systems. In: Houben, G.-J., McCalla, G., Pianesi, F., Zancanaro, M. (eds.) UMAP 2009. LNCS, vol. 5535, pp. 247–258. Springer, Heidelberg (2009)
3. Amatriain, X., Pujol, J.M., Tintarev, N., Oliver, N.: Rate it again: increasing recommendation accuracy by user re-rating. In: International ACM Conference on Recommender Systems, pp. 173–180 (2009)
4. Amatriain, X.: Mining large streams of user data for personalized recommendations. ACM SIGKDD Explorations Newsletter, vol. 14(2), pp. 37–48 (2013)
5. Bellogin, A., Said, A., de Vries, A.P.: The magic barrier of recommender systems – no magic, just ratings. In: Dimitrova, V., Kuflik, T., Chin, D., Ricci, F., Dolog, P., Houben, G.-J. (eds.) UMAP 2014. LNCS, vol. 8538, pp. 25–36. Springer, Heidelberg (2014)
6. Chang, S., Harper, F.M., Terveen, L.: Using groups of items to bootstrap new users in recommender systems. In: ACM Conference on Computer Supported Cooperative Work and Social Computing, pp. 1258–1269 (2015)
7. Ekstrand, M.D., Ludwig, M., Konstan, J.A., Riedl, J.: Rethinking the recommender research ecosystem: reproducibility, openness, and LensKit. In: ACM Conference on Recommender Systems, pp. 133–140 (2011)
8. Herlocker, J.L., Konstan, J.A., Terveen, L.G., Riedl, J.: Evaluating collaborative filtering recommender systems. ACM Trans. Inf. Syst. **22**(1), 5–53 (2004)
9. Hussein, T., Linder, T., Gaulke, W., Ziegler, J.: Hybreed: a software framework for developing context-aware hybrid recommender systems. User Model. User-Adap. Inter. **24** (1–2), 121–174 (2014)
10. Koren, Y., Bell, R., Volinsky, C.: Matrix factorization techniques for recommender systems. Computer **8**, 30–37 (2009)
11. Li, B., Chen, L., Zhu, X., et al.: Noisy but non-malicious user detection in social recommender systems. World Wide Web J. **16**(5–6), 677–699 (2013)

12. Liu, N.N., Meng, X., Liu, C., Yang, Q.: Wisdom of the better few: cold start recommendation via representative based rating elicitation. In: ACM Conference on Recommender Systems, pp. 37–44 (2011)
13. O'Mahony, M.P., Hurley, N.J., Silvestre, G.: Detecting noise in recommender system databases. In: International Conference on Intelligent User Interfaces, pp. 109–115 (2006)
14. Pham, H.X., Jung, J.J.: Preference-based user rating correction process for interactive recommendation systems. Multimedia Tools Appl. **65**(1), 119–132 (2013)
15. Rashid, A.M., Karypis, G., Riedl, J.: Learning preferences of new users in recommender systems: an information theoretic approach. ACM SIGKDD Explorations Newsletter, vol. 10(2), pp. 90–100 (2008)
16. Rendle, S., Gantner, Z., Freudenthaler, C., Schmidt-Thieme, L.: Fast context-aware recommendations with factorization machines. In: International Conference on Research and Development in Information Retrieval, pp. 635–644 (2011)
17. Sarwar, B., Karypis, G., Konstan, J.A., Riedl, J.: Item-based collaborative filtering recommendation algorithms. In: International ACM Conference on World Wide Web, pp. 285–295 (2001)
18. Schelter, S., Boden, C., Schenck, M., Alexandrov, A., Markl, V.: Distributed matrix factorization with map-reduce using a series of broadcast-joins. In: International ACM Conference on Recommender Systems, pp. 281–284 (2013)
19. Yu, P., Lin, L., Wang, F., Wang, J., Wang, M.: Improving recommendations with collaborative factors. In: Li, F., Li, G., Hwang, S.-W., Yao, B., Zhang, Z. (eds.) WAIM 2014. LNCS, vol. 8485, pp. 30–33. Springer, Heidelberg (2014)
20. Zhuang, Y., Chin, W.S., Juan, Y.C., Lin, C.J.: A fast parallel SGD for matrix factorization in shared memory systems. In: International ACM Conference on Recommender Systems, pp. 249–256 (2013)

RankMBPR: Rank-Aware Mutual Bayesian Personalized Ranking for Item Recommendation

Lu Yu[1], Ge Zhou[1], Chuxu Zhang[2], Junming Huang[3],
Chuang Liu[1(✉)], and Zi-Ke Zhang[1(✉)]

[1] Alibaba Research Centre for Complexity Sciences,
Hangzhou Normal University, Hangzhou, China
coolluyu@gmail.com, flyzhouge@outlook.com, liuchuang@hznu.edu.cn,
zhangzike@gmail.com
[2] Department of Computer Science, Rutgers University, Brunswick, USA
cz201@cs.rutgers.edu
[3] Web Sciences Centre, University of Electronic Science and Technology of China,
Chengdu, China
mail@junminghuang.com

Abstract. Previous works indicated that pairwise methods are state-of- the-art approaches to fit users' taste from implicit feedback. In this paper, we argue that constructing item pairwise samples for a fixed user is insufficient, because taste differences between two users with respect to a same item can not be explicitly distinguished. Moreover, the rank position of positive items are not used as a metric to measure the learning magnitude in the next step. Therefore, we firstly define a *confidence function* to dynamically control the learning step-size for updating model parameters. Sequently, we introduce a generic way to construct mutual pairwise loss from both users' and items' perspective. Instead of user-oriented pairwise sampling strategy alone, we incorporate item pairwise samples into a popular pairwise learning framework, *bayesian personalized ranking* (BPR), and propose *mutual bayesian personalized ranking* (MBPR) method. In addition, a rank-aware adaptively sampling strategy is proposed to come up with the final approach, called RankMBPR. Empirical studies are carried out on four real-world datasets, and experimental results in several metrics demonstrate the efficiency and effectiveness of our proposed method, comparing with other baseline algorithms.

1 Introduction

Building predictor for top-k recommendation by mining users' preferences from implicit feedback [1] can help to produce recommendations in a wide range of applications [9,14,15,17]. One significant challenge is that only positive observations are available. For example, we can only observe that a user bought a book, or a movie ticket from the web logs. Such issue is called "one-class" recommendation [4] and many works have offered solutions to measure relevance of a user-item pair by transforming traditional rating estimation problem to a ranking task.

© Springer International Publishing Switzerland 2016
B. Cui et al. (Eds.): WAIM 2016, Part I, LNCS 9658, pp. 244–256, 2016.
DOI: 10.1007/978-3-319-39937-9_19

To solve the one-class recommendation problem, Rendle *et al.* [2] proposed *bayesian personalized ranking* (BPR), a popular pairwise learning framework. Different from pointwise method dealing with a regression problem [5], BPR is built under assumption that everything that a user has not bought is of less interest for the user than those bought items. Empirically, pairwise ranking methods can perform better than pointwise approaches, and have been used as the workhorse by many recommendation approaches to tackle challenges from different applications, like tweet recommendation [22], tag recommendation [9], relation extraction [7], entity recommendation in heterogeneous information network [21] etc. In addition, many works aiming to optimise pairwise preference learning for one-class recommendation problems are recently proposed via leveraging novel information like social connection [19], group preference [8], or improving the strategy of selecting pairwise samples [3].

To the best of our knowledge, current pairwise learning methods [2,8,19, 20] construct pair examples from the user side. In this paper, we argue that unilaterally constructing samples from users' perspective is insufficient based on one main reason: taste differences between two users with respect to a same item can not be explicitly distinguished. Therefore, we introduce *mutual bayesian personalized ranking* (MBPR) to offer alternative idea to express the pairwise interactions, instead of the *user-based pairwise preference* alone. In addition, inspired by the recent literature [3], we optimize the sampling strategy of MBPR via utilising the rank position of positive samples to dynamically control the learning speed. Experimental results on four real-world datasets show that our proposed method significantly improves the performances comparing with other baseline algorithms on four evaluation metrics.

2   Preliminary

Let $\mathcal{U}$ and $\mathcal{I}$ denote the user and item set, respectively. We use $\mathcal{T}$ to denote the observed feedbacks from n users and m items. Each case $(u, i) \in \mathcal{T}$ means that user u ever clicked or examined item i. For a given user u, the relationship between user u and item i can be measured as x_{ui}, then a recommendation list of items could be generated from items $\mathcal{I} \setminus \mathcal{I}_u$, where $\mathcal{I}_u$ denotes the clicked or examined items by user u. In practise, only top-k recommendations can attract users' attention, and recommendation task in such a case can be modified as a learning to rank problem. In order to represent user u' preference over items with only "one-class" observations $\mathcal{T}$, pairwise learning approaches typically regard observed user-item pairs $(u, i) \in \mathcal{T}$ as a positive class label, and all other combinations $(u, j) \in (\mathcal{U} \times \mathcal{I} \setminus \mathcal{T})$ as a negative one. Then intermedia training samples $(u, i, j) \in D_{\mathcal{T}}$ are constructed according to Assumption 1.

Assumption 1. *For a given user u, unequal relationship exists between the examined item i and the unexamined item j, and user u would show more preference to item i than item j.*

2.1 Bayesian Personalised Ranking (BPR)

For a given user u, the relationship between user u and item i can be measured as x_{ui}. BPR will fit to the dataset D_T to correctly learn the ranks of all items via maximizing the following posterior probability of the parameter space θ.

$$p(\theta| >_u) \propto p(>_u |\theta)p(\theta), \tag{1}$$

where notation $>_u = \{i >_u j : (u, i, j) \in D_T\}$ denotes the pairwise ranking structure for a given u, and $p(\theta)$ is the prior probability of parameter space θ. Then BPR assumes that each case of $>_u$ is independent, and the above likelihood of pairwise preferences (LPP) can be modified as $LPP(u) = \prod_{(u,i,j) \in D_T} p(i >_u j|\theta)$, where $p(i >_u j|\theta) = \sigma(x_{uij}(\theta))$, and $\sigma(x) = \frac{1}{1+e^{-x}}$. In terms of $p(\theta)$, we define it as a normal distribution with zero mean and covariance matrix $\sum_{\theta} = \lambda_\theta I$, that is, $\theta \backsim \mathcal{N}(0, \sum_{\theta})$. Now we can infer the BPR by filling $p(\theta)$ into the maximum posterior probability in Eq. (1).

$$\ln LPP(u) + \ln p(\theta) = \sum_{(u,i,j) \in D_T} \ln \sigma(x_{uij}) - \frac{\lambda_\theta}{2}||\theta||^2, \tag{2}$$

where λ_θ are model regularization parameters. Here we choose $x_{uij}(\theta) = x_{ui} - x_{uj}$. The specific definition of the preference function used in this paper is $x_{ui} = b_u + b_i + W_u V_i^\top$, where $W \in \mathbb{R}^{|\mathcal{U}| \times d}$, $V \in \mathbb{R}^{|\mathcal{I}| \times d}$, and d is the number of latent factors. b_u and b_i are bias features for each user u and for each item i, respectively. Therefore, $x_{uij}(\theta) = b_i + W_u V_i^\top - b_j - W_u V_j^\top$. Generally, stochastic gradient descent based algorithms can be used as a workhorse to optimise the posterior distribution. More specifically, the parameters θ are randomly initialized. With iteratively traversing each observation $(u, i) \in \mathcal{T}$, a negative sample j is picked and parameters θ can be updated with gradients as shown in Eq. (3) until the stopping criterion is matched, and return the learned model.

$$\theta \leftarrow \theta + \alpha(\frac{\partial \ln LPP(u)}{\partial \sigma(x)} \cdot \frac{\partial x_{uij}}{\partial \theta} - \lambda_\theta \theta) \tag{3}$$

According to literature [2], the way to select a negative j could have significant impact on the performance of BPR. Typically, a bootstrap sampling approach with replacement is suggested, *i.e.* for a given (u, i), j is randomly selected out depending on an uniform distribution. However, Rendle et al. [3] argued that for a given user u, good negative samples which could make an effective updating could lie in a higher position than positive item i. While, uniform sampling could equally regard each negative item without taking the ranks of positive items i into consideration. Sequently, Rendle et al. [3] proposed a novel adaptive sampling strategy after giving insight into the learning diagram, especially the evolution of gradient distribution. Series of experiment results demonstrate that adaptive sampling a negative j according to their position in the rank list could improve the convergence speed and effectiveness of BPR. In more detail, the adaptive sampling strategy is described as follows:

For sampling a negative item j for the given (u, i), we firstly extend both user's and items' latent feature, that is, $W_u' = [1, W_u], V_i' = [b_i, V_i]$.

1. Sample a rank r from distribution $p(r) \propto \exp(\frac{-r}{\gamma_i})$, where γ_i is a hyperparameter controlling the expectation position of sampled item j. A small value of γ_i could generate a small value of r with a high probability. It indicates that a negative item j ranking in a high position could be probably sampled.
2. Sample a factor dimension d according to probability distribution $p(d|u) \propto |W'_{ud}|\delta_d$, where δ_d denotes the variance over all items' dth factor, and W'_{ud} denotes the value of user u's dth factor.
3. Sort items according to $V'_{\cdot,d}$, and return the item j on position r in the sorted list.

However, as we see, about $|\mathcal{I} \setminus \mathcal{I}_u|$ negative cases need be examined for each given (u,i). If the number of items in the database is enormous, the scale of negative candidates make it intractable to directly apply such strategy. In order to reduce the complexity, Rendle *et al.* proposed to pre-compute the ranks every $|\mathcal{I}|\log|\mathcal{I}|$ training cases.

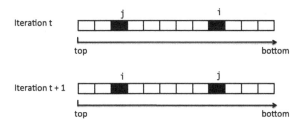

Fig. 1. Simple illustration to describe the item permutation for user u in different learning stages. The term "top" and "bottom" stands for the top and bottom position in the rank list, respectively.

3   Our Solution

3.1   Rank-Aware Pairwise Learning

In last Sect. 2.1, we can see that adaptively ranking the items could be implemented every a fixed steps in order to abate the computation cost. However, we argue that it still needs to sort $|\mathcal{I} \setminus \mathcal{I}_u|$ items for each $u \in \mathcal{U}$, which will dramatically increase the computation resources with the growth of the size of user and item set. Moreover, the rank position of the positive item i is not utilized as a significant criterion to measure the learning performance. One basic assumption of adaptive sampling is that selecting the negative sample j ranking in a higher position could help to find a good case to effectively update pairwise parameters. However, it omits that the ranks of positive item i when utilizing the benefits brought by the position information of negative item j. We believe that utilising the rank information of positive item i is crucial to offer an alternative way to improve the performance of pairwise learning algorithm based on BPR.

Let π_t^u denote the item permutation of user u in the iteration t, and $\pi_t^u(i)$ represents the position of item i. Suppose that for a given $(u, i) \in T$ in iteration t, $\pi_t^u(i)$ is very close to the bottom as it is shown Fig. 1. At that moment, user u's preference over positive item i is insufficiently learned under this situation. It still has a huge position gap between item i and j. If we sample the negative j which ranks much higher than a positive item i, we should update the corresponding parameters with a large magnitude due to the significant position gap. If positive item i ranks closely to the top position in iteration $t + 1$, we could say that preference function x_{ui} is learned well for this case, then we should update parameters with a slight step size. Therefore, we propose to transform the rank information as a crucial criterion to support the updating magnitude of model parameters.

To tackle the aforementioned challenge, we propose *rank-aware bayesian personalized ranking* (RankBPR) to utilize the rank position of item i for a given (u, i). We define a *confidence function* $C_u(rank_i)$, which serves as dynamic weight to control the updating step size of model parameters. Generally, $C_u(rank_i)$ is a monotonically increasing function. By integrating $C_u(rank_i)$ into Eq. (2), we obtain the *rank-aware bayesian personalised ranking* (RankBPR), and the objective function can be modified as follows:

$$\ln LPP(u) + \ln p(\theta) = \sum_{u \in \mathcal{U}} \sum_{i \in \mathcal{I}_u} C_u(rank_i) \sum_{j \in \mathcal{I} \backslash \mathcal{I}_u} \ln \sigma(x_{uij}) - \lambda_\theta ||\theta||^2. \quad (4)$$

Correspondingly, the flow of stochastic gradient as shown in Eq. (3) could be revised as follows:

$$\theta \leftarrow \theta + \alpha(C_u(rank_i) \cdot \frac{\partial \ln LPP(u)}{\partial \sigma(x)} \cdot \frac{\partial x_{uij}}{\partial \theta} - \lambda_\theta \theta) \quad (5)$$

As we see, the key part of the *confidence function* lies in the rank position of item i. In this work, we define $C_u(rank_i)$ as follows:

$$C_u(rank_i) = \sum_{s=1}^{rank_i} \gamma_s, with \ \gamma_1 \geq \gamma_2 \geq \gamma_3 \geq ... \geq 0.$$
$$rank_i = \sum_{j \in \mathcal{I} \backslash \mathcal{I}_u} \mathbb{I}[\beta + \hat{h}_{uj} \geq \hat{h}_{ui}] \quad (6)$$

where $\mathbb{I}$ is the indicator function, and $\hat{h}_{ui}$ measures the relevance between user u and item i. β is a margin parameter to control the gap between the positive item i and negative sample j. We choose $\gamma_s = 1/s$ as the weighting approach, which could assign large value to top positions with rapidly decaying weight for lower positions. Intuitively, a positive item i, which ranks in the highest position, could produce a low confidence to update the parameters. Oppositely, a bottom positive item i could offer a high confidence. In terms of the $\hat{h}_{ui}$, we give two different definitions under the assumption that ranks of a item list can be measured as the inner-product of user-item features or only according to a specific dimension of users' feature vector. For simplicity, one can denote $\hat{h}_{ui} = x_{ui}$ as a response to measure the relevance of user u and items i as the product of their feature vectors.

$$\hat{h}_{ui} = b_u + b_i + W_u V_i^\top \tag{7}$$

Besides this, we suppose that the value of each element of user u's feature vector W_u can represent u's specific taste, which drives us to rank all items only depending on the match score in a single dimension of feature vectors. To formulate our idea, we define a probability distribution for sampling a factor d as $p(d|u) \propto |W_{ud}|$. For each training case (u, i), we firstly sample a factor d according to $p(d|u) \propto |W_{ud}|$, then calculate $rank_i$ based on the match score $\hat{h}_{ui}$ between a user-item pair:

$$\hat{h}_{ui} = W_{ud} V_{id}, \tag{8}$$

where $W_{.d}$ and $V_{.d}$ denotes the value of dth element.

3.2 Fast Estimating Rank Position of Item i

Note that in each iteration, the exact value of $rank_i$ needs to be calculated for each observation is extremely expensive, when there are massive amount of items in the database. In this work, we employ a sampling process [10] to fast estimate the $rank_i$. More specifically, for a given sample (u, i), one uniformly draws a random negative item from $\mathcal{I}$ until finding a j, which satisfies $\beta + \hat{h}_{uj} \geq \hat{h}_{ui}$. Then the $rank_i$ can be approximated as $rank_i \approx \lfloor \frac{|\mathcal{I}|-1}{K} \rfloor$, where $\lfloor \cdot \rfloor$ denotes the floor function and K is the number of steps to find a item j. From this approximation of $rank_i$, we can see that the computations on seeking a negative item j could increase sharply if the positive item i happens to rank at the top of the list. To accelerate the estimation of $rank_i$, we define a parameter κ to limit the maximum sampling steps. Correspondingly, we also utilise an item buffer $buffer_{ui}$ whose size equals to κ to store every sampled negative item j. Finally, $rank_i$ can be approximated as $rank_i \approx \lfloor \frac{|\mathcal{I}|-1}{min(K,\kappa)} \rfloor$, and the negative item j will be selected from the top of the sorted $buffer_{ui}$ in descending order based on $\hat{h}_{uj}$. The complete procedure of RankBPR can be described in Algorithm 1.

3.3 Mutual Sample Construction for Pairwise Learning

Beside the optimized sampling strategy, we can see that pairwise learning approaches like BPR only focus on constructing samples (u, i, j) for a fixed user u. In this paper, we argue that only sampling negative samples w.r.t. a fixed user u is insufficient based on the following consideration:

- Recommendation task naturally involves with two basic types of entities, i.e. users and items. To better represent the relevance between them, mutual pairwise sampling is essential to not only capture a user's preferences over two different items, but also measure how different is a pair of users with respect

---

Algorithm 1. Optimizing models with RankBPR

Require: $\mathcal{T}$: Training Set
Ensure: Learned $\hat{\theta}$
 1: initialize θ
 2: **repeat**
 3:     randomly draw (u,i) from $\mathcal{T}$
 4:     estimate $C_u(rank_i)$ according to Equation (6)
 5:     select a negative sample j from the top of the sorted $buffer_{ui}$
 6:     $x_{uij} \leftarrow x_{ui} - x_{uj}$
 7:     $\theta \leftarrow \theta + \alpha(C_u(rank_i) \cdot (1 - \sigma(x_{uij}))\frac{\partial}{\partial\theta}x_{uij} - \lambda_\theta\theta)$
 8: **until** convergence.

---

to a same item. Intuitively, a piece of training sample (u, i, j) could help to distinguish the difference between item i and j from a user u's perspective, while two different users' preferences over the same item i could not be explicitly represented.

In order to tackle the aforementioned challenges, we propose to construct the pairwise loss from item's perspective instead of the construction of pairwise training samples from user's perspective alone. Specifically, another type of training samples $(i, u, v) \in D_{\mathcal{T}}$ for a given $(u, i) \in \mathcal{T}$ would be extracted under the proposed Assumption 2.

Assumption 2. *A group of users who ever selected the same items might have closer tastes than other users. For a given item $(u, i) \in \mathcal{T}$, user u could have a stronger preference than user v, who did not ever explicitly click or rate item i.*

We think that users purchasing the same item i could have much closer connection than those who do not purchase item i. Following the Assumption 2, mutual sampling construction could explicitly leverage the user connection information to distinguish the differences among users. Then two types of pairwise relationship are extracted for a given observation $(u, i) \in \mathcal{T}$. To formulate our idea, we propose *mutual bayesian personalized ranking* (MBPR) to incorporate mutual pairwise samples into BPR framework. As a response to our assumption, the basic idea of mutual pairwise sampling equals to user u- and item i-central pairwise construction. Therefore, two kinds of pairwise likelihood preferences, i.e. $p(i >_u j|\theta)$ and $p(u >_i v|\theta)$, are instantiated. The modified objective function we are going to maximise in this work is described as follows:

$$\ln LPP(u) + \ln LPP(i) + \ln p(\theta)$$

$$= \sum_{(u,i,j)\in D_{\mathcal{T}}} \ln \sigma(x_{uij}) + \sum_{(i,u,v)\in D_{\mathcal{T}}} \ln \sigma(x_{iuv}) - \frac{\lambda_\theta}{2}||\theta||^2. \qquad (9)$$

Different from standard BPR, mutual pairwise samples are constructed. In MBPR, for each observation $(u, i) \in \mathcal{T}$, a negative item $j \in \mathcal{I} \setminus \mathcal{I}_u$ and user $v \in \mathcal{U} \setminus \mathcal{U}_i$ are picked and parameters θ can be updated with the following gradients

$$\theta \leftarrow \theta + \alpha(\frac{\partial \ln LPP(u)}{\partial \sigma(x_{uij})} \cdot \frac{\partial x_{uij}}{\partial \theta} + \frac{\partial \ln LPP(i)}{\partial \sigma(x_{iuv})} \cdot \frac{\partial x_{iuv}}{\partial \theta} - \lambda_\theta \theta) \tag{10}$$

where $\mathcal{U}_i$ denotes the set of users who ever clicked item i. By employing rank-aware sampling approach, we further apply mutually dynamic sampling strategy to optimise the procedure of MBPR on selecting the negative samples and propose RankMBPR, in which parameters θ would be updated with the modified gradients:

$$\theta \leftarrow \theta + \alpha(C_u(rank_i)\frac{\partial \ln LPP(u)}{\partial \sigma(x_{uij})} \cdot \frac{\partial x_{uij}}{\partial \theta} + C_i(rank_u)\frac{\partial \ln LPP(i)}{\partial \sigma(x_{iuv})} \cdot \frac{\partial x_{iuv}}{\partial \theta} - \lambda_\theta \theta),\tag{11}$$

where $C_i(rank_u)$ can be calculated in the same way as $C_u(rank_i)$ with a slightly different definition as follows:

$$C_i(rank_u) = \sum_{s=1}^{rank_u} \gamma_s, with \ \gamma_1 \geq \gamma_2 \geq \gamma_3 \geq ... \geq 0$$
$$rank_u = \sum_{v \in \mathcal{U} \backslash \mathcal{U}_i} \mathbb{I}[\beta + \hat{h}_{iv} \geq \hat{h}_{iu}]. \tag{12}$$

Table 1. Statistics of the datasets

Datasets	#Users	#Items	#Observations	#Density
ML100K	943	1682	100000	6.3%
Last.fm	1892	17632	92834	0.28%
Yelp	16826	14902	245109	0.097%
Epinions	49289	139738	664823	0.02%

4   Experiments

Datasets and Settings: Four datasets[1] are used in this paper and statistics of them are summarized in Table 1. In this work, we adopt 5-fold cross validation method to demonstrate the performance of our proposed approach. More specifically, repeated validation experiments are conducted 5 times, in which we randomly select 80% of observed feedback as training set to train the ranking model, and the rest as the testing set. The final performance of each algorithm

---

[1] http://grouplens.org/datasets/movielens/
http://grouplens.org/datasets/hetrec-2011/
http://www.yelp.com/dataset_challenge
http://www.trustlet.org/wiki/Epinions.

is measured as the average results on four top-N metrics, which are used to compare the performance in measuring recall ratio (Recall@5, Recall@10), precision of top-10 recommendation list (MAP@10, MRR@10). We mainly compare our method with the following approaches:

- **PopRec**: A naive baseline that generates a ranked list of all items with respect to their popularity, represented by the number of users who ever rated, or clicked the target items.
- **WRMF**: This method defines a weight distribution for each $(u, i) \in \mathcal{U} \times \mathcal{I}$, then employs matrix factorization model to solve a regression problem via optimizing a square loss function [5]. It is the state-of-the-art one-class collaborative filtering method.
- **ICF**: Item-based CF is a classical collaborative filtering approach, and was initially explored in the literature [13]. It is a generic approach which could be used for rating prediction and item recommendation with only implicit feedback.
- **BPR**: This method uses a basic matrix factorization model as the scoring function and BPR as the workhorse to learn the ranks of items [2].
- **AdaBPR**: This method is slightly different from BPR-MF as an adaptive sampling strategy is proposed to improve the learning efficiency of BPR [3]
- **StaticBPR**: This method replaces the uniform sampling strategy with a static sampling strategy according to the popularity of items [3].

4.1 Performance Evaluation

Convergence: Initially, we empirically study the learning curves of BPR-based algorithms with different pairwise sampling strategies (see Fig. 2) on four datasets. Due to the limited space, we only list the learning curves on Last.fm and ML100K. From Fig. 2, we can see that all algorithms almost converge after a number of iterations, then fluctuate in a tiny range around the converging performance in the different metrics. We can find that RankBPR is superior to other baselines, and RankBPR-full outperforms RankBPR-dim, which proves the efficiency and effectiveness of rank-aware sampling strategy with definition of $\hat{h}_{u,i}$ in Eq. (7). Therefore, the way to calculate $\hat{h}_{u,i}$ for RankMBPR follows RankBPR-full. Turn to MBPR, its performance indicates that mutual construction of pairwise samples could also benefit for the pairwise learning methods. As incorporating both features of MBPR and RankBPR, RankMBPR achieves the best performance. From the learning curves we can see that stopping criterion could be defined as a fixed number of iterations for BPR-based algorithms. In this work, we set the number of iterations as 2000 for all BPR-based algorithms.

Performance Evaluation: Table 2 shows the average recommendation performance of different algorithms. We highlight the results of the best baseline and our proposed method respectively. Validation results show that RankMBPR is superior to all baseline algorithms on all four datasets. Note that MBPR also has comparative better performance than baseline approaches. One possible reason

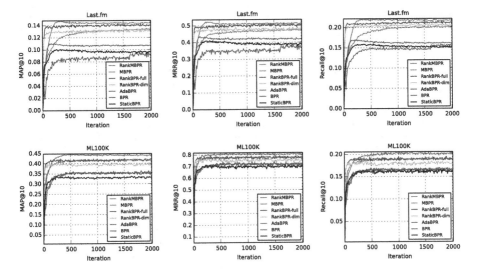

Fig. 2. Learning curves of different BPR-based algorithms in several evaluation metrics. In this case, we set dimension d to 30, and each algorithms iterated 2000 times. RankBPR-full takes Eq. (7) as the relevance function for RankBPR algorithm, while RankBPR-dim employs Eq. (8). In this work, we set $\beta = 1.0$ for both RankBPR-full and RankMBPR, and $\beta = 0.1$ for RankBPR-dim. Maximum size κ of $buffer_{ui}$ is fixed as 200 for RankBPR-full, RankBPR-dim, and RankMBPR.

may be that AdaBPR pays less attention to deeply explore the effects brought by the rank position of positive samples. In terms of WRMF, it is not directly to optimise ranking. Comparing with BPR, our proposed methods both utilise the differences among users by leveraging the construction of mutual pairwise samples, instead of the user-side solely, and the benefits brought by the rank-aware sampling method. From this table, we could find some interesting evidences. As the density of the dataset decreases, the performance gap between our proposed method and BPR-based methods will become larger. In particularly, AdaBPR performs not so much as ICF in the most sparse dataset, Epinions.

5   Related Work

Recently many works have began to adopt learning-to-rank idea to explore users' preference on items from ranking perspective [2,3,6,16,19,23,24]. As one of typical pairwise learning method, BPR is flexible to incorporate different contextual information to make BPR adaptive to different tasks. Riedel *et al.* [7] employs BPR to automatically extract structured and instructed relations. Zhao *et al.* [19] indicated out that users tend to have many common interests with their social friends, and proposed to leverage social connections to improve the quality of item recommendation. Pan *et al.* [8] pointed out that group features should be taken into account for further exploring users' preferences on items.

Table 2. Experimental results of all baseline algorithms. The last column shows the improvement of the proposed method compared with the best baseline method. The latent dimension is fixed as $d = 30$ for MF-based methods, like WRMF, BPR, AdaBPR.

Datasets	Metrics	PopRec	ICF	WRMF	BPR	AdaBPR	MBPR	RankMBPR	Improv
ML100K	Recall@5	0.0553	0.0882	0.1052	0.1037	**0.1123**	0.1142	**0.1202***	7.03 %
	Recall@10	0.0974	0.1473	0.1764	0.1708	**0.1823**	0.1876	**0.1921***	5.37 %
	MAP@10	0.1986	0.3041	0.3591	0.3712	**0.4024**	0.4188	**0.4312***	7.38 %
	MRR@10	0.5449	0.6843	0.7236	0.7389	**0.7682**	0.7788	**0.7963***	4.39 %
Last.fm	Recall@5	0.0476	0.1057	0.112	0.124	**0.133**	0.131	**0.147***	10.52 %
	Recall@10	0.0747	0.1546	0.169	0.191	**0.202**	0.201	**0.218***	7.92 %
	MAP@10	0.0413	0.1016	0.106	0.117	**0.128**	0.133	**0.145***	13.28 %
	MRR@10	0.1995	0.4260	0.431	0.452	**0.482**	0.471	**0.512***	6.22 %
Yelp	Recall@5	0.0156	0.0273	0.0262	0.0352	**0.0364**	0.0386	**0.0406***	11.53 %
	Recall@10	0.0289	0.0471	0.0428	0.0596	**0.0601**	0.0638	**0.0672***	11.81 %
	MAP@10	0.0098	0.0174	0.0172	0.0240	**0.0242**	0.0279	**0.0281***	16.11 %
	MRR@10	0.0286	0.0459	0.0498	0.0640	**0.0648**	0.0724	**0.0742***	14.51 %
Epinions	Recall@5	0.0143	**0.0293**	0.0227	0.0248	0.0261	0.0302	**0.0347***	18.43 %
	Recall@10	0.0217	**0.0459**	0.0369	0.0416	0.0441	0.0487	**0.0548***	19.39 %
	MAP@10	0.0097	**0.0199**	0.0159	0.0168	0.0178	0.0205	**0.0258***	29.64 %
	MRR@10	0.0351	**0.0617**	0.0547	0.0518	0.0542	0.0628	**0.0802***	29.98 %

Rendle *et al.* [9] extended BPR to learn tensor factorization method for recommending related tags to users with respect to a given item. Besides BPR, various methods inherit the pairwise learning idea. In [10], Jason *et al.* defined order pairwise ranking loss and develop online Weighted Approximate-Rank Pairwise (WARP) method, which can be applied to various top-k learning problems. Later, Jason *et al.* in [11,12] presented the possible applications of WARP in video recommendation and collaborative retrieval task. Zhao *et al.* [20] proposed a novel personalized feature projection method to model users' preferences over items. A boosting algorithm [18] is also proposed to build ensemble BPR for one-class recommendation task.

6   Conclusions and Future Work

In this paper, we propose to complementarily distinguish users' taste differences from items' side. Two kinds of pairwise preference likelihood are defined. An improved sampling strategy is customized to our proposed method (MBPR), then an rank-aware MBPR (RankMBPR) is introduced to help sufficiently learn users' preferences over items. The experimental results on four datasets show that RankMBPR perseveres the most efficacy than all baseline algorithms. In future, we incline to further explore the probable effects of the contextual information, or the network structure on helping to select the good pairwise samples.

Acknowledgments. Chuxu Zhang thanks to the assistantship of Computer Science Department of Rutgers University. This work was partially supported by the National Natural Science Foundation of China (No. 11305043), and the Zhejiang Provincial

Natural Science Foundation of China (No. LY14A050001), the EU FP7 Grant 611272 (project GROWTHCOM) and Zhejiang Provincial Qianjiang Talents Project (Grant No. QJC1302001).

References

1. Koren, Y.: Factorization meets the neighborhood: a multifaceted collaborative filtering model. In: Proceedings of the 14th ACM SIGKDD International Conference on Knowledge Discovery and Data Mining, pp. 426–434. ACM (2008)
2. Rendle, S., Freudenthaler, C., Gantner, Z., Schmidt-Thieme, L.: BPR: Bayesian personalized ranking from implicit feedback. In: Proceedings of the Twenty-Fifth Conference on Uncertainty in Artificial Intelligence, pp. 452–461. AUAI Press (2009)
3. Rendle, S., Freudenthaler, C.: Improving pairwise learning for item recommendation from implicit feedback. In: Proceedings of the 7th ACM International Conference on Web Search and Data Mining, pp. 273–282. ACM (2014)
4. Pan, R., Zhou, Y., Cao, B., Liu, N.N., Lukose, R., Scholz, M., Yang, Q.: One-class collaborative filtering. In: Eighth IEEE International Conference on Data Mining, pp. 502–511. IEEE (2008)
5. Hu, Y., Koren, Y., Volinsky, C.: Collaborative filtering for implicit feedback datasets. In: Eighth IEEE International Conference on Data Mining, pp. 263–272. IEEE (2008)
6. Shi, Y., Larson, M., Hanjalic, A.: List-wise learning to rank with matrix factorization for collaborative filtering. In: Proceedings of the Fourth ACM Conference on Recommender Systems, pp. 269–272. ACM (2010)
7. Sebastian, R., Limin, Y., Andrew, M.: Relation extraction with matrix factorization and universal schemas. In: Joint Human Language Technology Conference/Annual Meeting of the North Computational Linguistics (HLT-NAACL) (2013)
8. Pan, W., Chen, L.: GBPR: group preference based bayesian personalized ranking for one-class collaborative filtering. In: Proceedings of the Twenty-Third International Joint Conference on Artificial Intelligence, IJCAI, pp. 2691–2697 (2013)
9. Rendle, S., Schmidt-Thieme, L.: Pairwise interaction tensor factorization for personalized tag recommendation. In: Proceedings of the Third ACM International Conference on Web Search and Data Mining, pp. 81–90. ACM (2010)
10. Weston, J., Bengio, S., Usunier, N.: Large scale image annotation: learning to rank with joint word-image embeddings. Mach. Learn. $81(1)$, 21–35 (2010)
11. Weston, J., Yee, H., Weiss, R.J.: Learning to rank recommendations with the k-order statistic loss. In: Proceedings of the 7th ACM Conference on Recommender Systems, pp. 245–248. ACM (2013)
12. Weston, J., Wang, C., Weiss R., Berenzweig A.: In: Proceedings of the 29th International Conference on Machine Learning, pp. 9–16. ACM (2012)
13. Sarwar, B., Karypis, G., Konstan, J., Riedl, J.: Item-based collaborative filtering recommendation algorithms. In: Proceedings of International Conference on the World Wide Web, pp. 285–295. ACM (2001)
14. Celma, O.: Music Recommendation and Discovery in the Long Tail. Ph.D. thesis, Universitat Pompeu Fabra, Barcelona, Spain (2008)
15. Das, A.S., Datar, M., Garg, A., Rajaram, S.: Google news personalization: scalable online collaborative filtering. In: Proceedings of the 16th International Conference on World Wide Web, pp. 271–280. ACM (2007)

16. Qiu, H., Zhang, C., Miao, J.: Pairwise one class recommendation algorithm. In: Cao, T., Lim, E.-P., Zhou, Z.-H., Ho, T.-B., Cheung, D., Motoda, H. (eds.) PAKDD 2015. LNCS, vol. 9078, pp. 744–755. Springer, Heidelberg (2015)

17. Qu, M., Qiu, G., He, X., Zhang, C., Wu, H., Bu, J., Chen, C.: Probabilistic question recommendation for question answering communities. In: Proceedings of the 18th International Conference on World Wide Web, pp. 1229–1230. ACM (2009)

18. Liu, Y., Zhao, P., Sun, A., Miao, C.: A boosting algorithm for item recommendation with implicit feedback. In: Proceedings of International Joint Conference on Artificial Intelligence, IJCAI (2015)

19. Zhao, T., Mcauley, J., King, I.: Leveraging social connections to improve personalized ranking for collaborative filtering. In: Proceedings of the 23rd ACM International Conference on Information and Knowledge Management, pp. 261–270. ACM (2014)

20. Zhao, T., McAuley, J., King, I.: Improving latent factor models via personalized feature projection for one-class recommendation. In: Proceedings of the 24th International Conference on Information and Knowledge Management. ACM (2015)

21. Yu, X., Ren, X., Sun, Y., Gu, Q., Sturt, B., Khandelwal, U., Norick, B., Han, J.: Personalized entity recommendation: a heterogeneous information network approach. In: Proceedings of the 7th International Conference on Web Search and Data Mining, pp. 283–292. ACM (2014)

22. Chen, K., Chen, T., Zheng, G., Jin, O., Yao, E., Yu, Y.: Collaborative personalized tweet recommendation. In: Proceedings of the 35th International SIGIR Conference on Research and Development in Information Retrieval, pp. 661–670. ACM (2012)

23. Chen, C., Yin, H., Yao, J., Cui, B.: TeRec: a temporal recommender system over tweet stream. In: Proceedings of the 39th International Conference on Very Large Data Bases, pp. 1254–1257. ACM (2012)

24. Yin, H., Sun, Y., Cui, B., Hu, Z., Chen, L.: LCARS: a location-content-aware recommender system. In: Proceedings of the 19th ACM SIGKDD International Conference on Knowledge Discovery and Data Mining, pp. 221–229. ACM (2013)

Unsupervised Expert Finding in Social Network for Personalized Recommendation

Junmei Ding[1], Yan Chen[1], Xin Li[2], Guiquan Liu[1(✉)], Aili Shen[1], and Xiangfu Meng[3]

[1] University of Science and Technology of China, Hefei, China
{dingjm,ycwustc,gqliu,ailisheng}@mail.ustc.edu.cn
[2] IFlyTek Research, Hefei, China
xinli2@iflytek.com
[3] Liaoning Technical University, Fuxin, Liaoning, China
marxi@126.com

Abstract. Personalized Recommendation has drawn greater attention in academia and industry as it can help people filter out massive useless information. Several existing recommender techniques exploit social connections, i.e., friends or trust relations as auxiliary information to improve recommendation accuracy. However, opinion leaders in each circle tend to have greater impact on recommendation than those of friends with different tastes. So we devise two unsupervised methods to identify opinion leaders that are defined as experts. In this paper, we incorporate the influence of experts into circle-based personalized recommendation. Specifically, we first build explicit and implicit social networks by utilizing users' friendships and similarity respectively. Then we identify experts on both social networks. Further, we propose a circle-based personalized recommendation approach via fusing experts' influences into matrix factorization technique. Extensive experiments conducted on two datasets demonstrate that our approach outperforms existing methods, particularly on handing cold-start problem.

Keywords: Personalized recommendation · Experts · Social connections

1 Introduction

With the development of Internet, people suffer from useless *Information Blast*. Personalized Recommendation stands out and draws a lot of attention from both academia and industry as it can not only help people filter out massive useless information but also benefit merchants by increasing their revenue through virtual marketing, such as Amazon, Douban, Netflix, etc. In the aforementioned online services, users can review or give ratings, e.g. 1 to 5 to express their like and dislike on items. Besides, they can add someone as friends or follow others to be fans. These behaviors thus contribute to explicit and implicit social connections between users.

B. Cui et al. (Eds.): WAIM 2016, Part I, LNCS 9658, pp. 257–271, 2016.
DOI: 10.1007/978-3-319-39937-9_20

Fig. 1. Experts' interests regularize users' preferences on different categories.

It is well known that cold-start problem and data sparsity are two inevitable challenges in real-world recommendation scenarios. In order to alleviate these influence, several existing methods incorporate social connections into models under the assumption that a user's taste is similar to or affected by his social connections, which is known as *homophily* in sociology [1,2]. Despite of the success and improvement, the existing recommender algorithms hardly explore the influence of opinion leaders underlying the social connections, which are described as experts. In this paper, compared to those of friends with different tastes, we argue that sharing the same interest with opinion leaders has greater impacts on users in specific circles. Based on the aforementioned assumptions, we list our ideas as follows. Firstly, when we want to see a movie just for killing time, we become frustrated at choosing which one to see from millions of movies. In this case, we would like to follow opinion leaders who share the same interest in movie field rather than friends who have few interests in movies. Secondly, in a certain merchant's website, reviews of experts in such merchant's category are always ranked high, thus they have a greater influence on after-coming users. Thirdly, the rationale of conventional methods that leverage social connections to help to alleviate cold-start problems is that users with few ratings may have enough social connections. Thus we argue that users with few ratings seldom have enough social connections, and they are most probably affected by experts or opinion leaders. Hence, these ideas motivate us to incorporate the influence of experts into circle-based social connections to ease the cold-start problems to some extent.

In this paper, we propose a circle-based approach and use experts' interests to regularize users' preferences to improve the performance of personalized recommendation. As demonstrated in Fig. 1, we consider "friends based experts" and "similarity based experts" in our model. Explicit and implicit social networks are built by users' friendships and similarity, respectively. Then two types of experts are identified on both networks by using information transition theory and consumption records. Further, expert underlying the social connections are divided into several circles according to their rating items' category. Finally, a circle-based approach using matrix factorization technique, which is regularized by experts' interests, is proposed. Experiments on two datasets confirm the competitiveness of our algorithm over state-of-the-art methods, especially on handling the *cold-start* problem. In summary, the contributions are listed as follows:

- We use explicit and implicit social connections to build two kinds of social networks, then two unsupervised methods are proposed to identify the experts on both networks based on information transmission theory and user rating records.
- We propose a circle-based personalized recommendation approach via fusing experts' influence into matrix factorization technique.
- We conduct extensive experiments including user cold-start problem on two real-world datasets and compare the results with the existing algorithms.

The rest of the paper is organized as follows. Section 2 provides a brief review of related work. Section 3 details our approach for personalized recommendation. The performance is presented in Sect. 4, followed by the conclusion in Sect. 5.

2 Related Work

Collaborative Filtering (CF) is an popular and widely used technique for personalized recommendation. The technology can be categorized into two types, memory-based [3] and model-based [4,5]. The premise of recommendation is that users similar to target user's perspective or items are the same as users' historical records. The mechanism of model-based CF adopts the pure rating data to estimate. In this paper, we focus on matrix factorization model due to its efficiency and effectiveness. The basic low rank matrix factorization technique is illustrated by Ruslan [5] and is well applied by Yohuda [4] to win the Netflix Prize. The mechanism behind the idea is that a few primary latent factors are in control of users' and items' features, respectively. The model is trained by minimizing the square error objective function with regularization terms.

With the availability of social information, various models have been proposed by incorporating social connections to improve the recommendation performance, like SocialMF [2], SoReg [6] and hTrust [7]. Under the assumption that user's feature vectors should be close to their neighbors so that they could be learnt even for users who have few ratings, Mohsen in [2] incorporates trust propagation into the model, while Ma [6] and Tang [7] devise regularization terms for disparate objective function by investigating various user similarity measures and homophily coefficient respectively. In addition, interpersonal influence and social relation are well studied in [8,9]. Moreover, Ma et al. [10] discusses the influences on implicit social recommendation in detail.

However, the above methods ignore the influences of the opinion leaders underlying social connections, particularly on users who have different tastes with their friends. In [11], Yang proposes a circle-based recommendation by inferring category-specific social circles and devises several variants of weighting friends within circles. Following that, interpersonal influence and interests similarity are studied based on circles [12,13]. In [14], Ma considers both trust and distrust relationships and proposes a SVD signs based community mining method to discover the trust matrix. Lin et al. [15] exploits potential experts on virtual social networks, which are extracted from implicit feedbacks, and integrate the content-based methods by employing CF. Zhao et al. [16] utilizes the "following relations" between the users and topical interests to build the user-to-user graph, and proposes the Graph Regularized Latent Model (GRLM). In GRLM, expertise of users can be inferred based on both past question-answering activities and the inferred graph. What differentiates our approach to the aforementioned methods is that we explore the expert influences which is not only underlying explicit social connections, i.e., friend relationships, but also underlying implicit social connections, i.e., users' similarity preferences, in each circle.

3   Model Specification

In this section, we devise two methods identifying experts and propose a personalized recommendation algorithm via exploiting expert relationships underlying the circle-based social connections. Experts who have distinct influences on users in specific category are used to constrain matrix factorization objective function.

3.1   Problem Definition

Formally, let $U = \{u_1, u_2, ..., u_M\}$ and $V = \{v_1, v_2, ..., v_N\}$ be the set of users and items, where M and N refer to the number of users and items respectively. User relationship can be formulated as a matrix $F \in \mathbb{R}^{M \times M}$, where each entry f_{ij} equals to 1 when u_i and u_j are connected in social networks otherwise 0. According to the categories of items rated by each user, the user can be divided into corresponding circles. The definition is shown as follows:

Definition 1 (Circled Social Connection). *Two users u_i and u_j are said to be in the same circle C^c, i.e., $(u_i, u_j) \in C^c$, iff (1) $f_{ij} = 1$ (2) $V_{u_i}^c \neq \varnothing$ and $V_{u_j}^c \neq \varnothing$, where V_u^c denotes a set of items which belongs to category c and is rated by user u.*

3.2   Basic MF

As is mentioned above, the basic MF technique factorizes the entire rating matrix R into two matrices $U \in \mathbb{R}^{M \times D}$ and $V \in \mathbb{R}^{N \times D}$ where U and V are the user-feature and item-feature matrices respectively, $D \ll M, N$. Since we employ the concept of circle, the rating matrix R should be split into small pieces regarding to each category c, aka R^c and the objective function should be revised as follows accordingly:

$$\mathcal{L}^c = \frac{1}{2}||I \odot (R^c - U^c V^{cT})||_F^2 + \frac{\lambda}{2}(||U^c||_F^2 + ||V^c||_F^2) \tag{1}$$

where $\odot$ is hadamard product and U^c, V^c are the user-feature and item-feature matrices regarding to each category. I is an indicator matrix with the same dimension of R^c.

3.3   Identifying Experts

We consider "friends based experts" and "similarity based experts" in our model. Specifically, users' friendships and similarity are employed to construct two social networks. Then the experts are identified on both social networks via information transmission theory and user rating behaviors.

Building Social Networks. Building explicit social network with friendships, the user is a node in the network, and the friend relationship is the edge of the network. There is an edge between two users if they are friends. Finally, a social network is formed among these uses.

The implicit social network is constructed by the similarity between users. The construction is a little complex, which can be divided into the following three steps.

– Firstly we should calculate the similarity among users. The similarity between two users is calculated according to their ratings on items.
– Then the similarity threshold σ is selected and determined. If the similarity between u_i and u_j is greater than σ, there is an implicit edge e_{ij} between them.
– Finally, the weight ρ_{ij} of e_{ij} is given by the similarity between u_i and u_j.

To estimate the influence of parameter σ on the implicit social networks, we do some experiments in Sect. 4.4. An expert, generally, is a person with extensive knowledge or ability based on research, experience or occupation in a particular area. Due to experts' opinions are very constructive, there are a lot of users willing to follow them, so they may have abundant social connections. Hence, we define Eqs. 2 and 3 to represent the influence of each user on the explicit and implicit social networks respectively. By ranking those influences of users, we can identify the experts on the list.

Friends Based Expert and Similarity Based Expert. When both social networks are built, we can get matrix $F \in \mathbb{R}^{M \times M}$, where each entry f_{ij} equals to 1 when u_i and u_j are connected in social network otherwise 0. The user's influence is calculated by information transmission theory and user rating behaviors. According to [17], where a multi-step diffusion model with a shrinking parameter is proposed, we denote the transmission factor as ϕ_1, ϕ_2 and ϕ_3 for the first three layers. So u_i's influence under both social network can be calculated by:

$$P_{u_i}^1 = \phi_1 |V_{u_i}| + \phi_2 \sum_{f_{ij}=1} |V_{u_j}| + \phi_3 \sum_{f_{jk}=1} |V_{u_k}| \tag{2}$$

$$P_{u_i}^2 = \phi_1 |V_{u_i}| + \phi_2 \sum_{f_{ij}=1} \rho_{ij} |V_{u_j}| + \phi_3 \sum_{f_{jk}=1} \rho_{jk} |V_{u_k}| \tag{3}$$

where ρ_{ij} and ρ_{jk} are the weights of the corresponding edge. Using Eqs. 2 and 3, we can obtain the influence of each user on both social networks. Meanwhile, experts are identified by ranking the influence of each user.

3.4   Experts Regularization

Since users are more inclined to follow experts who have similar interests in a certain circle. This motivates us to assume that, experts preference has an important impact on users in their circles. So we incorporate the influence of experts into the regularization terms.

Influences from experts are two-folds: firstly, the precondition of an expert influencing an ordinary user is that they both rate the same merchant. Secondly, an expert has his own assertion and is unlikely to be influenced by other experts. Based on aforementioned two premises, we use "Friends Based Expert" and "Similarity Based Expert" to define the influence values in matrix E_1^c and E_2^c respectively. Expert matrix E^c is illustrated below:

$$E_{ij}^c = \begin{cases} t\dfrac{|V_{u_i}^c \cap V_{u_j}^c|}{\sum_{u_k \in C^c}|V_{u_j}^c \cap V_{u_k}^c|} + (1-t)\dfrac{|V_{u_i}^a \cap V_{u_j}^c|}{|V_{u_i}^c|} & V_{u_i}^c \cap V_{u_j}^c \neq \varnothing, u_i \notin E, u_j \in E \\ 0 & \text{otherwise} \end{cases}$$

(4)

where t is the trust trade-off between the expert u_j and himself u_i in circle c.

Note that E^c should be normalized throughout each row so that $\Sigma_j E^{*c}_{ij} = 1$ and naturally experts regularization is derived as:

$$\text{Reg}_{experts} = ||U^c - E^{*c}U^c||_F^2$$

(5)

3.5   Models and Training

With Eqs. 1 and 5 in hand, we propose Circle-based Recommendation with Experts Regularization (CRER). We adopt three different regularization terms to devise our methods. The details are as follows:

CRER1. We design the model CRER1 with friends and both kinds of experts regularization terms. S^{*c} adopt the method proposed in [11] to devise friends regularization for a user. CRER1 is devised as:

$$\mathcal{L}_1^c = \frac{1}{2}||I^c \odot (R^c - U^cV^{cT})||_F^2 + \frac{\lambda}{2}(||U^c||_F^2 + ||V^c||_F^2)$$
$$+ \frac{\alpha}{2}||U^c - S^{*c}U^c||_F^2 + \frac{\beta}{2}||U^c - E_1^{*c}U^c||_F^2 + \frac{\gamma}{2}||U^c - E_2^{*c}U^c||_F^2$$

(6)

where α, β and γ are tuning parameters for weighting the significance of these terms.

CRER2. Model CRER2 has Friends and Friends Based Expert Regulation terms. CRER2 only uses explicit social connections (friends) as auxiliary information, and does not use any implicit social connections.

$$\mathcal{L}_2^c = \frac{1}{2}||I^c \odot (R^c - U^cV^{cT})||_F^2 + \frac{\lambda}{2}(||U^c||_F^2 + ||V^c||_F^2)$$
$$+ \frac{\alpha}{2}||U^c - S^{*c}U^c||_F^2 + \frac{\beta}{2}||U^c - E_1^{*c}U^c||_F^2$$

(7)

Algorithm 1. Circle-based Recommendation with Experts Regularization

---

   input : R^c, E_1^{*c}, E_2^{*c}
   output: U^c, V^c
1 random U^c, V^c;
2 **for** $step = 1$ *to MAX STEP* **do**
3 | $U^c = U^c - \eta \frac{\partial L^c}{\partial U^c}$;
4 | $V^c = V^c - \eta \frac{\partial L^c}{\partial V^c}$;
5 | $\eta = \frac{\eta}{1+step/MAX STEP}$;
6 $\widehat{R}^c = U^c V^{cT}$;

---

where α and β are tuning parameters for weighting the significance of the two terms.

CRER3. Model CRER3 has dual Experts Regulation terms. It comprehensively uses the explicit social connections (friends) and implicit social connections (user similarity) as auxiliary information for personalized recommendation.

$$\mathcal{L}_3^c = \frac{1}{2}||I^c \odot (R^c - U^c V^{cT})||_F^2 + \frac{\lambda}{2}(||U^c||_F^2 + ||V^c||_F^2)$$
$$+ \frac{\beta}{2}||U^c - E_1^{*c}U^c||_F^2 + \frac{\gamma}{2}||U^c - E_2^{*c}U^c||_F^2 \qquad (8)$$

where β and γ are tuning parameters for weighting the significance of the two terms.

Training. The training process within each category will be conducted independently. Batched gradient descent can be employed here to minimize the objective function till convergency. $\mathcal{L}_3^c$ learning equations are shown below:

$$\frac{\partial L_3^c}{\partial U^c} = I^c \odot (U^c V^{cT} - R^c)V^c + \lambda U^c + \beta(\mathbf{I} - E_1^{*c})(U^c - E_1^{*c}U^c) + \gamma(\mathbf{I} - E_2^{*c})(U^c - E_2^{*c}U^c) \quad (9)$$

$$\frac{\partial L_3^c}{\partial V^c} = I^c \odot (U^c V^{cT} - R^c)^T U^c + \lambda V^c \qquad (10)$$

where $\mathbf{I}$ is an identity matrix. The corresponding algorithm is shown in Algorithm 1.

In Algorithm 1, η is learning steps, and it will be gradually reduced with the increase of iterations. After learning the user preference matrix U^c and the item characteristic matrix V^c via Algorithm 1, then the possible rating from the user u_i to the item v_j in category c will be predicted by $\widehat{r}_{ij}^c = u_i^c v_j^{cT}$.

3.6 Complexity Analysis

The complexity owes much to the computation of objective function $\mathcal{L}_3^c$ and its gradients. Let us assume $\overline{r}$ and $\overline{e}$ are the average number of ratings and influencing experts of a user respectively. The computational cost of $\mathcal{L}_3^c$ (Eq. 8) is $\mathcal{O}(M(\bar{r}+2\bar{e})D)$ and the cost of its corresponding gradients is $\mathcal{O}(M(\bar{r}+2\bar{e}^2)D)$. In real world $\bar{r}$ is relatively small. Besides due to the requirement that influences from experts only happens when a user have visited the same merchants with the experts, the scale of $\bar{e}$ is no larger than $\bar{r}$, i.e., $\bar{e} \leqslant \bar{r}$. Thus, the entire complexity is approximately linear with the number of users, which shows the computational efficiency of our proposed model.

4 Experiments

In this section, we conduct extensive experiments on our final model CRER3 on Yelp and Epinions datasets, and compare it with CRER1 and CRER2 proposed in Sect. 3 and several state-of-the-art methods.

Table 1. Statistic of each category on both datasets

(a) Yelp dataset

Category	User Count	Item Count	Rating Count	Sparsity
Activelife	16,004	2,379	29,800	7.83e-4
Arts	30,212	1,900	79,087	1.38e-3
Beauty	21,622	4,521	39,835	4.08e-4
Hotels	25,676	1,987	58,541	1.15e-3
Nightlife	50,258	4,253	161,165	7.54e-4
Restaurants	84,541	29,169	874,428	3.55e-4
Services	38,836	7,886	89,441	2.92e-4
Shopping	24,344	8,687	70,716	3.34e-4

(b) Epinions dataset

Category	User Count	Item Count	Rating Count	Sparsity
Services	11,022	808	39,861	4.48e-3
Games	7,576	1,577	34,143	2.86e-3
Books	8,416	2,064	29,373	1.69e-3
Music	7,194	2,758	37,014	1.87e-3
Hardware	6,805	1,020	1,6041	2.31e-3
Hotels	10,195	1,686	43,794	2.55e-3
Restaurants	7,447	1,059	25,982	3.29e-3
Cars	7,082	1,106	14,154	1.81e-3
Beauty	5,552	1,714	23,549	2.47e-3
Kids	7,810	2,224	54,385	3.13e-3

4.1 Datasets and Metrics

For Yelp dataset, we use Yelp Challenge round 5 dataset[1]. According to different categories, the items are divided into corresponding circles. The total user number is 366,715, while the number of users rating items less 4 is 280,969. Table 1(a) is a statistic of users and items on eight categories of Yelp dataset. Epinions is a website where users can review and rate items and we adopt the published version of the Epinions dataset[2]. It consists of 22,164 users and a total of 296,277 different items. The total number of ratings is 922,267. By dividing items into different categories, we also can get some circles. Table 1(b) is a statistic of users and items in ten categories of Epinions.

We perform 5-fold cross validation in our experiments. In each fold, we use 80 % of data as the training sets and the remaining 20 % as the test sets. The evaluation metrics we use in our experiments are Root Mean Square Error (RMSE)

---

[1] http://www.yelp.com/dataset_challenge.
[2] http://www.public.asu.edu/~jtang20/datasetcode/truststudy.htm.

and Mean Absolute Error (MAE), as these are the most popular accuracy measures in the literature of recommender systems. The RMSE is defined as

$$RMSE = \sqrt{\frac{\sum\limits_{(u,i)\in R_{test}} (R_{u,i} - \hat{R_{u,i}})^2}{|R_{test}|}}$$

where R_{test} is the set of all user-item pairs (u, i) in the test set and $R_{u,i}$ is the real rating value of user u on item i, $\hat{R_{u,i}}$ is the corresponding predicted rating value. The MAE is defined as

$$MAE = \frac{\sum\limits_{(u,i)\in R_{test}} |R_{u,i} - \hat{R_{u,i}}|}{|R_{test}|}$$

4.2 Baseline and Comparison

In this section, we compare the performance of the proposed approaches with the existing models including:

- BaseMF: This method is proposed in [5] by factorizing user-item matrix, which only utilizes the rating matrix without considering any social information.
- PRM: It is proposed in [12] which considers user's personality and social information. It takes into account the commodity and the correlation among users.
- CirCon3: The concept of circle-based recommendation system is presented in [11]. The new feature of "Friends Circles" and rating data combined with social network are used to infer "Trust Circles".
- SR: This method is proposed in [10], which not only uses the user rating records but also fuses the implicit user and item social information into social recommendation techniques, including similar and dissimilar relationships.
- CRER1: This model considers friends and two different experts regularizations to regularize user's preference. CRER1 pays attention to the relationship between friends and the influence of the experts on explicit and implicit social networks.
- CRER2: It contains experts inferred from friends' trust information and friends relationship regularizations. CRER2 not only identifies experts through the friends relationship, but also directly adopts friend relationships as a auxiliary information for matrix factorization.
- CRER3: This approach adopts two types of expert regularizations based on users' friend relationships and similarity preferences, respectively. It adequately considers the impacts of experts on users in each circle.

In all our experiments, we set the dimensionality for low-rank matrix factorization to be d=10 and the regularization constant to be λ=0.1.

Tables 2, 3, 4 and 5 show the performance on the Yelp and Epinions datasets based on two metrics. Parameters β and γ are set to 5, as they can make the best result on both datasets. From the four tables, we can see that CirCon3 performs better than PRM. But in [12], the author said that PRM is better than CirCon3 because, first, we found that, although PRM also uses Yelp data, but the amount of data we use is significantly bigger than theirs; second, PRM dataset shows less users, more items, and also has a lot of rating count, so a user may rate more items. The PRM method pays more attention to the relationship among users and the relationship between users and items, so the algorithm effect better when one can rate more items. And CirCon3 takes into account the user's trust friends, so in a large amount of users, CirCon3's performance will be better. In the results of two metrics on the Yelp and Epinions datasets, CirCon3 is nearly neck and neck with SR. Moreover, the performance of the whole SR is slightly better than that of CirCon3 on Yelp dataset while CirCon3 is slightly superior to SR on Epinions dataset. The reason may lie in different sparsity of the two datasets.

Table 2. MAE comparisons for eight categories on Yelp (dimensionality = 10).

Category	BaseMF	PRM	CirCon3	SR	CRER1	CRER2	CRER3
Activelife	1.086	0.981	0.866	0.794	0.787	0.759	**0.752**
Arts	1.017	0.980	0.851	0.849	0.748	0.743	**0.739**
Beauty	1.408	1.165	0.974	0.941	0.914	0.908	**0.907**
Hotels	0.938	0.933	0.850	0.837	0.763	0.754	**0.753**
Nightlife	0.985	0.974	0.852	0.851	0.777	0.774	**0.771**
Restaurants	0.857	0.791	0.739	0.754	0.735	0.736	**0.734**
Services	1.077	0.982	0.833	0.908	0.808	0.798	**0.793**
Shopping	1.201	0.975	0.951	0.845	0.798	0.784	**0.781**
Average	1.071	0.973	0.864	0.847	0.791	0.782	**0.779**

Table 3. RMSE comparisons for eight categories on Yelp (dimensionality = 10).

Category	BaseMF	PRM	CirCon3	SR	CRER1	CRER2	CRER3
Activelife	1.701	1.343	1.162	0.990	0.980	0.933	**0.922**
Arts	1.571	1.340	1.124	1.121	0.916	0.908	**0.899**
Beauty	2.721	1.889	1.458	**1.359**	1.491	1.489	1.483
Hotels	1.383	1.282	1.174	1.131	0.960	0.942	**0.938**
Nightlife	1.452	1.356	1.126	1.126	0.969	0.967	**0.959**
Restaurants	1.122	0.971	**0.873**	0.911	0.885	0.891	0.887
Services	1.813	1.392	1.105	1.314	1.037	1.018	**1.004**
Shopping	2.054	1.361	1.377	1.123	1.025	1.002	**0.994**
Average	1.727	1.367	1.175	1.134	1.033	1.019	**1.011**

In most cases, our models CRER1, CRER2 and CRER3 are superior to compared algorithms in two kinds of metrics on Yelp and Epinions datasets. The

Table 4. MAE comparisons for ten categories on Epinions (dimensionality = 10).

Category	BaseMF	PRM	CirCon3	SR	CRER1	CRER2	CRER3
Services	1.181	1.141	1.050	1.065	0.967	0.958	**0.955**
Games	1.121	1.046	0.890	0.915	0.779	0.788	**0.763**
Books	1.304	1.056	0.948	0.994	0.819	**0.789**	0.793
Music	1.157	1.039	0.875	0.915	0.749	0.744	**0.738**
Hardware	1.335	1.087	1.156	0.996	0.933	0.936	**0.876**
Hotels	1.148	1.049	0.876	0.912	0.765	**0.753**	0.755
Restaurants	1.190	1.075	0.975	0.919	0.909	0.905	**0.904**
Cars	1.301	1.071	1.011	1.035	0.913	0.892	**0.849**
Beauty	1.308	1.169	1.052	1.081	0.985	0.972	**0.971**
Kids	1.273	1.121	0.960	0.929	0.847	0.852	**0.836**
Average	1.540	1.357	1.224	1.220	0.867	0.859	**0.844**

Table 5. RMSE comparisons for ten categories on Epinions (dimensionality = 10).

Category	BaseMF	PRM	CirCon3	SR	CRER1	CRER2	CRER3
Services	2.029	1.831	1.670	1.704	1.472	1.449	**1.448**
Games	1.838	1.495	1.224	1.281	0.974	0.995	**0.949**
Books	2.349	1.496	1.323	1.429	1.041	**0.985**	0.996
Music	1.923	1.444	1.171	1.269	0.898	0.892	**0.881**
Hardware	2.392	1.576	1.875	1.429	1.361	1.327	**1.220**
Hotels	1.956	1.533	1.237	1.317	1.001	**0.976**	0.985
Restaurants	2.025	1.590	1.431	**1.307**	1.325	1.319	1.320
Cars	2.179	1.562	1.479	1.544	1.266	1.222	**1.114**
Beauty	2.350	1.817	1.626	1.684	1.461	1.441	**1.439**
Kids	2.278	1.649	1.416	1.338	1.160	1.169	**1.145**
Average	2.665	1.999	1.806	1.788	1.196	1.178	**1.150**

circle-based algorithm using elites information to assist the recommendation is effective. The performance of CRER1, CRER2 and CRER3 on both dataset are basically better than those of the first four methods according to MAE and RMSE evaluation metrics. And CRER3 outperforms CRER2 and CRER1, while CRER2 is slightly better than CRER1. We attribute the better performance to users' preferences captured by experts' influences in each circle. Besides, the unsupervised methods of finding experts are effective. Further, on Yelp dataset, in Beauty and Restaurants categories, SR and CirCon3 are slightly superior to our methods on RMSE metrics, respectively. Because users in Restaurants category is very large, so considering friendship may be more helpful to improve recommendation accuracy. While in Beauty category, with only a few users and less rating counts, so model with implicit similar and dissimilar relationship has better performance. On Epinions dataset, in Books and Hotels attributes, CRER2 performs best. So, in some case, we should consider friend and expert relationships together to capture users' preferences. We know CRER1 and CRER2 consider the regularization terms of friend and expert, but in most cases the performance of CRER2 is better than that of CRER1. It may be that friend and similarity based expert will weaken the influence of each other when they work at the same time, thus leading to the bad performance. However, compared to CRER1, without friends regularization term, the accuracy of CRER3 is significantly improved, the reason may lie in friends preference limiting the impact of similarity based experts. Overall, CRER3 method which makes full use of users' friends and similarity preferences to exploit the influence of experts on personalized recommendation demonstrates its competitiveness, and performs better.

4.3   The Cold Start Problem

The cold-start problem is a potential problem in personalized recommendation. It may cause the inaccuracy of prediction, as it does not gather abundant information of users or items, e.g. new users. In this section, we compare our methods with comparative algorithms on user cold-start problem. The test sets are divided into five groups according to the number of user rating items. Figure 2 shows the performance of the different algorithms under different user rating counts on both datasets, respectively.

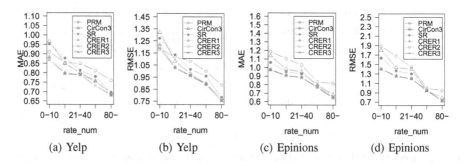

(a) Yelp   (b) Yelp   (c) Epinions   (d) Epinions

Fig. 2. Impact of user's rated number on MAE and RMSE metrics of both datasets.

The horizontal axis represents the number of user rating items, e.g. "0–10" means user rating items between 0 and 10, "11–20" means user rating items between 11 and 20, "80-" means user rating items more than 80, and so on. The vertical axis of Fig. 2(a) and (c) represent MAE values of Yelp and Epinions datasets. Figure 2(b) and (d) represent RMSE values of Yelp and Epinions datasets respectively. From these four graphs we can see that the accuracy of the algorithm will be higher when the number of rating increases. And in any case, our methods are superior to baseline algorithms on both datasets. Figure 2(a) and (b) show that when the rating number is very little, the SR and CirCon3 algorithms perform poor, but with the increase of the user ratings, their performances will be better. From Fig. 2(c) and (d), we can observe that PRM performs worst on Epinions dataset. SR outperforms PRM and CirCon3. The effect of CRER1, CRER2 and CRER3 is almost the same. In a whole, from Fig. 2 can be seen that the methods we proposed achieve the lowest MAE and RMSE on both datasets. This is because that our method not only considers users in different areas of expertise, but also exploit experts in the field to capture user's preferences.

Our methods are always the best in the five sets of the cold-start experiments. The reasons may lie in that both datasets we use have a large number of users, a small number of items, and the number of users rating items is small, so in this case elite groups that are identified on explicit and implicit social networks play an important role in personalized recommendation. From Fig. 2 we can see that the performance are getting better when the number of items that user ratings increase, at the same time, their accuracy is also getting closer. When user rates more items, the more helpful to improve the accuracy of the model, but the degree of improvement will be smaller. Hence, our methods fusing expert influences into circle-based personalized recommendation can alleviate the user cold-start problem to some extent.

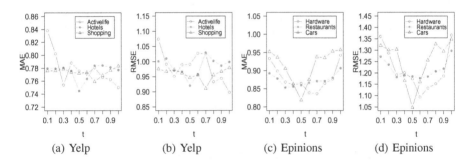

(a) Yelp   (b) Yelp   (c) Epinions   (d) Epinions

Fig. 3. The performance of CRER3 w.r.t t on three categories of both datasets.

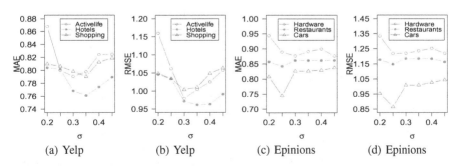

(a) Yelp   (b) Yelp   (c) Epinions   (d) Epinions

Fig. 4. The performance of CRER3 w.r.t σ on three categories of both datasets.

4.4   The Impact of σ and t

In order to study the influence of experts identified on the explicit and implicit social networks, we do the following experiments on the parameters σ and t. we choose Activelife, Hotels and Shopping categories of Yelp, and Hardware, Restaurants and Cars categories of Epinions to estimate the impact of σ and t. As aforementioned, σ is user similarity threshold (Sect. 3.3 Building Social Network) and t is the parameter in Eq. 4. σ is the weight of implicit social network edge and it controls which edge can be reserved in implicit social network built by users' similarity preference. t is the trust trade-off between experts and users in a specific circle, and it determines the influence of experts on a specific user. We vary the value of t as $\{0.1, 0.2, 0.3, 0.4, 0.5, 0.6, 0.7, 0.8, 0.9, 1.0\}$, and σ as $\{0.2, 0.25, 0.3, 0.35, 0.4, 0.5\}$. Figures 3 and 4 show the MAE and RMSE performance of parameters t and σ on CRER3. From Fig. 3 we can observe several peaks on the line chart which show that our method CRER3 is sensitive to t. From Fig. 3(a) and (b), for Yelp dataset, it is suggested that our method performs well when $t \in [0.3, 0.6]$. The optimal value of t is different for different categories. As for Epinions dataset, from the Fig. 3(c) and (d), they show the superiority of $t \in [0.3, 0.8]$ over other numerical value intervals. Cars category is more sensitive to the parameter t than Hardware and Restaurant categories. That illustrates the impact of experts on the user is obvious under Cars category. As for parameter σ, Fig. 4 shows the performance on Yelp and Epinions datasets. It is obvious that the results firstly become better with the increase of σ however they become worse when σ surpasses a certain value, i.e., 0.3 and 0.25 in our experiments for Yelp and Epinions datasets.

5   Conclusion

In this paper, we propose a personalized recommendation algorithm via incorporating experts influences into circle-based social networks to improve the accuracy of personalized recommendation. Specifically, we first construct two social networks by employing users' friend relationship and similarity preferences. Then based on information transmission theory and user rating records, we devise two unsupervised methods to identify experts on both social networks. Based on circle-based social networks, experts are divided into several circles according to their historical rating records, where items or merchants in each circle belong to the same category. Further, we devise two expert regularization terms to constrain matrix factorization objective function. Comparing with state-of-art algorithms, our methods can accurately capture users' preferences. Extensive experiments conducting on Yelp and Epinions datasets demonstrate the competitiveness of our methods, especially in handing user cold-start problem.

Acknowledgments. This research was partially supported by grants from the National Natural Science Fund Project of China (Grant No. 61232018 and 61325010).

References

1. Sun, J., Tang, J.: A survey of models and algorithms for social influence analysis. In: Aggarwal, C.C. (ed.) Social Network Data Analytics, pp. 177–214. Springer, US (2011)
2. Jamali, M., Ester, M.: A matrix factorization technique with trust propagation for recommendation in social networks. In: RecSys. pp. 135–142 (2010)
3. Linden, G., Smith, B., York, J.: Amazon. com recommendations: item-to-item collaborative filtering. IEEE Internet Comput. **7**(1), 76–80 (2003)
4. Koren, Y.: Factorization meets the neighborhood: a multifaceted collaborative filtering model. In: KDD. pp. 426–434 (2008)
5. Salakhutdinov, R., Mnih, A.: Probabilistic matrix factorization. In: NIPS. pp. 1257–1264 (2007)
6. Ma, H., Zhou, D., Liu, C., Lyu, M.R., King, I.: Recommender systems with social regularization. In: WSDM. pp. 287–296 (2011)
7. Tang, J., Gao, H., Hu, X., Liu, H.: Exploiting homophily effect for trust prediction. In: WSDM. pp. 53–62 (2013)
8. Huang, J., Cheng, X., Guo, J., Shen, H., Yang, K.: Social recommendation with interpersonal influence. In: ECAI. pp. 601–606 (2010)
9. Wang, F., Jiang, M., Zhu, W., Yang, S., Cui, P.: Recommendation with social contextual information. IEEE Trans. Knowl. Data Eng. **26**(11), 2789–2802 (2014)
10. Ma, H.: An experimental study on implicit social recommendation. In: The 36th International ACM SIGIR conference on research and development in Information Retrieval, SIGIR 2013, pp. 73–82. Dublin, Ireland, 28 July - 01 August 2013
11. Yang, X., Steck, H., Liu, Y.: Circle-based recommendation in online social networks. In: KDD. pp. 1267–1275 (2012)
12. Feng, H., Qian, X.: Recommendation via user's personality and social contextual. In: CIKM. pp. 1521–1524 (2013)
13. Zhao, G., Qian, X., Feng, H.: Personalized recommendation by exploring social users' behaviors. In: Gurrin, C., Hopfgartner, F., Hurst, W., Johansen, H., Lee, H., O'Connor, N. (eds.) MMM 2014, Part II. LNCS, vol. 8326, pp. 181–191. Springer, Heidelberg (2014)
14. Ma, X., Lu, H., Gan, Z.: Improving recommendation accuracy by combining trust communities and collaborative filtering. In: Proceedings of the 23rd ACM International Conference on Conference on Information and Knowledge Management, pp. 1951–1954. ACM (2014)
15. Lin, C., Xie, R., Guan, X., Li, L., Li, T.: Personalized news recommendation via implicit social experts. Inf. Sci. **254**, 1–18 (2014)
16. Zhao, Z., Wei, F., Zhou, M., Ng, W.: Cold-start expert finding in community question answering via graph regularization. In: Renz, M., Shahabi, C., Zhou, X., Cheema, M.A. (eds.) DASFAA 2015. LNCS, vol. 9049, pp. 21–38. Springer, Heidelberg (2015)
17. Song, X., Tseng, B.L., Lin, C.Y., Sun, M.T.: Personalized recommendation driven by information flow. In: Proceedings of the 29th Annual International ACM SIGIR Conference on Research and Development in Information Retrieval, pp. 509–516. ACM (2006)

An Approach for Clothing Recommendation Based on Multiple Image Attributes

Dandan Sha[1], Daling Wang[1(✉)], Xiangmin Zhou[2], Shi Feng[1],
Yifei Zhang[1], and Ge Yu[1]

[1] School of Computer Science and Engineering,
Northeastern University, Shenyang, People's Republic of China
shadd0830@gmail.com, {wangdaling, fengshi,
zhangyifei, yuge}@cse.neu.edu.cn
[2] School of Computer Science and Information Technology,
RMIT University, Melbourne, Australia
xiangmin.zhou@rmit.edu.au

Abstract. Currently, many online shopping websites recommend commodities to users according to their purchase history and the behaviors of others who have similar history with the target users. Most recommendations are conducted by commodity tags based similarity search. However, clothing purchase has some specialized characteristics, i.e. users usually don't like to go with the crowd blindly and will not buy the same clothing twice. Moreover, the text tags cannot express clothing features accurately enough. In this paper we propose a novel approach that extracts multi-features from images to analyze its content in different attributes for clothing recommendation. Specifically, a color matrix model is proposed to distinguish split joint clothing. ULBP feature is extracted to represent fabric pattern attribute. PHOG, Fourier, and GIST features are extracted to describe collar and sleeve attributes. Then, some classifiers are trained to classify clothing fabric patterns and split joint types. Experiments based on every attribute and their combinations have been done respectively, and have achieved satisfied results.

Keywords: Clothing recommendation · Multiple attributes · Classification

1 Introduction

Recently many online shopping websites provide personalized recommendation for facilitating users' selection. The websites can recommend relevant commodities to users based on their purchased, browsed, or collected products. The recommendations are mainly conducted using similarity search based on commodities' tags. In addition, some websites recommend the products purchased by others to a target user based on their popularity, and believe that users may buy the products that many others have purchased. This type of approaches is called as collaborative recommendation. Yet, for clothing recommendation, these approaches cannot fully meet users' specialized requirements due to the following reasons: (1) The tags-based recommendation may be inaccurate due to ambiguous and limited text expression and nonstandard tag names.

B. Cui et al. (Eds.): WAIM 2016, Part I, LNCS 9658, pp. 272–285, 2016.
DOI: 10.1007/978-3-319-39937-9_21

Same commodities may have different tags in different websites and different commodities may have the same tags. (2) Most people make a purchase based on personal preferences rather than others' behaviors [8], and they don't even like to go with the crowd in psychology. Even if two users buy the same clothes, they may do that for different purposes. (3) If a user bought a dress, generally it is impossible for the user to purchase the same dress again [16]. Thus, simple recommendation based on users' purchasing history and recommending the same one is irrational in this case.

Intuitively, if a user purchased a clothing (image), she (he) is supposed to like its some parts or characteristics at least. For example, in Fig. 1(a), a user may only like the fabric pattern of ink butterfly. In Fig. 1(b), a user may like its polo collar style. In Fig. 1 (c), a user may like its lace sleeves. In Fig. 1(d), a user may like its split joint style of sleeves parts different from the main body part.

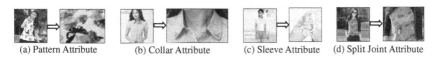

(a) Pattern Attribute   (b) Collar Attribute   (c) Sleeve Attribute   (d) Split Joint Attribute

Fig. 1. Possible clothing parts (attributes) liked by users

In text analysis, a word may have several different semantics. Similarly, an image interesting a user may contain more semantic context. If a user buys a clothing, she (he) won't buy the same one again. However, she (he) may buy other clothes which share similar attributes with it, including pattern, collar, sleeves, and split joint. Based on this intuition, we propose an approach based on pattern, collar, sleeves, split joint respectively and their freely customized combinations for clothing recommendation in online clothing shopping. In clothing recommendation, color, pattern, collar, sleeve, and so on are called as attributes [9, 18]. Thus our recommendation is made based on different attributes respectively.

The main contributions of this paper are summarized as follows: (1) We collect a large online clothing shopping dataset for investigating the content based online shopping recommendation task. Additionally, these datasets are thoroughly labeled with complete attributes. (2) We propose an accurate human body part segmentation method that detects global attributes like pattern and split joint, and detail attributes like collar and sleeve. A novel color matrix model is proposed for the split joint clothing judgment issue. (3) Besides pattern attribute, we consider the split joint, collar, and sleeves attributes which are usually not be considered in the clothing recommendation systems, and provide diversified results to meet different users' requirements. (4) We evaluate the effectiveness of our approach by experiments over real datasets.

The rest of the paper is organized as follows. Section 2 introduces related work. Section 3 describes our research problem and recommendation framework based on different clothing attributes. Section 4 describes the attribute detection of clothing image. Section 5 shows the attribute content analysis and recommendation. Section 6 shows our experimental results and the conclusion is given in Sect. 7.

2 Related Work

For clothing recommendation, Huang et al. [4] model users' preferences based on the users' search history, and recommend the most relevant results by analyzing type, color, and appearance features of clothes. Lin et al. [8] recommend an appropriate combination from the user's available clothing options based on the day's weather and the user's schedule. Ajmani et al. [1] utilize semantic analysis of clothing features, media features, and user profile for personalized clothing recommendation.

Clothing retrieval is also one of the support techniques. Wang et al. [18] retrieve clothing on photo collections by combining color matching and attribute (e.g. T-shirt, collar, button, horizontal stripes, etc.) learning. Fu et al. [3] encode semantic context into the bag-of-word model to solve the cross-scenario clothing retrieval problem. Liu et al. [9] search clothing from online shops similar to a human photo captured on street. Chen et al. [2] proposed a way of clothing fabric image retrieval based on impressions. Li et al. [6] search clothes from websites based on text and image provided by users. Mizuochi et al. [13] retrieve clothing based on local similarity with multiple images. Lin et al. [7] implemented clothing retrieval by using a hierarchical deep search technique. Yamaguchi et al. [19] proposed to pre-train global models of clothing and learn local models of clothing on the fly from retrieved examples for parsing clothing. The above recommendations and retrievals are mainly based on global features or attributes. We focus on different attributes respectively and synthetically.

In clothing image process domain, Lu et al. [11] implemented clothing image's main color and attribution annotation by segmenting image into regions, and removing background and noise. Yamazaki and Inaba [20] use image features derived from clothing fabrics, wrinkles, and clothing overlaps to classify clothes. Wang and Ai [17] combine blocking models to address the person-wise occlusions for highly occluded images. Zhou et al. [24] create 3D garment models from a single image. Huang et al. [5] use both global and local features to improve the estimation accuracy of cloth fabric images. Ramisa et al. [14] use depth and appearance features for informing robot grasping of highly wrinkled clothes. Yang et al. [21, 22] classify clothes patterns into plaid, striped, patternless, and irregular, and developed a feature combination to form a local image descriptor. Manfredi et al. [12] developed a system for clothing segmentation and color classification. Liu [10] proposed image salient region segmentation based on the color and texture of the region. In [23], an attribute-augmented semantic hierarchy (A^2SH) was proposed to organize the semantic concepts into multiple levels. Our multiple image attributes-based recommendation is just inspired by A^2SH [23] and upper-/lower-body online shopping clothing [9].

3 Our Clothing Recommendation Framework

Given a clothing image dataset I_{cloth}, for any clothing image $i_c \in I_{cloth}$, let $A = \{A_1, A_2, A_3, A_4, A_5\}$ be the attributes set of i_c, where A_1, A_2, A_3, A_4, and A_5 are color, fabric pattern, collar, sleeve, and split joint attributes, respectively. We describe these attributes using descriptor $dpr_{c,Ai}$ ($i = 1, 2,..., 5$). For an image i_u from user u, we describe it with descriptor $dpr_{u,Ai}$ ($i = 1, 2,..., 5$), i.e.:

$$i_c \sim <dpr_{c,1}, dpr_{c,2}, dpr_{c,2}, dpr_{c,4}, dpr_{c,5}>$$
$$i_u \sim <dpr_{u,1}, dpr_{u,2}, dpr_{u,3}, dpr_{u,4}, dpr_{u,5}> \tag{1}$$

We will recommend a clothing set I_r with k images based on the free combinations of these attributes, i.e.:

$$I_r = \text{Top}_k(\underset{i_c \in I_{cloth}}{\operatorname{argmin}} dist(i_c, i_u)) \tag{2}$$

where $dist(i_c, i_u)$ is the distance between i_c and i_u. When we perform recommendation based on any attribute A_i, we only search i_c in the corresponding subset $I_i \subseteq I_{cloth}$, and represent them as the descriptors $dpr_{c,i}$ of i_c and $dpr_{u,i}$ of i_u to A_i. Based on these descriptors, the distance between i_c and i_u is calculated. As such, the recommendation can be performed based on the function as below:

$$dist(i_c, i_u) = \sum_{i=1}^{5} w_i dist(dpr_{c,i}, dpr_{u,i}) \tag{3}$$

where w_i is the weight of an attribute distance. If the ith attribute is chosen in recommendation, $w_i = 1$, otherwise $w_i = 0$. For an attribute choice A_i, we have to calculate all images $i_c \in I_i$ and $I_i \subseteq I_{cloth}$. To tackle this task, we must address the following problems:

- Get the most representative part of the human body to each clothing attribute.
- Selecting the best image features to describe the characteristics of each attribute.
- Design a recommendation algorithm based on multiple attributes.

The framework of our clothing recommendation is shown as Fig. 2. The framework includes modeling stage and recommendation stage. In modeling stage, we conduct 6 operations: (a) download clothing data (include images, tags, URL, and so on) from shopping websites; (b) remove the background of every clothing image; (c) segment each image; (d) get the most representative part of the human body corresponding to each attribute; (e) extract features from the human body parts; (f) obtain related descriptors expressing the clothing images. In the recommendation stage, we first obtain the descriptor of a user selected image. If a clothing is uploaded by the target user, a number of preprocessing operations (h) ~ (l) will be done as the operations (b) ~ (f). Otherwise, we can get the descriptor of a clothing image browsed by this user from the dataset directly in (g). Then, we search similar clothing from clothing datasets as in (m), and give the recommendation results using different attributes. Finally, in (n), we re-rank the recommendation results and return them based on user's feedback on which attributes are interesting. The operations in (a) ~ (f) are performed offline, while those in (g) ~ (m) can be done either online or offline. In fact, not all attributes are necessary for a recommendation, because they are not always characteristic. Thus we provide 5 attribute choices for users' declarative recommendation.

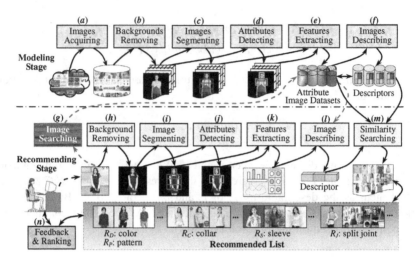

Fig. 2. Clothing recommendation based on different attribute

4   Clothing Image Segmentation and Attribute Detection

4.1   Clothing Image Segmentation and Upper Body Clothing Detection

As the clothing images on the shopping websites contain clean backgrounds having different colors, we will remove the interference from background. Considering the Grabcut [15] is efficient in eliminating image background and has been widely used in clothing retrieval and recommendation, we employ it to remove the backgrounds. Figure 3(a) and (b) show how to apply Grabcut for clothing segmentation.

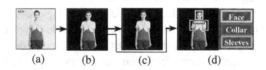

Fig. 3. Background removing, upper body detecting, face, collar and sleeves with bounding boxes (Color figure online)

Following that, to get the representative clothing parts, we utilize the face location in a clothing image to predict a bounding box below the head, which contains the clothing of the upper-body. Given an image in the dataset, once faces are detected and located, we can get the initial estimation of the main clothing parts. Because most original images are 600 × 600 pixels in our dataset, we set the bounding box 300 × 300 pixels and partition it into 9 blocks (Fig. 3(c)) for further pattern attribute extraction.

4.2   Collar and Sleeves Attribute Detection in Clothing Image

Strong relationships exist among the human body parts, especially in the clothing models. We can utilize the proportion and location of a face in a clothing image to locate the collar and sleeves. Denote H_h and H_w as the height and width of the model's head bounding box. For obtaining the most appropriate locations of collar and sleeve, we set different parameters and observe the percentage of "containing and only containing collar (sleeve)" images in all images. We call it as *Discrimination*. The *Discrimination*s of collar and sleeve are shown in Tables 1 and 2, respectively. If the parameters are set too small, the relevant parts cannot be contained. If they are set too large, other parts may also be contained. From Table 1, the best results can be obtained when the collar bounding box's height C_h is $0.6 \times H_h$ and its width C_w is $1.4 \times H_w$. Similarly, from Table 2, the sleeve bounding box's height S_h should be equal to H_h and its width S_w should be $0.7 \times H_w$. The results are shown in the Fig. 3(d).

Table 1. Discrimination of collar partitioning

$C_w \backslash C_h$	$0.4H_h$	$0.5H_h$	$0.6H_h$	$0.7H_h$	$0.8H_h$
$1.2H_w$	0.15	0.19	0.25	0.27	0.10
$1.3H_w$	0.18	0.25	0.65	0.56	0.13
$1.4H_w$	0.23	0.67	**0.81**	0.75	0.21
$1.5H_w$	0.15	0.65	0.73	0.65	0.15
$1.6H_w$	0.11	0.33	0.56	0.37	0.12

Table 2. Discrimination of sleeve partitioning

$S_w \backslash S_h$	$0.8H_h$	$0.9H_h$	$1.0H_h$	$1.1H_h$	$1.2H_h$
$0.5H_w$	0.08	0.10	0.15	0.17	0.20
$0.6H_w$	0.21	0.26	0.58	0.33	0.29
$0.7H_w$	0.28	0.52	**0.79**	0.65	0.38
$0.8H_w$	0.24	0.32	0.68	0.43	0.21
$0.9H_w$	0.18	0.21	0.38	0.23	0.15

4.3   Split Joint Clothing Judgment Based on Color Matrix

Our split joint clothing judgment mainly includes two parts: *color matrix construction* and *color matrix transformation*. In color matrix construction, for each upper-body clothes, it is difficult to recognize sleeves and main body parts accurately. Thus, to construct a color matrix for a clothing, we perform several operations. First, we partition the upper body image into 10×10 equal size blocks because fewer blocks cannot show the relationships of split joint attribute and 100 blocks is enough for training. The result is as shown in Fig. 4(a). We utilize their color and positional relationship to estimate split joint attribute. Then we extract the domain color of each block and describe it as a number. As such, each upper-body clothing is described as a 10×10 matrix. Finally, an RGB color histogram is computed over each block to obtain its main color. By dividing each of R, G, and B components into 5 parts in order to distinguish the color of skin, each color histogram is described as a 125-dimensional vector. To avoid the problem with split joint color clothing for short sleeves blouse, we remove the blocks that are most filled with skin before color matrix transformation. We analyze skin blocks' color distribution and find the color of skin and its threshold. A skin block is reset to black. A color matrix has $10 \times 10 \times 125$ dimensions. Based on the preliminary experiments on our clothing dataset, directly using color matrixes for SVM training produces inaccurate results of color split joint determination.

We conduct *color matrix transformation* to reduce the matrix dimensionality, so better SVM training classifier can be achieved. Given a color matrix, the number of each color's occurrence is firstly counted and colors are sorted in descending order within $0 \sim n$. Background and skin blocks are reset to 0. Then the matrix elements are replaced with n. Figure 4(b) and (c) show an example of color matrix transformation, representing the original color matrix of Fig. 4(a) and its final one respectively.

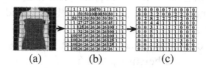

(a)   (b)   (c)

Fig. 4. Clothing upper body partitioning and its color Matrix (Color figure online)

The new color matrix generated from transformation can be treated as a descriptor of an image. We consider attribute classification as a supervised learning and adopt SVM classifier for this problem. By training we can determine whether a clothing is split joint or not, and whether it is upper-lower joint or sleeves-body (middle-sides) joint. Then the split joint attribute of ith clothing can be described as a 3-dimensional vector $sp_i = <sp_{i,1}, sp_{i,2}, sp_{i,3}>$, where $sp_{i,j}$ ($j = 1 \sim 3$) is binary. In detail, $sp_{i,1} = 0$ represents the ith clothing is split joint clothing, $sp_{i,2} = 1$ means the ith clothing is upper-lower split joint, and $sp_{i,3} = 1$ means the ith clothing is sleeves-body split joint. For example, the split joint attribute value of a upper-lower split joint clothing is <0, 1, 0> , and the split joint attribute value of a sleeves-body split joint clothing is <0, 0, 1>.

5   Clothing Recommendation Based on Attribute Content

5.1   Color and Texture Feature Extraction for Clothing Pattern

We use the upper body clothing as shown in Fig. 3(c) and extract the RGB color histogram of each clothing to represent its color of the whole clothing. Then the ith clothing image can be described as a 125-dimensional vector $<rgb_{i,1}, rgb_{i,2}, ..., rgb_{i,125}>$.

We detect the **ULBP** feature of images to describe the patterns of clothes. A set of clothing-specific pattern attributes are defined based on empirical study on clothing catalog, which includes vertical, plaid, horizontal, floral print, drawing, and plain as shown in Fig. 5. For eliminating the interference of collar and sleeve, we segment the upper-body cloth into 9 blocks and only use the center part in Fig. 3(c). The block's ULBP feature is a 59-dimensional vector. Using ULBP, the ith clothing image is described as $<ulbp_{i,1}, ulbp_{i,2}, ..., ulbp_{i,59}>$. We use SVM to train the 6 pattern categories.

5.2   PHOG, Fourier, and GIST Feature Extraction for Collar and Sleeves

There are several types of collars, including round collar, v-neck, doll collar, polo color, and so on. Two collars belonging to the same type may have differences in detail.

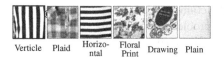

Verticle	Plaid	Horizo-ntal	Floral Print	Drawing	Plain

Fig. 5. Clothing pattern categories

For example, the first 5 images shown in Fig. 6 are all "doll collar", but with differences. The next 3 images are all "polo collar", and the difference among them is obvious in collar shape. The ninth image is "round collar", while it is hard to clearly describe the last image using text tag. Similarly, sleeves also have different shapes. In Fig. 7, the first 3 images are "puff sleeve", but with different "puff" shapes. The fourth and fifth images are "lace sleeve", but the "lace" is distributed in sleeves and other parts as well. Thus, tags cannot describe collar and sleeves shape accurately.

Doll Collar   Polo Collar   Round Collar   ? Collar

Fig. 6. Some clothing collar examples

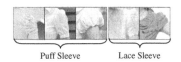

Puff Sleeve   Lace Sleeve

Fig. 7. Some clothing sleeve examples

Different from fabric pattern, collar and sleeve images have clear border and shape in visual inspection, thus their shape features are more important. Our collars and sleeves attributes based recommendation takes shape features into consideration. We have got collar and sleeve parts using the parts detection of human body as shown in Fig. 3(d), so the shape features are extracted from images to describe the attributes. As illustrated in [9], PHOG feature and body parts close to the neck are most effective for retrieving similar collar shapes. In addition, Fourier and GIST features are extracted for collar and sleeve description.

PHOG Feature Extraction. HOG (Histogram of Oriented Gradient) is an object detection feature descriptor used in computer vision and image processing within an image sub-region that is quantized into k bins. In our case $k = 12$, because each angle is 15°. The appearance and shape of the partial object can be well described by gradient or edge density distribution of an image. Moreover, PHOG (Pyramid Histogram of Oriented Gradients) can describe feature from more levels ($L = 3$ in our work) based on [8], and the final Pyramid Histogram of Oriented Gradients (PHOG) descriptor (1020 features) for the image is a concatenation of all the HOG vectors.

Fourier Feature Extraction. Description or discrimination of boundary curves (shapes) is an important problem in image processing and pattern recognition. Fourier descriptors (FD's) have interesting properties in this respect. Here we define

$$\phi * (t) = \mu_0 + \sum_{k=1}^{\infty} A_k \cos(kt - \alpha_k).$$ The set $\{A_k, \alpha_k; k = 1, 2,..., \infty\}$ includes the Fourier descriptors (FD) for curve γ (9 features).

GIST Feature Extraction. The GIST descriptor has been proved to be powerful in scene categorization and retrieval. Each clothing image is convolved with Gabor filters at 8 scales and 8 orientations. The filter responses are averaged within each of 4×4 divisions of the image (512 features) [8].

Given a clothing part image x_j^i ($j = 1$ for collar, $j = 2$ for sleeve), we have the unified feature $cs_j^i = \left[p_j^i, f_j^i, g_j^i \right]$, which is a concatenation of PHOG feature p_j^i, Fourier feature f_j^i, and GIST feature g_j^i. After normalization of the PHOG, GIST, and Fourier features, the 3 features are combined to describe collars and sleeves.

5.3 Clothing Recommendation Based on Attribute Content

We show the 'Recommendation Stage' in Fig. 2 in Algorithm 1.

Algorithm1: Recommendation Clothing Based on Detailed Attribute Content

---

Input: user image i_u, clothing image dataset $I_{cloth}=\{I_1, I_2(I_{21}\sim I_{26}), I_3, I_4, I_5 (I_{51}, I_{52})\}$;
Output: Recommended set I_r with top-k clothing image;
Describe:

    1) generate descriptor of i_u ($dpr_{u,1}\sim dpr_{u,5}$) and recommendation list A_i ($i=1, 2, ..., 5$);
    2) obtain user's selection A_i for setting $w_i=0$ or 1; //by interacting with users
    3) for $i=1$ to 5
    4) {if $w_i=1$ calculate $dist(i_c, i_u)=dist(dpr_{c,i}, dpr_{u,i})$ $\forall i_c \in I_i \subseteq I_{cloth}$; //see Formula (3)
    5) $I_r^i = \text{Top}_{\lambda k} (\underset{i_c \in I_i}{\arg\min} \, dist(i_c, i_u))$ }; //generate recommendation i^{th} subset

    6) return $I_r = Top_k \bigcap_{i=1}^{5} I_r^i$; //Formula (1)

Using the methods proposed in Sects. 4, 5.1, and 5.2, we store all downloaded clothing images with different attributes and their descriptors including color, split joint, pattern, collar, and sleeves. The clothing image dataset I_{cloth} is partitioned into $I_1 \sim I_5$ that store pattern, collar, sleeves, and split joint attribute images respectively. For split joint and pattern attribute, we construct classifiers C_J and C_P, by which the images are classified. In addition, I_2 is partitioned into $I_{21} \sim I_{26}$, which corresponds to vertical, plaid, horizontal, floral print, drawing and plain pattern. I_5 is further partitioned into I_{51} and I_{52}, which correspond to sleeves-body and upper-lower split joint. Given a clothing image i_c in I_{cloth}, corresponding to Formula (1), its descriptors include: $dpr_{c,1} = <rgb_1$, ..., $gb_{125}>$, $dpr_{c,2} = <ulbp_{c,1},...$, $ulbp_{c,59}>$, $dpr_{c,3} = \left[p_1^r, f_1^r, g_1^r \right]$, $dpr_{c,4} = \left[p_2^r, f_2^r, g_2^r \right]$, and $dpr_{c,5} = <sp_{r,1}, sp_{r,2}, sp_{r,3}>$. Similarly, i_u can be described as descriptors

$dpr_{u,1} \sim dpr_{u,5}$ (in Line 1). Since I_{cloth} is partitioned into 5 subsets, Formula (3) is varied based on users' attribute choices (in Line 4). The returned I_r contains k images, while for every attribute, we give λk images that have minimal distances with i_u ($\lambda = 5, 10,...$) (in Line 5). When a user selects more than one attributes, the returned I_r with top-k images are the intersection of λk images for every attribute (in Line 6).

6 Experiments

6.1 Dataset and Experimental Setting

We build a clothing dataset consisting of 8000 clothing images downloaded from Tmall (https://www.tmall.com/) online shopping website. We select training dataset for attribute classification, including 1000 images for split joint, 1000 images for fabric pattern, and 6000 images for color, pattern, collar, sleeve, split joint attributes' recommendation test. The clothing recommendation performance is measured based on the evaluation criterion of [3, 9]. We evaluate the top k recommended datum by a precision which is computed by: $\mathrm{Pr}\,ecision@k = \sum_i^k Rel(i)\Big/N$, where $Rel(i)$ is the ground truth relevance between the user's image i_u and i^{th} ranked image in clothing dataset, and N is normalization constant used to ensure the correct ranking result in a precision score of 1. We have conducted a subjective user study. 5 evaluators majored in computer science participated in the user study. Each evaluator is given the recommended clothes returned by our algorithm. After viewing these clothes, they were asked to give a score from 1 to k indicating how much the recommended clothes are relevant to current source clothing based on specified attributes. Then we get the precision of our recommendation algorithm.

6.2 Recommendation Results Based on Single Attribute

Figure 8 shows the recommendation results based on single attribute. In Fig. 8(a) ~ (f), every first image (with red frame) is from users, and the next 5 images are top-5 recommendation results. Figure 8(a) only considers pattern without other attributes. We can see that results have similar pattern (floral print), but different from first image in collar, sleeves, even color. In Fig. 8(b), as color is used, the results become more ac-curate. Figure 8(c)–(d) consider only split joint attribute. Here "Split Joint" is selected by users, and detail type (upper-lower or sleeves-body) is detected automatically by our algorithm. The first image in Fig. 8(c) is upper-lower split joint, thus the returned results only contain the images with upper-lower split joint. So does Fig. 8(d) for sleeves-body split joint. Figure 8(e)–(f) consider collar and sleeves, respectively.

6.3 Recommendation and Evaluation Results Based on Different Attributes

After showing recommendation list, a user can customize different attributes and their combination. In Fig. 9, the first image is from a user, and next images are the top-6

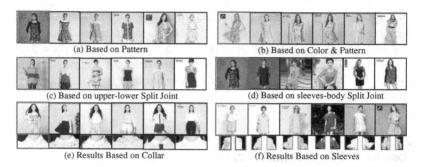

(a) Based on Pattern   (b) Based on Color & Pattern

(c) Based on upper-lower Split Joint   (d) Based on sleeves-body Split Joint

(e) Results Based on Collar   (f) Results Based on Sleeves

Fig. 8. Recommendation results based on different attributes (Color figure online)

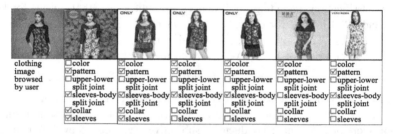

Fig. 9. Recommendation results based on different attributes

returned results when the user selects all attributes (we set $\lambda = 10$ by experiment analysis). The more matched attributes an image has, the more anterior it is ranked.

We extract color feature, detect collar and sleeves attribute from the first image. For pattern and split joint attribute, the image is recognized as "floral print" and "sleeve-body" by classifiers. Thus users' selections concern with the attributes and their free combination. For first image, Fig. 10 shows some recommendation results from Tao-Bao (https://www.taobao.com/) in 2016.1.13.

Fig. 10. Recommendation results based on tag and "Similarity" in taobao.com/

In Fig. 10, the first row shows search results based on tags, and the second row gives recommended results based on "similarity". Combining Fig. 9 with Fig. 10, we can see the difference between them. The results in Fig. 10 consider the same tag (in first row) or similar style (in second row), i.e. global feature, and ours considers the detailed attributes. From global style, the recommendation results in Fig. 10 are significant, and our recommendation can be a supplementary for users' customized requirements.

Finally, we measure the clothing retrieval performance by the evaluation criterion *Precision@k* of [3, 9] and show it in Fig. 11. Here we only compare our proposed alternatives for different attributes and their combinations. By human evaluation for determining whether the returned top-*k* results match given images or not, we can compute *Precision@k* (*k* = 1, 3, 5, 7, 9) as shown in Fig. 11. From a global view, *Precision@k* decreases with *k* increasing. It shows that our recommendation can rank the best match result first. In the single attribute based recommendations, our evaluation only considers the corresponding attribute. Given an image, if we recommend the images based on color attribute, we only see whether the returned results are consistent with the given image in color. In this case, it is not strange that the recommendation based on single attribute can achieve better result than attribute combination. However, based on all attributes, we can also achieve 65 % (top-9) ∼ 75 % (top-1) *Precision@k* results. On the other hand, the results based on collar and sleeve have lower *Precision@k* than pattern and split joint attributes, this is caused by the trained classifiers for pattern and split joint attributes.

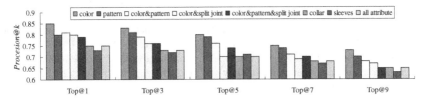

Fig. 11. Recommendation precision based on different attributes combinations (Color figure online)

7 Conclusion

In this paper, we have addressed the clothing recommendation problem by analyzing the different attributes of clothing image including fabric pattern, collar, sleeves, and split joint with various features. We have proposed classification based recommendation over multiple image attributes. Extensive experimental results have proved that our approach is a supplementary of current clothing recommendation for users' special requirements, and can be extend to the recommendation based on different attributes from multi-clothing rather than single clothes easily.

Acknowledgment. The project is supported by National Natural Science Foundation of China (61370074, 61402091), the Fundamental Research Funds for the Central Universities of China under Grant N140404012.

References

1. Ajmani, S., Ghosh, H., Mallik, A., Chaudhury, S.: An ontology based personalized garment recommendation system. In: Web Intelligence/IAT Workshops, pp. 17–20 (2013)
2. Chen, Y.-W., Sobue, S., Huang, X.: KANSEI based clothing fabric image retrieval. In: Trémeau, A., Schettini, R., Tominaga, S. (eds.) CCIW 2009. LNCS, vol. 5646, pp. 71–80. Springer, Heidelberg (2009)
3. Fu, J., Wang, J., Li, Z., Xu, M., Lu, H.: Efficient clothing retrieval with semantic-preserving visual phrases. In: Lee, K.M., Matsushita, Y., Rehg, J.M., Hu, Z. (eds.) ACCV 2012, Part II. LNCS, vol. 7725, pp. 420–431. Springer, Heidelberg (2013)
4. Huang, C., Wei, C., Wang, Y.: Active learning based clothing image recommendation with implicit user preferences. In: ICME Workshops, pp. 1–4 (2013)
5. Huang, X., Chen, D., Han, X., Chen, Y.: Global and local features for accurate impression estimation of cloth fabric images. In: SII, pp. 486–489 (2013)
6. Li, R., Wang, D., Zhang, Y., Feng, S., Yu, G.: An approach of text-based and image-based multi-modal search for online shopping. In: Gao, H., Lim, L., Wang, W., Li, C., Chen, L. (eds.) WAIM 2012. LNCS, vol. 7418, pp. 172–184. Springer, Heidelberg (2012)
7. Lin, K., Yang, H., Liu, K., Hsiao, J., Chen, C.: Rapid clothing retrieval via deep learning of binary codes and hierarchical search. In: ICMR, pp. 499–502 (2015)
8. Lin, Y., Kawakita, Y., Suzuki, E., Ichikawa, H.: Personalized clothing-recommendation system based on a modified Bayesian network. In: SAINT, pp. 414–417 (2012)
9. Liu, S., Song, Z., Liu, G., Xu, C., Lu, H., Yan, S.: Street-to-shop: cross-scenario clothing retrieval via parts alignment and auxiliary set. In: CVPR, pp. 3330–3337 (2012)
10. Liu, Z.: A garment image segmentation method based on salient region and JSEG. JSW **10**(11), 1274–1282 (2015)
11. Zhaolao, L., Zhou, M., Xuesong, W., Yan, F., Xiaohui, T.: Semantic annotation method of clothing image. In: Kurosu, M. (ed.) HCII/HCI 2013, Part V. LNCS, vol. 8008, pp. 289–298. Springer, Heidelberg (2013)
12. Manfredi, M., Grana, C., Calderara, S., Cucchiara, R.: A complete system for garment segmentation and color classification. Mach. Vis. Appl. (MVA) **25**(4), 955–969 (2014)
13. Mizuochi, M., Kanezaki, A., Harada, T.: Clothing retrieval based on local similarity with multiple images. In: ACM Multimedia, pp. 1165–1168 (2014)
14. Ramisa, A., Alenyà, G., Moreno-Noguer, F., Torras, C.: Using depth and appearance features for informed robot grasping of highly wrinkled clothes. In: ICRA, pp. 1703–1708 (2012)
15. Rother, C., Kolmogorov, V., Blake, A.: "GrabCut": interactive foreground extraction using iterated graph cuts. ACM Trans. Graph. (TOG) **23**(3), 309–314 (2004)
16. Wang, J., Zhang, Y.: Opportunity model for e-commerce recommendation: right product; right time. In: SIGIR, pp. 303–312 (2013)
17. Wang, N., Ai, H.: Who blocks who: simultaneous clothing segmentation for grouping images. In: ICCV, pp. 1535–1542 (2011)
18. Wang, X., Zhang, T., Tretter, D., Lin, Q.: Personal clothing retrieval on photo collections by color and attributes. IEEE Trans. Multimedia (TMM) **15**(8), 2035–2045 (2013)
19. Yamaguchi, K., Kiapour, M., Ortiz, L., Berg, T.: Retrieving similar styles to parse clothing. IEEE Trans. Pattern Anal. Mach. Intell. (PAMI) **37**(5), 1028–1040 (2015)
20. Yamazaki, K., Inaba, M.: Clothing classification using image features derived from clothing fabrics, wrinkles and cloth overlaps. In: IROS, pp. 2710–2717 (2013)
21. Yang, X., Yuan, S., Tian, Y.: Assistive clothing pattern recognition for visually impaired people. IEEE Trans. Hum.-Mach. Syst. (THMS) **44**(2), 234–243 (2014)

22. Yang, X., Yuan, S., Tian, Y.: Recognizing clothes patterns for blind people by confidence margin based feature combination. In: ACM Multimedia, pp. 1097–1100 (2011)
23. Zhang, H., Zha, Z., Yang, Y., et al.: Attribute-augmented semantic hierarchy: towards bridging semantic gap and intention gap in image retrieval. In: ACM Multimedia, pp. 33–42 (2013)
24. Zhou, B., Chen, X., Fu, Q., Guo, K., Tan, P.: Garment modeling from a single image. Comput. Graph. Forum (CGF) **32**(7), 85–91 (2013)

SocialFM: A Social Recommender System with Factorization Machines

Juming Zhou$^{(\boxtimes)}$, Dong Wang$^{(\boxtimes)}$, Yue Ding$^{(\boxtimes)}$, and Litian Yin$^{(\boxtimes)}$

School of Software, Shanghai Jiao Tong University, Shanghai, China
{zhoujuming,wangdong,dingyue,yinlitian}@sjtu.edu.cn

Abstract. Exponential growth of web2.0 makes social information be an indispensable part for recommender systems to solve cold start and sparsity problems. Most of the existing matrix factorization (MF) based algorithms for social recommender systems factorize rating matrix into two low-rank matrices. In this paper, we propose an improved factorization machines (FMs) with social information, called SocialFM. Our approach can effectively simulate the influence propagation by estimating interactions between categorical variables and specifying the input feature vectors. We combine user trust value with similarity to compute the influence value between users. We also present social regularization and model regularization to impose constraint on the objective function. Our approach is a general method, which can be easily extended to incorporate other context like user mood, timestamp, location, etc. The experiment results show that our approach outperforms other state-of-the-art recommendation methods.

Keywords: Recommender system · Factorization machines · Social network · Regularization

1 Introduction

Recommender System (RS) [1,16] aims to recommend fitting items to users. Typically, Collaborative Filtering (CF) [1] relying on the theory that similar users have similar interest makes a great success on RS. However, CF suffers from the inherent weaknesses: sparsity and cold start. Nearly all of the CF algorithms can not handle users who never rated any items. Moreover, large numbers of methods [12,17] assume users are i.i.d. (independent and identically distributed) during the learning process. In social network, users behaviors influence each other. When we see our friends send a tweet "Titanic is great", we may want to go to cinema. Recently, with the exponential expending of web2.0, social apps and websites bring large volume of social information. Jiang et al. [4] apply LDA on context and extract item features to model the interpersonal influence and individual preference. They find individual preference and interpersonal influence have low correlation and both of them have contributions on rating. So, it is important to incorporate the social information into recommender system.

© Springer International Publishing Switzerland 2016
B. Cui et al. (Eds.): WAIM 2016, Part I, LNCS 9658, pp. 286–297, 2016.
DOI: 10.1007/978-3-319-39937-9_22

Matrix factorization (MF) is a general and effective method, which supposes that in the user-item rating matrix only a small number of factors influence preferences and factorizes rating matrix into user feature matrix and item feature matrix. Many approaches [3,8–10] fuse the basic method with social information to improve the accuracy. Recently, Rendle [14] proposes a novel method factorization machines (FMs) modeling all interactions between variables using factorized parameters. It can mimic several specialized factorization methods and outperforms MF. In this paper, we propose an improved factorization machines (FMs) with social information, called SocialFM. More specifically, our approach presents four domains: user, item, trustee and other rated item, and specifies the input feature vectors by taking rating and influence value into account. We calculate influence value between users by taking trust value and user similarity into account. During the learning process, motivated by the intuition that users' favor is close to their similar trustees' favor, we impose a social regularization to constrain the objective function. Moreover, considering the inner connection between domains, we propose another FMs regularization. At last, we use stochastic gradient descent (SGD) method to learn the parameters. Our approach is quite general, can be easily extended to incorporate contextual information like user mood, timestamp, location, etc. Experiment on Filmtrust and Epinions datasets shows that our approach outperforms the state-of-the-art algorithms.

The rest of paper is organized as follows. In Sect. 2, we discuss the related work on social network, some classic social RS and basic FMs. In Sect. 3, we first present the improved Social FM with two regularization and demonstrate an example of our approach with a small dataset, then we describe how to compute influence value and show the final algorithmic procedure. Experimental results are presented in Sect. 4, followed by the conclusion and future work in Sect. 5.

2   Related Work

In this section, we first show the difference between social and trust relationship. Then we review some related work on recommender systems especially in social network. At last, the basic FMs model is presented.

2.1   Trust and Social Network

There are two kinds of relationship in social information, trust relationship and social friendship. They are close but not the same. User a trusts user b when a is willing to rely on the actions of b [11], a is trustor, b is trustee. But a does not need to know b in real life. So, it more like stars in WeiBo and official accounts in WeChat. Social friendship indicates users trust each other and they are more likely classmates, colleagues or relatives in real life, like friends in QQ and WeChat. Both of two relationship fit the assumption that users' favor is close to users who they trust. However, social friendship is more sensitive to data, users are not always willing to rely on their friends, some of them may have different tastes.

Trust network can be defined as a directed graph in which vertices represent users in network, edges are corresponding to trust relationship, the weight on the edge represents trust value for a pair of vertices. Usually, we transform trust network to an asymmetric matrix S. $s_{a,b}$ donates the trust value from trustor a to trustee b. Social network is specific trust network, which is an undirect graph. Its matrix is symmetric, if b is a friend of a, that means a trusts b, b also trusts a and $s_{a,b} = s_{b,a}$.

2.2 Social Recommender Systems

Matrix factorization (MF) [7] is a basic method, which assumes only a small number of factor influences preferences and factorizes the rating matrix into user feature matrices and item feature matrices. Ma et al. [8] propose Social Trust Ensemble (RSTE) which combines MF and social based approach with the notion that predicted rating of user u on item i should reflect favors of u self and u's trustees. Jamali et al. [3] incorporate trust propagation in the matrix factorization approach relying on the assumption that users have the same taste of other users they trusts, called SocialMF. This approach keep the user feature vectors close to the features of their direct neighbors. Its runtime is much faster than STE. Ma et al. [10] coin the term social regularization to represent constraint on recommender systems. In [20], Yang et al. focus on inferring category-specific social trust circles and outlines three variants of weight with friends based on expertise levels to represent trust value. In [19] Yang et al. propose a hybrid model TrustMF combining both a truster model and a trustee model from the perspectives of trusters and trustees.

2.3 Factorization Machines

Factorization machines [13,14] are quite generic approaches since they can mimic several factorization models, like PITF [15], SVD++ [6] and BPTF [18]. In FMs, each rating behavior with other information generates a transaction described by vector $\mathbf{x}$ with p real-value variables representing the profile like user, item, friend etc. A FM (degree = 2) models all the interactions between pairs of variables to learn better predicted value. The model equation for a factorization machine of degree $d = 2$ is defined:

$$\hat{y}(x) = w_0 + \sum_{j=1}^{p} w_j x_j + \sum_{j=1}^{p} \sum_{j'=j+1}^{p} x_j x_{j'} \sum_{f=1}^{k} v_{j,f} v_{j',f} \tag{1}$$

where k is the number of factors, p is the variable number of the vector $\mathbf{x}$, w_0 is the global bias, w_j is specific bias for j-th variable. So, model parameters that have to be estimated are:

$$w_0 \in R, \quad \mathbf{w} \in R^p, \quad \mathbf{v} \in R^{p \times k}. \tag{2}$$

In Eq. 1, $\sum_{f=1}^{k} v_{j,f} v_{j',f}$ can be viewed as the similarity of two columns by factorizing them. This allows the FM model to deal sparse data more reliable. It has been proved that the Eq. 1 can be computed very efficiently as it is equivalent to:

$$\hat{y}(x) = w_0 + \sum_{j=1}^{p} w_j x_j + \frac{1}{2} \sum_{f=1}^{k} \left[\left(\sum_{j=1}^{p} v_{j,f} x_j \right)^2 - \sum_{j=1}^{p} v_{j,f}^2 x_j^2 \right] \tag{3}$$

In Eq. 3, the FM are multilinearity. For each model parameter $\theta \in \Theta$, it can be divided into two functions g_θ and h_θ, which are independent of the value θ:

$$\hat{y}(\mathbf{x}) = g_\theta(\mathbf{x}) + \theta h_\theta(\mathbf{x}) \tag{4}$$

with:

$$h_\theta(\mathbf{x}) = \frac{\partial \hat{y}(\mathbf{x})}{\partial \theta} = \begin{cases} 1, & \text{if } \theta \text{ is } w_0 \\ x_l, & \text{if } \theta \text{ is } w_l \\ x_l \sum_{j \neq l} v_{j,f} x_j, & \text{if } \theta \text{ is } w_{l,f} \end{cases} \tag{5}$$

For least squares regression $l(\hat{y}, y)$, its derivative is defined:

$$\frac{\partial}{\partial \theta} l(\hat{y}(\mathbf{x}|\theta), y) = 2(\hat{y}(\mathbf{x}|\theta) - y) h_\theta(\mathbf{x}) \tag{6}$$

3 SocialFM

In this section, we describe the details of social factorization machine (SocialFM) model, here degree $d = 2$.

3.1 Model

The rating and social information can be transformed into four categorical domains, user U, item I, trustee T, other rated item RI. The user domain U and item domain I are transformed into a real-valued vector using indicator variable. E.g. vector $(1,0,0)$ indicates the active user is the first user in user domain. SocialFM specifies domain T and domain RI by taking the rating value and influence value into account and normalizing their values to indicate their values and relationship. Moreover, domain RI does not contain the current item. For current item doesn't influence itself and FISM [5] has shown better performance than the SVD++. Domain RI can be viewed as the implicit influence of rated items, domain T hold the implicit influence of trustees.

Example 1. Here, we take an example assuming data comes from a movie review system. The system records users rate movies and their trust relationship. We select some of them to be a small dataset. Let U, I be:

$$U = \{\textbf{A}\text{lice, } \textbf{B}\text{ob, } \textbf{C}\text{harlie}\}$$

$$I = \{\textbf{TI}\text{tanice, } \textbf{N}\text{otting } \textbf{H}\text{ill, } \textbf{S}\text{tart } \textbf{W}\text{ars, } \textbf{S}\text{tart } \textbf{T}\text{rek}\}$$

We use the abbreviated word overstriking before. Observed rating and trust relationship (assume influence value has been computed) are:

Rating Record = {{A,TI,5}, {A,NH,3}, {B,NH,4}, {B,SW,1},
{B,ST,1}, {C,SW,4}, {C,ST,2}, {C,TI,5}}
Trust NetWork={{A,B,0.1}, {A,C,0.4}, {B,C,0.5}}

where the set {A,TI,5} means user A rates TI 5 point. Set {A,B,0.1} indicates the trust relationship that A trusts B with 0.1 trust value. Figure 1 shows the final result feature vectors **x** that are the input for SocialFM. x^1 means the first transaction that Alice rates Titanic with 5 point having 20 % trusting Bob and 80 % trusting Charlie. She only have rated Notting Hill except Titanic.

Objective Function. Base on the FMs, we define a global loss function over all training data **D** to be:

$$
\text{OPT}(\mathbf{D}, \mathbf{\Theta}) = \underset{\theta}{\arg\min} \left(\sum_{(\mathbf{x},y) \in \mathbf{D}} l\left(\hat{y}\left(\mathbf{x}|\theta\right) - y\right) + \sum_{\theta \in \mathbf{\Theta}} \lambda_\theta \|\theta\|_F^2 \right.
$$

$$
+ \beta \sum_{a=1}^{n} \sum_{b \in C_a^+} \sum_{f=1}^{k} \left(g\left(a,b\right) \|v_{a,f} - v_{b,f}\|_F^2 \right) \tag{7}
$$

$$
\left. + \alpha \sum_{j=1}^{p} \sum_{f=1}^{k} \|v_{j,f} - v_{j',f}\|_F^2 \right)
$$

	Feature vector **x**														Target y	
x^1	1	0	0	1	0	0	0	0	0.2	0.8	0	1	0	0	5	y^1
x^2	1	0	0	0	1	0	0	0	0.2	0.8	1	0	0	0	3	y^2
x^3	0	1	0	0	1	0	0	0	0	1	0	0	0.5	0.5	4	y^3
x^4	0	1	0	0	0	1	0	0	0	1	0	0.8	0	0.2	1	y^4
x^5	0	1	0	0	0	0	1	0	0	1	0	0.8	0.2	0	1	y^5
x^6	0	0	1	0	0	1	0	0	0	0	5/7	0	0	2/7	4	y^6
x^7	0	0	1	0	0	0	1	0	0	0	5/9	0	4/9	0	2	y^7
x^8	0	0	1	1	0	0	0	0	0	0	0	0	2/6	4/6	5	y^8
	A	B	C	TI	NH	SW	ST	A	B	C	TI	NH	SW	ST		
		User			Movies				Trust Users			Other rated movies				

Fig. 1. Example for feature vectors x created from the small dataset mentioned before. The i-th row represents feature vector x^i with its corresponding target rating y^i. The first 3 columns represent indicator variables for the active user who rates the item, the next 4 columns are the indicator variables for current item, the next 3 columns represent the indicator variables for trustees and influence value, the last 4 columns are valuable indicators for rated items and their rating

where n is the number of users, k and p are the number of factors and variable number of the vector $\mathbf{x}$, $|| \cdot ||_F^2$ denotes the Frobenius norm, $g\,(a,b)$ is the influence value, C_a^+ is the user set who user a trusts, the superscript means the profile in corresponding domain. λ_θ is a L2 regularization parameter to avoid overfitting. We use the same regulation parameters for similar model parameters(i.e. λ_w for w_j and λ_v for $v_{j,f}$).

Here, we propose two kinds of regularization. Motivated by the assumption that user's favor is close to similar trustees, we impose constraints between active users and their trustees individually. The influence is concerned with trust value and user similarity. This regularization handles two situations. When user a and his trustee b have totally opposite favor, the small influence value will have a little effect on objective function. If we use $\sum_{a=1}^{n} \sum_{f=1}^{k} ||v_{a,f} - \sum_{b \in C_a^+} s_{a,b} v_{b,f}||_F^2$, it may cause the information loss when users' trustees have diverse favor. Moreover, domain U represents the features of users, domain T represents the features of trustees. If, they are the same person, their features should be similar, so are items. The best way is to use cosine similarity. However, it's hard to calculate the partial derivative, we just simplify it to make their value be close.

3.2 Influence value

In Sect. 3.1, the social regularization requires the influence value between friends which is consist of trust value $s_{a,b}$ and user similarity $sim\,(a,b)$. In trust network, if user a trusts user b, $s_{a,b} = 1$, vice versa conversely. We compute trust value:

$$s_{a,b} = s_{a,b} \times \sqrt{\frac{d^-\,(\nu_b)}{d^+\,(\nu_a) + d^-\,(\nu_b)}} \tag{8}$$

where $d^+\,(\nu_a)$ is the outdegree of the node ν_a indicating the number of users that user a trusts. $d^-\,(\nu_b)$ is the indegree of the node ν_b representing the number of users who trust user b.

In social network, for the outdegree and indegree are same for each node. The trust value $s_{a,b} = s_{b,a} = \sqrt{\frac{1}{2}}$, if user a is user b's friend. So, we can't roughly regard trust value as final influence value. When user a and user b are friends and their features are similar which means they have similar taste, we can deem they have deep influence to each other. Motivated by this, we take user similarity into account calculated by Pearson Correlation Coefficient(PCC) [2]. PCC can deal with the situation that each user has own rating styles. The similarity between user a and user b with PCC method is:

$$sim\,(a,b) = \frac{\sum\limits_{i \in I(a) \bigcup I(b)} (r_{a,i} - \bar{r}_a)\,(r_{b,i} - \bar{r}_b)}{\sqrt{\sum\limits_{j \in I(a) \bigcup I(b)} (r_{a,i} - \bar{r}_a)^2} \sqrt{\sum\limits_{i \in I(b) \bigcup I(b)} (r_{b,i} - \bar{r}_b)^2}} \tag{9}$$

where $r_{a,i}$ is rating user a gave item i, $\bar{r}_a$ denotes the average rating of user a, i is the subset of items which user a and user b both rated. To avoid negative influence, we use the function $f(x) = (x+1)/2$ to restrict the PCC similarities in [0,1]. So, the final influence value between user a and user b is:

$$g(a,b) = s_{a,b} * sim(a,b) \tag{10}$$

At last, we normalize values as $g(a,b) = g(a,b)/\sum_{b \in C_a^+} g(a,b)$.

3.3   Social FMs Training

We use stochastic gradient descent(SGD) algorithm to optimize our model for it's simple and works well for the loss functions. Algorithm 1 shows how to optimize Social FMs by SGD.

Algorithm 1. Learn Social FMs by SGD

Input:
  Training data S, Regularization parameters λ, learning rate η, initialization σ
Output:
  Model
  $w_0 \leftarrow 0; \mathbf{w} \leftarrow (0, ..., 0); \mathbf{v} \sim N(0, \sigma)$;
  repeat
    for $(\mathbf{x}, y) \in D$ **do**
      $w_0 \leftarrow w_0 - \eta \left(\frac{\partial}{\partial w_0} l\left(\hat{y}(\mathbf{x}|\theta), y\right) + 2\lambda_0 w_0 \right)$
      for $j \in [1, ..., p] \wedge x_j \neq 0$ **do**
        $w_j \leftarrow w_j - \eta \left(\frac{\partial}{\partial w_j} l\left(\hat{y}(\mathbf{x}|\theta), y\right) + 2\lambda_w w_j \right)$
        for $f \in [1, ..., k]$ **do**
          if $j \in$ domain U and active user is a **then**
            $v_{j,f} \leftarrow v_{j,f} - \eta \Bigg(\frac{\partial}{\partial v_{j,f}} l\left(\hat{y}(\mathbf{x}|\theta), y\right) + 2\lambda_v v_{j,f}$
            $+ 2\beta \bigg(\sum_{b \in C_a^+} g(a,b)(v_{a,f} - v_{b,f}) + \sum_{b \in C_a^-} g(b,a)(v_{a,f} - v_{b,f}) \bigg)$
            $+ 2\alpha \left(v_{a,f} - v_{a',f}\right) \Bigg)$
          else
            $v_{j,f} \leftarrow v_{j,f} - \eta \left(\frac{\partial}{\partial v_{j,f}} l\left(\hat{y}(\mathbf{x}|\theta), y\right) + 2\lambda_v v_{j,f} + 2\alpha \left(v_{j,f} - v_{j',f}\right) \right)$
          end if
        end for
      end for
    end for
  until stopping criterion is met

If $j \in$ domain U, j represents the active user a in domain U, we need to compute all the differences between user a and his trustees in set C_a^+ and trustors in set C_a^-. In the each step of learning, users' feature vectors $\mathbf{v}$ are updated by the influence of their trustees and trustors, which can be viewed as the process of influence propagation. When the values of the objective function as Eq. 7 shown converges in the learning phase, the propagation of influence will reach a harmonic status.

In [14] Rendle shows FMs are efficient. Because in real-word data, vectors $\mathbf{x}$ of transactions are huge sparse, almost all of elements are zero. The computational complexity of evaluating each predicted rating and learning parameters are $O\left(k\bar{m}\right)$ and $O\left(1\right)$, where $\bar{m}\left(x\right)$ is the average of m(x) for x∈ all transactions. In SocialFM, the computational complexity of evaluating each predicted rating doesn't change, computational complexity of learning is $O\left(|C_u|\right)$, $|C_u|$ is the average number of trustees and trustors for all users.

4   Experiment

4.1   Datasets

We evaluate our method on two famous datasets: Filmtrust and Epinions. Fimtrust is a movie rating dataset taken from www.librec.net/datasets.html, which consists of ratings from 1508 users who rated 2071 item. The total rating number is 35497 and total trust statements is 1853 issued by 609 users. The density is 1.13 % and 0.42 % in terms of ratings and trust relations. Epinions.com is a well-known knowledge sharing and review site established in 1999. We randomly select some of the data from www.trustlet.org/wiki/Epinions. It has 1528 user and 1999 different items. The number of ratings and trust relationship are 12057 and 78308. Its density is 0.394 % and 3.35 % in terms of ratings and trust relations respectively. Both of the datasets contain item rating and social relationship. Compare to Filmtrust, active users in Epinion dataset trust more people but rate less items on average.

4.2   Metrics

In the experiment, We use two popular metrics: Root Mean Square Error (RMSE) and Mean Absolute Error (MAE) to evaluate the prediction quality. The metric MAE and RMSE are defined as:

$$MAE = \frac{\sum_{i,j}|r_{i,j} - \hat{r_{i,j}}|}{N}, \qquad RMSE = \sqrt{\frac{\sum_{i,j}\left(r_{i,j} - \hat{r_{i,j}}\right)^2}{N}} \qquad (11)$$

Where $r_{i,j}$ is the rating value in validation set, $\hat{r_{i,j}}$ is the predicted rating value. N is the number of rating in validation set. Smaller values of MAE and RMSE mean better performance.

4.3   Comparisons and Parameters Settings

In our experiments, in order to show the improvement of our Social FM, we compare six methods with some social state-of-the-art algorithms:

- PMF [12]: this method is well known in recommender system. It models the Base MF method in a probabilistic way but doesn't consider social relationship
- RSTE [8]: this method combines MF and social based approach with the notion that predicted rating of user u on item i should reflect favors of u self and u's trustees.
- SoRec [9]: this method fuses a user-item rating matrix with the users social network using probabilistic matrix factorization.
- SocialMF [3]: this method makes users feature vectors be dependent on the feature vectors of their direct neighbors in the social network.
- SoReg [10]: this method proposes an social regularization to constrain the matrix factorization objective function and improve the prediction accuracy.
- TrustMF [19]: this method proposes a hybrid model combining trustor model and trustee model from the perspectives of trustors and trustees.

For all methods to be validated, we set respective optimal parameters based on the our experiments or suggested by previous work. The number of latent factor d=5/10 and $\lambda = 0.001$ for L2 regularization. Moreover we set other parameters: in RSTE $\alpha = 0.4$ for Epinions and 1 for Filmtrust, in SoRec $\lambda_c = 1$ for Epinions and 0.01 for Filmtrust, in SoReg $\beta = 0.1$, in SocialMF $\lambda_t = 5$ for Epinions 1 for Filmtrust, in TrustMF $\lambda_t = 1$.

4.4   Result

For each experiment discussed below, we use a 5-fold cross-validation of learning and testing and take the mean as the final result. We randomly select 80 % of data as the training set and the remaining 20 % as the validation set, where the random seed is 1. We set $\lambda_0 = -0.005$, $\lambda_w = 0.0003$, $\lambda_v = 0.01$, $\alpha = 0.1$, $\beta = 0.1$ for both of two datasets. We validate the MAE and RMSE of SocialFM and compare the performance with the methods mentioned before.

Tables 1 and 2 report the MAE and RMSE values of all comparison partners on Flimtrust and Epinions datasets. From the results, we can observe that our approach consistently outperforms other methods and improve performance (MAE and RMSE) 3.3 %, 0.7 %, 4.2 %, 1.5 % in term of d=5 and d=10 respectively with Filmtrust dataset, 2.0 %, 2.1 %, 1.7 %, 1.3 % with Epinions dataset. Moreover, PMF method in Filmtrust get much better results, because the rating data in Epinions dataset is much sparser than Filmtrust. On the contrary, trust data in Filmtrust dataset is sparser than Epinions, that's why SocialMF and TrustMF method get better performance in Epinions comparing to other methods.

In our method, the parameters α and β play an important role to constrain the loss function. Parameter α control how much the inner relationship of domains, if it's too big, the model will be overfitting. Parameter β control how much our model consider social information into account. Figure 2 how α and β affect the MAE and RMSE. We observe that α impact more obviously than β, because social regularization is concerned about influence value which in

Table 1. Experimental results on Filmtrust

Dataset	Metrics	PMF	RSTE	SoRec	SocialMF	SoReg	TrustMF	SocialFM
d = 5	MAE	0.714	0.628	0.628	0.638	0.674	0.631	**0.605**
	RMSE	0.949	0.810	0.810	0.837	0.878	0.810	**0.798**
d = 10	MAE	0.735	0.640	0.638	0.642	0.6668	0.631	**0.604**
	RMSE	0.968	0.835	0.831	0.8844	0.875	0.819	**0.797**

Table 2. Experimental results on Epinions

Dataset	Metrics	PMF	RSTE	SoRec	SocialMF	SoReg	TrustMF	SocialFM
d = 5	MAE	1.710	0.907	0.943	0.853	0.879	0.847	**0.836**
	RMSE	2.130	1.209	1.261	1.112	1.203	1.095	**1.074**
d = 10	MAE	1.569	0.956	0.902	0.845	0.873	0.845	**0.835**
	RMSE	1.985	1.294	1.191	1.089	1.154	1.087	**1.072**

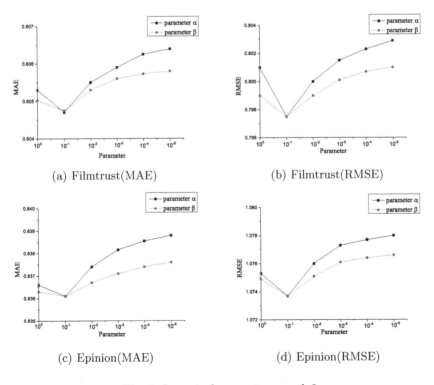

(a) Filmtrust(MAE)   (b) Filmtrust(RMSE)

(c) Epinion(MAE)   (d) Epinion(RMSE)

Fig. 2. Impact of parameter α and β

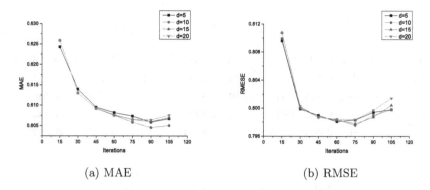

(a) MAE   (b) RMSE

Fig. 3. Performance comparison on Filmtrust with different kinds of dimensionality

our method is not so accurate. Moreover, with the value decreasing, MAE and RMSE will decrease at first, get the best result at 0.1 and then increase.

Figure 3 shows the performance (RMSE and MAE) for different kinds of dimensionality on Filmtrust where learning rate $\eta = 0.001$. We can observe that our method get the best performance on RMSE and MAE when d (dimensionality) = 10, while other kinds of dimensionality degrade performance a little for MAE.

5   Conclusion and Future Work

In this paper, we propose a social-based recommender system with factorization machines which can effective simulate the influence propagation by estimating interactions between categorical variables and specifying the input feature vectors with four domains: user, item, trustee and rated item. We also impose the social regularization and model inner regularization to constrain the objective function. Our approach is quite general, can be easily extended to incorporate contextual information. Experiment on two datasets shows that our approach outperforms the state-of-the-art algorithm.

In our proposed model, we only utilize user rating data to calculate the user similarity, which is sparse and may generate some mistakes. We plan to incorporate the context information like tag into our model and compute user similarity by user bias on tags. We also decide to design a more accurate user influence function.

References

1. Adomavicius, G., Tuzhilin, A.: Toward the next generation of recommender systems: a survey of the state-of-the-art and possible extensions. IEEE Trans. Knowl. Data Eng. **17**(6), 734–749 (2005)
2. Breese, J.S., Heckerman, D., Kadie, C.: Empirical analysis of predictive algorithms for collaborative filtering. In: Proceedings of the Fourteenth Conference on Uncertainty in Artificial Intelligence, pp. 43–52. Morgan Kaufmann Publishers Inc. (1998)

3. Jamali, M., Ester, M.: A matrix factorization technique with trust propagation for recommendation in social networks. In: Proceedings of the Fourth ACM Conference on Recommender Systems, pp. 135–142. ACM (2010)
4. Jiang, M., Cui, P., Liu, R., Yang, Q., Wang, F., Zhu, W., Yang, S.: Social contextual recommendation. In: Proceedings of the 21st ACM International Conference on Information and Knowledge Management, pp. 45–54. ACM (2012)
5. Kabbur, S., Ning, X., Karypis, G.: Fism: factored item similarity models for top-n recommender systems. In: Proceedings of the 19th ACM SIGKDD International Conference on Knowledge Discovery and Data Mining, pp. 659–667. ACM (2013)
6. Koren, Y.: Factorization meets the neighborhood: a multifaceted collaborative filtering model. In: Proceedings of the 14th ACM SIGKDD International Conference on Knowledge Discovery and Data Mining, pp. 426–434. ACM (2008)
7. Koren, Y., Bell, R., Volinsky, C.: Matrix factorization techniques for recommender systems. Computer **8**, 30–37 (2009)
8. Ma, H., King, I., Lyu, M.R.: Learning to recommend with social trust ensemble. In: Proceedings of the 32nd International ACM SIGIR Conference on Research and Development in Information Retrieval, pp. 203–210. ACM (2009)
9. Ma, H., Yang, H., Lyu, M.R., King, I.: Sorec: social recommendation using probabilistic matrix factorization. In: Proceedings of the 17th ACM Conference on Information and Knowledge Management, pp. 931–940. ACM (2008)
10. Ma, H., Zhou, D., Liu, C., Lyu, M.R., King, I.: Recommender systems with social regularization. In: Proceedings of the Fourth ACM International Conference on Web Search and Data Mining, pp. 287–296. ACM (2011)
11. Mayer, R.C., Davis, J.H., Schoorman, F.D.: An integrative model of organizational trust. Acad. Manag. Rev. **20**(3), 709–734 (1995)
12. Mnih, A., Salakhutdinov, R.: Probabilistic matrix factorization. Adv. Neural Inf. Process. Syst. **20**, 1257–1264 (2007)
13. Rendle, S.: Factorization machines. In: 2010 IEEE 10th International Conference on Data Mining (ICDM), pp. 995–1000. IEEE (2010)
14. Rendle, S.: Factorization machines with libFM. ACM Trans. Intell Syst. Technol. (TIST) **3**(3), 57 (2012)
15. Rendle, S., Schmidt-Thieme, L.: Pairwise interaction tensor factorization for personalized tag recommendation. In: Proceedings of the Third ACM International Conference on Web Search and Data Mining, pp. 81–90. ACM (2010)
16. Resnick, P., Varian, H.R.: Recommender systems. Commun. ACM **40**(3), 56–58 (1997)
17. Salakhutdinov, R., Mnih, A.: Bayesian probabilistic matrix factorization using Markov chain Monte Carlo. In: Proceedings of the 25th International Conference on Machine Learning, pp. 880–887. ACM (2008)
18. Xiong, L., Chen, X., Huang, T.K., Schneider, J.G., Carbonell, J.G.: Temporal collaborative filtering with Bayesian probabilistic tensor factorization. In: SIAM SDM, vol. 10, pp. 211–222. (2010)
19. Yang, B., Lei, Y., Liu, D., Liu, J.: Social collaborative filtering by trust. In: Proceedings of the Twenty-Third International Joint Conference on Artificial Intelligence, pp. 2747–2753. AAAI Press (2013)
20. Yang, X., Steck, H., Liu, Y.: Circle-based recommendation in online social networks. In: Proceedings of the 18th ACM SIGKDD International Conference on Knowledge Discovery and Data Mining, pp. 1267–1275. ACM (2012)

Identifying Linked Data Datasets for sameAs Interlinking Using Recommendation Techniques

Haichi Liu[1]($\boxtimes$), Ting Wang[1], Jintao Tang[1], Hong Ning[1],
Dengping Wei[1], Songxian Xie[1], and Peilei Liu[2]

[1] College of Computer, National University of Defense Technology,
Changsha, Hunan Province, China
liuhaichi@nudt.edu.cn
[2] Academy of National Defense Information, Wuhan, Hubei Province, China

Abstract. Due to the outstanding role of owl:sameAs as the most widely used linking predicate, the problem of identifying potential Linked Data datasets for sameAs interlinking was studied in this paper. The problem was regarded as a Recommender systems problem, so serveral classical collaborative filtering techniques were employed. The user-item matrix was constructed with rating values defined depending on the number of owl:sameAs RDF links between datasets from Linked Open Data Cloud 2014 dump. The similarity measure is a key for memory-based collaborative filtering methods, a novel dataset semantic similarity measure was proposed based on the vocabulary information extracted from datasets. We conducted experiments to evaluate the accuracy of both the predicted ratings and recommended datasets lists of these recommenders. The experiments demonstrated that our customized recommenders outperformed the original ones with a great deal, and achieved much better metrics in both evaluations.

Keywords: Linked data datasets · Interlinking · sameAs links · Recommender systems

1 Introduction

In order to be considered as Linked Data, the datasets published on the web have to be connected, or linked, to other datasets [1]. The RDF links such as owl:sameAs between datasets are fundamental for Linked Data as they connect data islands into a global data space so-called Web of Data. Data linking [2] can be formalized as an operation, which takes two Linked Data dataset as input and produces a collection of links between entities of the two datasets as output. When a new dataset was published as Linked Data, the publisher should check all the datasets in the Web of Data to identify the possible links, which is very time-consuming. So if there are some technology can be utilized, being recommended based on known links and focusing on those datasets most likely to link, one can sharply reduce the computational costs if the recommendations are accurate enough.

B. Cui et al. (Eds.): WAIM 2016, Part I, LNCS 9658, pp. 298–309, 2016.
DOI: 10.1007/978-3-319-39937-9_23

In the Web of Data, an increasing number of owl:sameAs[1] statements have been published to support merging distributed descriptions of equivalent RDF resources from different datasets. The owl:sameAs property is part of the Web Ontology Language (OWL) ontology [3], the official semantics of owl:sameAs is: *an owl:sameAs statement indicates that two URI references actually refer to the same thing.* When all of these owl:sameAs statements are taken together, they form a very large directed graph connecting Linked Data datasets to each other. Due to the outstanding role of owl:sameAs as the most widely used linking predicate [4], we focus on recommendation of datasets for sameAs interlinking. Previous works [5–8] mostly did not distinguish RDF link types when identifying datasets for interlinking, and experiments were conducted on the experimental data constructed from RDF links of various types, while the graphs formed from various types of RDF links exhibit different characters [4]. Previous works would be of less help for real application scenarios, as dataset publishers still do not know what kinds of RDF links can be established furthermore how to configure the data linking algorithms. Due to the limitations of previous methods, it is necessary to find better ways.

In this paper we try to tackle the problem of identifying more datasets that can be established owl:sameAs links with, when the publisher's dataset has already linked to a few datasets. This is the scenario that the Recommender systems [11] techniques can be applied. We construct user-item matrix with rating values depending on the number of owl:sameAs RDF link triples between datasets from newly updated LOD Cloud 2014 dump [4]. Several classical collaborative filtering methods of Recommender systems are applied. Utilizing the semantic schema information extracted from Linked Data datasets, we define dataset semantic similarity to replace the original similarity component of memory-based collaborative filtering methods to develop our customized recommenders. To evaluate the recommenders, we conduct two experiments for assessing rating and top n recommendation accuracy. Experimental results demonstrate our customized recommenders perform much better than the original ones. The MAEs are only half of the original ones, the values are low to the range of (0.3, 0.5) on a rating scale of 1 to 7. The F-Measures are almost twice higher, the values are within the range of (0.2, 0.5), which are promising given the large set of datasets to recommend from. This drastic improvement are liable on the peculiar properties of the merging of dataset semantic similarity and memory-based collaborative filtering recommenders. The source codes and experimental data have been uploaded to Github[2].

The rest of the paper is organized as follows. In Sect. 2, at first we describe the framework which consider the dataset identification problem as a Recommender systems problem and how we construct user-item rating matrix. Then we describe the collaborative filtering technologies we used upon the problem. At last we define a dataset semantic similarity algorithms used for injecting domain-specific information. In Sect. 3, we describe the experiments data,

---

[1] http://www.w3.org/2002/07/owl#sameAs.

[2] https://github.com/HaichiLiu/Recommending-Datasets-for-Interlinking.

evaluation methodology, and results. In Sect. 4, we present related works. Finally, in Sect. 5 we conclude the paper.

2   Recommender Systems Techniques

We model the problem of identifying target dataset for sameAs interlinking as a Recommender systems problem, and we describe how to construct user-item rating matrix which is necessary for recommendation algorithms in Sect. 2.1. Several representative collaborative filtering algorithms we employed are briefly described in Sect. 2.2. Also we define a dataset semantic similarity algorithm as the similarity computation component of memory-based recommenders in Sect. 2.3.

2.1   Recommendation Framework

Recommender systems are personalized information agents that attempt to predict which items out of a large pool a user may be interested in. The user's interest in an item is expressed through the rating the user gives the item. Generally, the interaction between user and item is represented with a user-item rating matrix. A recommender system has to predict the ratings for items that the user has not yet seen. With these estimated ratings the system can recommend the items that have the highest estimated rating to the target user. Note that item is a general term used to denote what the system recommends to users, and can be of any type, like movies, books, websites, or news articles. In our case, these items are Linked Data datasets available in the Web of Data. We use $U = \{u_1, u_2, u_3, ..., u_n\}$ to denote the set of dataset publishers (users), $D = \{d_1, d_2, d_3, ..., d_m\}$ for the set of datasets (items). We view that each dataset d_i is published by a unique publisher u_i, this makes $n = m$. This may not be hold in real world, but actually u_i is merely an identifier of dataset d_i in the publishers set U, which makes the representation to be understood easily in a Recommender systems scenario. And we denote R as an $n \times n$ matrix of ratings $r_{i,j}$, with $i \epsilon \{1, ..., n\}, j \epsilon \{1, ..., n\}$. Recommender algorithms are used to predicting the rating values of a certain dataset publisher for the datasets he or she has not linked, or recommending a ranked list of datasets he or she might want to link according to the rating values predicted.

We aggregate all owl:sameAs RDF links by dataset, meaning that we consider dataset publisher (user) of dataset a has a rating for dataset b if there exists at least one owl:sameAs RDF link triple from dataset a which contains the subject of the triple to the dataset b which contains the object. We find that some Linked Data dataset publisher did not choose the standard http://www.w3.org/2002/07/owl#sameAs as linking predicates, but use terms from proprietary vocabulary, such as http://www.abes.fr/owlsameAs, even mistakenly used http://www.w3.org/2002/07/owlsameAs, http://www.w3.org/2000/01/rdf-schema#sameAs, we also extract links defined by these predicates. Since we can view that a dataset is sameAs interlinked to itself, the number of RDF link triples equals to the number of entities defined in the dataset. Rating values are set based on number of owl:sameAs RDF link triples, the rating value equals to the number of digits of link triples count. We illustrate the construction of rating matrix from datasets interlinking with the example as Fig. 1. For example, dataset d_1 has 243 RDF links to dataset d_2, the corresponding matrix entry $r_{1,2}$ equals to 3.

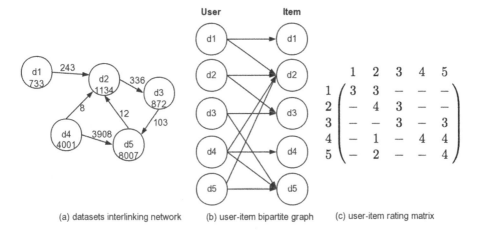

(a) datasets interlinking network   (b) user-item bipartite graph   (c) user-item rating matrix

Fig. 1. An example to illustrate how to construct rating matrix from datasets interlinking network. (a) is an example of interlinking network of five datasets, identified by $d1, d2, ..., d5$. The number inside the circle is the entities number of each dataset. The arrows represents owl:sameAs RDF links between datasets with a number of RDF links triples count. (b) is an example of user-item bipartite graph constructed from (a), each user has a link set to itself in item set. Generated rating matrix is shown in (c), and it is a 5×5 matrix.

2.2 Collaborative Filtering Recommendation

Collaborative filtering is widely implemented and the most mature recommendation technique. The concept is to make correlations between users or between items. There are memory-based and model-based techniques [11].

Memory-Based Recommendation. Memory-based recommenders can be divided into: user-based and item-based recommenders. The main idea of user-based algorithms is simply as follows: given a ratings matrix and the ID of the current (active) user as an input, identify nearest users that had similar preferences to those of the active user in the past. Then, for every item i that the active user has not yet seen, a prediction is computed based on the ratings for i made by the nearest users:

$$r_{a,i} = \sum_{b \in N_a} sim(a, b) \times r_{b,i} \qquad (1)$$

The similarity measure between user a and b, $sim(a,b)$, is essentially a distance measure and is used as a weight. Different algorithms can be used to compute the similarity, such as cosine, Pearson, Spearman, Euclidean distance, et al.

The main idea of item-based algorithms is to compute predictions using the similarity between items rather than the similarity between users. An item-based algorithm computes a weighted average of these other ratings as the following with $sim(i,j)$ is computed similar to what we did in user-based recommender:

$$r_{a,i} = \sum_{j \epsilon S_i} sim(i,j) \times r_{a,j} \tag{2}$$

Matrix Factorization Models. Matrix factorization models [12] is a model-based method which maps both users and items to a joint latent factor space of dimensionality f. Accordingly, each item i is associated with a vector $q_i \epsilon R_f$, and each user u is associated with a vector $p_u \epsilon R_f$. For a given item i, the elements of q_i measure the extent to which the item possesses those factors, the same is true for a user. The resulting dot product, $q_i^T p_u$, approximates the rating of user u for item i. To learn the factor vectors (p_u and q_i), the system minimizes the regularized squared error on the set of known ratings:

$$\min_{p^*,q^*} = \sum_{u,i \epsilon K} (r_{u,i} - q_i^T p_u)^2 + \lambda(|p_u|^2 + |q_i|^2) \tag{3}$$

Here, K is the set of (u,i) pairs for which $r_{u,i}$ is known (the training set). Simon Funk popularized a stochastic gradient descent optimization of Eq. 3 wherein the algorithm loops through all ratings in the training set. It modifies the parameters by a magnitude proportional to γ in the opposite direction of the gradient, yielding:

$$e_{u,i} = r_{u,i} - q_i^T p_u \tag{4}$$

$$q_i \leftarrow q_i + \gamma \cdot (e_{u,i} \cdot p_u - \lambda \cdot q_i) \tag{5}$$

$$p_u \leftarrow p_u + \gamma \cdot (e_{u,i} \cdot q_i - \lambda \cdot p_u) \tag{6}$$

2.3   Injecting Domain-Specific Information

Both user-based and item-based recommender rely on similarity component. In user-based recommender the similarity between two users is based on their ratings of items that both users have rated, likewise in item-based recommender the similarity between two items is based on ratings of users that rated both items. As a Linked Data dataset is a collection of RDF triples describing entities with RDFS vocabularies or OWL ontologies, this motivates us to extract its semantic schema features to define dataset semantic similarity, and make it as the similarity component of memory-based recommenders to develop our customized recommenders. In this section, we describe how we model a Linked Data dataset with vector space model (VSM) using semantic features, and how to calculate similarity between datasets based on the model further.

Vector Space Model for Linked Data Dataset. Vector space model is an algebraic model for representing text documents (and any objects, in general) as vectors of identifiers, such as, for example, index terms. In VSM each document is represented by a vector in a m-dimensional space, where each dimension corresponds to a term from the overall vocabulary of a given document collection. VSM was adapted to model the dataset in this paper. A Linked Data dataset uses one or more RDFS vocabularies or OWL ontologies. The vocabulary provides the terms (classes and properties) for expressing the data. A vocabulary URI, a class URI or a property URI used in triples of the dataset can be seen as *semantic features* of the dataset, they are called *vocabulary feature*, *class feature* and *property feature* respectively. Let $T = \{t_1, t_2, ..., t_m\}$ be the dictionary, that is the set of semantic features of datasets in the corpus. Formally, each dataset $d_i = \{w_{1i}, w_{2i}, ..., w_{mi}\}$, where w_{kj} is the weight for feature t_k in dataset d_i.

Dataset representation in the VSM raises two issues: weighting the terms and measuring the feature vector similarity. The most commonly used term weighting scheme is TF-IDF (Term Frequency-Inverse Document Frequency) weighting:

$$TF - IDF(t_k, d_j) = \frac{f(t_k, d_j)}{|\{t_i \epsilon d_j\}|} \cdot log \frac{n}{|\{d_j \epsilon D : t_k \epsilon d_j\}|} \tag{7}$$

$f(t_k, d_j)$ is the number of times that feature t_k occurs in dataset d_j.

Dataset Semantic Similarity. As stated earlier, a similarity measure is required to determine the closeness between two datasets. Many similarity measures have been derived to describe the proximity of two vectors; among those measures, cosine similarity is most widely used:

$$sim(d_i, d_j) = \frac{\sum_k w_{ki} \cdot w_{kj}}{\sqrt{\sum_k w_{ki}^2} \cdot \sqrt{\sum_k w_{kj}^2}} \tag{8}$$

As we assume each dataset is published by a unique publisher, the dataset semantic similarity can be used as user similarity component in user-based recommender as well as item similarity component in item-based recommender. Using dataset semantic similarity in memory-based methods also helps to relieve the cold start problem of recommender systems, which is common in our scenario, as the number and the variety of Linked Data datasets are increasing rapidly. A "new" dataset with few interlinkings to other datasets cannot easily be recommended in pure memory-based recommenders, because the similarity was computed on past ratings. While dataset semantic similarity is computed utilizing only the information of datasets themselves, recommendations can also be made for new datasets.

3 Experiments

3.1 Experimental Data

We construct the experimental data from the LOD Cloud 2014 dataset published in [4]. The data is a crawl of the Web of Linked Data conducted in

April 2014, which contains 8,038,396 resources crawled from 900,129 documents. The crawled data is provided for download as a single N-Quads formatted 2.6 GB zipped dump file. Using the dataset URIs published by the authors, we managed to divide the dump into 990 separated dataset dumps, in which the quads whose subject' URI defined under the same dataset URI are grouped together. For each dataset dump, we extract semantic features of property, class and vocabulary types. The property features of a dataset are obtained by grouping all the predicate URIs of RDF triples in the dataset dump. The class features are obtained by grouping the object URIs of RDF triples with (s rdfs:type o) pattern. Since the namespace of class or property URI are the URI of the vocabulary where the class and property were defined, by grouping all the namespace of class and property URIs of a dataset, we can get the vocabulary features of the dataset. Using the method we describe in Sect. 2.1, We manage to construct our experimental user-item matrix with 990 users, 990 items and 1641 ratings from users to items.

3.2 Evaluation Methodology

Evaluating Rating Accuracy. Mean Absolute Error (MAE) is used to measure the closeness of predicted ratings to the true ratings. It is defined as the average absolute difference between the n pairs $< p_h, r_h >$ of predicted ratings and real ratings:

$$MAE = \frac{\sum_{h=1}^{n} |p_h - r_h|}{n} \qquad (9)$$

In our experiments, for each user, we take a certain percentage of the ratings as "training data" to produce recommendations, and the rest of the ratings is compared against estimated rating values to compute MAE. The results may differ as the data set is split randomly, hence for each algorithm, we run the test for 10 times and take the average score for final presentation.

Evaluating Top N Recommendations. It's not always essential to present estimated rating values to users. In many cases, an ordered list of recommendations, from best to worst, is sufficient. So we could apply classic information retrieval metrics F1-Measure to evaluate recommenders. We adopt *leave-one-out* strategy, for a user we remove his top n ratings, and use his left ratings and all the other users' ratings as training set. The final scores are calculated by averaging all the users' test results. As the test rating records are selected in descending order of its value rather than randomly, for memory-based algorithms we do not need to repeat the experiments. For matrix factorization algorithm in which calculation was started from randomly initialized vectors, we run the test 10 times and present the average results.

3.3 Results and Discussion

The recommendation algorithms and evaluation methods are implemented with the help of a Java machine learning library called Mahout [13]. To inject domain-specific information as stated in Sect. 2.3, we implement a customized class that extends ItemSimilarity and UserSimilarity of Mahout.

To the best of our knowledge, there are rare works applying Recommender systems techniques to the problem of identifying target dataset for sameAs interlinking. For comparision, we have chosen three simple recommenders: Random, ItemAverage and ItemUserAverage and three original collaborative filtering recommenders: Item-based, User-based and Rating SGD. Random recommender produces random recommendations and preference estimates. ItemAverage recommender is a simple recommender that always estimates rating for an item to be the average of all known preference values for that item. ItemUserAverage recommender is like ItemAverage recommender, except that its estimated ratings are adjusted for the users' average rating value. Item-based recommender is the original one implemented in Mahout, Item-{Vocabulary, Class, Property} recommenders are our customized recommenders in which the similarity components of original item-based algorithm are replaced by our dataset semantic similarity components with vocabulary, class and property features respectively. This is the same for User-based recommenders. For User-based recommenders, there are two ways for choosing neighborhoods: fixed-size neighborhoods (noted with n as the neighborhoods size parameter) and threshold-based neighborhoods (noted with t as the threshold parameter). We explored a range of possible choices of both parameters for both evaluation. For fixed-size neighborhoods, n is in the range of $[1,10]$ with 1 as step size, for threshold-based neighborhoods, t is in the range of $[0.1,0.9]$ with 0.1 as step size. The best results are shown in the Tables 1 and 2, and the optimized parameters are noted in cells. There are three parameters can be tuned for RatingSGD recommender, f is the number of factors used to compute this factorization, γ is the learning rate, and i is the number of iterations. These parameters also have been tuned for optimum performance.

When evaluating rating accuracy, we vary the percent of rating records used for training from 50 % to 90 %. In Table 1 we can see that the MAEs are around 2.5 for Random recommender. With some simple intuitions, the MAEs are lower to about 1.0 or 1.2 for ItemUserAverage and ItemAverage recommenders. Original item-based recommender in Mahout has further lower MAEs about 0.8. Our Item-{Vocabulary, Class, Property} recommender shows better performance in MAEs at the training percent 50 %, 60 % and 70 %, but worse at 80 % and 90 %. The MAEs of original user-based recommender with fixed-size neighborhoods are in the range of $(0.9, 1.0)$, the MAEs of original user-based recommender with threshold-based neighborhoods are also in the range of $(0.9, 1.0)$ but lower. Both User-based recommenders have better performance than Item-based recommender. RatingSGD recommender is generally better than original item and user based recommenders, but not as good as our customized recommenders. User-{Vocabulary, Class, Property} recommenders with fixed-size neighborhoods mostly have better performance than the original User-based recommender with fixed-size neighborhoods. User-{Vocabulary, Class, Property} recommenders with threshold-based neighborhoods have much better performance than the original User-based recommender with threshold-based neighborhoods, actually the MAEs of User-Class recommender with threshold-based neighborhoods are the lowest of all tested recommenders at all training percent. The values are around $(0.3, 0.5)$, which are only half of the MAEs achieved by the best original recommender, i.e., the Item-based recommender.

Table 1. MAE comparison of various recommenders

	50 %	60 %	70 %	80 %	90 %
Random	2.4591	2.5758	2.4587	2.5194	2.5293
ItemAverage	1.0517	1.0702	1.0793	1.0867	1.0351
ItemUserAverage	1.2407	1.2429	1.2380	1.2671	1.2265
Item-based	0.8611	0.8698	0.8361	0.7700	0.8148
Item-Vocabulary	0.7754	0.8049	0.7978	0.8359	0.8454
Item-Class	0.7329	0.7757	0.7882	0.8021	0.8246
Item-Property	0.7775	0.8191	0.7981	0.8678	0.8863
User-based $n = 8$	1.0151	0.9890	1.0951	0.9683	0.9138
User-Vocabulary $n = 4$	1.0043	0.9930	0.9077	0.9399	0.8584
User-Class $n = 2$	0.8796	0.9125	1.0679	1.1103	0.8546
User-Property $n = 10$	0.9720	1.0057	1.0761	1.1061	0.9074
User-based $t = 0.6$	0.9620	0.9403	0.9995	0.9660	0.9268
User-Vocabulary $t = 0.6$	0.7934	0.7177	0.7537	0.7002	0.6627
User-Class $t = 0.9$	**0.3669**	**0.4607**	**0.4153**	**0.4102**	**0.3904**
User-Property $t = 0.9$	0.6149	0.5794	0.5709	0.8153	0.5667
RatingSGD $f = 20\ i = 50\ \gamma = 0.01$	0.8649	0.8252	0.8419	0.8518	0.8607

For evaluating the Top N recommendation, we evaluate top 1 to top 10 recommendation performance of various recommenders for comprehensive comparisons. The results are shown only for top 1 to 5 in Table 2 due to the space limit, the trend of results for top 6 to 10 tests are similar. Random recommender failed completely in this test, since randomly recommending a few number of datasets out of 990 datasets can hardly hit the right answers. The other two simple recommenders also performed very poorly. Original item and user based recommenders performed better. Item-based recommender achieved F1-Measures larger than 0.1 for top $\{3, 5, 6, 7, 8\}$ test cases. The F1-Measures of User-based recommender with fixed-size neighborhoods are higher and within the range of $(0.2, 0.5)$. The F1-Measures of User-based recommender with threshold-based neighborhoods are within the range of $(0.1, 0.2)$. RatingSGD recommender performed worse than original item and user based recommenders, the F1-Measures are always below 0.05. Our customized Item-{Vocabulary, Class, Property} recommenders have close performance compared with Item-based recommender. While User-{Vocabulary, Class, Property} recommenders all achieve much better performance compared with the original user-based ones. The User-Vocabulary recommender with fixed-size neighborhoods achieves the best F1-Measures in all the top n test cases except top 1, the F1-Measures are within the range of (0.2, 0.5), almost twice higher than the best original recommender, i.e., User-based recommender with fixed-size neighborhoods.

Table 2. F1-measure comparison for Top N Recommendation

	top 1	top 2	top 3	top 4	top 5
Random	NaN	NaN	NaN	NaN	NaN
ItemAverage	0.0156	0.0072	NaN	NaN	NaN
ItemUserAverage	0.0156	0.0072	NaN	NaN	NaN
Item-based	0.0272	0.0874	0.1310	0.0667	0.1400
Item-Vocabulary	0.0066	0.0504	0.1071	0.2000	0.1600
Item-Class	0.0132	0.0360	0.0952	0.1000	0.1200
Item-Property	NaN	0.0504	0.0952	0.1500	0.1000
User-based $n = 3$	**0.0468**	0.2009	0.2130	0.2171	0.2188
User-Vocabulary $n = 10$	0.0353	**0.2437**	**0.3709**	**0.4710**	**0.4316**
User-Class $n = 8$	0.0400	0.2102	0.3538	0.4437	0.3716
User-Property $n = 10$	0.0266	0.2336	0.2811	0.2648	0.2387
User-based $t = 0.9$	0.0054	0.0942	0.131	0.2000	0.1800
User-Vocabulary $t = 0.5$	0.0203	0.1956	0.2519	0.3984	0.3568
User-Class $t = 0.5$	0.0301	0.1892	0.2736	0.4193	0.3124
User-Property $t = 0.5$	0.0203	0.1956	0.2519	0.3984	0.3568
RatingSGD $f = 20$ $i = 50$ $\gamma = 0.01$	0.0156	0.0072	0.0119	0.0500	0.0200

4 Related Works

Identifying relevant datasets for interlinking is a novel research area. There are a few approaches developed specifically for this purpose, which can be categorized into two groups.

In the first category, the problem is tackled in a retrieval way, these methods try to retrieve datasets that can be interlinked with for a given dataset. Nikolov et al. [9] proposed a method that depends on an third-party semantic web search service. They use labels of randomly selected individuals from a dataset to query the search service and aggregated the results by datasets. They conducted experiments on three datasets as examples. Also not all datasets have instances with labels, Ell et al. [10] show that only 38.2 % of the non-information resources have a label. Lopes et al. [6] proposed a probabilistic approach based on Bayesian theory, they defined rank score functions that exploit vocabulary features of dataset and the known dataset links. Liu et al. [5] modeled the problem in an Information Retrieval way, they have developed a ranking function called collaborative dataset similarity which is proven to be quite effective. Using learning to rank algorithm to incorporate these ranking functions, they can further improve the performance and achieve the best MAP (Mean Average Precision) compared with previous works.

In the second category, the problem is tackled using link prediction approaches. The graphs of datasets and interlinking between them are constructed and link prediction measures are used to rank potential dataset pairs. Lopes et al. [6] represented the data space as a directed graph, and used the Preferential Attachment and the Resource Allocation to measure the likelihood that two datasets can be connected. The linear combine of these two score is used to rank the dataset pair. But when computing Preferential Attachment score, instead of using out-degree of source dataset, they used the size of similarity set of source dataset. Similarity set is defined as the set of datasets which have vocabulary features in common with source dataset. Mera et al. [7] developed a dataset interlinking recommendation tool called TRT. They implemented most of state of art local and quasi-local similarity indices, but these indices are not combined in any way. They also developed a tool called TRTML [8]. In TRTML, the interlinking of datasets was represented as an undirected graph, and four link prediction measures were implemented and three supervised classification algorithms were used. They balanced the percentage of unlinked tripleset pairs considered for better performance when comparing various algorithms, in this way the testing settings can no longer reflect the real challenges as the real-world distribution is extremely imbalanced.

5   Conclusion

The Web of Data is constantly growing, in order to be considered as Linked Data, the datasets published on the web have to be interlinked to other datasets. Data linking between two given datasets is a time-consuming process, if there are some techiques can bepublisher, it will substantially reduce the need to perform exploratory search. As the ubiquitous owl:sameAs property is used to connect these datasets, we focus on this type of link, and try to solve the problem of identifying target dataset for sameAs interlinking, when the publishers dataset has linked to a few datasets. This is the scenario that the Recommender systems techniques can be applied. We construct user-item matrix with rating values depending on the number of RDF link triples between datasets. We extract vocabulary features of dataset, and define a dataset semantic similarity algorithm as the similarity component of memory-based recommenders. The experiments show that Recommender systems techniques is effective for the problem and our customized recommenders perform better than original collaborative filtering recommenders. For future work, we plan to exploit more advanced recommendation techniques and develop more effective features focusing on the topical aspect of datasets.

Acknowledgements. This material is based on work supported by the National Natural Science Foundation of China (61200337, 61202118, 61472436)

References

1. Bizer, C., Heath, T., Berners-Lee, T.: Linked data - the story so far. Int. J. Semantic Web Inf. Syst. **5**, 1–22 (2009)
2. Ferrara, A., Nikolov, A., Scharffe, F.: Data linking for the semantic web. Int. J. Semantic Web Inf. Syst. **7**, 46–76 (2011)
3. Bechhofer, S., van Harmelen, F., Hendler, J., Horrocks, I., McGuinness, D., Patel-Schneider, P., Stein, L.A.: OWL web ontology language reference. W3C Recommendation (2004). www.w3.org/TR/owl-ref
4. Schmachtenberg, M., Bizer, C., Paulheim, H.: Adoption of the linked data best practices in different topical domains. In: Mika, P., et al. (eds.) ISWC 2014, Part I. LNCS, vol. 8796, pp. 245–260. Springer, Heidelberg (2014)
5. Liu, H., Tang, J., Wei, D., Liu, P., Ning, H., Wang, T.: Collaborative datasets retrieval for interlinking on web of data. In: Presented at the Proceedings of the 24th International Conference on World Wide Web Companion, WWW 2015, Florence, Italy, 18–22 May 2015, Companion Volume (2015)
6. Lopes, G.R., Leme, L.A.P.P., Nunes, B.P., Casanova, M.A., Dietze, S.: Two approaches to the dataset interlinking recommendation problem. In: Benatallah, B., Bestavros, A., Manolopoulos, Y., Vakali, A., Zhang, Y. (eds.) WISE 2014, Part I. LNCS, vol. 8786, pp. 324–339. Springer, Heidelberg (2014)
7. Caraballo, A.A.M., Nunes, B.P., Lopes, G.R., Leme, L., Casanova, M.A., Dietze, S.: TRT - a tripleset recommendation tool. In: Presented at the Proceedings of the ISWC 2013 Posters & Demonstrations Track, Sydney, Australia, 23 October 2013
8. Caraballo, A.A.M., Arruda Jr., N.M., Nunes, B.P., Lopes, G.R., Casanova, M.A.: TRTML - a tripleset recommendation tool based on supervised learning algorithms. In: Presutti, V., Blomqvist, E., Troncy, R., Sack, H., Papadakis, I., Tordai, A. (eds.) ESWC Satellite Events 2014. LNCS, vol. 8798, pp. 413–417. Springer, Heidelberg (2014)
9. Nikolov, A., d'Aquin, M., Motta, E.: What should I link to? identifying relevant sources and classes for data linking. In: Pan, J.Z., Chen, H., Kim, H.-G., Li, J., Horrocks, I., Mizoguchi, R., Wu, Z., Wu, Z. (eds.) JIST 2011. LNCS, vol. 7185, pp. 284–299. Springer, Heidelberg (2012)
10. Ell, B., Vrandečić, D., Simperl, E.: Labels in the web of data. In: Aroyo, L., Welty, C., Alani, H., Taylor, J., Bernstein, A., Kagal, L., Noy, N., Blomqvist, E. (eds.) ISWC 2011, Part I. LNCS, vol. 7031, pp. 162–176. Springer, Heidelberg (2011)
11. Adomavicius, A.: Toward the next generation of recommender systems: a survey of the state-of-the-art and possible extensions. IEEE Trans. Knowl. Data Eng. **17**, 734–749 (2005)
12. Koren, Y., Bell, R., Volinsky, C.: Matrix factorization techniques for recommender systems. IEEE Comput. Soc. **42**, 30–37 (2009)
13. Owen, S., Anil, R., Dunning, T., Friedman, E.: Mahout in Action. Manning Publications Co., Shelter Island (2011)

Query-Biased Multi-document Abstractive Summarization via Submodular Maximization Using Event Guidance

Rui Sun[1], Zhenchao Wang[1], Yafeng Ren[2], and Donghong Ji[1(✉)]

[1] Computer School, Wuhan University, Wuhan, China
{ruisun,dhji}@whu.edu.cn
[2] Singapore University of Technology and Design, Singapore, Singapore

Abstract. This paper proposes an abstractive multi-document summarization method. Given a document set, the system first generates sentence clusters through an event clustering algorithm using distributed representation. Each cluster is regarded as a subtopic of this set. Then we use a novel multi-sentence compression method to generate K-shortest paths for each cluster. Finally, some preferable paths are selected from these candidates to construct the final summary based on several customized submodular functions, which are designed to measure the summary quality from different perspectives. Experimental results on DUC 2005 and DUC 2007 datasets demonstrate that our method achieves better performance compared with the state-of-the-art systems.

1 Introduction

Summarization approaches can be categorized as extractive and abstractive (Mani 2001). Except of presenting the summarized information, the summary should be readable. Extractive approaches use sentences as the basic units for content selection and summary generation. In standard benchmark, most of them can achieve better performance than abstractive approaches. However, abstractive approaches receive much attention because their generated results are more close to manual summary. Some studies show that there is an empirical limit intrinsic to pure extraction (Genest and Lapalme 2011).

Abstractive approaches require a deeper text analysis and the ability of generating new sentences to convey the important content from text. A few abstractive approaches (Genest and Lapalme 2011; Bing et al. 2015; Li et al. 2015) construct new sentences by exploring more fine-grained syntactic units than sentences, namely, noun/verb phrases. However, sentence synthesis is still very inaccurate based on a set of phrases with little grammatical information (Zhang 2013). Most of recent abstractive approaches focus on rewriting techniques (Cheung and Penn 2014; Banerjee et al. 2015).

In this paper, we propose a model for abstractive multi-document summarization based on multi-sentence compression. This model aims to generate the summary using submodular function maximization. The framework of this model

© Springer International Publishing Switzerland 2016
B. Cui et al. (Eds.): WAIM 2016, Part I, LNCS 9658, pp. 310–322, 2016.
DOI: 10.1007/978-3-319-39937-9_24

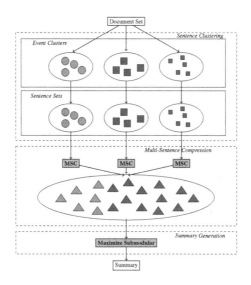

Fig. 1. Our system framework.

is shown in Fig. 1. Different from previous approaches based on multi-sentence compression (Mehdad et al. 2014; Banerjee et al. 2015), our sentence clustering and multi-sentence compression are guided by event information. In the sentence clustering stage, the cluster of similar sentences are generated for each subtopic of the document set. Previous clustering algorithm are conducted based on the similarity between sentences. Differently, our clusters are generated through event clustering based on distributed representation. The clustering algorithm mainly has two advantages. Firstly, the structured event contains necessary grammatical information, which is much less noise than sentences. Moreover, we use the distributed representation to alleviate the limitation of morphology or word sense. In the second stage, we exploit a multi-sentence compression model to generate K candidate paths for each sentence cluster. SalMSC (Sun et al. 2015) is an aggressive multi-sentence compression model. It generates arbitrary length compressions by using a special graph construction strategy. Our event information can be naturally introduced into this model. In the third stage, we hope to search the preferable sentences from the candidate paths of each cluster to form a summary. Recently, some studies perform summarization by maximizing submodular function (Dasgupta et al. 2013; Lin and Bilmes 2011). Following this direction, we design a combination of three submodular functions to search the final summary. These functions consider the topic coverage, query relevance and diversity of sentences in summary simultaneously. We propose a greedy algorithm to generate the final summary under the constraint of length.

We conduct the experiments on standard datasets DUC 2005 and DUC 2007 to demonstrate the effectiveness of our method. The results show that the event information can facilitate not only the sentence clustering, but also the compression generation. These submodular functions measure the quality of summary from different perspective during the process of path selection. Our method outperforms the state-of-art abstractive systems on both datasets.

2 Related Work

Abstractive summarization approach has received more attention in recent years. Some studies focus on sentence synthesis exploiting fine-grained syntactic units. Genest and Lapalme (2011) define information items as the basic units for abstract representation, and then integrate these items into a sentence structures according to a text-to-text generation process. Bing et al. (2015) first constructs a pool of concepts and facts represented by noun/verb phrases to construct new sentences. They employ integer linear optimization to construct the new sentence under some constraints. Li et al. (2015) propose a sparse-coding-based method to calculate the salience of noun/verb phrases and simultaneously perform the sentence compression and mention rewriting via a unified optimization method.

While other researchers pay more attention to the structure with more semantic information. Li (2015) introduce Basic Semantic Unit (BSU) to construct a semantic link network, and generate the abstractive summary through summary structure planning. Liu et al. (2015) make use of Abstract Meaning Representation (AMR) and focus on a graph-to-graph transformation to generate the final summary. Cheung and Penn (2013) regard the caseframe as a shallow approximation of semantic role, and conduct a series of studies comparing the human-written model summaries to system summaries at the semantic level of caseframes. In addition, Cheung and Penn (2014) present sentence enhancement for text-to-text generation in abstractive summarization by allowing the combination of dependency subtrees. Obviously, the rewriting techniques are commonly used in these methods, such as sentence compression (Mehdad et al. 2014; Li et al. 2014; Banerjee et al. 2015), sentence fusion (Barzilay and Mckeown 2005) or sentence enhancement (Cheung and Penn 2014).

In this paper, we use the multi-sentence compression to generate the summary. Multi-sentence compression based on word graph (MSC) is first proposed by Filippova (2010). Sun et al. (2015) modify the construction of word graph and introduce the salience information into the calculation of the weights of vertexes. The linguistic quality of each path is ensured by the guidance of structured event and a trigram language model. Their results demonstrate that the modified multi-sentence compression algorithm (SalMSC) achieve better performance in headline generation.

Mehdad et al. (2014) introduce MSC into meeting summarization. Similar to our framework, Banerjee et al. (2015) also employ MSC for abstractive multi-document summarization. Their method mainly contains the following differences from our method: (1) They conduct the sentence clustering by initializing cluster centers using each sentence from the most important document. The sentences in other documents are assigned to the corresponding cluster based on cosine similarity. The number of clusters is limited by the identified document. (2) They generate K-shortest paths from the sentences in each cluster using MSC. Here, we use more aggressive SalMSC to generate the paths for each cluster. (3) They employ an integer linear programming (ILP) model with the objective of maximizing information content and readability of the final summary, considering the length of path, information score and linguistic quality. We combine three submodular functions to generate the final summarization.

3   Our Method

Our method consists of three steps, namely sentence clustering, sentence compression and summary generation.

3.1   Sentence Clustering

Given a document set, the summarization task aims at finding the important subtopics in these documents and organizing the representative semantic units, e.g., sentences and phrases, to form the summary. Most of sentence clustering methods reply on the calculation of sentence similarity. Events are meaningful structures, which contain the necessary grammatical information, and yet are much less sparse than sentences (Sun et al. 2015). We propose to regard the event as basic unit to perform sentence clustering. Thus, our sentence clustering consists of the following three components: (1) Extracting events in sentences; (2) Generating event clusters; (3) Producing candidate sentences.

Extracting Events. We exploit events as the basic units for sentence clustering. Here an event is a tuple (S, P, O), where S is the subject, P is the predicate and O is the object. For example, for the sentence *"Ukraine Delays Announcement of New Government"*, the event is *(Ukraine, Delays, Announcement)*. This type of event structures has been used in open information extraction (Fader et al. 2011), and has a range of applications (Ding et al. 2014; Ng et al. 2014; Sun et al. 2015). We apply an simple open event extraction approach. Different from traditional event extraction, open event extraction does not have a closed set of entities and relations (Fader et al. 2011). We follow Hu et al. (2013) to extract events.

Given a text, we first use the Stanford dependency parser[1] to obtain the Stanford typed dependency structures of the sentences. Then we mainly focus on two relations, *nsubj* and *dobj*, for extracting event arguments. Event arguments which have the same predicate are merged into one event, represented by tuple *(Subject, Predicate, Object)*. For example, given the sentence, *"Egypt would cancel the Cairo economic summit"*, we can obtain two event arguments: *nsubj(cancel, Egypt)* and *dobj(cancel, summit)*. The two event arguments are merged into one event: *(Egypt, cancel, summit)*.

---

[1] http://nlp.stanford.edu/software/lex-parser.shtml.

Clustering Events. We aim at clustering the events into different subtopics for different document set. Each event is represented using a feature vector. In recent years, a large number of studies show that the distributed representation gives the strong performance in various NLP applications. Instead of discrete feature, we use a distributed feature vector to represent each event. Grefenstette and Sadrzadeh (2011) provide a distributional version of the type-logical meaning of the event:

$$\overrightarrow{Sub, Verb, Obj} \;=\; \overrightarrow{Verb} \odot (\overrightarrow{Sub} \otimes \overrightarrow{Obj}) \tag{1}$$

The compositional semantic of each event can be calculated by point-wise multiplication of the meaning of predicate to the Kronecker product of its subject and object. The semantic vector is learned using the CBOW/Skip model of word2vec tool[2] from a large-scale WikiPedia corpus[3].

The Chinese Whispers clustering algorithm[4] (Biemann 2006) is used to cluster the events. This graph-based algorithm is highly scalable and nonparametric. In the event graph, the weight of each edge is denoted as the cosine similarity $sim(e_i, e_j)$ of semantic vectors of events.

Producing Candidate Sentences. In this section, our goal is to prepare candidate sentences for multi-sentence compression. Because a sentence may contain two or more events, we propose to extract the candidate according to the constituent parsing subtree of the event. We employ a simple heuristic method to extract the candidate sentences. First, we identify the predicate node of each event in the full constituent tree. Then, we locate the nearest sentence node $\langle S \rangle$ or $\langle SBAR \rangle$ to this predicate node after traversing from it to $\langle ROOT \rangle$ node. This sentence fragment corresponding to this node is extracted as the candidate.

It is noted that some fragments are uncompleted structures, i.e. infinitive structure. In other words, these fragments may lost important syntactic constituent, namely *Subject*. We sample 5 events for each news article, and extract their sentence fragments according to above-mentioned heuristic rules. The infinitive structures appear 14.27 % times in the total fragments. To address this problem, the event structure is beneficial to get the lost syntactic constituent back. A reasonable interpretation is that the dependency relations in event participants may be beyond the boundary of fragments. In addition, the co-reference mentions in the candidate sentences are replaced with the representative word of corresponding lexical chain. It is significant to construct the word graph and improve the performance of multi-sentence compression. As an example, *"...Arafat... to postpone the declaration of statehood on May 4...",* we can extract the fragment, *"Arafat postpone the declaration of statehood on May 4"*, as the candidate.

---

[2] https://code.google.com/p/word2vec/.
[3] http://dumps.wikimedia.org/enwiki/20140102/.
[4] http://wortschatz.informatik.uni-leipzig.de/~cbiemann/software/CW.html.

3.2 Sentence Compression

Word-Graph Construction. In our word graph, the edges are added for **all** word pairs of one sentence in written order, not for the adjacent word pairs in Filippova (2010). If a word w_j occurs after w_i in one sentence, then an edge $w_i \rightarrow w_j$ is added for the graph. Similarly, we add two artificial words, $\langle S \rangle$ and $\langle E \rangle$, to the start position and end position of all sentences, respectively. The word graph represents collection of sentences efficiently by mapping the identical words to a single vertex. The common vertexes shared by different candidates increase the fusion probability of sentence fragments, which are regarded as a condensed representation of candidates for each cluster. So, we conduct coreference resolution for each candidate before the construction of word graph. Figure 2 gives an simple example of the word graph.

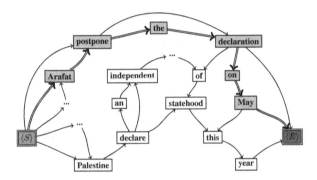

Fig. 2. Word graph generated from candidates and a possible compression path.

Sentence Compression. Given a graph $G = (\mathcal{V}, \mathcal{E})$, where $\mathcal{V}$ and $\mathcal{E}$ denote nodes and edges, respectively. Sun et al. (2015) firstly introduce the salience information into the calculation of the vertex weights. The vertex V_i is salient when its corresponding word is close to the salient events. We propose that the centroid of event cluster should be representative. The salience score of each event can be calculated according to the distance between each event and centroid event. The vertex weight is computed as follows:

$$w(V_i) = \sum_{e \in CE} sal(e) \exp\{-dist(V_i.w, e)\}$$

$$sal(e) = 1 - sim(e, e_{cent})$$

(2)

where $V_i.w$ denotes the word of vertex V_i, and $dist(w, e)$ denotes the distance from the word w to the event e. We use e_{cent} to represent the centroid event of each cluster.

The search space is large because of the redundant edges in the graph. Consequently, we also exploit beam search algorithm to generate compressions.

The paths are measured from two aspects, namely the sum edge score and fluency score. The trigram language model is trained using SRILM[5] on English Gigaword (LDC2011T07).

3.3 Summary Generation

We combine three submodular functions, namely, topic contribution, query contribution and diversity. Then we design a greedy search algorithm to generate the final summary.

Submodular Functions. Given a set of objects $\mathcal{T} = \{T_1, \cdots, T_n\}$ and a function $\mathcal{F} : 2^{\mathcal{T}} \rightarrow \mathbb{R}$ that returns a real value for any subset $\mathcal{S} \subset \mathcal{T}$, $f(.)$ is a submodular if for any $\mathcal{R} \subset \mathcal{S} \subset \mathcal{T}$ and $s \in \mathcal{T} \backslash \mathcal{S}$ satisfy:

$$f(\mathcal{R} \cup \{s\}) - f(\mathcal{R}) \geq f(\mathcal{S} \cup \{s\}) - f(\mathcal{S}) \tag{3}$$

As we can see, the incremental value of s decreases as the context in which s is considered grows from $\mathcal{R}$ to $\mathcal{S}$. The monotone nondecreasing submodular functions can simply be referred to as monotone submodular. In this paper, we use three functions to measure the information coverage of each paths.

(1) **Topic Coverage.** Each cluster represents a subtopic in the document set. The subtopics have different salience to the document set. We use the document coverage to measure the salience of each subtopic. We believe that the compressed paths should have different topic contributions according to the salience of corresponding subtopics. The topic contribution can be calculated by the following formula:

$$f\_topic(S) = \sum_{s \in S} score(s) wcluster(s) \tag{4}$$

where $wcluster$ is refer to the cluster weight: $\frac{\sum_{d \in D, e \in s} 1\{e \in d\}}{N}$. Intuitively, the higher the cluster coverage, the stronger the topic contribution of the path in this cluster. The paths in salient cluster should be considered early.

(2) **Query Relevance.** Query-biased summarization requires the generated summary could not only convey the main content of documents, but also bias to the information needs of a specific query. To identify the summary sentence with the highest coverage of query content, we propose a score that counts the number of query terms that appear in the path. In order to reward the score to cover more salient terms in the query content, we also consider the *tf.idf* score of query terms in this coverage formulation.

$$f\_query(S) = \sum_{s \in S} \frac{\sum_{q_i \in s} tf.idf(q_i)}{\sum_{q_i \in Q} tf.idf(q_i)} \tag{5}$$

where Q denotes the query of this document set, and q_i is the query term.

---

[5] http://www.speech.sri.com/projects/srilm/.

(3) **Diversity.** The cluster has a property of diversity naturally. Traditional extractive summary tends to select a sentence with the top score for each cluster. All candidate sentences are considered in submodular-based approaches. To diminish the benefit of choosing elements from the same cluster, a diversity function is introduced to measure the diversity of a summary set S. Traditional diversity function employs MMR to penalize the redundancy. Here, we define a diversity function as:

$$f\_diver(S) = \frac{\sum_k (\sum_{S_i, S_j \in S \cap P_k} dissim(S_i, S_j))}{\sum_k (\sum_{S_i, S_j \in P_k} dissim(S_i, S_j))} \tag{6}$$

where the function $dissim$ refers to the dissimilarity of paths S_i and S_j in same cluster. It is calculated as: $1 - sim(S_i, S_j)$.

Summary Generation. As above mentioned, three function are non-negative, monotone, submodular. Our objective function consists of these functions, which aim at finding a summary that maximizes

$$f(S) = f\_topic(S) + \alpha f\_query(S) + \beta f\_diver(S) \tag{7}$$

where the parameters α and β can be used to scale the range of f_query and f_diver, respectively.

Input: *Paths set P; L, parameters in f are fixed*
Output: *Generated Summary S*
$S \leftarrow \{\,\}, U \leftarrow P$
while $U \neq null$ **do**
   $s \leftarrow \arg\max_{s \in U} f(S + s)$
   $U \leftarrow U \setminus \{\, s\,\}$
   $S \leftarrow S \cup \{\, s\,\}$
   $U \leftarrow \{\, p | p \in U, \sum_{s \in S} len(s) + len(p) \leq L\,\}$
end while

Fig. 3. Greedy maximize submodular algorithm.

Our summary generation algorithm is shown in Fig. 3. During the search process, we use S to save the paths in our generated summary. For each iteration, we find path s with the largest value of objective function, and maintain the set of candidate paths U for the next iteration. Once the set U is *null*, the greedy searching algorithm terminates and the paths in the set S is composed of our final summary.

Previous extractive work showed that finding the optimal summary can be viewed as a combinatorial optimization problem which is NP-hard to solve (McDonald 2007). Our searching algorithm is still suboptimal. Furthermore, the parameters are important to search the summary sentences. In this paper, we fix these parameters in search process.

4   Experiment

4.1   Settings

Datasets. We conduct the experiments on the DUC 2005–2007 datasets[6] on multi-document summarization. DUC 2005 and DUC 2007 are used as test datasets to evaluate our methods. They contain 50 and 45 topics respectively. We use DUC 2006 as our development set. The length of the generated summary is limited by 250 words.

Baseline Systems. We select a few baseline and state-of-the-art abstractive systems working on DUC 2005 and DUC 2007, respectively. `MSC+ILP` denotes the system proposed by Banerjee et al. (2015). `TopBSU` denotes the best system in Li (2015). These two systems achieve the best performance on two datasets, respectively. In addition, we design several baselines to validate the influence of event guidance and multi-sentence compression algorithm to summary generation.

Evaluation Metric. ROUGE (Lin 2004) has been proven to be effective in measuring qualities of summaries and correlates well to human judgments. We use it for automatic evaluation of summaries in this work. Following Zheng et al. (2014), we report Rouge-L and Rouge-SU4 scores of various systems on DUC 2005. We use the scores of baseline systems on DUC 2005 as reported in their work. On DUC 2007, we report Rouge-1, Rouge-2 and Rouge-SU4 scores following Li (2015). The scores of baseline systems reported in their work are directly used in this paper.

In addition, we conduct human evaluations, using the same method as Banerjee et al. (2015). Four evaluators are asked to rate the generated summaries according to two different factors, namely informativeness and readability. Each summary is given a subjective score from 1 to 5, with 1 being the worst and 5 being the best. The 10 document sets from DUC 2007 and their summaries are selected for human rating.

4.2   Results

To fine-tune the parameters of our model, we conduct development experiment on DUC 2006. For efficiency, these parameters are optimized separately. In this paper, we set the values of parameters as shown in Table 1.

Automatic Evaluation. Table 2 shows the final results on the testing set. We design several baselines to validate the influences of event guidance (`Event`), multi-sentence compression algorithm (`SalMSC`) and submodular maximization (`SubM`) to summary generation. `Sent` is refer to the sentence clusters based on the similarity[7] between sentences. `Event+SalMSC+SubM` is refer to our full model.

---

[6] http://duc.nist.gov/data.html.
[7] We use Sent2Vec, which code is available at https://github.com/klb3713/sentence2vec, to learn the vectors of sentences.

Table 1. Parameter setting in our method.

Parameter	Description	Value
B	The beam size	128
λ	The fluency weight	0.8
α	The weight of query relevance	1.3
β	The weight of diversity	2.5

Table 2. Performance comparison for automatic evaluation on DUC 2005 and DUC 2007.

Method	DUC 2005			DUC 2007		
	Rouge–L	Rouge–SU4	Method	Rouge–1	Rouge–2	Rouge–SU4
Sent+MSC	0.33235	0.10927	Sent+MSC	0.38237	0.08351	0.13179
Sent+MSC+SubM	0.33875	0.11042	Sent+MSC+SubM	0.39435	0.08863	0.14079
Event+MSC	0.34237	0.11428	Event+MSC	0.40128	0.09072	0.14745
Event+MSC+SubM	0.35023	0.12075	Event+MSC+SubM	0.41902	0.09932	0.15756
Event+SalMSC	0.34853	0.11775	Event+SalMSC	0.41679	0.09736	0.15422
Event+SalMSC+SubM	0.35689[a]	0.12566[a]	Event+SalMSC+SubM	0.42208[a]	0.10203[a]	0.15761[a]
MSC+ILP	0.35772	0.12411	TopBSU	0.42145	0.11016	0.15632
RDMS	0.35376	0.12297	RA–MDS	0.403	0.092	0.146
FGB	0.35018	0.12006	MASC	0.41224	0.10650	0.15767
KM	0.29107	0.10806	MultiMR	0.41967	0.10302	0.15385
NMF	0.28716	0.11278	TTG	0.39268	0.09645	0.14553
DUC Best	0.34764	0.10012	RankBSU	0.39123	0.08742	0.14381
LexRank	0.33179	0.12021	Avg–2007	0.39684	0.09495	0.14671
Centroid	0.32562	0.11007	NIST–baseline	0.33126	0.06425	0.11114

[a] denotes that the result is significantly better with a p-value below 0.05.

By comparison with different clustering algorithm, sentence compression algorithm, we can examine the effectiveness of our model. As shown in Table 2, the model achieves better Rouge scores on both DUC 2005 and DUC 2007, demonstrating the structured event is beneficial to guide the subtopic splitting and sentence compression. Further, we validate our submodular maximization algorithm. Those baseline systems without SubM, simply choose the path with the top score from each cluster. As shown in Table 2, better performance is achieved by SubM, demonstrating the effectiveness of our algorithm. We report results of previous state-of-the-art systems as well. Compared with MSC+ILP and TopBSU, our full model achieves comparable performance. If with more accurate dependency parser for event extraction, our system will be expected to achieve better performance. As shown in Table 2, our full model is significantly better than other state-of-the-art systems.

Manual Evaluation. Following Banerjee et al. (2015), we conduct human evaluation. The results are shown in Table 3, which contains two aspects: informativeness, fluency. Similar to the automatic evaluation results, Event+SalMsc+SubM achieves better performance than other systems. There are two possible reasons. Firstly, the candidate sentences are extracted with event guidance, hence Event methods reduce noises from long sentences; Secondly, in multi-sentence

Table 3. Results from the manual evaluation.

Method	Info.	Flu.
Sent+MSC	3.81	3.73
Event+MSC+SubM	4.05	3.95
Event+SalMSC+SubM	4.43	4.21

Table 4. Generated summaries for the topic "Oslo Accords" (D0720E) by the different methods.

Method	Rouge–1	Rouge–2	Rouge–SU4	Method	Rouge–1	Rouge–2	Rouge–SU4
Event+SalMSC +SubM	0.46374	0.14792	0.21562	Sent+MSC	0.42887	0.14474	0.17551

To implement the oslo peace accords, israel and the palestinians Under the oslo principles, he has the right to **declare an independent palestinian state**. The united states, seeking to persuade the palestinians not to declare a state. The u.s. would help move the peace process forward after israel's may 17 elections. The final-status issues that the two sides have to tackle include the status of jerusalem, the borders of the **palestinian entity**, the fate of **palestinian refugees**, the future of **jewish settlements** in the west bank and gaza strip and water. To persuade the palestinians not to declare a state. Israel to implement the oslo accords and the hebron agreement in a bid to move the mideast peace process forward. As invited, a delegation of the palestinian islamic resistance movement headed by its spiritual leader sheik ahmed yassin attended the whole world, Particularly the united states and israel, are anxiously to see what decision the council would make on the statehood issue. To renew intifada if the **peace process collapses**. The meeting is in progress, hundreds of palestinians from the democratic front for the liberation of palestine, are demonstrating in front of the office. A majority of israelis were overwhelmed by oslo, the rabin assassination and the blowing up of israeli and palestinian self-rule areas. The two sides to implement the agreement in may 4 to avert being exploited by hard-line prime minister benjamin netanyahu in the coming elections. To facilitate that process, to bring israeli and palestinian leaders together within six month after israeli elections to review the status of their efforts and to help **reach an agreement**	Palestinian leader yasser arafat *has stated many times in the past that* under the oslo principles, he has the right to **declare an independent palestinian state** *on may 4 this year*. Under the oslo accords, israel opposes this idea, threatening to annex parts of the west bank. The oslo, the israeli and palestinian negotiators reached agreement with israel. President yasser arafat said thursday urged israel and the palestinian national authority -lrb- pna minister benjamin netanyahu's policies on peace process collapses. Israel has handed over about 43 percent of the west bank to the palestinians two years ago. Following a series of suicide bombing attacks, the peace process and led to the current likud government came to power in israel last year. He urged the u.s. and other countries to put pressure on the israeli document concerning israel's troop redeployment in the west bank. Palestinian leader yasser arafat has stated many times in the past that he has the right to declare an independent palestinian state on may 4 this year. Under the oslo accords and the hebron agreement signed last israeli troop redeployment in the west bank. The oslo accords have to **reach a framework agreement** with israel and the palestinians. The united states, seeking to persuade the palestinians not to declare a state, urged israel if the **peace process collapses**. So far has handed over about 43 percent of the west bank to the palestinians two years ago. Following a series of islamic suicide bombing attacks, the peace process and led to the current likud government came to power in israel last year

compression stage, the structured event information is introduced to effectively restrict the choice of nodes.

Example Outputs. Table 4 presents a comparison of summaries from the different systems for the topic "*Osla Accords*". Based on this example, we can know that our full model tends to generate more acceptable results. Our method can identify the important phrases, such as "*palestnian entity*", "*palestinian refugees*" and "*jewish settlements*". Sent+MSC lose some important information due to ignoring the salience information. For the event "*Arafat declare an independent palestinian state*", Sent+MSC brings more redundant information "*on may 4 this year*". Conversely, our method can generate more succinct results. In addition, through we do not select the compression from each cluster respectively, our submodular functions ensure the informativeness and the diversity of final result simultaneously. The above observation and analysis show the effectiveness of our proposed model, which is designed using event guidance and submodular maximization.

5 Conclusion and Future Work

We proposed a novel model for abstractive summarization, introducing a combination of three submodular functions to generate the final summary, based on a set of events. Our model can generate sentence clusters through an event clustering algorithm based on distributed representation. We designed a set of submodular functions to search the preferable paths which are generated by a multi-sentence compression algorithm. The summary quality is ensured through maximizing the submodular-based objective function from topic coverage, query relevance and diversity. Experimental results on DUC 2005 and DUC 2007 demonstrated that our model achieved better performance comparing with the state-of-the-art systems.

For future work, we plan to explore two directions. Firstly, we plan to introduce the event information into discourse structure analysis. In addition, we plan to investigate supervised methods for abstractive summarization.

Acknowledgments. We thank all reviewers for their detailed comments. This work is supported by the State Key Program of National Natural Science Foundation of China (Grant 61133012), the National Natural Science Foundation of China (Grant 61373108, 61373056), the National Philosophy Social Science Major Bidding Project of China (Grant 11&ZD189). The corresponding author of this paper is Donghong Ji.

References

Banerjee, S., Mitra, P., Sugiyama, K.: Multi-document abstractive summarization using ILP based multi-sentence compression. In: Proceedings of IJCAI 2015, pp. 1208–1214 (2015)

Barzilay, R., McKeown, K.R.: Sentence fusion for multidocument news summarization. Comput. Linguist. **31**(3), 297–328 (2005)

Biemann, C.: Chinese whispers: an efficient graph clustering algorithm and its application to natural language processing prob-lems. In: Proceedings of the First Workshop on Graph Based Methods for Natural Language Processing, pp. 73–80 (2006)

Bing, L., Li, P., Liao, Y., Lam, W.: Abstractive multi-document summarization via phrase selection and merging. In: Proceedings of ACL 2015, pp. 1587–1597 (2015)

Cheung, J.C.K., Penn, G.: Towards robust abstractive multi-document summarization: a caseframe analysis of centrality and domain. In: Proccedings of ACL 2013, pp. 775–786 (2013)

Cheung, J.C.K., Penn, G.: Unsupervised sentence enhancement for automatic. In: Proccedings of EMNLP 2014, pp. 775–786 (2014)

Dasgupta, A., Kumar, R., Ravi, S.: Summarization through submodularity and dispersion. In: Proccedings of ACL 2013, pp. 1014–1022 (2013)

Ding, X., Zhang, Y., Liu, T., Duan, J.: Using structured events to predict stock price movement: an empirical investigation. In: Proceedings of EMNLP 2014, pp. 1415–1425 (2014)

Fader, A., Soderland, S., Etzioni, O.: Identifying relations for open. In: Proceedings of EMNLP 2011, pp. 1535–1545 (2011)

Filippovai, K.: Multi-sentence compression: finding shortest paths in word graphs. In: Proceedings of Coling 2010, pp. 322–330 (2010)

Genest, P.-E., Lapalme, G.: Framework for abstractive summarization using text-to-text generation. In: Proceedings of the Workshop on Monolingual Text-To-Text Generation, pp. 64–73 (2011)

Grefenstette, E., Sadrzadeh, M.: Experimental support for a categorical compositional distributional model of meaning. In: Proceedings of EMNLP 2011, pp. 1394–1404 (2011)

Hu, Z., Rahimtoroghi, E., Munishkina, L., Swanson, R., Walker, M.A.: Unsupervised induction of contingent event pairs from film scenes. In: Proceedings of EMNLP 2013, pp. 369–379 (2013)

Li, C., Liu, Y., Liu, F., Zhao, L., Weng, F.: Improving multi-documents summarization by sentence compression based on expanded constituent parse trees. In: Proceedings of EMNLP 2014, pp. 691–701 (2014)

Li, P., Bing, L., Lam, W., Li, H., Liao, Y.: Reader-aware multi-document summarization via sparse coding. In: Proceedings of IJCAI 2015, pp. 30–35 (2015)

Li, W.: Abstractive multi-document summarization with semantic information extraction. In: Proceedings of EMNLP 2015, pp. 1908–1913 (2015)

Lin, H., Bilmes, J.: A class of submodular functions for document summarizatio. In: Proccedings of ACL 2011, pp. 510–520 (2011)

Lin, C.-Y.: Rouge: a package for automatic evaluation of summaries. In: Text Summarization Branckes Out: Proceedings of the ACL-04 Workshop, pp. 74–81 (2004)

Liu, F., Flanigan, J., Thomson, S., Dadeh, N., Smith, N.A.: Toward abstractive summarization using semantic representations. In: Proceedings of NAACL 2015, pp. 1077–1086 (2015)

Mani, I.: Automatic Summarization. Natural Language Processing, vol. 3. John Benjamins Publishing Company, Amsterdam (2001)

McDonald, R.: A study of global inference algorithms in multi-document summarization. In: Amati, G., Carpineto, C., Romano, G. (eds.) ECiR 2007. LNCS, vol. 4425, pp. 557–564. Springer, Heidelberg (2007)

Mehdad, Y., Carenini, G., Ng, R.T.: Abstractive summarization of spoken and written conversations based on phrasal queries. In: Proceedings of ACL 2014, pp. 1220–1230 (2014)

Ng, J.-P., Chen, Y., Kan, M.-Y., Li, Z.: Exploiting timelines to enhance multi-document summarization. In: Proceedings of ACL 2014, pp. 923–933 (2014)

Sun, R., Zhang, Y., Zhang, M., Ji, D.: Event-driven headline generation. In: Proceedings of ACL 2015, pp. 462–472 (2015)

Zhang, Y.: Partial-tree linearization: generalized word ordering for text synthesis. In: Proceedings of IJCAI 2013, pp. 2232–2238 (2013)

Zheng, H.-T., Gong, S.-Q., Chen, H., Jiang, Y., Xia, S.-T.: Multi-document summarization based on sentence clustering. In: Neural Information Processing, pp. 429–436 (2014)

Graph Data Management

Inferring Diffusion Network on Incomplete Cascade Data

Peng Dou, Sizhen Du, and Guojie Song[(✉)]

Key Lab of Machine Perception (MOE), Peking University, Beijing, China
billydou20@gmail.com,
{sizhen,gjsong}@pku.edu.cn

Abstract. Inferring the diffusion network based on observed cascades is fundamental and of interest in the field of information diffusion on the network. All the existing methods which aim to infer the network assume that the cascades are complete without any missing nodes. In real world, not every infection between nodes can be easily observed or acquired. As a result, there are some missing nodes in the real cascades, which indicates that the observed cascades are incomplete and makes more challenges for solving this problem. Being able to recover the incomplete cascades is critical to us since inferred network based on the incomplete cascades can be inaccurate. In this paper, we tackle the problem by developing a two-stage framework, which finds the paths that contain the missing nodes at first and then estimate the location and infection time of missing nodes. Experiments on real and synthetic data show the accuracy of our algorithm to finding the missing node on the network, as well as the infection time of the missing nodes.

1 Introduction

The problem of diffusion process has gained much attentions in recent years [1–4]. Inferring the diffusion process is one of the hotly debated topics, which covers lots of fields in real life. For example, microblogs can be created or reposted by many people everyday, thus the information will be shared rapidly on the network. However we may not be able to know how the information diffuses on the network. Some users did not cite the source when publishing the information, which means that only the infection time of nodes on the network is recorded. Apart from the information diffusion, epidemiology is another application of inferring diffusion process. The infection time of a patient can be easily observed while who has infected the patient is often difficult to find. Therefore, inferring the diffusion process is a problem demanding prompt solution.

Some existing methods have been proposed to tackle the problem of inferring the diffusion networks based on the complete observed cascades [5–7]. Even if we can handle this problem, it is based on the condition that all the nodes on the network can be observed so that we are able to record their infection time. However, some microblogs may be deleted by their users and it will lead to an incomplete observed data. Unfortunately, there is no guarantee that the cascades

© Springer International Publishing Switzerland 2016
B. Cui et al. (Eds.): WAIM 2016, Part I, LNCS 9658, pp. 325–337, 2016.
DOI: 10.1007/978-3-319-39937-9_25

are totally complete since some nodes cannot be observed in real world. Once the malicious information spread all over the network, can we infer the diffusion process of it only based on the incomplete cascades? Since our purpose is to infer how and when the information diffuses, finding the missing nodes becomes the crux of solving the problem.

In order to achieve this purpose, we face the following three challenges: first of all, given a set of cascades, we have to tell whether there exists at least one missing node. Inferring the network based on incomplete cascades will lead to an inaccurate result. Secondly, the infection time of the missing nodes has to be estimated accurately for the reason that the estimated infection time determines the inferred network. Inaccurate infection time will definitely result in a wrong network. Thirdly, the diffusion process can be inferred by the cascades and it is the last task we have to handle. As mentioned above, inferring diffusion network on intact cascade can be solved with the existing methods. However, the first two challenges, which are also the most vital and thorny, are what we care about and eagle to solve.

To the best of our knowledge, this is the first paper that deals with the problem of inferring networks with incomplete data. In this paper, we present a method to tackle these two challenges in terms of accuracy and efficiency. In more detail, we find the paths which include the missing nodes by analyzing the inferred network based on the incomplete cascades. Once finding those paths, we locate the missing nodes on the network and then estimate the infection time of the missing nodes according to their neighbors. We conduct the experiment on synthetic data to evaluate the performances of our method, which reveals that we can recover the missing nodes decently. We also apply our algorithm to the real world data and achieve an excellent result.

Related Work. There are several works related to our work. Zhang et al. focus on the Bayesian network inference by exploiting causal independence [8]. Network structure learning for estimating the dependency structure of directed probabilistic model is proposed by Getoor et al. [9]. However, this problem is intractable while heuristic solutions cannot guarantee the efficiency and performance. The problem of inferring the diffusion network has been considered recently. The proposed algorithm NetInf [5] assumes the propagation among edges is homogeneous and it uses submodular optimization to solve the problem. Meyer and Leskovec propose CONNIE [6], which uses convexity and heuristics. Gomez-Rodriguez et al. propose NetRate algorithm [7] to tackle the problem of underlying dynamics and estimating the transmission rates. Myers et al. [10] focus on the information diffusion through the influence external out-of-network sources. All the methods above make the assumption that the observed cascades are complete, thus they cannot be applied to the proposed problem. The recent related work [11] is concerning finding the source of diffusion network based on observed infection, while it still cannot handle the problem we propose above.

2 Problem Formulation

2.1 Transmission Model

The information spreads over the network, which leaves a trace called cascade. There is at least one source of the information, of which the infection time is 0, which means that the source node is infected at time 0. Each time when a node i is infected, it subsequently infects all of its neighbors independently on the network, and the infection is homogenous among all the edges on the network. Let Δt_{ij} denote the time that node i takes to infect j, thus $t_j = t_i + \Delta t_{ij}$, where t_i, t_j stand for the infection time of node i and j respectively.

2.2 Problem Statement

The observed cascades $C = \{t^1, t^2, \ldots, t^c\}$ is the trace of the spread process in the diffusion network G, which records the time when each single node on the network gets infected. The transmission between edges obeys a specific distribution, which is a prior information. For a specific observed cascade, namely t^i, consists of infection time of the nodes. For node k, the infection time of it in cascade t^i is denoted as t_k^i and $t_k^i \in [0, T^c] \cup \{+\infty\}$, where T^c is the size of observation window and $+\infty$ means the node is not infected during the observation window. Apart from all the observed nodes, there is a set $M = \{u_1, u_2, \ldots, u_m\}$, which represents missing nodes on the network, of which the infection time cannot be observed. We aim to find those missing nodes on the diffusion network and recover their infection time as well as the network structure. Table 1 presents the important notations used in this paper.

Table 1. Notations

Notations	Descriptions
G	The diffusion network
$\widetilde{G}^C$	The inferred network based on the cascades C
(x, y)	The directed edge from x to y
$\overrightarrow{xy}$	The directed path from x to y
$\overrightarrow{(x, y)}$	The potential short path from x to y
$l\overrightarrow{xy}$	The distance from x to y
C	The set of the cascades $\{t^1, t^2, \ldots, t^c\}$
d	The time delay of infection over one edge
$\overline{d}$	The expected time delay of infection
$\overline{v}$	The variance on time delay of infection
t_i	The infection time of node i
M	The set of missing nodes of G

3 Proposed Method

In this section, we propose a method to tackle the problem of inferring network with missing nodes. In order to find the missing node, we trace specific edges called **P**otential **S**hort **E**dge (**PSE**) where the missing node is located at first. There is a remarkable gap between the time delay of the potential short edges and time delay given by the prior distribution, which we adopt as the evidence to find the potential short edges. Then, we aim to locate these missing nodes on the network and estimate the infection time of the missing node to recover the whole cascades, which is based on the time delay given by the incomplete cascades.

3.1 Find the Potential Short Edges with Missing Nodes

The major concern is that the infection time of missing nodes cannot be observed which leads to incomplete cascades. Thus the inferred network $\widetilde{G^C}$ based on the incomplete cascades is not accurate consequently when applying the existing method. It is remarkable that some edges appear in $\widetilde{G^C}$ which not included in G because of the missing nodes. Based on the prior distribution, the time delay of these edges can be seen abnormal. Therefore, our method firstly take a look at those edges as defined below.

Definition 1 (Short Edge). *For an inferred graph $\widetilde{G^C}$, we define the short edge (x, y) as follows:*

$$(x, y) \in \widetilde{G^C}, (x, y) \notin G, and \overrightarrow{xy} \in G$$

If (x, y) is one of the short edges in $\widetilde{G^C}$, it takes more time for x to infect y for the reason that some other nodes on the path $\overrightarrow{xy}$ should be infected before y is infected. According to the observed cascades, we can unearth those short edges by counting and analyzing the time delay between (x, y) in $\widetilde{G^C}$. For a path $\overrightarrow{xy}$ whose length is $l\overrightarrow{xy}$, since the infection among edges is independent, the expected time delay from x to y, denoted as $\overline{t_y - t_x}$, can be expressed by $\bar{d}$ as follows:

$$\overline{t_y - t_x} = l\overrightarrow{xy} \cdot \bar{d} \tag{1}$$

Equation (1) provides us with a simple approach to tell whether (x, y) is likely to be a short edge when we have t_x and t_y, on the condition that $t_x - t_y > \bar{d}$ remarkably. However, we cannot ignore the fact that, it is likely that x infects y with a long time delay even if they are neighbors. Let n_{xy} be the number of cascades where $t_y > t_x$ and $m_{xy,\gamma}$ be the number of cascades where $t_y - t_x > (1+\gamma)\bar{d}$. The probability that $t_y - t_x > (1+\gamma)\bar{d}$ is denoted as p_γ. Note that γ is the threshold which controls the size of $m_{xy,\gamma}$ and p_γ is smaller as γ increasing. Intuitively, a bigger $m_{xy,\gamma}$ indicates higher probability that (x, y) is a short edge. In other words, (x, y) is unlikely to be an edge from the true diffusion network G. Here, we have the following lemma:

Lemma 1. *Given an arbitrary integer k ranged from 0 to n_{xy}, the probability that $m_{xy,\gamma} \geq k$ can be presented as:*

$$P(m_{xy,\gamma} \geq k) \leq 2^{n_{xy}} p_\gamma^k (1 - p_\gamma)^{n_{xy} - k}$$

Proof.

$$P(m_{xy,\gamma} \geq k) = \sum_{i=k}^{n_{xy}} P(m_{xy,\gamma} = i)$$

$$< \sum_{i=k}^{n_{xy}} C_{n_{xy}}^i p_\gamma^i (1 - p_\gamma)^{n_{xy} - i} \leq \sum_{i=k}^{n_{xy}} C_{n_{xy}}^i p_\gamma^k (1 - p_\gamma)^{n_{xy} - k}$$

$$\leq p_\gamma^k (1 - p_\gamma)^{n_{xy} - k} \sum_{i=k}^{n_{xy}} C_{n_{xy}}^i$$

$$\leq 2^{n_{xy}} p_\gamma^k (1 - p_\gamma)^{n_{xy} - k} \qquad \square$$

We aim to distinguish the short edges from all the edges, the probability of $t_y - t_x > (1 + \gamma)\bar{d}$ is smaller when (x, y) is not a short edge as mentioned above. Note that $P(m_{xy,\gamma} \geq k)$ decreases when k is larger. Therefore, given a probability threshold η, k can be computed so that $P(m_{xy,\gamma} \geq k) < \eta$, which indicates that the probability that (x, y) is a short edge is more than 1-η. The theorem below presents us how to compute the k, which also tells us how to judge whether an edge in $\widetilde{G^C}$ is *potential short edge*.

Theorem 1. *The edge (x,y) is a short edge with the probability over $1 - \eta$ when $m_{xy,\gamma} \geq k$:*

$$k = \frac{\ln \eta - n_{xy} \ln 2(1 - p_\gamma)}{\ln p_\gamma - \ln(1 - p_\gamma)}$$

*and we call (x, y) **PSE**(potential short edge).*

Proof.

$$P(m_{xy,\gamma} \geq k) \leq 2^{n_{xy}} p_\gamma^k (1 - p_\gamma)^{n_{xy} - k} < \eta$$
$$\ln(2^{n_{xy}} p_\gamma^k (1 - p_\gamma)^{n_{xy} - k}) < \ln \eta$$
$$n_{xy} \ln 2 + k \ln p_\gamma + (n_{xy} - k) \ln(1 - p_\gamma) < \ln \eta$$
$$k \geq \frac{\ln \eta - n_{xy} \ln 2(1 - p_\gamma)}{\ln p_\gamma - \ln(1 - p_\gamma)}$$

$$\square$$

Since we are interested in finding the missing node, what PSE tells us is that there can be at least one missing node between x and y if (x, y) is PSE. The remaining question is how to determine the infection time of the missing node and its position in network G. In the following sections, an efficient algorithm for solving this question is proposed and elaborated.

3.2   Estimating the Infection Time of Missing Nodes

We have proved that the missing nodes lie in the PSE with large probability, namely at least $1 - \eta$. Figure 1 shows an example for us: (a) represents the inferred graph, (z,x) and (z,y) are both PSE, which can be inferred by both (b) and (c). z is actually a root of a tree and x and y are the leaves. In (b), x and y share a father while in (c), x and y have their own father. The purpose of our work is equivalent to finding out the nodes inside the tree rooted in z. The example of (b) and (c) reminds us that we should determine whether two nodes share a common ancestor. Intuitively, the branch on the tree makes the difference of infection time between the nodes. As the branch appears earlier, the difference is greater. For example, the difference of infection time between x and y in (b) is smaller than that in (c). Let $v_{t_x-t_y}$ represent the variance of the infection time of x and y, and $\bar{v}$ is the variance of the distribution on the diffusion. We build the relationship of them on the variance as follows.

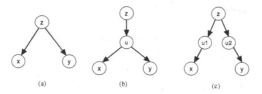

Fig. 1. Missing nodes

Lemma 2. *Given* $\overrightarrow{(z,x)}$ *and* $\overrightarrow{(z,y)}$, *let* u *be the Lowest Common Ancestor(LCA) of* x *and* y, *we have the following equation:*

$$v_{t_x-t_y} = (\overrightarrow{lux} + \overrightarrow{luy}) \cdot \bar{v}$$

Proof.

$$v_{t_x-t_z} = v_{t_u-t_z} + v_{t_x-t_u}, v_{t_y-t_z} = v_{t_u-t_z} + v_{t_y-t_u}$$
$$v_{t_x-t_y} = v_{(t_x-t_z)-(t_y-t_z)} = v_{t_x-t_u} + v_{t_y-t_u}$$

Since the infection among edges is independent

$$v_{t_x-t_y} = \overrightarrow{lux}\cdot\bar{v} + \overrightarrow{luy}\cdot\bar{v} = (\overrightarrow{lux}+\overrightarrow{luy})\cdot\bar{v} \qquad \square$$

It is likely that the LCA of x and y is just z, thus it is necessary for us to determine whether the LCA of x and y is one of the missing nodes instead of z, as presented below.

Theorem 2. *Given* $\overrightarrow{(z,x)}$ *and* $\overrightarrow{(z,y)}$, *there exists a node* u *which* $u \neq z$ *and* $u \in M$, *when*

$$\frac{v_{t_x-t_y}}{\bar{v}} < \frac{t_x + t_y - 2t_z}{\bar{d}}$$

Proof. According to Eq. (1), it is easy to get

$$\overrightarrow{lzx} = \frac{t_x - t_z}{\bar{d}}, \overrightarrow{lzy} = \frac{t_y - t_z}{\bar{d}}$$

Let u be the LCA of x and y, as induced in Eq. (1),

$$(\overrightarrow{lux} + \overrightarrow{luy}) = \frac{v_{t_x - t_y}}{\overline{v}}$$

Because u is infected by z by the path $\overrightarrow{zu}$, it means

$$\overrightarrow{lux} + \overrightarrow{luy} < \overrightarrow{lzx} + \overrightarrow{lzy}$$

So, $\dfrac{v_{t_x - t_y}}{\overline{v}} < \dfrac{\overline{t_x - t_z}}{\overline{d}} + \dfrac{\overline{t_y - t_z}}{\overline{d}} = \dfrac{\overline{t_x + t_y - 2t_z}}{\overline{d}}$ ☐

Corollary 1. *When two nodes x and y share a common father, which means* $\overrightarrow{lux} = \overrightarrow{luy} = 1$, *we have* $\frac{v_{t_x - t_y}}{\overline{v}} = \frac{\overline{t_x + t_y - 2t_u}}{\overline{d}} = 2$.

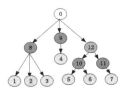

Fig. 2. A possible hidden tree based on PSE

For the node z and all of PSE headed on it, as mentioned above, there exists a hidden tree rooted in z and its leaves are the tails of these PSE. In order to find the missing nodes on the tree, we can manage to find the father of the nodes in tree and recover it from bottom to up. For example, Fig. 2 shows a possible hidden tree rooted in node 0 and nodes 1 to 7 are the leaves of the tree, while means (0,1), (0,2), (0,3), (0,4), (0,5), (0,6) and (0,7) are the PSE we find. At the first round, we aim to find the father of nodes 1 to 7. Since several nodes may share the same father, we deal with this situation by applying the corollary1. Thus, we discover that node 8, 9, 10 and 11 are the fathers of these leaves. Then, node 12, which is father of 10 and 11, is easily to be detected at the second round. However, not only the missing nodes are what we have to discover but the infection time of them has to be determined. We let z to be the root of the tree as mentioned above, and the leaves which share the same father u, denoted as $x_1, x_2, \ldots, x_k$. It is simple to get the average of the infection time of these leaves, which is $\overline{t_x} = \dfrac{\sum\limits_{i=1}^{k} \overline{t_{x_i}}}{k}$. As Theorem 2 shows, $\overrightarrow{lzx}$ can be estimated by $\frac{v_{t_z - \overline{t_x}}}{\overline{v}}$. And we have know that $\overrightarrow{lzu} = \overrightarrow{lzx} - 1$, thus combined with the prior distribution, we can estimate the infection time of u as follows:

$$t_z + (\frac{v_{t_z - \overline{t_x}}}{\overline{v}} - 1)(\overline{t_x} - t_z)$$

We start to find the missing nodes from a given node, denoted as $\tilde{z}$, which is the head of at least one PSE. Adopting the conclusion of the theorems and

---

Algorithm 1. Estimating the infection time of missing nodes

Require: $\widetilde{z}, PSE, C\{t^1, t^2, \ldots, t^c\}$;

1: $S = \emptyset$;
2: **for** $x \in V(\widetilde{G^C})$ **do**
3:      **if** $(\widetilde{z}, x) \in PSE$ **then**
4:          $S = S \bigcup x$;
5: **while** S is not $\emptyset$ **do**
6:      choose an arbitrary node x_1 in S, $X = x_1$;
7:      **for** $y \in S$ and $y \neq x$ **do**
8:          **if** $\frac{v_{t_x} - t_y}{v} \approx 2$ **then**
9:              $X = X \bigcup y$;
10:     t = 0;
11:     create a missing node u;
12:     **for** c = 1 to $|C|$ **do**
13:         **for** $x \in X$ **do**
14:             $t = t + t_x^c$;
15:         $\overline{t_x} = t/|C|$;
16:         $t_u^c = t_z + (\frac{v_{t_z} - \overline{t_x}}{v} - 1)(\overline{t_x} - t_z)$;
17:     add the missing node u to $\widetilde{G^C}$;
18:     S = S \ X;

---

equations above, we are able to estimate all the missing nodes related to $\widetilde{z}$ and their infection time in the cascades. Our method to find the missing node is presented in Algorithm 2. Lines 2–4 are intent to get all the leaves of the tree whose root is $\widetilde{z}$, and the loop from lines 5–16 manage to find the fathers of these leaves and estimate their infection time. Note that Algorithm 2 functions as one round to find the missing nodes in the tree, a higher tree requires more calls of Algorithm 2.

Algorithm 2. Inferring the network with missing cascades.

Require: $C\{t^1, t^2, \ldots, t^c\}$;

1: **while** true **do**
2:      $\widetilde{G^C} = NetRate(C)$;
3:      $PSE = Finding\_PSE(\widetilde{G^C}, \eta, \gamma)$;
4:      **if** PSE is $\emptyset$ **then**
5:          return $\widetilde{G^C}$;
6:      $\widetilde{z} = \underset{x \in \widetilde{G^C}}{\arg\max} |\{y|(x, y) \in PSE\}|$;
7:      $C = Finding\_missing\_nodes(\widetilde{z}, PSE, C)$;

The whole algorithm combines the two algorithms presented above. Firstly, we infer the network based on missing data and try to find all the PSE to ensure that there is at least one missing node in $\widetilde{G^C}$. Secondly, we select a node which is the head of one of all PSE. The selected node is the root of the hidden tree and we find the corresponding missing nodes are located on the tree. Thirdly, we estimate the infection time of the missing node to build the new cascades for use later. These three steps keeps working until no PSE is detected, as Algorithm 3 presented.

4 Experimental Evaluation

In this section, we will evaluate the performances of our method on a range of datasets and networks: synthetic networks which simulates the structure of social networks and real diffusion networks.

4.1 Evaluation Criteria

To evaluate the accuracy of our method, we consider three evaluation criteria: the number of missing nodes, the structure of network and the infection time of missing nodes. As stated above, $\widetilde{G^C}$ is the inferred network and M' is the inferred set of missing nodes. Since we have to know the relationship between the inferred missing nodes and real missing nodes, we build the mapping between the inferred missing nodes and real missing nodes to ensure the feasibility of evaluation, and the mapping is denoted as $\tilde{f}$. The evaluation is based on the comparison between inferred ones and real ones.

- **The inferred missing nodes.**
  $f(M') = \frac{|\{m|m \in M \cap M'\}|}{|M'|}$, which is the precision of the inferred missing nodes.
- **The inferred network structure.**
  Let $E(M)$ be the set of edges in real networks which are related to the real missing nodes, where $E(M) = \{(x,y)|x \in M \text{ or } y \in M\}$. The evaluation function, $f(\widetilde{G^C}) = \frac{\left|\{(x,y)|(x,y) \in \widetilde{G^C}, \tilde{f}(x) \in M \text{ or } \tilde{f}(y) \in M\}\right|}{|E(M)|}$, which is the precision of the inferred network.
- **The inferred cascades.**
  We first compute the accuracy of infection time of a single node x, which is $h(x) = \frac{\sum_{i=1}^{c} \frac{\min\{t^i_{\tilde{f}(x)}, \tilde{t}^i_x\}}{\max\{t^i_{\tilde{f}(x)}, \tilde{t}^i_x\}}}{c}$. Then, we consider all the missing nodes so that $f(\widetilde{C}) = \frac{\sum_{x \in M' \cap M} h(x)}{|M' \cap M|}$, which is the average accuracy of the missing nodes on infection time.

4.2 Experiments on Synthetic Data

Experimental Setup. We consider the Kronecker Graph model to generate the diffusion networks, and we consider core-periphery [0.9,0.5;0.5,0.3] network structure. We then simulate the cascades on the generated network with the transmission model and a given transmission rate. Each time we pick the starting nodes uniformly at random so that at least 95 % nodes are recorded in the cascades. Once a node gets infected, we record the time of the infection and mark it as infected status in case that it will be infected later. The transmission model varies from exponential, power law and rayleigh, while the size of the network varies from 64, 128 and 256. The parameters for finding the potential paths are set as follows: $\gamma = 1, \eta = 0.05$.

Varying the Network Size. Intuitively, the larger the network is, the more accurately our method is to work as other factors are fixed. From the Fig. 3(a), we can see that the it is obvious the accuracy declines as the network gets larger. On the network with 64 nodes, the accuracy of all the criteria is over 0.7. While the network is larger, the accuracy is relatively lower. For example on the network with 256 nodes, we can observe that the accuracy of inferred nodes and network structure decrease by 15 %. This is due to the fact that, with the same number of cascades, smaller network makes it easier to infer.

Varying the Cascade Size. It is easier for us to find the missing nodes when the amount of cascades is larger, which can also make the result more accurate. We generate the cascades at the beginning of the experiment carefully, and now we examine the dependence on cascade size. Figure 3(b) plots the relationship between the accuracy of our method and the cascade size. As expected, the accuracy can be seen increasing when the number of cascades gets bigger.

Varying the Number of Missing Nodes. We also examine the results when varying the number of missing nodes. Once we generate the network and the corresponding cascades, we eliminate some nodes from the cascades, which are the missing nodes we aim to find. We select nodes randomly from the networks to be the missing nodes, which account for a specific proportion of all the nodes. The proportion varies from 10 %, 20 % and 30 %. If there are few missing nodes on the network, since we choose them randomly, they will disperse on the network. When we have more missing nodes, two nodes may connected by one edge are both missing. It is more difficult to infer the connected two nodes, compared with one missing nodes between two known nodes. Figure 3(c) shows the accuracy when varying the proportion of the missing nodes on the network.

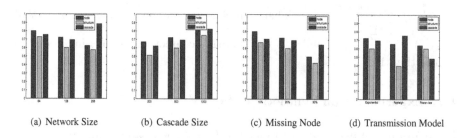

(a) Network Size   (b) Cascade Size   (c) Missing Node   (d) Transmission Model

Fig. 3. Evaluation on synthetic data (Color figure online)

Varying the Transmission Model. As mentioned above, we consider three types of the transmission model: exponential, power law, rayleigh. The transmission rate is 1 for exponential and rayleigh model and 0.5 for the power law model. In Fig. 3(d), we observe that the accuracy of exponential model is better than that of other two models. The accuracy of all the three criteria is over 65 %. Notice that given the same transmission rate, the variance of rayleigh distribution is larger than exponential distribution, it makes the inference much harder and affects the accuracy in this case.

4.3 Experiments on Real Data

Dataset Description. The real data we use is extracted from the quotes and phrases online that appear most frequently from August, 2008 to April, 2009. The method we apply to trace information is MemeTracker Metholody. For the experiment, we use the network which has 500 nodes and over 3000 edges, and the nodes we choose are the top-500 documents.

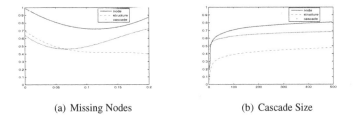

(a) Missing Nodes   (b) Cascade Size

Fig. 4. Evaluation on real data

Evaluation on Real Data. We range the proportion of missing nodes from 0 to 0.2 and examine how the performances of our method depends on the number of missing nodes. We also take the three criteria into account as elaborated above to evaluate our method, as the Fig. 4(a) shows. The accuracy of nodes is above 75 % in all cases, and comes to nearly 90 % when the proportion is 0.2. There is also a rise on the accuracy of inferred cascade, which is eventually over 70 %. Apart from these two criteria, we observe that the accuracy of inferred network structure declines. Because the inferred structure is more complex when adding the missing nodes thus it is more difficult for us to infer. Even though, we are able to infer 40 % of the real network structure. However, note that some isolated missing nodes may not be infected during on most of the cascades, adding one of these nodes to the set of missing nodes will make the accuracy drop immediately. Thus the smoothing effect is carried on the curve of the accuracy. Because of these uninfected missing nodes, we observe a decline on the accuracy at the beginning, rather than rising all the time. With more missing nodes added, the accuracy is tending towards stability. At the same time, we also examine the relationship of our method with the cascade size. From Fig. 4(b), we can observe that the accuracy increases as cascade size becomes larger. The accuracy is comparatively lower when the number of cascade is small, because too few cascades are not able to provide the information about the missing nodes. However, the increment speed of accuracy is high at the beginning, and the accuracy can be kept at a fair level over 50 cascades. The accuracy for nodes, structure and cascade is 68 %, 37 %, 61 % respectively. After a period of spurt, the accuracy increases slowly. When the cascade size is 500, the accuracy of inferred missing nodes reaches 85 %.

5   Conclusion

We have proposed a new problem of inferring diffusion network with missing data and developed a method to handle it. Evaluation on both synthetic data and real data shows our algorithm can accurately recover the missing nodes, network structure and cascades. Meanwhile, our method can handle various transmission model, such as exponential, power law, rayleigh and other models. There are some interesting problems for future work. Further concentration can be focus on heterogeneous diffusion network. Moreover, an efficient and accurate algorithm on inferring diffusion network is expected to improve the efficiency.

Acknowledgement. This work was supported by the National Science and Technology Support Plan(2014B AG01B02), the National Natural Science Foundation of China (61572041), the National High Technology Research and Development Program of China (2014AA015103), and Beijing Natural Science Foundation (4152023).

References

1. Rogers, E.M.: Diffusion of innovations. Gen. Inf. **51**(6), 866–879 (1995)
2. Kempe, D., Kleinberg, J., Tardos, E.: Maximizing the spread of influence through a social network. In: Proceedings of the Ninth ACM SIGKDD International Conference on Knowledge Discovery and Data Mining, pp. 137–146 (2003)
3. Adar, E., Adamic, L.A.: Tracking information epidemics in blogspace. In: Proceedings of the 2005 IEEE/WIC/ACM International Conference on Web Intelligence, pp. 207–214. IEEE Computer Society (2005)
4. Watts, D.J., Dodds, P.S.: Influentials, networks, and public opinion formation. J. Consum. Res. **34**(4), 441–458 (2007)
5. Gomez-Rodriguez, M., Leskovec, J., Krause, A.: Inferring networks of diffusion and influence. CoRR, vol. abs/1006.0234 (2010). http://arxiv.org/abs/1006.0234
6. Myers, S.A., Leskovec, J.: On the convexity of latent social network inference. Eprint Arxiv, pp. 1741–1749 (2010)
7. Gomez-Rodriguez, M., Balduzzi, D., Scholkopf, B.: Uncovering the temporal dynamics of diffusion networks. In: Proceedings of the 28th International Conference on Machine Learning (ICML 2011), pp. 561–568 (2011)
8. Zhang, N.L., Poole, D.: Exploiting causal independence in Bayesian network inference. J. Artif. Intell. Res. **5**, 301–328 (1996)

 9. Getoor, L., Friedman, N., Koller, D., Pfeffer, A.: Learning probabilistic relational models of link structure. J. Mach. Learn. Res. 1300–1309 (2002)
10. Myers, S.A., Zhu, C., Leskovec, J.: Information diffusion and external influence in networks. In: Proceedings of the 18th ACM SIGKDD International Conference on Knowledge Discovery and Data Mining, pp. 33–41. ACM (2012)
11. Farajtabar, M., Gomez-Rodriguez, M., Du, N., Zamani, M., Zha, H., Song, L.: Back to the past: source identification in diffusion networks from partially observed cascades. In: Proceedings of Eighteenth International Conference on Artificial Intelligence and Statistics, pp. 232–240 (2015)

Anchor Link Prediction Using Topological Information in Social Networks

Shuo Feng, Derong Shen[✉], Yue Kou, Tiezheng Nie, and Ge Yu

College of Information Science and Engineering, Northeastern University,
Shenyang, China
shendr@mail.neu.edu.cn

Abstract. People today may participate in multiple social networks (Facebook, Twitter, Google+, etc.). Predicting the correspondence of the accounts that refer to the same natural person across multiple social networks is a significant and challenging problem. Formally, social networks that outline the relationships of a common group of people are defined as aligned networks, and the correspondence of the accounts that refer to the same natural person across aligned networks are defined as anchor links. In this paper, we learn the problem of Anchor Link Prediction (ALP). Firstly, two similarity metrics (Bi-Similarity BiS and Reliability Similarity ReS) are proposed to measure the similarity between nodes in aligned networks. And we prove mathematically that the node pair with the maximum BiS has higher probability to be an anchor link and a correctly predicted anchor link must have high ReS. Secondly, we present an iterative algorithm to solve the problem of ALP efficiently. Also, we discuss the termination of the algorithm to give a tradeoff between precision and recall. Finally, we conduct a series of experiments on both synthetic social networks and real social networks to confirm the effectiveness of our approach.

Keywords: Anchor link prediction · Social network · Topological information · Aligned networks

1 Introduction

Social networks are becoming ubiquitous in our daily life. Formally, social networks that outline the relationships of a common group of people are defined as aligned networks. In practice, there are many aligned networks, e.g., Google+ and Facebook. And the accounts that refer to the same natural person across aligned networks are defined as anchor links, e.g., the account of Lady GaGa in Twitter and that in Facebook form an anchor link. In this paper, we focus on predicting anchor links in aligned networks, and we name this problem as Anchor Link Prediction (ALP).

ALP is an important task in social network analysis. Firstly, ALP is helpful in constructing a portrait of a natural person, e.g., Facebook is a good representation of people's personal profile in life and LinkedIn is better to provide people's job information. Reconciling Facebook and LinkedIn may have a full outline of people on different aspects. Secondly, leveraging anchor links, many business applications may become easier, e.g., when people are registering new accounts in the social network,

B. Cui et al. (Eds.): WAIM 2016, Part I, LNCS 9658, pp. 338–352, 2016.
DOI: 10.1007/978-3-319-39937-9_26

the system may recommend information to their friends in other social networks. Thirdly, ALP is useful in solving the problem of conflict detection. e.g., many people may assign incorrect personal information such as birthday in different networks. Reconciling the networks can help improve the consistency.

In real life, topology information based ALP is useful in network De-anonymizing problem [11]. Network owners often share networks with many third parties. To alleviate privacy concerns, most attributes are not available. For the third parties, de-anonymizing the network is an important work. Thus, in this paper, we only use topology information to solve the problem of ALP. [6, 7] are much related to ours. They use common neighbor as node pair similarity and regard the unlinked node pairs with the most common neighbors as anchor links. However, they mainly have two disadvantages. Firstly, many unlinked node pairs may have the most common neighbors at the same time. As a result, they may hesitate in choosing the anchor links that conflict to each other. Secondly, they conduct a one-time prediction for each anchor link. In other words, they do not consider the correctness of the previously predicted anchor links in the following iteration. Following, we conduct the contributions of our work.

Firstly, two similarity metrics, BiS and ReS, are proposed to measure the similarity of node pairs. Compared with the other similarity metrics, we have two advantages: (I) The unlinked node pair with the maximum BiS has higher probability to be an anchor link. (II) A correct anchor link must have higher ReS.

Secondly, we present an algorithm (BiSP_ReSD) using BiS to do Prediction and using ReS to do incorrect link Detection. In the algorithm, two processes (anchor link prediction and incorrect link detection) are conducted iteratively. In anchor link prediction step, the node pair with the maximum BiS is predicted as an anchor link. And in incorrect link detection step, the anchor links with low ReS are regarded as the incorrect ones. The time complexity of BiSP_ReSD is $O(D^K D'^K + |L|)$ in each iteration, where D (or D') is the average degree of the aligned network, L is the set of the anchor links and K is a positive constant number. Also, we discuss the termination of the algorithm.

Finally, we conduct a serious of experiments on both synthetic social networks and real social networks to confirm the effectiveness of our algorithm.

The rest of the paper is organized as follows. In Sect. 2 we discuss related works. In Sect. 3 we give the preliminary. In Sect. 4 we propose two similarity metrics. In Sect. 5 we propose an algorithm to solve the problem of ALP. In Sect. 6 we show the experimental results. In Sect. 7 we conclude the paper.

2 Related Work

ALP is firstly formalized as connecting corresponding identities across communities [10]. They supposed that people prefer to use similar username across multiple networks. Further, [1, 2] utilize additional textual attributes (e.g., email address, image, name) to improve the accuracy. Recently, [14] aims at identifying a certain user using limited information. However, statistics in [3] shows that the significant attributes might be set unobservable or protected by people in purpose to protect self-privacy. Thus, behavior information is recognized to solve the problem.

Recently, several behavior information based methods are proposed to solve ALP [4, 5, 12]. [12] solves the problem mainly through people's writing style, language pattern and username modification. [5] supposes that people's daily habits will perform on both of the aligned networks. They focus on check-in time and check-in location. And [4] conducts a sensor to detect the frequent behavior pattern in a fixed-length time period. Traditionally, behavior information based ALP methods are based on machine learning models and always require large amount of training data.

ALP can be regarded as maximum common subgraph problem [9, 13], whose goal is to find a one-to-one mapping between the nodes in the aligned networks and maximum the number of the overlapped edges. Network alignment problem is a NP-hard problem. And most methods are heuristic and provide near optimal solution. However, the time complexity of these methods is still high, which can be hardly used to social networks.

A different work is studied in [11]. They study the problem of de-anonymizing social networks, which is much related to ours. They iteratively predict new anchor links and keep all the score of the anchor links higher than a fixed threshold. However, this method is a heuristic method, there is no theoretical guarantees for their algorithm.

Inspired by [11], Korula and Lattanzi [7] formulate the problem mathematically and they are the first to solve ALP with theoretical guarantees. [6] observes a strong sensitivity of their algorithm to the number of the pre-linked anchor links: if the pre-linked anchor links are too little, their algorithm would face the cold start problem. Further [8] solves the cold start problem by reducing the precision of the result.

3   Preliminary

In this paper, we suppose that the aligned networks are subgraphs of an invisible underlying network [7]. We use $G^*(V^*, E^*)$ to denote the underlying network, where V^* denote the set of the nodes and E^* denote set of the relationships between the nodes in V^*. We assume that the underlying network G^* fits power law distribution. In other words, each node prefers to attach to the nodes with high attractiveness. Moreover, the attractiveness of each node would be proportional to its degree. Thus, we approximately regard that each node in G^* prefers to attach to the nodes with high degree. In other words, given $i^* \in V^*$, $j^* \in V^*$, $k^* \in V^*$, the edge e_{i^*, j^*} is more likely to exist than e_{i^*, k^*}, if $|N(j^*)| > |N(k^*)|$, where $N(j^*)$ denotes the set of the neighbors of j^* and $|N(j^*)|$ denotes the size of the set $N(j^*)$.

We denote the couple of aligned networks that generated from G^* as $G(V, E)$ and $G(V', E')$, where $V \subseteq V^*$, $V' \subseteq V^*$, $E \subseteq E^*$ and $E' \subseteq E^*$. Especially, we use ' or nothing on the top-right corner of a lowercase to distinguish a node in different aligned networks, e.g., $i \in V$ and $i' \in V'$ denote the nodes in G and G' respectively. For each node $i \in V$ (or $i' \in V'$), we use $\tilde{i}$ (or $\tilde{i}'$) to denote the original node of i (or i') in G^*. So we say an anchor link (i, i') is correctly predicted if $\tilde{i} = \tilde{i}'$. We assume that the aligned networks G

and G' are generated by selecting the edges randomly from G^* with the probability S and S' respectively, where both S and S' are larger than 0.5. We would further discuss the case of selecting nodes randomly in Sect. 6.

4   Similarity Metrics

4.1   Bi-Similarity BiS

We use BiS to measure similarity of the unlinked node pairs. Given a couple of aligned networks $G(V, E)$, $G(V', E')$ and an anchor link set L, we regard BiS of an unlinked node pair (a, a') as

$$BiS(a, a') = \sum_{(x,x') \in L} T_{(x,x')}(a, a'), \tag{1}$$

where (x, x') is an anchor link in L. $T_{(x,x')}(a, a')$ is regarded as the similarity brought by (x, x'). We denote $T_{(x,x')}(a, a')$ as transitive similarity from (x, x') to (a, a').

$$T_{(x,x')}(a, a') = \sum_{k=1}^{K} Pr(x, k, a) Pr(x', k, a'), \tag{2}$$

where K is a constant number. $Pr(x, k, a)$ is the probability that a random surfer starts from x and reaches at a after k steps without passing by the visited nodes. $Pr(x, k, a)$ could be obtained through the paths from x to a.

For example, as shown in Fig. 1, G and G' are a couple of aligned networks. We regard (x, x') as an anchor link and we take $T_{(x,x')}(c, c')$ as an example. For $k = 1$, we have $Pr(x, 1, c) = Pr(x', 1, c') = 1/3$. For $k = 2$, we have $Pr(x, 2, c) = 1/3$ and $Pr(x', 2, c') = 1/6$. For $k \geq 3$, there is no path from x to c without passing by the visited nodes. So we have $T_{(x,x')}(c, c') = 1/6$.

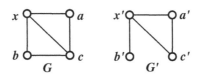

Fig. 1. A simple example of aligned networks

For a fixed anchor link (x, x'), it may spread transitive similarity to many other unlinked node pairs. In our intuition, the more degree a and a' have, the more likely transitive similarity from (x, x') to (a, a') would exist.

Lemma 1. Each anchor link (x, x') has higher probability to spread transitive similarity to (a, a') than (b, b'), if $\min(|N(\tilde{a})|, |N(\tilde{a}')|) > \max(|N(\tilde{b})|, |N(\tilde{b}')|)$.

Prove. When $K = 1$, the probability that (x, x') spreads transitive similarity to (a, a') can be denoted as $Pr(e_{x,a}, e_{x',a'})$, which is the probability of existence of the edge $e_{x,a}$

and $e_{x',a'}$. We have $Pr\left(e_{x,a}, e_{x',a'}\right) = SS' Pr\left(e_{\tilde{x},\tilde{a}}, e_{\tilde{x}',\tilde{a}'}\right)$. For any other node pair (b, b'), where $\max\left(|N(\tilde{b})|, |N(\tilde{b}')|\right) < \min\left(|N(\tilde{a})|, |N(\tilde{a}')|\right)$, we have $Pr\left(e_{\tilde{x},\tilde{b}}\right) < Pr\left(e_{\tilde{x},\tilde{a}}\right)$ and $Pr\left(e_{\tilde{x}',\tilde{b}'}\right) < Pr\left(e_{\tilde{x}',\tilde{a}'}\right)$. This is because each node in G^* prefers to attach to the nodes with high degree. So we have $Pr\left(e_{x,a}, e_{x',a'}\right) > Pr\left(e_{x,b}, e_{x',b'}\right)$. And the proof is complete. For the case $K > 1$, the proof is similar with the case of $K = 1$.

Theorem 1. The node pair with maximum BiS is an anchor link with high probability.

Proof. According to Lemma 1, if $\tilde{a} = \tilde{a}'$ and the degree of $\tilde{a}$ is maximum in G^*, each anchor link would have higher probability to spread transitive similarity to (a, a') than any other node pair. Thus, when the number of anchor links tends to be extremely large, $BiS(a, a')$ equals the sum of the probability that anchor links spread transitive similarity to (a, a'). Thus, we say the node pair with maximum BiS must be an anchor link.

4.2   Reliability Similarity ReS

Although Theorem 1 guarantees that the node pairs with the maximum BiS would always be an anchor link, the real-world networks are not always well-designed. So we propose ReS to measure whether an anchor link in hand is correct.

Definition 1 (Similarity Witness). Given a couple of aligned networks G, G' and an anchor link (x, x'). We denote (x, x') as a Positive Similarity Witness (PSW) of (a, a'), if x is a neighbor of a and x' is a neighbor of a' simultaneously. And on the other hand, we denote (x, x') as a Negative Similarity Witness (NSW) of (a, a'), if either x is a neighbor of a or x' is a neighbor of a'.

For example, as shown in Fig. 1, suppose (b, b') is an anchor link. We regard (b, b') as a PSW of (c, x'), because b is a neighbor of c and b' is a neighbor of x'. We regard (b, b') as a NSW of (x, a'), because b is a neighbor of x and b' is not a neighbor of a'.

Formally, we present the formula of ReS as follow.

$$\Re(a, a') = |PSW(a, a')| / (\xi + |NSW(a, a')|), \tag{3}$$

where $PSW(a, a')$ is the set of PSW of (a, a') and $NSW(a, a')$ is the NSW set. ξ is an extremely small positive real number to avoid denominator to be zero.

Theorem 2. Given an anchor link set L, we suppose that all the anchor links in L are correct. For any anchor link (a, a') in L, no other nodes $i \neq a, i' \neq a'$ would have

2.1 $|PSW(a, i')| \geq |PSW(a, a')|$ and $|PSW(i, a')| \geq |PSW(a, a')|$.
2.2 $|NSW(a, i')| \leq |NSW(a, a')|$ and $|NSW(i, a')| \leq |NSW(a, a')|$.
2.3 $\Re(a, i') \geq \Re(a, a')$ and $\Re(i, a') \geq \Re(a, a')$.

Proof. We firstly prove Theorem 2.1. We assume (v, v') is an anchor link in L, where $v \neq a$ and $v' \neq a'$. And the probability that (v, v') is a PSW of (a, a') can be denoted as $Pr(e_{v,a}, e_{v',a'}) = SS'Pr(e_{\tilde{v},\tilde{a}}, e_{\tilde{v}',\tilde{a}'})$. And because both (v, v') and (a, a') are correctly predicted, we have $Pr(e_{\tilde{v},\tilde{a}}, e_{\tilde{v}',\tilde{a}'}) = Pr(e_{v',a'})$. For the other node $i' \neq a'$, $Pr(e_{v,a}, e_{v',i'}) = SS'Pr(e_{\tilde{v},\tilde{a}}, e_{\tilde{v}',\tilde{i}'}) \leq SS'Pr(e_{\tilde{v},\tilde{a}}) = Pr(e_{v,a}, e_{v',a'})$. And when the number of anchor links in L tends to be extremely large, we have

$$|PSW(a, a')| = \sum_{(v,v') \in L} Pr(e_{v,a}, e_{v',a'}) \tag{4}$$

Thus, $|PSW(a, a')|$ gets maximum when $\tilde{a} = \tilde{a}'$, and Theorem 2.1 is proved. Next, we prove Theorem 2.2. We give the expansion of $|NSW(a, a')|$

$$
\begin{aligned}
|NSW(a, a')| &= \sum_{(v,v') \in L} Pr(e_{v,a}) + Pr(e_{v',a'}) - 2Pr(e_{v,a}, e_{v',a'}) \\
&= \sum_{(v,v') \in L} SPr(e_{\tilde{v},\tilde{a}})\left(1 + S'Pr(e_{\tilde{v}',\tilde{a}'}|e_{\tilde{v},\tilde{a}})\left(1/SPr(e_{\tilde{v},\tilde{a}}|e_{\tilde{v}',\tilde{a}'}) - 2\right)\right),
\end{aligned}
\tag{5}
$$

where $Pr(e_{\tilde{v},\tilde{a}}|e_{\tilde{v}',\tilde{a}'})$ is the probability of the existence of the edge $e_{\tilde{v},\tilde{a}}$ in the case that the edge $e_{\tilde{v}',\tilde{a}'}$ exists. Based on the assumption $S > 0.5$, we find only when both $Pr(e_{\tilde{v},\tilde{a}}|e_{\tilde{v}',\tilde{a}'})$ and $Pr(e_{\tilde{v}',\tilde{a}'}|e_{\tilde{v},\tilde{a}})$ equal 1, the formula gets minimum. In other words, for any other nodes $i \neq a, i' \neq a'$, $|NSW(i, a')|$ and $|NSW(a, i')|$ would be larger than $|NSW(a, a')|$. And we recognize that only when $\tilde{a}' = a'$, $Pr(e_{\tilde{v},\tilde{a}}|e_{\tilde{v}',\tilde{a}'})$ and $Pr(e_{\tilde{v}',\tilde{a}'}|e_{\tilde{v},\tilde{a}})$ equal 1.

Theorem 2.3 can be proved easily based on Theorems 2.1 and 2.2.

5 Algorithm BiSP_ReSD

In this section, a novel algorithm BiSP_ReSD is presented to solve the problem of ALP. We use BiS to do Prediction (BiSP) and use ReS to do incorrect link Detection (ReSD).

Figure 2 shows the outline of BiSP_ReSD. Firstly, we compute each node pair's BiS and ReS brought by the pre-linked anchor links, and record them in a similarity index M. Then we iteratively conduct our prediction and detection. In each iteration, we firstly search for the unlinked node pair with the maximum BiS and regard this unlinked node pair as a new anchor link. And then we compute the incremental similarity brought by this new anchor link. Simultaneously, we carry out incorrect link detection to check the incorrect anchor links.

All through the process, we focus on three issues. (1) How similarity index M works and how to update M. Given a newly predicted anchor link, we would introduce the incremental similarity computation in Sect. 5.1. (2) How to detect the incorrect anchor links. In Sect. 5.2, we would describe the method to detect incorrect anchor

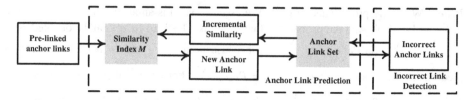

Fig. 2. Outline of BiSP_ReSD

links using ReS. (3) When to terminate the predicting process. In Sect. 5.3, we introduce an early termination strategy to give a tradeoff between precision and recall.

5.1 Incremental Similarity Computation

Before describing incremental BiS and ReS computation, we would introduce the similarity index M. As shown in Fig. 3, we record BiS of each node pair in each cell in shadow. For ReS, we know ReS is consist of the number of PSW and NSW from formula 3. And we have the following formula

$$|NSW(a, a')| = |PN(a)| + |PN(a')| - 2|PSW(a, a')|, \qquad (6)$$

where $PN(a)$ is the subset of the neighbors of a that are linked nodes, and we denote $PN(a)$ as positive neighbors (PN) of a. For example, in Fig. 1, we regard (x, x') and (a, a') as two anchor links. We have $PN(b) = \{x\}$ and $PN(c') = \{x', a'\}$.

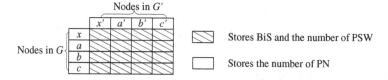

Fig. 3. Similarity index M

From formula 6, we could record ReS into similarity index M easily. We separately record PSW and PN of each node pair instead of ReS. For PSW, we would record it in each cell in shadow just as BiS. And for PN, we would record it in the first row or the first column of M in blank. So far, we have explained how M works.

Next, we introduce the method to compute incremental BiS. From formula 1, we realize that BiS is the sum of transitive similarity from each anchor link. Thus, the incremental BiS can be regarded as the transitive similarity spreading from the newly predicted anchor link. Firstly, we would seek for the paths starting from the anchor link. Then, we need to check whether the paths are the ones without passing by the visited nodes. Finally, we would compute the incremental BiS using formula 2. We realize that the computation cost of incremental BiS is proportional to the number of the paths from the anchor link $O(D^K D'^K)$, where D and D' are the average degree of the aligned network respectively.

For incremental ReS, the newly predicted anchor link just influences the ReS of its neighbors. For example, as shown in Fig. 1, we assume (a, a') is the new anchor link. (a, a') would increase PSW of the node pairs in $\{x, c\} \times \{x', c'\}$, and (a, a') would increase PN of $\{x, c\} \cup \{x', c'\}$, which would further change ReS of the node pairs related to x, c, x' and c'. Therefore, for the newly predicted anchor link (u, u'), we firstly increase PSW of the node pairs that are in $N(u) \times N(u')$. And then we would increase PN of the nodes in $N(u) \cup N(u')$. We realize that the computation cost of incremental ReS is $O(DD')$.

5.2 Incorrect Link Detection

According to Theorem 2, we consider an anchor link is correctly predicted if no other related node pair has higher ReS than this anchor link. Thus, in incorrect link detection, we should iteratively check each anchor link. For each anchor link, there may be $O(|V| + |V'|)$ related node pairs at most. Therefore, the time complexity of incorrect link detection is $O(|L|(|V| + |V'|))$ in each iteration. We notice that the computation cost is extremely high, and we would use PSW instead of ReS to conduct incorrect link detection approximately.

Firstly, we collect the node pairs with top-φ_w numbers of PSW among the ones related to the anchor link. And we think the node pair with the maximum ReS must be in these node pairs. Then we compare ReS of these φ_w node pairs with that of the anchor link in check. Using this approximate method, the time complexity is reduced to $O(|L|)$.

In total, the time complexity of our algorithm is $O(D^K D'^K + |L|)$ in each iteration.

5.3 Early Termination

In realization, we notice that nodes with high degree are predicted with high accuracy. But for the nodes with low degree, the prediction is not always satisfying. In this section, an early termination is presented to give a tradeoff between precision and recall.

Definition 2 (Reliability of Anchor Link Set). B and L' are subsets of anchor link set L. And all the anchor links in L' are correct. We denote the reliability of set B as

$$\Re(B, L') = \frac{\sum_{(a,a') \in B} |PSW_{L'}(a, a')|}{\sum_{(a,a') \in B} |NSW_{L'}(a, a')|},$$

where $PSW_{L'}(a, a') = PSW(a, a') \cap L'$ and $NSW_{L'}(a, a') = NSW(a, a') \cap L'$.

Theorem 3. We assume both B and L' are subsets of anchor link set L and all the anchor links in L' are correctly predicted. If all the anchor links in B are also correct, we have $\Re(B, L') = SS'/(S + S' - 2SS')$.

Proof. We have

$$\Re(B, L') = \frac{\sum_{(a,a')\in B} \sum_{(v,v')\in L'} SS' Pr\left(e_{\tilde{v},\tilde{a}}, e_{\tilde{v}',\tilde{a}'}\right)}{\sum_{(a,a')\in B} \sum_{(v,v')\in L'} SPr\left(e_{\tilde{v},\tilde{a}}\right) + S' Pr\left(e_{\tilde{v}',\tilde{a}'}\right) - 2SS' Pr\left(e_{\tilde{v},\tilde{a}}, e_{\tilde{v}',\tilde{a}'}\right)}$$

.

Because the node pair (a, a') in B and (v, v') in L' are correctly predicted, we have $\tilde{v} = \tilde{v}', \tilde{a} = \tilde{a}'$. So $Pr\left(e_{\tilde{v},\tilde{a}}, e_{\tilde{v}',\tilde{a}'}\right) = Pr\left(e_{\tilde{v},\tilde{a}}\right) = Pr\left(e_{\tilde{v}',\tilde{a}'}\right)$. Theorem 3 is proved.

We use Theorem 3 to measure the quality of the recently predicted result. We assume at time t the anchor link set is $L_t = \{l_1, l_2 \ldots l_n\}$, where l_i denote the predicted anchor link at iteration i. And we construct a box $B_t = \{l_{n-\varphi_B} \ldots l_n\}$ with size φ_B to record the recently predicted anchor links. If $\Re(B_t, L') < \varphi_k SS'/(S + S' - 2SS')$, we recognize that the recently predicted anchor links may be not satisfactory. And we need to terminate the algorithm. $\varphi_k \in (0, 1)$ is a parameter to control the precision and recall of the algorithm. When φ_k is large, the result has high precision and low recall. And when φ_k is small, the result has low precision and high recall. Traditionally, φ_B is 100. φ_B cannot be set too large or small. When φ_B is large, many incorrect predictions are predicted. And when φ_B is small, the early termination is too sensitive and the algorithm may be terminated too early.

We notice that $SS'/(S + S' - 2SS')$ is an invisible parameter and the correctly predicted subset L' is unavailable. But we notice that the nodes with high degree are predicted early and with high accuracy. Thus, at each time t, we also construct a box H_t

$$H_t = \{(a, a') | (a, a') \in L, |N(a)| > D, |N(a')| > D'\}$$

The nodes in H_t are all with high degree. And we approximatively regard H_t as the correctly predicted subset and we have $\Re(H_t, H_t) = SS'/(S + S' - 2SS')$. So the algorithm terminates when we have $\Re(B_t, H_t) < \varphi_k \Re(H_t, H_t)$.

6 Experiments

6.1 Experiment Setup

Environment. All approaches are implemented using JAVA. All the experiments are performed on the computer equipped with Intel i5-4590 CPU at 3.30 GHz and 8 GB of main memory. The operating system is a 64-bit installation of Windows 7.

Datasets. As shown in Table 1, Facebook (we use F in short) is the communication network between users in Facebook. DBLP (D) is a network recording the co-author relationship between authors. Enron (E) is an email communication network of Enron. Facebook, DBLP and Enron are all real-world networks labeled with timestamp. For Facebook and Enron network, we generate aligned networks by taking edges in disjoint time intervals. And we would select different set of papers to generate co-author

Table 1. Datasets

Network	Number of nodes	Number of edges
Facebook (F)	23,629	240,992
DBLP (D)	36,692	183,831
Enron (E)	317,080	1,049,866
RG	20,000	400,000
PA	20,000	399,791

aligned networks in DBLP. RG and PA are synthetic datasets. RG denotes the underlying network fits random graph model, and PA denotes the underlying network fits preference attach model [15]. All the synthetic datasets are generated by igraph.

6.2 Effectiveness Evaluation

Performance on Real-World Networks. We firstly analyze the effectiveness of our algorithm on real-world datasets. We compare our methods with the method in [7] (CN) and the method in [13] (MCS). For each dataset, we select different type of pre-linked anchor links: high, low and random. In high (low) type, we select 10 % anchor links whose degree are higher (lower) than average degree as the pre-linked anchor links previously. We use FMeasure to evaluate performance of each approach. As shown in Fig. 4, our method has higher precision and recall than CN and MCS in most datasets. For different type of pre-linked anchor links, our method performs better when the pre-linked anchor links have low degree. This is because the nodes with low degree is hard to predict. When the pre-linked nodes have low degree, the prediction become easier.

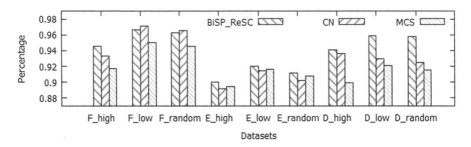

Fig. 4. Performance on real-world datasets

Performance on Synthetic Datasets. We conduct experiments on synthetic datasets to prove the correctness of theoretical analysis. We regard RG and PA as the underlying network. Moreover, we generate aligned networks by deleting edges (nodes) randomly with the probability of 0.8 (S and S' are both 0.8). We use SEdge (SNode) to denote different aligned network generation: selecting edges (nodes) randomly from the underlying network. In the experiments, we randomly select 10 % underlying anchor links as the pre-linked anchor links. As shown in Table 2, we have good performance on synthetic datasets, especially on RG_ SNode and RG_SEdge. This is because the degree of the nodes in RG fit average distribution. Compared with RG, many nodes with low degree are in PA, which may cause incorrect prediction. Also, we recognize the precision is lower than recall in RG_SNode and PA_SNode. This is because there are many unique nodes that exist in only one of the aligned networks. And the simi-

Table 2. Performance on synthetic datasets

	RG_SNode	RG_SEdge	PA_SNode	PA_SEdge
Precision	98.70 %	100 %	92.36 %	96.99 %
Recall	99.83 %	100 %	94.53 %	93.55 %

larity between these nodes may be large enough to be an anchor link. Thus, we may unavoidably regard them as anchor links.

Performance w.r.t. S, S' and Pre-linked Anchor Links. In this experiment, we use PA to conduct our experiment and generate aligned networks by selecting edge randomly. We assume $S = S'$, which means the overlap of the aligned networks is only S^2. As shown in Fig. 5, our algorithm performs better with the increase of the pre-linked anchor links. And our algorithm performs better with the increase of S. That means the more similar the two aligned networks are, the better performance we have. We also notice that our algorithm has good performance even with little pre-linked anchor links.

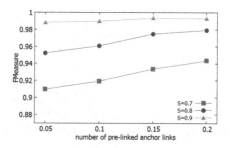

Fig. 5. Performance w.r.t. S and pre-linked anchor links

6.3 Efficiency Evaluation

In this section, we mainly discuss the variation of the parameter in our algorithm. And we will also show some characteristics of our algorithm.

Table 3. Performance w.r.t. K

	$K = 1$	$K = 2$	$K = 3$
Time(ms)	161,498	643,985	1,918,166
FMeasure	92.53 %	93.51 %	93.66 %

Performance w.r.t. K. In our algorithm, K controls the accuracy of BiS, which will further influences the prediction of a new anchor link. We conduct our experiment on Facebook and randomly select 5 % underlying anchor links as the pre-linked anchor links. As shown in Table 3, we notice the time cost increases exponentially with the increase of K. But there is not obvious change in FMeasure, this is because we conduct incorrect link detection to figure out the incorrect ones. Especial, even when K equals 1, our method has good performance. So the optimal value of K is 1.

Performance w.r.t. φ_w. φ_w is much related to the efficiency of the incorrect link detection. We conduct our experiment on Facebook and randomly select 10 % underlying anchor links as the pre-linked anchor links. As shown in Fig. 6, we vary φ_w from 4 to 12. We notice that both precision and recall have an obvious increase with φ_w varied from 4 to 8. When φ_w gets larger than 8, recall remains unchanged and precision increase slowly. So we conclude that φ_w is suitable to be 8.

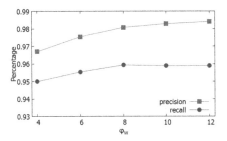

Fig. 6. Performance w.r.t. φ_w

Performance w.r.t. φ_k and φ_B. We conduct experiments on Facebook and randomly select 10 % underlying anchor links as the pre-linked anchor links. As shown in Fig. 7, both precision and recall have obvious decrease with the increase of φ_k. And precision gets maximum when φ_k equals 0.6. Although a lower φ_k may improve the recall, the precision may decrease obviously. And in Fig. 8, we discuss the optimal setting of φ_B. Recall increases hardly with φ_B varied from 50 to 100. This means that when $\varphi_B = 50$, many anchor links cannot be identified. And both precision and recall tend to be steady with φ_B varied from 100 to 200. So the optimal value of φ_B is 100.

Efficiency of Early Termination. In early termination, we employ B_t and H_t to record the recently predicted anchor links and anchor links with high degree at iteration t. We record $\Re(B_t, H_t)$ and $\Re(H_t, H_t)$ in each iteration. We conduct our experiment on Facebook and randomly select 10 % underlying anchor links as the pre-linked anchor

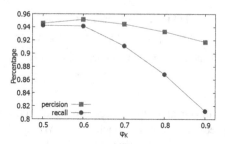

Fig. 7. Performance w.r.t. φ_k

links. As shown in Fig. 9, we use B_t and H_t to denote $\Re(B_t, H_t)$ and $\Re(H_t, H_t)$ respectively in each iteration t. We find that H_t is relatively stable. And in the previous 19000 iteration B_t varies around H_t. But in the latter iteration, B_t is much lower than H_t. And during these iteration, nodes with low degree are predicted incorrectly. So early termination is useful in improving the precision.

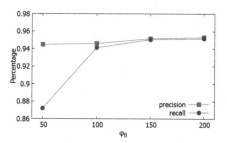

Fig. 8. Performance w.r.t. φ_B

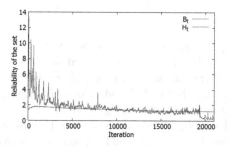

Fig. 9. Efficiency of early termination (Color figure online)

7 Conclusion

In this paper, we firstly propose two similarity metrics to evaluate the similarity between the node pairs in aligned networks. Then we present an algorithm BiSP_ReSD to solve the problem ALP. Finally, we test the effectiveness of our method on both synthetic social networks and real social networks. Comparing with other methods, we conduct incorrect link detection to detect the incorrect ones. In the future, we would further take user community information into consideration to answer ALP.

Acknowledgment. This work is supported by the National Basic Research 973 Program of China under Grant No. 2012CB316201, the National Natural Science Foundation of China under Grant No. 61472070.

References

1. Vosecky, J., Hong, D., Shen, V.Y.: User identification across multiple social networks. In: 2009 First International Conference on Networked Digital Technologies, NDT 2009. IEEE (2009)
2. Raad, E., Chbeir, R., Dipanda, A.: User profile matching in social networks. In: 2010 13th International Conference on Network-Based Information Systems (NBiS), pp. 297–304. IEEE (2010)
3. Labitzke, S., Taranu, I., Hartenstein, H.: What your friends tell others about you: low cost linkability of social network profiles. In: Proceedings of 5th International ACM Workshop on Social Network Mining and Analysis, San Diego, CA, USA (2011)
4. Liu, S., Wang, S., Zhu, F., et al.: Hydra: large-scale social identity linkage via heterogeneous behavior modeling. In: Proceedings of the 2014 ACM SIGMOD International Conference on Management of Data, pp. 51–62. ACM (2014)
5. Kong, X., Zhang, J., Yu, P.S.: Inferring anchor links across multiple heterogeneous social networks. In: Proceedings of the 22nd ACM International Conference on Information and Knowledge Management, pp. 179–188. ACM (2013)
6. Yartseva, L., Grossglauser, M.: On the performance of percolation graph matching. In: Proceedings of the First ACM Conference on Online Social Networks, pp. 119–130. ACM (2013)
7. Korula, N., Lattanzi, S.: An efficient reconciliation algorithm for social networks. Proc. VLDB Endowment **7**(5), 377–388 (2014)
8. Kazemi, E., Hamed, S.H., Grossglauser, M.: Growing a graph matching from a handful of seeds. In: Proceedings of the VLDB Endowment International Conference on Very Large Data Bases, 8(EPFL-ARTICLE-207759) (2015)
9. Zhu, Y., Qin, L., Yu, J.X., et al.: High efficiency and quality: large graphs matching. VLDB J. Int. J. Very Large Data Bases **22**(3), 345–368 (2013)
10. Zafarani, R., Liu, H.: Connecting corresponding identities across communities. In: ICWSM (2009)
11. Narayanan, A., Shmatikov, V.: De-anonymizing social networks. In: 2009 30th IEEE Symposium on Security and Privacy, pp. 173–187. IEEE (2009)
12. Zafarani, R., Liu, H.: Connecting users across social media sites: a behavioral-modeling approach. In: Proceedings of the 19th ACM SIGKDD International Conference on Knowledge Discovery and Data Mining, pp. 41–49. ACM (2013)

352 S. Feng et al.

13. Bayati, M., Gerritsen, M., Gleich, D.F., et al.: Algorithms for large, sparse network alignment problems. In: 2009 Ninth IEEE International Conference on Data Mining, ICDM 2009, pp. 705–710. IEEE (2009)
14. Vesdapunt, N., Garcia-Molina, H.: Identifying Users in Social Networks with Limited Information (2014)
15. Barabási, A.L., Albert, R.: Emergence of scaling in random networks. Science **286**(5439), 509–512 (1999)

Collaborative Partitioning for Multiple Social Networks with Anchor Nodes

Fenglan Li$^{(\boxtimes)}$, Anming Ji, Songchang Jin, Shuqiang Yang, and Qiang Liu

College of Computer, National University of Defense Technology, Changsha, China
{lifenglan92,jssyjam,jsc04,sqyang9999,lq_zj1981}@126.com

Abstract. Plenty of individuals are getting involved in more than one social networks, and maintaining multiple relationships of social networks. The value behind the integrated information of multiple social networks is high. Howerver, the research of multiple social networks has been less studied. Our work presented in this paper taps into abundant information of multiple social networks and aims to resolve the initial phase problem of multi-related social network analysis based on MapReduce by partition the mutli-related social networks into non-intersecting subsets. To concretize our discussion, we propose a new multilevel framework (CPMN), which usually proceed in four stages, Merging Phase, Coarsening Phase, Intial Partitioning Phase and Uncoarsening Phase. We propose a modified matching strategy in the second stage and a modified refinement algorithm in the fourth stage. We prove the effective of CPMN on both synthetic data and real datasets. Experiments show that the same node in different social networks is assigned to the same partition by 100 % without sacrificing the load balance and edge-cut too much. We believe that our work will shed light on the study of multiple social networks based on MapReduce.

1 Introduction

With the rapid development of computer science and technology, online social networks like Twitter, Facebook, QQ, Wechat, Microblog, etc., have become ubiquitous in our daily life. Undoubtedly, Facebook is replacing our address books and Skype calls have shorten our distance and have become an important communication medium. To enjoy more social services, an army of individuals is getting involved in various social networks and maintaining distinct social circles in differnt networks simultaneously. The users own more than one account in various networks are called **anchor nodes**.

Given the facts above, the studies of multiple social networks begin to capture researchers's attention [1–3]. The information behind single social networks isn't so sufficient as multiple social networks. The utilization of integration information from multiple networks enhances the accuracy and efficiency of social networks applications. An intuitive example is friendship recommandation, through co-processing, the information of multi-related networks can be integrated and the one with higher probability is more likely to be recommended.

© Springer International Publishing Switzerland 2016
B. Cui et al. (Eds.): WAIM 2016, Part I, LNCS 9658, pp. 353–364, 2016.
DOI: 10.1007/978-3-319-39937-9_27

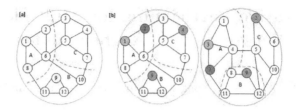

Fig. 1. A comparison of graph partitioning for a single network and for multiple networks. The edges cut off with red dotted line are the edge-cut. The nodes filled by different colors represent the anchor nodes, the nodes with the same color in different networks (see[b]) are the same users.

However, as the network continues to expand, the analysis of the whole social networks is a significant challenge, especially when the relationships of different social networks are taken into account. Traditional analysis methods based on single machines are not suitable for the requirement of performance anymore. MapReduce, a progromming paradigm proposed by Google, gives us a new approach to solve large-scale social network analysis problem by making use of the power of multi-machines [4–6]. Social networks analysis based on MapReduce needs to divide the whole social networks into multiple non-intersecting sub networks. Consequently, the sub networks in different server can be analysed in Map phase separately. The result of Map phase can be transmitted to Reduce phase to form the final result. On that account, the quality of partitioning is the key factor affecting social networks analysis based on MapReduce.

In this paper, we attemped to address the collaborative partitioning for multiple social networks that integrates heterogeneous information of different networks through the anchor nodes. For simplicity, the initial weight of all the edges is equal 1. Although a large body of literature has appeared in the graph partitioning (GP) for a single network, the GP for multiple networks has been less studied. The GP for a single network try to ensure the load balance and the edge-cut minimization. In addition to the above two goals, the collaborative partitioning for different related networks has the requirement to put the same nodes of multiple networks into the same paritions. With that, data locality can be further improved and communication overhead among servers can be reduced. When dealing with multiple networks, the load balance means the weight of nodes of different networks in the same partitions is roughly equivalent (Fig. 1).

A novel multilevel k-way partitioning for multiple networks is proposed to try to address those problems. In this paper, (1) we propose a new framework (CPMN) to tackle the multiple graph partitioning problem. (2) We propose a modified coarsen algorithm for multiple networks. (3) We develop a modified refinement algorithm to generate a k-way partition and always guarantee the same node in the same partition without sacrificing load balance and edge-cut.

The rest of paper is organized as follows: Sect. 2 is about related work. In Sect. 3, we formulate the problem. Section 4 is about the detail of our framework and strategies. Some experiments are conducted to show the performance of our framework in Sect. 5. We make a conclusion in Sect. 6.

2   Related Work

Mathematically, the problem of social networks partitioning can be solved by graph partitioning. Typically, graph partitioning problem falls under the category of NP-complete problem [7]. Thus, it is unlikely to find an optimal partition algorithm in polynomial-time. Many traditional algorithms are developed based on heuristics to find reasonable partitions. In this section, we will briefly introduce some classical algorithms related to our work.

We group algorithms into four clusters based on the idea for graph partitioning. The first class is moving algorithm that includes group migration algorithm, genetic algorithm (GA) [8], simulated annealing (SA) [9] etc. Group migration algorithm is the most commonly used among them. Kernighan-Lin algorithm (KL) [10] and Fiduccia-Mattheyses method (FM) [11] are the ground-breaking network partition methods which aim to divide a network into 2 balanced partitions. Given an initial bisection, the KL method tries to find a sequence of node pair exchanges that lead to a reduction of the cut size. A single iteration of the KL leads $O(|N|^3)$ time consumed. KL is so important that plenty of algorithms are inspired by it. FM is one of variations of KL algorithm that tries to reduce the time cost. Different from KL which chooses node pair to exchange, FM selects single node to move. Therefore, the time cost can be reduced to $O(|E|)$.

Another class algorithm is base on the idea of analytic geometry. One of the representative algorithms is spectral partitioning method [12,13]. Given a graph with adjacency matrix A and degree matrix D, the basic idea of spectral partitioning is computing the eigenvector of laplacian of matrix that is defined as $L = D - A$. The graph is divided into bisection through the sign of the corresponding vector entry. However, the price of this method is very high due to the computation of the engenvector corresponding to the second smallest eigenvalue [14].

The third class of graph partitioning algorithms makes full use of the geometric information of the graph nodes which can be called Geometric partitioning algorithms. Taking recursive coordinate bisection (RCB) [15] and inertial method [16] as example, RCB selects a coordinate axis and then tries to find a plane, orthogonal to the selected axis, that can bisect the nodes into equal subsets.Instead of selecting a coordinate axis, the axis of minimum angular momentum of the set of nodes is choosen in Inertial method.

The last class of algorithms is called multilevel graph partitioning schemes [17–21]. First, the graph is first coarsened down to a small number of vertices, then a reasonable initial partitions of the smallest graph is computed, and then the partitioning is sucessively refined level by level with refinement algorithms. Therefore, partitions of original graph are obtained. Multilevel algorithms have been shown to be good iteratives to both spectral and geometric algorithms [12].

The graph partitioning for multiple networks has been less studied. To our best known, synergistic partitioning in multiple large scale social networks [22]

is the first and the only one to study. It divides the first datum network with
multilevel k-way partitioning method into k partitions, then applies a modified
label propagation algorithm (LPA) to divide aligned network.

3   Problem Formulation

In this section, the concepts of multiple social networks are described and the
problem that this paper foces on is formulated (Fig. 2).

Definition 1 *(Multiple Social networks). Mathematically, a network can be
described as $G^i = (V^i, E^i)(i = 1, 2, ..., t)$, where V^i and E^i represent the set of
nodes and edges of the G^i, respectively. We use $n = |V^i|$ and $m = |E^i|$ to record
the number of nodes and edges. The sequence of graphs $(G^1, G^2, ..., G^t)$ corresponds to the multiple social networks. The networks mentioned in this paper are
undirected and unweighted networks.*

Definition 2 *(Anchor nodes and Non-anchor nodes). The same users
invoved in different social networks are called anchor nodes. The set constructed
by the anchor nodes are anchor nodes set A, and $A = \cap V^i$, $1 < i < t$. Besides
the anchor nodes, the left nodes are non-anchor nodes. These non-anchor nodes
constitute the non-anchor node set NA, $NA = \cup na^i$, where na^i is the non-
anchor nodes set of the ith graph. It just shows the relationship of anchor nodes
set A and non-anchor node NA: $A \cap NA = \oslash$, $A \cup NA = \cup V^i$, $0 < i < t$.*

Definition 3 *(Merged network). When the anchor nodes are known, the
graphs can be merged through the links. That means the merged graph of
$(G^1, G^2, ..., G^t)$ can be described by $G = (V, E)$ where $V = \cup V^i$, $E = \cup E^i$,
$0 < i < t$. When combined, the nodes are assigned new numbers.*

There are three concerns that need to be thought highly in collaborative
partitioning for multiple social networks with anchor nodes. Anchor nodes come
to be the first. If all the anchor nodes are allocated to the same partitions

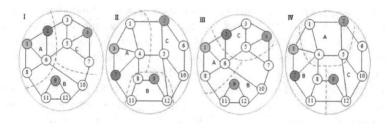

Fig. 2. Diagrammatic sketch of CPMN. The same color in different networks represents
the corresponding anchor pair. [I,II] shows the result of CPMN taking the anchor
nodes into consideration. [III,IV] is the result of an independent partitioning method.
Compared the partition results of networks, [I,II] obtains better partitions than [III,IV]
in the load balance and edge-cut minimazation.

respectively, the data locality will be improved and the communication between partitions is reduced simultaneously. The partitions should ensure the equal task of each processor to avoid the waiting delay and reduce the time cost. Meanwhile, more communication means worse efficiency and more time consumed. Add it all up, we can summarize our goals of collaborative partitioning for multiple social networks with anchor nodes as follows:

1. To make sure the same nodes in different networks could be aligned to the same partitions respectively.

2. As far as possible to ensure that balance of multiple social networks.

3. Minimize edge-cut of different partitions.

Mathematically, the graph partitioning for multiple networks with anchor nodes could be described as: For a given running configuration $Rp = (G^1, G^2, ..., G^t, A, k)$, and try to find k subsets $P_1, P_2, ..., P_k$ of V such that:

$$max(\frac{\sum_{v \in A} count(v)}{|A|}) \quad count(v) = 1 \text{ if } P_i[A_i[v]] \text{ is the same.} \tag{1}$$

In (1), $P_i[v]$ is the partition the node v in the $^i th$ network belongs to. And $A_i[v]$ is the label of anchor node v in the $^i th$ network.

$$min(\frac{max(\sum_{v \in G^i \ and \ v \in P_j} W_{P_j}(v))}{\frac{\sum_{v \in G^i} W(v)}{k}}) \quad i \in [1, t] \text{ and } j \in [1, k] \tag{2}$$

In (2), $W(v)$ represents the weight of node v and P_j is one of the partitions, G^i is one of the multiple social networks.

$$min(\sum_{i=1}^{t} Cut^i) \tag{3}$$

In (3), Cut^i is the number of edge crossing partitions in the $^i th$ network.

4 Proposed Methods

In this section, we describe a framework for collaborative partitioning for multiple social networks with anchor nodes, and the framework is called CPMN.

SPMN proposed in [22] divides the processing into two stages: datum generation stage and network alignment stage. Different from that, this paper intends to put forward a strategy of merging multiple networks into a larger graph firstly on GP for multiple networks. And then, inspired by the multilevel k-way partitioning proposed by Karypis and Kumar [23,24], we try to take advantage of the multilevel methods to divide the merged graph into k partitions. Therefore, the framework of CPMN falls into four phases: merging phase, coarsening phase, initial partitioning phase and uncoarsening phase. The paradigm can be illustrated in Fig. 3. Next we will describe each of these phases in more detail.

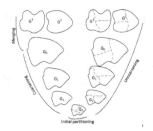

Fig. 3. Diagrammatic sketch of the framework of collaborative partitioning for multiple networks with anchor nodes: Merging, Coarsening, Initial partitioning and Uncoarsening.

4.1 Merging Phase

During the merging phase, For two (G^1, G^2), the merging phase will apply a merging method to generate a merged and larger graph $G = (V, E)$ where $V = V^1 \cup V^2$, $E = E^1 \cup E^2$ according to the anchor nodes set A. In order to be good with respect to the original graphs, the adjacent nodes of anchor nodes are merged, while the repeated edges are excluded. We add an array to identify the networks that nodes belong to. Thus the information of different networks is preserved. Also the weight and size of the nodes are held, while the non-anchor nodes and the connectivity information remain unchanged. However, the number of nodes from multiple networks is reallocated. The process is the preparations of collaborative partitioning for multiple networks so that the same node of different networks can always be aligned into the same partition.

4.2 Coarsening Phase

The phase aims to reduce the size of merged graph $G_0 = (V_0, E_0)$ and generate a sequence of smaller graphs $G_1, G_2, ..., G_m$, such that $|V_1| > |V_2| > ... > |V_m|$. The stage is achieved by collapsing the nodes and edges. The node of G_{i+1} consists of a set of nodes of G_i. Let V_i^x be the set of nodes of G_i combined to form node v of G_{i+1}, and E_i^y be the set of adjacent edges of V_i^x combined to form edges e that node v of G_{i+1} is adjacent to. In order to be good respect to the original graph and preserve the connectivity information in the coarser graph, the three formulas are used.

$$\begin{cases} W(v) = \sum_{u \in V_i^x} W(u) \\ W(e) = \sum_{d \in E_i^y} W(d) \\ W(I) = \sum_{u \in V_i^x} I(u) \end{cases} \tag{4}$$

Here, the $I(v)$ means the identity of the node v that used to differentiate which networks the node v belongs to.

The core of the compression is to find a matching [18]. A matching is an edge set that no more than two edges are incident on the same node. Considering a graph $G_{i+1} = (V_{i+1}, E_{i+1})$ is produced by $G_i = (V_i, E_i)$ through a matching M_i. Hence

$$W(E_{i+1}) = W(E_i) - W(M_i) \tag{5}$$

Consequently, the key point of coarsening phase is looking for the maximal matching that contains all the edges that satisfy the condition of matching. This method is called heavy edge matching (HEM) [25]. We proposed a modified heavy edge matching method (MHEM) to adapt to the new multiple networks environment.

For a given node v, instead of choosing a maximal weight simply, we choose the node that belongs to different networks with v and the weight of link is maximized so that the load balance of different networks is easier to satisfy. If such node does not exist, the anchor node of adjacent to v is chosen. However, if both the conditions are not satisfied, we choose the maximal weight as alternative choice. Overall, the nodes from different networks have the highest priority. The anchor nodes follow. The nodes only with maximal weight have the lowest priority. Note that this algorithm does not guarantee the coarsest graph to own equal nodes of different networks, but our experiments show that it works well.

4.3   Initial Partitioning Phase

The phase is to compute a k-way partitions of the coarsest graph $G_m = V_m, E_m$, in which the abundant information of different networks are contained to enforce the balance and minimization of edge-cut of multiple networks. Each partition contains roughly $|V_0^1|/k$ and $|V_0^2|/k$ node weight of the original multiple networks. In our algorithm, the k-way partition of G_m is computed using the multilevel bisection algorithm [25].

4.4   Uncoarsening Phase

The last phase of our framework CPMN is to project the partition P_m of the coarser graph G_m back to the original graph through the intermediate graphs $G_m, G_{m-1}, ..., G_m$. Meanwhile, the refinement algorithm are applied to optimize the result of partition so that the load balance and minimal edge-cut are achieved with an acceptable error.

The Greedy Refinement (GR) method which is the variation of FM is proved to be an efficiency method. Inspired by the idea of GR, we proposed a modified GR algorithm (MGR) to refine the partitions.

In particular, MGR works as follows. Consider a graph $G_i = (V_i, E_i)$, and its partitions vector P_i. First, the nodes are visited in a random order. $ED[v]_b$ and $ID[v]$ are the external degree of v to partition b and the internal degree of v, respectively. If the node v is one of the boundary nodes set B, and the moving of v in partition to partition b doesnt violate the balance condition.

$$\begin{cases} W_{G^1}(a) - W_{G^1}(v) > W_{G^1}^{min} \\ W_{G^1}(b) + W_{G^1}(v) < W_{G^1}^{max} \\ W_{G^2}(a) - W_{G^2}(v) > W_{G^2}^{min} \\ W_{G^2}(b) + W_{G^2}(v) < W_{G^2}^{max} \end{cases} \tag{6}$$

Table 1. Data sets of basic test on synthetic data

Graph1		Graph2	
n	m	n	m
5000000	50774773	5100000	55148024

Table 2. Data sets of merging graph of Table 1 through different anchor nodes set

	D1	D2	D3
n	8430000	8850000	9100000
m	105922786	105922786	105922788
Anchor	1670000	1250000	100000

Table 3. Data sets of scalability test

	D4	D5	D6	D7
n	2025000	3600000	5350000	7100000
m	17613495	32550359	66771683	88170368

Table 4. Data sets on real networks

Foursquare		Twitter		Anchor
n	m	n	m	n
1063989	28199331	6487891	12227875	413178

$W_{G^i}^{min}$ and $W_{G^i}^{max}$ are the minimal and maximal weight the network G^i allowed. Also the node $|v|$ to be moved should satisfy one of the conditions:

1. $ED[v]_b > ID[v]$ and $ED[v]_b$ is the maximum among all $b \in B$.

2. $ED[v]_b = ID[v]$ and $\begin{cases} W_{G^1}(a) - W_{G^1}(b) > W_{G^1}(v) \\ W_{G^2}(a) - W_{G^2}(b) > W_{G^2}(v) \end{cases}$.

That is, the MGR moves v to a partition that results in the highest reduction in edge-cut without violating the balance condition or leads to the improvement of load balance. When the moving finished, the external degree and internal degree of adjacent nodes of node v should be updated. Therefore, the quality of partitions of collaborative partitioning for multiple networks can be promoted greatly during the uncoarsening phase.

5   Experiment Result

5.1   Experiment Environment and Data Set

In this paper, all the experiments are running on a server (Intel(R) Xeon(R) CPU E5-2403 0 @ 1.80 GHz, memory 64 GB) with 64-bit CentOS.

Data sets used can be divided into two categories: virtual networks and real social networks. Virtual networks are generated by LFR benchmark by changing parameters that LFR needs. Data sets used in basic experiment can be found in Tables 1 and 2. Table 3 shows the data sets we use in experiments on scalability. The two real social networks used in experiments are Foursquare and Twitter, which are crawled from May 7th to June 9th. 1,064,011 users are crawled in foursquare, about 413,182 users among them have registered in twitter as well. Number of followers and number of these 413182 users follow in Twitter are 1,030,855,018 and 187,295,465 respectively. To compare the performance of CPMN and Metis, we choose a subset of twitter whose size is roughly equal to foursquare. Detail information is llustrated in Table 4.

SPMN is based on MapReduce, it is unfair to compare CPMN and SPMN. Therefore, the baseline method used to compare with CPMN framework is Metis, the state-of-the-art method for a single network. To envaluate the accuarcy and balance performance of the framework CPMN, NMI (Normalized Mutual Information) [26] of anchor nodes and and edge-cut in multiple networks is computed by (1) and (3).

5.2   Experiments and Results

In this section, experiments results are shown to evaluate the performance of CPMN. The results of Metis is the addition of results of multiple networks so that the experiments results can be compared with CPMN.

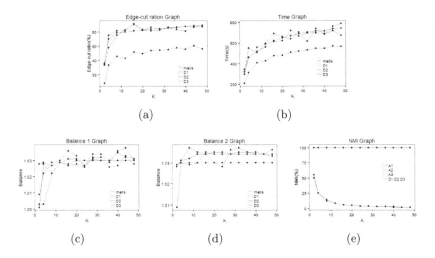

Fig. 4. Experiment results of CPMN on basic test of performace. (a): Edge-cut; (b): Time test; (c)(d): balance of the two graph; (e): NMI test. All the results are shown on the partitions of different anchor nodes of multiple networks. $A1, A2, A3$ represent the results of different anchor nodes on metis.

Performance of CPMN: Extensive empirical study shows our framework works well. The data sets can be seen in Table 3. From Figs. 4(a) and (b), we can see that curves of the Metis and CPMN changes are similar. However, the edge-cut ratio (time consumptions) of CPMN is higher than Metis during partition process. The reasons are as follows. First, the structure of merged graph is more complex and contain more intricate interactions. Second, in each phase of CPMN takes the anchor nodes into consideration. The results with more anchor nodes tend to generate more edge-cut and time consumption. It indicates that anchor nodes are the main reason for edge-cut sets and time consumption.

Figures 4(c) and (d) show the load balance of multiple networks. Although the unbalance of CPMN is a bit higher than Metis. Obviously, the unbalance is

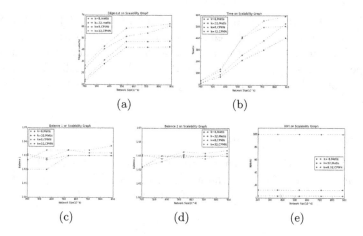

(a)   (b)

(c)   (d)   (e)

Fig. 5. Experiment results of CPMN on scalability test of performace. (a): Edge-cut; (b): Time test; (c)(d): balance of the two graph; (e): NMI test. All the results are shown on different size of multiple networks.

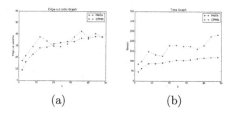

(a)   (b)

Fig. 6. Experiment results of CPMN on real networks test. (a): Edge-cut; (b): Time test. All the results are tested on two real social networks.

limited to 5 % that meets the requirement of collaborative partitioning. Because CPMN takes the anchor nodes and the relationships of multiple networks into account. From Fig. 4(e), we can see the NMI of CPMN is perfect compared with Metis. No matter what k and anchor nodes size is, the NMI remains unchanged at 100 % while NMI of Metis is falling down from 50 %. That's because Metis performs the division processes without considering any anchor nodes, but the merging phase of CPMN makes NMI keeping 100 % level.

Scalability of CPMN: Table 3 shows the datasets for the experiment. From Figs. 5(a) and (b), the comparison of CPMN and Metis on scalability is shown. The growth of edge-cut ratio and time consumption of CPMN appear to be similar with Metis. In Fig. 5(c) and (d), balance graph of multiple networks show non regularity owing to the random of networks. Similar with the basic experiments, the unbalance is limited to 5 %. From Fig. 5(e), NMI of CPMN and Metis stay fairly constant while the size of networks increase. Thus, we can see NMI is not related to the network size but k. The NMI of CPMN stays 100 % while Metis remains at a low level which is less than 50 %.

Performance on real social networks: The detail information about the two real social networks can be seen in Table 4. Here, we show parts of the experiments results as representative for the tendency of other results are similar with basic experiments. Figure 6 shows that our CPMN is comparable with Metis. The result of CPMN cuts off more edges than Metis and consumes more time because of the existence of anchor nodes. Despite all that, the same nodes in two real social networks are always divided into the same partition.

6 Conclusion

From the experiments on both synthetic and real data, we can see CPMN is an effective method to partition multiple social networks. The improvement of NMI of anchor nodes can be seen in Sect. 5. Actually, the anchor nodes pairs are assigned to the same partitions at 100 % level while it still keeps reasonable load balance of multiple networks. For a variety of reasons, the edge-cut and time consumption need to be further improved which may be the next potential research. Currently, our framework meets the basic requirements of collaborative partitioning of multiple networks. Making full use of the partitions of multiple related networks, it is easier to access to the social networks analysis based on MapReduce.

Acknowledgments. The work described in this paper is partially supported by National Key fundamental Research and Development Program (No. 2013CB329601, No. 2013CB329606) and National Natural Science Foundation of China (No. 61502517, No. 61372191, No. 61572492). The author is grateful to the anonymous referee for a careful checking of the details and helpful comments that improved this paper.

References

1. Zhang, J., Yu, P.S.: MCD: Mutual clustering across multiple heterogeneous networks. In: IEEE BigData Congress (2015)
2. Zhang, J., Yu, P.S.: Mutual community detection across multiple partially aligned social networks. arXiv preprint arXiv:1506.05529 (2015)
3. Zhang, J., Philip, S.Y.: Integrated anchor and social link predictions across social networks. In: Proceedings of the 24th International Conference on Artificial Intelligence, pp. 2125–2131. AAAI Press (2015)
4. Liu, G., Zhang, M., Yan, F.: Large-scale social network analysis based on mapreduce. In: 2010 International Conference on Computational Aspects of Social Networks (CASoN), pp. 487–490. IEEE (2010)
5. Gong, N.: Using map-reduce for large scale analysis of graph-based data (2011)
6. Jin, S., Yu, P.S., Li, S., Yang, S.: A parallel community structure mining method in big social networks. Math. Prob. Eng. **2015** (2015)
7. Feder, T., Hell, P., Klein, S., Motwani, R.: Complexity of graph partition problems. In: Proceedings of the Thirty-First Annual ACM Symposium on Theory of Computing, pp. 464–472. ACM (1999)

8. Golberg, D.E.: Genetic Algorithms in Search, Optimization, and Machine Learning. Addison-Wesley, Reading (1989)
9. Kirkpatrick, S., Gelatt, C.D., Vecchi, M.P., et al.: Optimization by simulated annealing. Science **220**(4598), 671–680 (1983)
10. Kernighan, B.W., Lin, S.: An efficient heuristic procedure for partitioning graphs. Bell Syst. Tech. J. **49**(2), 291–307 (1970)
11. Fiduccia, C.M., Mattheyses, R.M.: A linear-time heuristic for improving network partitions. In: 19th Conference on Design Automation, pp. 175–181. IEEE (1982)
12. Hendrickson, B., Leland, R.: A multilevel algorithm for partitioning graphs. In: Proceedings of the 1995 ACM/IEEE Conference on Supercomputing, p. 28. ACM (1995)
13. Pothen, A., Simon, H.D., Liou, K.P.: Partitioning sparse matrices with eigenvectors of graphs. SIAM J. Matrix Anal. Appl. **11**(3), 430–452 (1990)
14. Karypis, G., Kumar, V.: Multilevel graph partitioning schemes. In: ICPP (3), pp. 113–122 (1995)
15. Berger, M.J., Bokhari, S.H.: A partitioning strategy for nonuniform problems on multiprocessors. IEEE Trans. Comput. **100**(5), 570–580 (1987)
16. Farhat, C., Lesoinne, M.: Automatic partitioning of unstructured meshes for the parallel solution of problems in computational mechanics. Int. J. Numer. Meth. Eng. **36**(5), 745–764 (1993)
17. Smith, C.: The planet's 24 largest social media sites, and where their next wave of growth will come from. Business Insider **8** (2013)
18. Bui, T.N., Jones, C.: A heuristic for reducing fill-in in sparse matrix factorization. Technical report, Society for Industrial and Applied Mathematics (SIAM), Philadelphia, PA (United States) (1993)
19. Cheng, C.K., Wei, Y.C., et al.: An improved two-way partitioning algorithm with stable performance [VLSI]. IEEE Trans. Comput. Aided Des. Integr. Circ. Syst. **10**(12), 1502–1511 (1991)
20. Garbers, J., Promel, H.J., Steger, A.: Finding clusters in VLSI circuits. In: 1990 IEEE International Conference on Computer-Aided Design, ICCAD 1990. Digest of Technical Papers, pp. 520–523. IEEE (1990)
21. Hagen, L., Kahng, A.B.: A new approach to effective circuit clustering. In: 1992 IEEE/ACM International Conference on Computer-Aided Design, ICCAD 1992. Digest of Technical Papers, pp. 422–427. IEEE (1992)
22. Jin, S., Zhang, J., Yu, P.S., Yang, S., Li, A.: Synergistic partitioning in multiple large scale social networks. In: 2014 IEEE International Conference on Big Data (Big Data), pp. 281–290. IEEE (2014)
23. Karypis, G., Kumar, V.: Multilevel k-way partitioning scheme for irregular graphs. J. Parallel Distrib. Comput. **48**(1), 96–129 (1998)
24. Karypis, G., Kumar, V.: Parallel multilevel series k-way partitioning scheme for irregular graphs. SIAM Rev. **41**(2), 278–300 (1999)
25. Karypis, G., Kumar, V.: A fast and high quality multilevel scheme for partitioning irregular graphs. SIAM J. Sci. Comput. **20**(1), 359–392 (1998)
26. Studholme, C., Hill, D.L., Hawkes, D.J.: An overlap invariant entropy measure of 3d medical image alignment. Pattern Recogn. **32**(1), 71–86 (1999)

A General Framework for Graph Matching and Its Application in Ontology Matching

Yuda Zang[1(✉)], Jianyong Wang[1], and Xuan Zhu[2]

[1] Tsinghua University, Beijing, China
zangyd13@mails.tsinghua.edu.cn,
jianyong@tsinghua.edu.cn
[2] Samsung R&D Institute, Beijing, China
xuan.zhu@samsung.com

Abstract. Graph matching (GM) is a fundamental problem in computer science. Two issues severely limit the application of GM algorithms. (1) Due to the NP-hard nature, providing a good approximation solution for GM problem is challenging. (2) With large scale data, existing GM algorithms can only process graphs with several hundreds of nodes.

We propose a matching framework, which contains nine different objective functions for describing, constraining, and optimizing GM problems. By holistically utilizing these objective functions, we provide GM approximated solutions. Moreover, a fragmenting method for large GM problem is introduced to our framework which could increase the scalability of the GM algorithm.

The experimental results show that the proposed framework improves the accuracy when compared to other methods. The experiment for the fragmenting method unveils an innovative application of GM algorithms to ontology matching. It achieves the best performance in matching two large real-world ontologies compared to existing approaches.

Keywords: Graph matching · Optimization · Ontology matching · Algorithm

1 Introduction

Structured data has to be searched, identified and recognized in many fields such as artificial intelligence, computer vision, and knowledge engineering. It is most efficient to represent structured data using graphs, because computers can handle graphs efficiently and effectively. This prevalent way of depicting data gives rise to a lot of powerful graph based algorithms. A notable one among them is graph matching (GM). The GM problem can be described as finding the optimal correspondence between two given graphs. To make it more clear, the alignment result is the best fit for graph structures and labels of vertices according to specific criteria. The GM can be formulated as a quadratic assignment problem (QAP) known to be NP-hard [1].

There are no universally efficient methods for solving the general NP-hard GM problems. Even simple problems with as few as ten variables can be extremely challenging to solve, while problems with a few hundred variables can be utterly intractable. Existing methods tend to solve GM using approximation techniques.

© Springer International Publishing Switzerland 2016
B. Cui et al. (Eds.): WAIM 2016, Part I, LNCS 9658, pp. 365–377, 2016.
DOI: 10.1007/978-3-319-39937-9_28

This paper investigates the existing GM algorithms in detail. It presents a standard framework containing multiple object functions for describing, constraining, and optimizing GM problems and reducing the computational cost.

Applying our framework has three benefits: First, the representation is natural, succinct, and practical. Its simplicity endows the framework with the quality that it can be easily adjusted to different settings and developed to process more complex problems. Second, it generalizes and improves the state of the art Path-following (PATH) [4] algorithm and its extensions, by providing alternative objective functions to enhance the matching performance. Third, a novel fragmented form of GM greatly improves the computational scalability of GM by partitioning the graphs according to previous knowledge. We will illustrate the benefits of our GM framework in synthetic and benchmark datasets and real Ontology Matching problems.

2 Related Work

The development of GM algorithms has been explored in several surveys [1, 2]. Traditionally, three groups of approximation strategy are adopted by GM algorithms: spectral relaxation, semi-definite programming (SDP), and doubly-stochastic relaxation. Previous surveys have proved lots of approaches of GM methods to be theoretically or practically outdated. Recent research on approximation methods mostly adopts the doubly-stochastic relaxation algorithms [3], with justifiable reasons. First, the drawback of spectral relaxation methods is that the spectral embedding of graph vertices is not uniquely defined [2]. Second, although SDP methods have the theoretical guarantee [11] to solve a polynomial time relaxation problem by approximating the original problem to at least 87.9 percent. For many NP-hard problems, it is too expensive to use SDP approaches in practice, because the method squares the problem size. In this paper, we only focus on the recent progress made through doubly-stochastic relaxation.

Usually, stochastic relaxation methods adopt a two-step scheme: firstly, solving a continuously relaxed problem instead of the original discrete problem; secondly, rounding the approximate solution to a binary one, i.e. converting the global optimal solution of relaxed problems to feasible solutions of the original problem. The limitation exists because the rounding step greatly reduce accuracy because it is independent of the cost function. However, a set of PATH and its extended algorithms have been proposed to overcome these shortcomings. By converting the global optima of the relaxed problem to a feasible solution of the original problem through a carefully designed path, the algorithms greatly reduce the precision loss in the former "rounding" step. This set of algorithms leads to top quality performances on GM problems.

The original PATH [4] introduces the initial algorithm and provides a solution for undirected same-vertex-sized GM. [5] further extends the algorithm to process directed graphs and in [8], researchers finally tackled the well-defined directed partial matching problem. [6, 7] adopt a more general formulation, as was introduced by Lawler [9].

Excluding these PATH-based strategies, recent research shows heuristic approaches can be developed to solve large scale GM problems. Authors in [10] designed a greedy mechanism to optimize the GM objective by maximizing the gain of objective value at each time when a matching pair is selected. The approximate algorithm named SiGMa leverages both individual and structural information in a scalable manner which can handle million-vertex-sized GM. However, it is widely accepted that heuristic is deficient in GM area. Heuristic methods are flawed when the data and information is scattered globally, rather than accumulated around local regions. In order to get a satisfying solution on universal situations, we mainly focus on the approximation methods in this paper.

Moreover, a notable trend arises in the GM field. In addition to optimization-based work, probabilistic frameworks show a promising ability to solve traditional problems [12, 13].

3 Graph Matching

In this Sect. 3.2 we formulate GM problem and propose a set of objective functions as a result. Then we analyze and prove their equivalence to each other. Finally in Sects. 3.3 and 3.4 we present our algorithm by introducing the convex/concave relaxation of the objective functions and the mechanism of PATH.

3.1 Notation

The following symbols are used, without specific declaration:

Capital letters denote a matrix X;
Lower-case letters with indices $x_{i_1 i_2}$ denote the element in i_1^{th} row and i_2^{th} column of X;
$A \otimes B$ is the Kronecker products of A and B;
Bold lower-case letters denote a column vector $\boldsymbol{u}$;
$\mathbb{R}^*$ denotes real number $\mathbb{R} \geq 0$.

3.2 Problem Formulation

The graphs in GM may have associated labels to all its vertices and the objective is to find an alignment that fits both the labels and graph structures well at the same time.

A graph $\mathcal{G} = (V, E)$ is denoted by a finite set of vertices $V = \{v_1, \ldots, v_n\}$ and a set of directed edges $E = \{e_1, \ldots, e_m\}, e_i \in V \times V$. The edges can be equivalently presented by the weighted incidence matrix $A \in \mathbb{R}^*_{n \times n}$, where a non-zero element a_{i_1, i_2} indicates an edge from the i_1^{th} vertex to the i_2^{th} vertex of V.

A pair of graphs to be matched are given by $\mathcal{G}_1 = (V_1, E_1), |V_1| = n_1$ and $\mathcal{G}_2 = (V_2, E_2), |V_2| = n_2$ along with their weighted incidence matrices $A_1 \in \mathbb{R}^*_{n_1 \times n_1}$ and $A_2 \in \mathbb{R}^*_{n_2 \times n_2}$. In order to quantify the similarities between graph labels, a vertex affinity matrix is also given by $K_v \in \mathbb{R}^*_{n_1 \times n_2}$ with $k^v_{i_1, i_2}$ to measure the similarity between the i_1^{th} vertex of V_1 and the i_2^{th} vertex of V_2.

Vertex matching matrix $X \in \mathcal{P}_{n_1 \times n_2}$ defines the possible mapping between vertices of $\mathcal{G}_1$ and vertices of $\mathcal{G}_2$, i.e. V_1 and V_2. It is constrained to be, at most, one-to-one mapping, i.e., $\mathcal{P}_{n_1 \times n_2}$ is the set of the following matrices:

$$\mathcal{P}_{n_1 \times n_2} = \left\{ X | X \in \{0, 1\}^{n_1 \times n_2}, X 1_{n_2} = 1_{n_1}, X^T 1_{n_1} \le 1_{n_2} \right\}$$

GM is to find the optimal vertex correspondence matrix X, such that overall compatibility of the vertex and edge is maximized. The principle of measurement is to maximize the similarity between weighted edges and the weighted vertices in matched graphs, in other words, to minimize the difference between the matched graphs. The compatibility of a possible matching can be described as:

$$\min_{X \in \mathcal{P}} -\mathrm{tr}\left(K_v^T X\right) + \left\|A_1 - XA_2X^T\right\|_F^2. \tag{1}$$

$$\min_{X \in \mathcal{P}} -\mathrm{tr}\left(K_v^T X\right) + \left\|X^T A_1 X - A_2\right\|_F^2. \tag{2}$$

$$\min_{X \in \mathcal{P}} -\mathrm{tr}\left(K_v^T X\right) + \left\|A_1 X - XA_2\right\|_F^2. \tag{3}$$

$$\min_{X \in \mathcal{P}} -\mathrm{tr}\left(K_v^T X\right) + \left\|X^T A_1 - A_2 X^T\right\|_F^2. \tag{4}$$

$$\min_{X \in \mathcal{P}} -\mathrm{tr}\left(K_v^T X\right) - \left\|A_1 + XA_2X^T\right\|_F^2. \tag{5}$$

$$\min_{X \in \mathcal{P}} -\mathrm{tr}\left(K_v^T X\right) - \left\|X^T A_1 X + A_2\right\|_F^2. \tag{6}$$

$$\min_{X \in \mathcal{P}} -\mathrm{tr}\left(K_v^T X\right) - \left\|A_1 X + XA_2\right\|_F^2. \tag{7}$$

$$\min_{X \in \mathcal{P}} -\mathrm{tr}\left(K_v^T X\right) - \left\|X^T A_1 + A_2 X^T\right\|_F^2. \tag{8}$$

Firstly, $\mathrm{tr}\left(K_v^T X\right) = \sum_{i_1=1}^{n_1} \sum_{i_2=1}^{n_2} k_{i_1,i_2}^v x_{i_1,i_2}$ denotes the similarity between vertices. Secondly, $\|.\|_F$ denotes the Frobenius norm.

Previous research measured compatibility as minimizing the difference between the transformed matrix and another matrix. For example in Function (1), minimizing $\|A_1 - XA_2X^T\|_F^2$ reflects the compatibility of the matching result, because XA_2X^T is the elementary transformation product of A_2 which aims to match A_1.

We notice that another way to measure the compatibility is to maximize the concurrence of the transformed matrix and another matrix. Function (5) denotes this by $\|A_1 + XA_2X^T\|_F^2$: the larger the value, the more closely matched the two graphs. The mathematical equivalence between Functions (1) and (5) is given:

Apply Cauchy–Schwarz inequality to $\|A_1 - XA_2X^T\|_F^2$ and $\|A_1 + XA_2X^T\|_F^2$:

$$\left\|A_1 + XA_2X^T\right\|_F^2 \le \|A_1\|_F^2 + \left\|XA_2X^T\right\|_F^2 + 2\sqrt{\|A_1\|_F^2 \left\|XA_2X^T\right\|_F^2}$$

$$\left\|A_1 - XA_2X^T\right\|_F^2 \geq \left\|A_1\right\|_F^2 + \left\|XA_2X^T\right\|_F^2 - 2\sqrt{\left\|A_1\right\|_F^2\left\|XA_2X^T\right\|_F^2}$$

Notice that in the above inequities, $\left\|A_1\right\|_F^2$ and $\left\|XA_2X^T\right\|_F^2$, are considered to be constant (elementary transformation does not change the Frobenius value). As illustrated in the inequity, equilibrium is reached when A_1 and XA_2X^T are identical. In other words, Functions (1) and (5) owns the same optimization target. The same goes for any pair of the objectives.

3.3   Convex and Concave Relaxation

By adopting doubly-stochastic matrices, i.e. the convex hull of the permutation matrices $\mathcal{P}_{n\times m}$ as the domain of GM problem:

$$\mathcal{D}_{n\times m} = \left\{X|X \in [0, 1]^{n\times m}, X1_m = 1_n, X^T1_n \leq 1_m\right\}$$

The GM problem is relaxed to its corresponding continuous version. The region containing feasible solutions is included by the domain of the relaxed problem. We prove that the relaxed form of Functions (1), (2), (3) and (4) are convex and those of Functions (5), (6), (7) and (8) are concave (see Sect. 4.2), while doubly-stochastic relaxed from of Function (9) is neither a concave nor convex relaxation. Section 4.2 outlines mathematical proof where the fragmented form of objectives incorporates the original form.

3.4   Path Algorithm

The key to PATH's success lies in the leverage of the objective function. We denote the original objective function as J_{gm} which needs an integer solution. Stochastic relaxation relaxes the objective function to a convex one denoted as J_{gm_convex} and to a concave one denoted as $J_{gm\_concave}$. Note that the global minimum of J_{gm_convex} can be found using convex optimization which is a real number solution and that a local minimum of $J_{gm\_concave}$ can be got when an initial value is given which is an integer solution. Once the global optimum of J_{gm_convex} has been found, PATH gradually converts the continuous optimum to an extreme point of $J_{gm\_concave}$ which is an integer solution through the following way [4]:

A series of combined objective functions is constructed at first.

$$J_\gamma = (1 - \gamma)J_{gm\_convex} + \gamma J_{gm\_concave}$$

γ is initially set to 0, to make the combined objective function equal to J_{gm_convex}. The next step is to increase γ by a small interval to find a local minimum of $J_{\gamma + \Delta\gamma}$, initialized as the local minimum of J_γ, using the Frank-Wolfe [14] algorithm which is not limited by convex and concave functions. As intervals are added, at last $\gamma = 1$, J_γ

becomes $J_{gm\_concave}$ and an integer solution is acquired as an approximation of the original problem.

PATH and its extensions benefits from the state-of-the-art accuracy. However, it has been widely accepted that the convex and concave relaxation is critical in all the related works which is very difficult to construct [3].

In our work, the abundance of relaxed objective functions may help us to seek a satisfying solution. In Sect. 5, we describe a comparison experiment concerning all possible combination usage of objective functions.

4 Fragmenting Method

In Sect. 4.1, we propose a fragmenting method to deal with large GM problem, and in Sect. 4.2, we uniformly prove the convex/concave property of the relaxed objective functions either original or fragmented.

4.1 Method Description

Usually, the matched vertices only exist between two equivalent subsets of V_1 and V_2 partitions. A partition of V_1 as $V_1^1, V_2^1, \ldots, V_t^1$ holds the condition that $V_1^1 \cap V_2^1 \cap \ldots$ $\cap V_t^1 = \emptyset, V_1^1 \cup V_2^1 \cup \ldots \cup V_t^1 = V_1$. Previous knowledge assures the correct matching lies exactly within the subset pairs $V_1^1 \leftrightarrow V_1^2, V_2^1 \leftrightarrow V_2^2, \ldots, V_t^1 \leftrightarrow V_t^2$. In some domains such as ontology matching, the data is organized in a hierarchical manner. By partitioning the data properly, we can significantly improve the performance.

Further, we denote the size of the vertices $V_1^1, V_2^1, \ldots, V_t^1$ as $|V_1^1| = n_1^1, |V_2^1| = n_2^1, \ldots, |V_t^1| = n_t^1$, and $n_1^1 + n_2^1 + \ldots + n_t^1 = n_1$, the vertices $V_1^2, V_2^2, \ldots, V_t^2$ are denoted similarly. The correspondence matrix X is fragmented as $X_1, X_2, \ldots X_t$, satisfying $X_1 \in \mathcal{P}_{n_1^1 \times n_1^2}, X_2 \in \mathcal{P}_{n_2^1 \times n_2^2}, \ldots, X_t \in \mathcal{P}_{n_t^1 \times n_t^2}$ where X_i is the matching of V_i^1 and V_i^2.

The vertex affinity matrix K_v should be fragmented as $K_i^v (i = 1, \ldots, t; K_i^v \in \mathbb{R}_{n_i^1 \times n_i^2})$. These fragmented vertex affinity matrices only comprise vertices of corresponding partition subset pairs, ensuring lower computing costs.

In addition, the incidence matrix A_1 is fragmented as $A_{ij}^1 \in \{0, 1\}^{n_i^1 \times n_j^1} (i, j = 1, 2, \ldots, t)$. A_2 is fragmented similarly. Figure 1 exemplifies the fragmenting form of the incidence matrix. The partitions are supposed to be $\{\{a, b, c\}, \{d, e, f\}\}$ and $\{\{1, 2, 3\}, \{4, 5\}\}$.

Using this method, the objective Functions (1), (2), (3), (4), (5), (6), (7), (8) and (9) must be fragmented into the new formulations. For example, Functions (3) and (7) can be fragmented in this manner:

$$\min_{X \in \mathcal{P}} J_{gm}(X_1, X_2, \ldots X_t) = -\sum_{i=1}^{t} tr\left(K_i^{vT} X_i\right) + \sum_{i=1}^{t} \sum_{j=1}^{t} \left\| A_{ij}^1 X_j - X_i A_{ij}^2 \right\|_F^2.$$

$$(9)$$

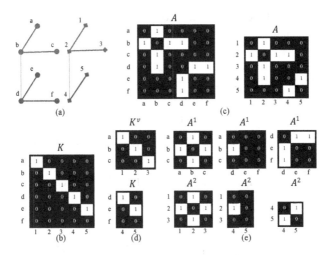

Fig. 1. Example of the fragmenting method. (a) Two graphs to be matched. (b) Vertex affinity matrix K_v. (c) The incidence matrices A_1 and A_2. (d) Fragmented vertex affinity matrices. (e) Fragmented incidence matrices.

$$\min_{X \in \mathcal{P}} J_{gm}(X_1, X_2, \ldots X_t) = -\sum_{i=1}^{t} tr\left(K_i^{vT} X_i\right) - \sum_{i=1}^{t} \sum_{j=1}^{t} \left\| A_{ij}^1 X_j + X_i A_{ij}^2 \right\|_F^2. \tag{10}$$

Functions (10) and (11) are equivalent to the original Functions (3) and (7), respectively, since all the label and structure information of the original formulations is contained, as illustrated in Fig. 1. By partitioning the vertex sets properly, we reduce the computing costs in two aspects.

(a) It fragments the vertex affinity matrix. The number of pairwise labels whose similarity need to be measured is reduced.

(b) It fragments the incidence matrices. The scale is significantly reduced.

4.2 Proof of Convex and Concave Property

In former section, we claimed the relaxed form of Functions (1), (2), (3) and (4) are convex and those of Functions (5), (6), (7) and (8) are concave. In this section, we prove that not only the original functions but also their fragmented functions have the property.

We relax the domain of Functions (10) and (11) from $\mathcal{P}$ to $\mathcal{D}$:

$$J_{vex} = \min_{X \in \mathcal{D}} J_{gm}(X_1, X_2, \ldots X_t). \tag{11}$$

$$J_{cav} = \min_{X \in \mathcal{D}} J_{gm}(X_1, X_2, \ldots X_t). \tag{12}$$

The Hessians of these fragmented forms of objective functions reveal their convex/concave nature:

$$\text{Hessian}(J_{vex}) = \sum_{i=1}^{t}\sum_{j=1}^{t}\alpha_{ij}\begin{bmatrix} I_{n_i^2 \times n_i^2} \otimes A_{ij}^{1}{}^{T}A_{ij}^{1} & -A_{ij}^{2}{}^{T} \otimes A_{ij}^{1}{}^{T} \\ -A_{ij}^{2} \otimes A_{ij}^{1} & A_{ij}^{2}A_{ij}^{2}{}^{T} \otimes I_{n_j^1 \times n_j^1} \end{bmatrix}$$

$$\text{Hessian}(J_{cav}) = -\sum_{i=1}^{t}\sum_{j=1}^{t}\alpha_{ij}\begin{bmatrix} I_{n_i^2 \times n_i^2} \otimes A_{ij}^{1}{}^{T}A_{ij}^{1} & A_{ij}^{2}{}^{T} \otimes A_{ij}^{1}{}^{T} \\ A_{ij}^{2} \otimes A_{ij}^{1} & A_{ij}^{2}A_{ij}^{2}{}^{T} \otimes I_{n_j^1 \times n_j^1} \end{bmatrix}$$

Denote $P_{ij} = I_{n_i^2 \times n_i^2} \otimes A_{ij}^{1}$, $Q_{ij} = A_{ij}^{2} \otimes I_{n_j^1 \times n_j^1}$. Ignoring the scalars, the matrices are all semi-definite, because for any vector $x^T = (x_1^T, x_2^T)$:

$$(x_1^T, x_2^T)\begin{bmatrix} P_{ij}^T P_{ij} & -P_{ij}^T Q_{ij}^T \\ -Q_{ij}P_{ij} & Q_{ij}Q_{ij}^T \end{bmatrix}\begin{pmatrix} x_1 \\ x_2 \end{pmatrix} = (P_{ij}x_1 - Q_{ij}^T x_2)^T(P_{ij}x_1 - Q_{ij}^T x_2)$$

$$(x_1^T, x_2^T)\begin{bmatrix} P_{ij}^T P_{ij} & P_{ij}^T Q_{ij}^T \\ Q_{ij}P_{ij} & Q_{ij}Q_{ij}^T \end{bmatrix}\begin{pmatrix} x_1 \\ x_2 \end{pmatrix} = \left(P_{ij}x_1 + Q_{ij}^T x_2\right)^T(P_{ij}x_1 + Q_{ij}^T x_2)$$

So Function (12) is convex and Function (13) is concave. When $t = 1$, it equals to the original definition. The proof also holds for Functions (4) and (8) which are convex and concave, respectively.

Algorithm 1 summarizes the workflow of the matching process. The computational complexity is $O(N^3)$ where N is the number of vertices because our approach integrates PATH thus it has the same complexity.

Algorithm 1: Graph matching algorithm

---

1 **input:**   graph incidence matrices A_1, A_2
                 and vertex affinity matrix K_v

2 **output:**   vertex matching matrix X

3 Fragmenting step (if necessary);

4 Relax the original problem to a convex optimization one;
  Initialize X as the global minimum of it;
  Initialize $\gamma = 0$;

5 **while** $\gamma + \Delta\gamma < 1$ **do** PATH

6 $\qquad\qquad\qquad\qquad \gamma' \leftarrow \gamma + \Delta\gamma$;

7 $\qquad\qquad\qquad\qquad$ Initialize $X' = X$;

8 $\qquad\qquad\qquad\qquad$ Find $X' = \arg\min_{X' \in \mathcal{D}} J_{\gamma'}$ by Frank

9 $\qquad\qquad\qquad\qquad \gamma \leftarrow \gamma', X \leftarrow X'$;

---

5 Experimental Study

We conduct several experiments on different datasets to evaluate several aspects of our approach. Since Functions (1), (2), (5) and (6) are biquadratic, we only use the quadratic Functions (3), (4), (7) and (8) to avoid the more complex high dimensional calculations.

Firstly we compare different methods of combining objective functions on synthetic datasets. Secondly, we compare our algorithm with other GM algorithms using the benchmark datasets. Finally, we creatively apply GM to the Ontology Matching problem and compare GM with other leading OM algorithms.

5.1 Test on Synthetic Datasets

The relaxed quadratic objective functions set is listed below:

$$\min_{X \in \mathcal{D}} -\mathrm{tr}\left(K_v^T X\right) + \|A_1 X - X A_2\|_F^2. \tag{13}$$

$$\min_{X \in \mathcal{D}} -\mathrm{tr}\left(K_v^T X\right) + \left\|X^T A_1 - A_2 X^T\right\|_F^2. \tag{14}$$

$$\min_{X \in \mathcal{D}} -\mathrm{tr}\left(K_v^T X\right) - \|A_1 X + X A_2\|_F^2. \tag{15}$$

$$\min_{X \in \mathcal{D}} -\mathrm{tr}\left(K_v^T X\right) - \left\|X^T A_1 + A_2 X^T\right\|_F^2. \tag{16}$$

On synthetic datasets we judge the effectiveness of all possible combinations for PATH algorithm. Every combination comprises both a convex and concave function. In all, there are four combinations in our framework, and we use random synthetic graphs [15] to test their performance. Datasets Syn1-Syn5 contain 150 vertices with different edge densities ranging from 0.1 to 0.5 at each interval.

The values of objective functions can be treated as evaluation criteria for approximate solutions. Note that all of the Functions (13), (14), (15) and (16) participate calculations in our algorithm. In order to objectively compare the combinations, we adopt the unused Function (9) as the evaluation objective.

Table 1 shows that the combinations (14, 16) and (15, 17) outperform the remaining combinations by providing all the best solutions. Both the combination pairs have a supplementary nature. In the following experiments, we will use both combinations in calculation. Only the best solution is selected and compared.

5.2 Test on Benchmark Datasets

There exists a QAP benchmark library[1] which is widely used for testing datasets with other GM algorithms [4–6]. Some of the datasets in the benchmark library were

---

[1] http://opt.math.tu-graz.ac.at/qaplib/.

Table 1. Performance comparison of different combination of relaxed objective functions

Data	(15, 16)	(14, 16)	(14, 17)	(15, 17)
Syn1	8.397E + 06	8.025E + 06	8.402E + 06	**7.980E + 06**
Syn2	2.291E + 07	2.197E + 07	2.278E + 07	**2.189E + 07**
Syn3	4.174E + 07	**4.061E + 07**	4.166E + 07	4.062E + 07
Syn4	6.108E + 07	**5.864E + 07**	6.053E + 07	5.875E + 07
Syn5	1.093E + 08	1.066E + 08	1.089E + 08	**1.063E + 08**

selected to compare our algorithm with the graduated assignment algorithm (GA) [16], PATH [4] and GNCCP [8].

We tested the algorithms on both symmetric and asymmetric datasets. Tables 2 and 3 illustrate the performance of the above algorithms. Here, 'GM' represents our own GM framework, OPT denotes the best known results provided by the benchmark library. All the results came from the corresponding research paper, and the OPT result was updated from the benchmark library.

Table 2. Test on symmetric benchmark datasets

Data	Opt	GA	PATH	GNCCP	GM
chr15a	9896	30370	19088	10840	**10064**
chr15c	9504	23686	16206	14890	**9504**
chr20b	2298	6290	5560	3164	**2618**
chr22b	6194	9658	8500	6918	**6808**
tai10a	135028	168096	152534	138306	**135028**
tai20a	703482	871408	753712	736710	**712134**
tai30a	1818146	2077958	1903872	1856666	**1837860**
tai40a	3139370	3668044	3281830	3180740	**3076921**

Table 3. Test on asymmetric benchmark datasets

Data	Opt	GA	PATH	GNCCP	GM
lipa20a	3683	3909	3885	3789	**3778**
lipa30a	13178	13668	13577	13459	**13390**
lipa40a	31538	32590	32247	32012	**31940**
lipa50a	62093	63730	63339	62901	**62835**
lipa60a	107218	109809	109168	108445	**108171**
lipa70a	169755	173172	172200	171421	**171180**
lipa80a	253195	258218	256601	255546	**254968**
lipa90a	360630	366743	365233	363480	**363062**

Note that in symmetric data, our approach outperforms all the baseline algorithms in all of the datasets. For the chr15c and tai10a datasets, we achieved the stated optimal solutions, and for tai40a dataset; our solution is more precise than OPT. For asymmetric datasets, our algorithm also shows outstanding performance. For all of the datasets, our algorithm achieves the best performance.

5.3   Ontology Matching

The matching of mouse and human anatomies is a critical problem. Most of the research is conducted by matching of NCI Thesaurus (human anatomy) to the Adult Mouse Anatomical Dictionary [17].

The human anatomy ontology contains 3298 hierarchical concepts, organized by "sub-class" and "part-of" structural properties. The mouse anatomy ontology, containing 2737 concepts, is similarly organized. From the perspective of data scale, this ontology matching problem is almost impossible to be solved by GM algorithms. However, the predefined hierarchy of ontology concepts, allows us to apply them to the fragmented form of GM.

We extract the top-level concepts that are incorporated by both ontologies. Since top-level anatomical concepts are quite simple, we manually aligned them to fragment them into several groups.

The concepts can be treated as nodes in GM. By lexical approaches, we quantify the labels of concepts. Specifically in our experiment, we used UMLS Terminology Services[2] to acquire synonymous names. Then we built a vertex affinity matrix K_v^T using the unit '1' to mark the anchored concepts by UMLS. Other concepts are quantified by the edit distance computation.

$$J_{gm} = \min_{X \in \mathcal{P}} -\alpha \sum\nolimits_{i=1}^{t} tr\left(K_i^{vT}X_i\right) + (1 - \alpha) \sum\nolimits_{i=1}^{t} \sum\nolimits_{j=1}^{t} \left\| A_{ij}^1 X_j - X_i A_{ij}^2 \right\|_F^2$$

There is a tuning problem for objective functions, denoted by the trade-off parameter $\alpha \in [0, 1]$. Small α highlights the matching of structures and large α highlights the matching of labels. In practice, we use $\alpha = 0.5$.

This Ontology Matching problem is formulated into the fragmented GM:

(a) Fragmenting the vertices. The aligned ten groups of concepts provide a natural partition of vertices. We use ten permutation matrices to denote the matching result:$X_1, X_2, \ldots X_{10}$.
(b) Fragmenting the vertex affinity matrix. We produce vertex affinity matrix $K_i^v (i = 1, \ldots, 10)$ for each of the 10 groups using lexical computation.
(c) Fragmenting the incidence matrix $A_{ij}^1(i, j = 1, \ldots, 10)$ and $A_{ij}^2(i, j = 1, \ldots, 10)$.

Once the matching result is acquired, for example, concept m is matched with concept n, ΔJ_{gm} is calculated as the contribution to the objective function. Finally, we ranked and acquired pairs with high contribution values as a result.

---

[2] https://uts.nlm.nih.gov/home.html#apidocumentation.

$$\Delta J_{gm} = J_{gm}(X_1, X_2, \ldots X_{10}) - J_{gm}(X_1, X_2, \ldots X_{10} | x_{mn} = 0)$$

In Table 4, we compare our fragmented GM algorithm, namely FGM, with other leading methods: AML, AOT and LogMap. The three systems show the best performances in recent OAEI evaluation [18]. The "StringEquiv" method computes identical labels.

Table 4. Performance comparison of the fragmented GM algorithm and other ontology matching algorithms on human and mouse anatomical ontologies.

Matcher	Runtime(s)	Precision	Recall	F-measure
FGM	25	0.956	**0.957**	**0.9565**
AML	28	0.956	0.932	0.944
AOT	896	0.436	0.775	0.558
LogMap	12	0.918	0.846	0.881
StringEquiv	–	**0.997**	0.622	0.766

6 Conclusion

In this paper, we proposed an approximation algorithm framework for formulating, reducing, and solving practical problems with GM methods. Among all possible combinations of objective functions, there are two supplementary pairs which achieve the best approximation performance. On benchmark datasets, our algorithm shows the best performance in comparison with other leading algorithms. Finally, we applied GM to an Ontology Matching scenario, and the experimental results confirm GM algorithms are promising to solve large scale ontology matching problems.

Acknowledgements. This work was supported in part by National Basic Research Program of China (973 Program) under Grant No. 2014CB340505, National Natural Science Foundation of China under Grant Nos. 61532010 and 61272088, and Tsinghua University Initiative Scientific Research Program.

References

1. Livi, L., Rizzi, A.: The graph matching problem. Pattern Anal. Appl. **16**(3), 253–283 (2013)
2. Conte, D., Foggia, P., Sansone, C., et al.: Thirty years of graph matching in pattern recognition. Int. J. Pattern Recognit. Artif. Intell. **18**(03), 265–298 (2004)
3. Liu, Z.Y., Qiao, H., Yang, X., et al.: Graph matching by simplified convex-concave relaxation procedure. Int. J. Comput. Vis. **109**(3), 169–186 (2014)
4. Zaslavskiy, M., Bach, F., Vert, J.P.: A path following algorithm for the graph matching problem. Pattern Anal. Mach. Intell. **31**(12), 2227–2242 (2009)
5. Liu, Z.Y., Qiao, H., Xu, L.: An extended path following algorithm for graph-matching problem. Pattern Anal. Mach. Intell. **34**(7), 1451–1456 (2012)
6. Zhou, F., De la Torre, F.: Factorized graph matching. In: CVPR 2012, pp. 127–134 (2012)

7. Zhou, F., De la Torre, F.: Deformable graph matching. In: CVPR 2013, pp. 2922–2929 (2013)
8. Liu, Z.Y., Qiao, H.: GNCCP—graduated nonconvexity and concavity procedure. Pattern Anal. Mach. Intell. **36**(6), 1258–1267 (2014)
9. Lawler, E.L.: The quadratic assignment problem. Manag. Sci. **9**(4), 586–599 (1963)
10. Lacoste-Julien, S., Palla, K., Davies, A., et al.: Sigma: simple greedy matching for aligning large knowledge bases. In: KDD 2013, pp. 572–580 (2013)
11. Goemans, M.X., Williamson, D.P.: Improved approximation algorithms for maximum cut and satisfiability problems using semidefinite programming. J. ACM **42**(6), 1115–1145 (1995)
12. Zass, R., Shashua, A.: Probabilistic graph and hypergraph matching. In: CVPR 2008, pp. 1–8 (2008)
13. Suchanek, F.M., Abiteboul, S., Senellart, P.: Paris: Probabilistic alignment of relations, instances, and schema. Proc. VLDB Endowment **5**(3), 157–168 (2011)
14. Jaggi, M.: Revisiting Frank-Wolfe: projection-free sparse convex optimization. In: ICML 2013, pp. 427–435 (2013)
15. Newman, M.E.J., Strogatz, S.H., Watts, D.J.: Random graphs with arbitrary degree distributions and their applications. Phys. Rev. E **64**(2), 026118 (2001)
16. Gold, S., Rangarajan, A.: A graduated assignment algorithm for graph matching. Pattern Anal. Mach. Intell. **18**(4), 377–388 (1996)
17. Bodenreider, O., Hayamizu, T.F., Ringwald, M., et al.: Of mice and men: aligning mouse and human anatomies. In: AMIA 2005 Annual Symposium Proceedings, p. 61 (2005)
18. Mika, P., et al. (eds.): The Semantic Web–ISWC 2014. Information Systems and Applications, incl. Internet/Web, and HCI, vol. 8796. Springer, Switzerland (2014)

Internet Traffic Analysis in a Large University Town: A Graphical and Clustering Approach

Weitao Weng[1], Kai Lei[1(✉)], Kuai Xu[2], Xiaoyou Liu[3], and Tao Sun[3]

[1] School of Electronics and Computer Engineering (SECE),
Peking University, Beijing, China
wengwt@pku.edu.cn, leik@pkusz.edu.cn
[2] School of Mathematical and Natural Sciences,
Arizona State University, Tempe, USA
kuai.xu@asu.edu
[3] The Network Information Center, Shenzhen University Town, Shenzhen, China
{liuxy,suntao}@utsz.edu.cn

Abstract. Campus networks consist of a rich diversity of end hosts including wired desktops, servers, and wireless BYOD devices such as laptops and smartphones, which are often compromised in insecure networks. Making sense of traffic behaviors of end hosts in campus networks is a daunting task due to the open nature of the network, heterogeneous devices, high mobility of end users, and a wide range of applications. To address these challenges, this paper applies a combination of graphical approaches and spectral clustering to group the Internet traffic of campus networks into distinctive traffic clusters in a divide-and-conquer manner. Specifically, we first model the data communication between a particular subnet of campus networks and the Internet on a specific application port via bipartite graphs, and subsequently use the one-mode projection to capture behavior similarity of end hosts in the same subnet for the same network applications. Finally we apply a spectral clustering algorithm to explore the behavior similarity to identify distinctive application clusters within each subnet. Our experimental results have demonstrated the benefits of our proposed method for analyzing Internet traffic of a large university town to discover anomalous behaviors and to uncover distinctive temporal and spatial traffic patterns.

Keywords: Traffic analysis · Bipartite graph · Spectral clustering · Campus network

1 Introduction

In the recent years, campus networks have witnessed tremendous growth of devices, applications and Internet traffic. Making sense of Internet traffic in

---

L. Kai—This work has been financially supported by Shenzhen General Research project No: JCYJ20150626111057728 and Key Research Project No: JCYJ2015 1014093505032, JSGG20140516162852628 and JCYJ20151030154330711.

B. Cui et al. (Eds.): WAIM 2016, Part I, LNCS 9658, pp. 378–389, 2016.
DOI: 10.1007/978-3-319-39937-9_29

campus networks remains a daunting task for network operators due to hetero-geneous devices, applications and the open nature of campus networks. Although a rich body of literature has focused on traffic analysis for Internet data centers, backbone networks, and wireless and cellular networks [1–4], little effort has been devoted to characterizing behavior patterns of end hosts and applications of campus networks. More importantly, the difficulty of managing and securing campus networks, an important component of Internet ecosystem, allows cyber attackers actively to explore and compromise end hosts and mobile devices in these networks. It becomes extremely important to develop effective techniques to understand traffic patterns and behavior dynamics of end hosts and applications in campus networks.

In this paper, we propose to use graph and cluster analysis techniques to analyze Internet traffic of a large campus network with thousands of users. Due to the wide diversity of applications and heterogeneous devices, we first employ a divide-and-conquer manner to separate Internet traffic based on subnets and applications, which allows for a more fine-grained traffic characterization. Subsequently, we apply bipartite graphs to model Internet traffic between end hosts in each subnet and the Internet on a specific application port, and obtain one-mode projections of bipartite graphs for locating behavior similarity of end hosts in each subnet. Finally, we leverage spectral clustering algorithms to group end hosts in each subnet/application into different behavior clusters based on their behavior similarity measured with Jaccard index.

Our experimental results show that distinctive traffic clusters from each subnet/application combination reveal critical insights of traffic patterns and host behaviors of campus networks such as the power-law distributions of traffic volumes, popular traditional applications and mobile apps, scanning activities and DDoS attacks. More importantly, these traffic clusters have a wide range of applications for managing and securing campus networks, since these clusters reflect common and unusual traffic behaviors of each subnet and each application. In particular, we demonstrate the applications of traffic clusters in detecting traffic anomalies, identifying compromised end hosts in the campus networks, and inferring emerging and disruptive applications. Moreover, our proposed method is generic which can be applied in other types of networks (e.g. backbones or access).

The contributions of this paper can be summarized as follows:

- We introduce a divide-and-conquer method to divide Internet traffic of a large campus network into subnets and applications, leading to fine-grained traffic characteristics.
- We apply a combination of graphical approaches and spectral clustering to model and analyze Internet traffic between campus networks and the Internet, and discover inherent traffic clusters with distinctive behavior patterns for each subnet/application.
- We demonstrate a broad range of applications of subnet/application traffic clusters for detecting traffic anomalies, locating compromised internal hosts, and identifying emerging and disruptive applications.

2   Background and Data Collections

In this section, we first briefly describe the background and motivations for characterizing behavior diversity of end hosts in campus networks and subsequently our data collection effort and the data-sets used in this study.

Campus networks typically support a rich diversity of end hosts including wired desktops and servers, mobile devices such as laptops, smartphones and tablets. Unlike other edge networks such as enterprise networks, data center networks, campus networks are open to thousands of students, faculty and staff for accessing the Internet through wired connections or wireless networks. Such a wide diversity of end hosts and applications create a variety of challenges and difficulties for characterizing traffic behaviors of campus networks.

Towards this end, this paper collects, analyzes and makes sense of Internet traffic of end hosts in a large university town. In particular, in this study we focus on Shenzhen university town (UTSZ) that hosts graduate schools of three major universities: Harbin Institute of Technology, Peking University, and Tsinghua University. Combining together, three graduate schools have a total of 1567 faculty and 7805 graduate students as of Spring 2015. To support the research and instructions, the UTSZ campus network consists of six subnets. Three subnets are assigned with Internet-routable addresses, and are mainly reserved for wired desktops and servers, while the other three subnets, configured with the private IP blocks, are for both wired desktops and wireless devices such as smartphones, tablets, and laptops.

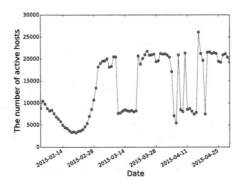

Fig. 1. The number of active hosts per day in the campus networks within three months

In this study, we collect real-time traffic flows from the border routers of UTSZ campus network which have a total Internet bandwidth of 4.4 Gbps. The router samples IP packets with a sampling ratio of 1 %, and exports 5-tuple network flows, identified by source IP address, destination IP address, source port, destination port and protocol, to our dedicated flow collection facility. Figure 1 illustrates the number of active hosts per day in the campus networks from February 4, 2015 to April 30, 2015. As shown in Fig. 1 , the number of active hosts varies at different times. The number of active hosts per day in the winter vacation is usually below 10,000, while there are about 20 thousand individual end devices in most school days, reflecting the challenges of characterizing behavior diversities of these devices and discovering anomalous traffic patterns.

3   Traffic Characteristics of University Town

In this section we first present the traffic patterns of the entire campus network and separate subnets and uncover the benefits of studying traffic for individual subnets. Subsequently, we shed light on the advantages of characterizing traffic behaviors for individual applications for each subnets.

3.1   Traffic Patterns of Campus Networks and Individual Subnets

Campus networks typically serve thousands of students, faculty and staff with very open policies, in which end users are free to use a wide diversity of applications. Figure 2(a) shows the total number of bytes for the outgoing traffic from all devices to the Internet during a 24-h time window. As the campus network consists of six different subnets, the aggregated traffic as shown in Fig. 2(a) does not tell the traffic pattern for individual subnets. Thus, we separate the outgoing traffic into six groups based on the subnets of the source IP addresses. Figure 2(b) illustrates the number of bytes for six different subnets, respectively. As shown in Fig. 2(b), the separate traffic statistics for each subnets could clearly tell the difference between each subnet, reflecting the diversity of end hosts, applications and traffic behavioral patterns of these subnets.

The subnets not only have different traffic patterns, but also have different patterns of active hosts over time. Figure 3 shows the number of unique source IP addresses which are observed from the outgoing traffic during a 24-h time window. It is interesting to see that these six subnets can be broadly classified

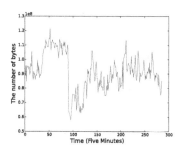
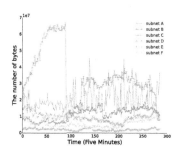

(a) The total number of bytes for the entire university town

(b) The total number of bytes for different subnets

Fig. 2. The total number of bytes for outgoing traffic from the entire university town and 6 different subnets during a 24-h time window

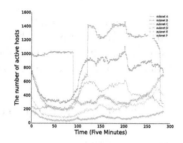

Fig. 3. The number of active hosts in six different subnets during a 24-h time window

into two groups based on similar temporal patterns: the first three subnetworks, i.e., subnets A–C, and the last three subnetworks, subnets D–F.

For the first group of subnets assigned to academic buildings, the number of unique source IP addresses grows during the working hours, but decreases over night. In contrast, the numbers of active hosts in the second group reach peak in the evening because these subnetworks are assigned to students dormitories. The aggregated counts of active hosts reveal less diurnal patterns than individual subnets shown in Fig. 3, thus suggesting the benefits of charactering traffic characteristics for individual subnets.

3.2   Traffic Patterns of Applications

The last few years have witnessed dramatical changes of Internet traffic landscape due to the rising popularity of web, video and mobile applications [2]. Our analysis on campus networks also observe a large number of applications across all subnets. We observe that the number of destination ports from the outgoing traffic from a single subnet with a private IP address block during a 5-min time window is over 2000. The subnet is not routable from the Internet, thus all communication flows are initiated from end hosts in the subnet. In other words, each destination port could potentially represent a unique Internet application.

Figure 4(a) illustrates the number of bytes for the subnet over the same time period, while Fig. 4(b) shows the same traffic statistics for the top 5 destination ports, i.e., 80/TCP, 53/UDP, 443/TCP, 8080/TCP and 8000/TCP. As shown in Fig. 4, we find that the temporal patterns of traffic statistics for individual applications reveal more insights than the aggregated traffic since the latter often mix the traffic for all applications. More importantly, a number of interesting traffic anomalies are easy to discover for individual applications, since a small deviation for a single application might be significant for the aggregated traffic. Thus, these empirical results successfully demonstrate the benefits and advantages of dividing campus traffic into individual subnets and specific applications.

In summary, our preliminary analysis of campus networks reveals that separating Internet traffic from campus networks into individual subnets and specific applications leads to fine-grained traffic characteristic and in-depth insights of behavior patterns of end hosts and mobile devices in the campus network. In

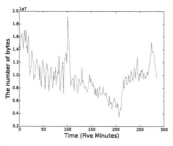 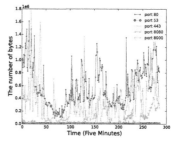

(a) The number of bytes for all appli- (b) The number of bytes for the top
cations   5 applications

Fig. 4. The total number of bytes for all applications and top 5 applications for a
private subnet during a 24-h time window

the next section, we will introduce a divide-and-conquer approach for discovering
traffic clusters from a specific subnet on a particular application with bipartite
graphs, one-mode projections and spectral clustering algorithms.

4   Spectral Clustering Analysis of Internet Traffic for Subnet/Application

In this section we describe how to apply graphical approaches and spectral clustering to discover traffic clusters for subnet/application in campus networks. The
first step of the method is to leverage bipartite graphs and one-mode projections
to model data communication between each subnet and the Internet for one
application port. Subsequently, we apply spectral clustering to divide end hosts
in each subnet into distinct traffic clusters based on their behavior similarity on
one application port.

4.1   Model Traffic Patterns with Bipartite Graphs and One-Mode Projections

The data communication between end hosts of each subnet in the university
town and end hosts of the Internet could be naturally modeled as a bipartite
graph, i.e., $B = (U, V, E)$, which has two node sets U, V and edges in E. Each
edge connects one node, e.g., u, v from each set U, V, respectively [5]. In this
study, all the hosts from the university town are assigned to the set U, while all
the hosts from the Internet form the set V. Therefore, edges in bipartite graphs
only represent traffic flows between the university town and the Internet.

In order to study the behavior similarity of outgoing traffic for end hosts in
the university town, we project the bipartite graph B onto the source nodes set
U with weights representing the Jaccard index between the neighborhoods of
the two nodes in the original bipartite graph [6]:

$$w_{u_i, u_j} = \frac{|N(u_i) \bigcap N(u_j)|}{|N(u_i) \bigcup N(u_j)|} \tag{1}$$

where $|N(u_i) \bigcup N(u_j)|$ presents the union of unique destination IP addresses of hosts u_i and u_j, and $|N(u_i) \bigcap N(u_j)|$ denotes the interaction of destination IP addresses of hosts u_i and u_j. Clearly, the weight of an edge is $0 \leq w \leq 1$. If two hosts share the same set of the destination IP addresses on the Internet, the two hosts will have a weight of 1 in the one-mode projection. In other words, the weight matrix of the one-mode projection essentially captures the similarity of outgoing traffic behavior among end hosts in the same subnet.

4.2 Spectral Clustering Algorithm for Discovering Traffic Clusters for Subnets/Applications

The weight matrix carrying behavior similarity allows us to explore cluster analysis to discover traffic clusters of end hosts in the same subnet. In this paper, we adopt a simple yet effective spectral clustering algorithm [7] to cluster end hosts into different behavior clusters. Given a similarity matrix from the weighted adjacency matrix of one-mode projection graph W, we first need to compute the diagonal matrix D whose diagonal entries consist of the sum of each $W's$ row and then find the normalized Laplacian matrix $L = D^{-1/2} W D^{-1/2}$. Next, the second step is to compute the largest k eigenvalues of the normalized Laplacian matrix L.

To find the optimal number of clusters k, we follow the same automated cluster number selection method in [8]. Compute the largest k eigenvalues $\lambda_1, \lambda_2, ..., \lambda_k$ of L until satisfying $\sum_{i=1}^{k} \lambda_i \geq \alpha \times \sum_{j=1}^{n} \lambda_j$ and $(\lambda_k - \lambda_{k+1}) \geq \beta \times (\lambda_{k-1} - \lambda_k)$. Based on our empirical observations from the data-sets used in the experiments we choose $\alpha = 0.8$ and $\beta = 2.0$. Next, we use the corresponding k eigenvectors as columns to construct a matrix and normalize each row to a unit length to obtain the matrix E, where each row of matrix E is considered as a source host of the subnet in campus network. The final step is to use K-means to extract the final partition and to assign the original $host_i$ to cluster j if and only if row i of E was assigned to cluster j.

The spectral clustering algorithm runs on the similarity matrix of one-mode projection for discovering traffic clusters for each subnet on a particular application. End hosts assigned into the same cluster share similar traffic pattern, thus the clustering results could potentially provide invaluable insights for traffic profiling and anomaly detection.

5   Exploring Subnet/Application Traffic Clusters

Dividing application traffic for each subnet into distinctive behavioral clusters is beneficial. First, it lies in the reduction of clusters from thousands of end hosts in a large scale network, which is of critical values for network operators and security analysts to manage, monitor and secure campus networks of the university town and thousands of end hosts in the network. Second, we find that the clusters often reflect distinct behavior groups in each application of one subnet and often tell interesting events. In this section we explore subnet/application traffic clusters for a broad range of applications. In particular, we will demonstrate how the traffic clusters detect anomalous traffic, uncover compromised internal hosts of campus networks, identify emerging unknown applications and facilitate traffic and performance engineering of the campus networks.

5.1 Anomaly Detection

Detecting traffic anomalies in campus networks is a challenging task due to the open nature of the network to students, faculty and staff, and the wide diversity of network applications and mobile apps. To demonstrate the benefit of our divide-and-conquer and clustering approach in anomaly detection, we use the traffic data-set from a subnet on destination port 80/TCP during one 5-min time window, in which 155 source hosts are grouped into 38 distinctive clusters. Figure 5 illustrates the log scale distribution of traffic volumes for all clusters during this time window. As shown in Fig. 5, one cluster significantly differs from any other clusters. Upon close examination, we find that this cluster consists of two hosts in the subnet talking to 15 destination hosts on the destination port 80/TCP with 61 Gb during 5 min.

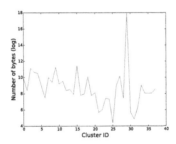

Fig. 5. The log scale distribution of traffic volumes of one subnet on destination port 80/TCP.

Our traffic clusters could also facilitate the detection of pervasive scanning activities towards campus networks or to the Internet. For example, we observe a cluster, containing one single source IP, sends data packets to hundreds of random destination IP addresses on random destination ports from two fixed source ports. These findings could be very useful for network managers to effectively detect and filter such malicious scanning activities.

5.2 Detecting Compromised End Hosts

The traffic clusters can also detect compromised end hosts in campus networks. For example, Fig. 6 illustrates the distribution of traffic volumes across all clusters of one subnet on one destination port. Clearly, cluster 4 contains a much higher traffic volume than the others. Our in-depth examination finds out that

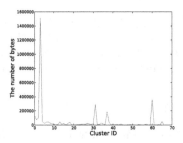

Fig. 6. The distribution of traffic volumes across all clusters of a subnet on destination port 80/TCP.

13 source hosts in this cluster send over 25,000,000 packets with random source ports to two networks on the Internet address space, indicating these source hosts are likely to be compromised end hosts, and are used by attackers to launch massive cyber attacks. Similarly, for another instance we find that all source hosts from one cluster of a subnet send a large number of TCP SYN packets to only one destination IP, accounting for over 99.25 % total traffic flows of the entire subnet during that time window. In addition, there is no response returned from the destination IP addresses, confirming the anomalous traffic behaviors of these internal hosts.

5.3   Identifying Emerging Applications

Another important application of traffic clusters in each subnet is to identify emerging and disruptive applications. Specifically, we run our proposed clustering method on the Internet traffic originating from one subnet on all the destination ports in order to group destination ports into distinct clusters based on the behavior similarity of these applications. For example, two unknown destination ports 10100/TCP and 10101/TCP always appear in the same clusters with destination ports 80/TCP and 443/TCP. We examine the top 10 destination ports with different subnets and two random chosen days, respectively. It is interesting to find that the unknown port 10100 or 10101 almost always appears in the top 10 destination ports, regardless of different subnet or time. Thus the traffic clusters provide critical insights for understanding emerging or disruptive applications and aid network operators or security analysts in gaining in-depth knowledge on such traffic.

5.4   Traffic Engineering of Campus Networks

As the users, devices, traffic and applications continue to grow in campus networks, network operators are often faced with challenges on mitigating the impact of sudden traffic spikes. Towards this end, we explore our proposed method to facilitate such tasks. Figure 7(a) shows the total bytes towards destination port 53/UDP from one subnet in the campus network over time. From

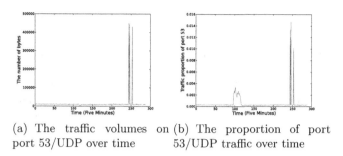

(a) The traffic volumes on port 53/UDP over time

(b) The proportion of port 53/UDP traffic over time

Fig. 7. The traffic volumes and the proportion of a subnet on port 53.

the aggregate traffic on the same subnet, it is very difficult to know which hosts or application contribute to such traffic anomalies. Figure 7(b) illustrates the proportion of destination port 53/UDP originating from the same subnet. Combined with Fig. 7(a), we can find out that the port 53/UDP contributes to sudden traffic spikes. Thus dividing Internet traffic of campus networks into each subnet and application and then clustering source hosts into smaller number of clusters allows network operators to effectively and quickly locate the root causes of traffic spikes and take corresponding actions to mitigate such events.

6   Related Work

Campus networks are an important part of Internet ecosystem due to their wide diversity of end hosts and devices as well as heterogeneous network applications. In addition, campus networks are often the early adopters of Internet innovations, thus a rich set of literature analyze traffic, applications, security and users of campus networks [9–12]. For example, [9] proposed a novel open flow architecture on campus switches which allows researchers to reuse controllers for experiments under heterogeneous environments. Similarly, [10] conducted a measurement study of YouTube traffic in a large university campus network to reveal interesting traffic footprint of YouTube video traffic, while [11] focused on UDP traffic in a campus network and found a significant growth of UDP traffic for the recent years.

Traffic characterizations and network security of campus networks have also attracted a number of research studies [13–15]. For example, [14] introduced a signal analysis methodology for classifying traffic anomalies in campus networks, and used deviation score techniques to detect anomaly traffic, while [15] applied entropy concepts from information theory to distinguish anomaly patterns from baseline distributions. Different from these prior work, our study focused on analyzing behavior similarity of end hosts in the campus networks and on exploring traffic clusters in the same subnets for the same network applications for improved security monitoring and network management in campus networks.

Graphical approaches, combined with data mining algorithms, have recently been used to model and analyze the Internet traffic behaviors and characteristics

[8,16,17]. For example, [16] adopted a K-means algorithm to classify Internet traffic with the first five packets of a data communication between two end hosts, while [17] characterized users' behaviors with bipartite graphs and clustering algorithms, and also demonstrates the practical applications of user behavioral characterization. Similar to these studies, our work also exploits bipartite graphs and one-mode projections to model data communication between end hosts, and applies spectral clustering algorithm to extract the inherent traffic clusters of each subnet for each network application in the campus network.

7   Conclusions and Future Work

Making sense of massive Internet traffic in a large scale campus network is a daunting task due to a rich diversity of end hosts and heterogeneous network applications. Given the size and complexity of hosts and applications, we explore a divide-and-conquer approach via focusing on behavior similarity of end hosts in the same subnets of campus networks for the same network applications. Subsequently, we leverage a combination of graphical approach and cluster analysis techniques to group end hosts of each subnet on each application into distinctive clusters. By applying spectral clustering algorithm, we extract distinctive clusters from each subnet, and use these clusters to discover anomaly traffic patterns and compromised end hosts in the campus network and identify emerging and disruptive applications for improved traffic engineering. Our future work lies in developing a real-time traffic anomaly detection algorithm that could handle massive traffic data and automatically detect the anomalous behavior of a given cluster.

References

1. Borgnat, P., Dewaele, G., Fukuda, K., Abry, P., Cho, K.: Seven years and one day: sketching the evolution of internet traffic. In: Proceedings of IEEE INFOCOM, pp. 711–719 (2009)
2. Labovitz, C., Iekel-Johnson, S., McPherson, D., Oberheide, J., Jahanian, F.: Internet inter-domain traffic. ACM SIGCOMM Comput. Commun. Rev. **41**(4), 75–86 (2011)
3. Kihl, M., Odling, P., Lagerstedt, C., Aurelius, A.: Traffic analysis and characterization of internet user behavior. In: Proceedings of ICUMT, pp. 224–231 (2010)
4. Xu, K., Zhang, Z.L., Bhattacharyya, S.: Profiling internet backbone traffic: behavior models and applications. ACM SIGCOMM Comput. Commun. Rev. **35**(4), 169–180 (2005)
5. Strogatz, S.H.: Exploring complex networks. Nature **410**(6825), 268–276 (2001)
6. Borgatti, S.P., Halgin, D.S.: Analyzing affiliation networks. In: Carrington, P., Scott, J. (eds.) The Sage Handbook of Social Network Analysis, pp. 417–433. Sage Publications, Thousand Oaks (2011)
7. Ng, A.Y., Jordan, M.I., Weiss, Y., et al.: On spectral clustering: analysis and an algorithm. Adv. Neural Inf. Process. Syst. **2**, 849–856 (2002)
8. Xu, K., Wang, F., Gu, L.: Behavior analysis of internet traffic via bipartite graphs and one-mode projections. IEEE/ACM Trans. Netw. **22**(3), 931–942 (2014)

9. McKeown, N., Anderson, T., Balakrishnan, H., et al.: Openflow: enabling innovation in campus networks. ACM SIGCOMM Comput. Commun. Rev. **38**(2), 69–74 (2008)

10. Zink, M., Suh, K., Gu, Y., Kurose, J.: Characteristics of youtube network traffic at a campus network-measurements, models, and implications. Comput. Netw. **53**(4), 501–514 (2009)

11. Lee, C., Lee, D.K., Moon, S.: Unmasking the growing UDP traffic in a campus network. In: Taft, N., Ricciato, F. (eds.) PAM 2012. LNCS, vol. 7192, pp. 1–10. Springer, Heidelberg (2012)

12. Henderson, T., Kotz, D., Abyzov, I.: The changing usage of a mature campus-wide wireless network. Comput. Netw. **52**(14), 2690–2712 (2008)

13. Olsen, F.: The growing vulnerability of campus networks. Chron. High. Educ. **48**(27), A35 (2002)

14. Barford, P., Kline, J., Plonka, D., Ron, A.: A signal analysis of network traffic anomalies. In: Proceedings of ACM SIGCOMM Workshop on Internet Measurment, pp. 71–82 (2002)

15. Gu, Y., McCallum, A., Towsley, D.: Detecting anomalies in network traffic using maximum entropy estimation. In: Proceedings of ACM SIGCOMM Conference on Internet Measurement, pp. 32–32 (2005)

16. Bernaille, L., Teixeira, R., Akodkenou, I., Soule, A., Salamatian, K.: Traffic classification on the fly. ACM SIGCOMM Comput. Commun. Rev. **36**(2), 23–26 (2006)

17. Qin, T., Guan, X., Wang, C., Liu, Z.: MUCM: multilevel user cluster mining based on behavior profiles for network monitoring. IEEE Syst. J. **PP**(99), 1–12 (2014)

Conceptual Sentence Embeddings

Yashen Wang[(⊠)], Heyan Huang, Chong Feng, Qiang Zhou,
and Jiahui Gu

Beijing Engineering Research Center of High Volume Language Information
Processing and Cloud Computing Applications, School of Computer,
Beijing Institute of Technology, Beijing, China
{yswang,hhy63,fengchong,qzhou,gujh}@bit.edu.cn

Abstract. Most sentence embedding models typically represent each sentence only using word surface, which makes these models indiscriminative for ubiquitous homonymy and polysemy. In order to enhance discriminativeness, we employ concept conceptualization model to assign associated concepts for each sentence in the text corpus, and learn conceptual sentence embedding (CSE). Hence, the sentence representations are more expressive than some widely-used document representation models such as latent topic models, especially for short text. In the experiments, we evaluate the CSE models on two tasks, text classification and information retrieval. The experimental results show that the proposed models outperform typical sentence embedding models.

Keywords: Sentence embedding · Conceptualization · Text representation

1 Introduction

The success of natural language processing tasks crucially depend on text representation, of which sentence representation is very important. Perhaps the most common fixed-length vector representation for texts is the bag-of-words or bag-of-n-grams [1]. However, they suffer from data sparsity and high dimensionality, and have very little sense about the semantics words or the distances between the words. This means that it could not discriminate whether word `apple` indicates a fruit or an IT company.

Recently, much the work with deep learning methods has involved learning sentence vector representations [2–6]. Despite of their usefulness, recent sentence embeddings face several challenges: (1) Most sentence embedding models represent each sentence only using word surface, which makes these models indiscriminative for ubiquitous homonymy and polysemy; (2) For short text, however, neither parsing nor topic modeling works well because there are simply not enough signals in the input. To solve the problem, we must derive more semantic signals from the input sentence, e.g., concepts.

In this paper, we proposed Conceptual Sentence Embedding (CSE), an unsupervised framework that learns continuous distributed vector representations for sentence. Specially, by innovatively introducing concept information, the semantic vector representations are learned to predict the surrounding words in contexts sampled from the sentence. Our technique is inspired by the recent work in learning vector representations of words using deep neural networks (DNN) [2, 7]. More precisely, we first

B. Cui et al. (Eds.): WAIM 2016, Part I, LNCS 9658, pp. 390–401, 2016.
DOI: 10.1007/978-3-319-39937-9_30

obtain concept distribution of the sentence, and generate corresponding concept vector. Then we concatenate the sentence vector, several word vectors with concept vector from a sentence, and predict the following word in the given context. All of the word vectors, sentence vectors and concept vectors are trained by the stochastic gradient descent and backpropagation [8]. While sentence vectors and concept vector are unique among sentences, the word vectors are shared. At prediction time, sentence vectors are inferred by fixing the word vectors, and training the new sentence vector until convergence.

In summary, the basic idea of CSE is that, we allow each word to have different embeddings under different concepts. For example, the word `apple` indicates a fruit under the concept *food*, and indicates an IT company under the concept *information technology*. Hence, concept information significantly contributes to the discriminative of sentence vector. Moreover, an important advantage of conceptual sentence vectors is that they are learned from unlabeled data. Meanwhile, it also addresses some of the key weaknesses of bag-of-words models and topic model-based DNN models (e.g., [2, 3]). Because, they enhance the semantical representation of the words with associated concepts, and could tolerate sparsity. The second advantage is that we take the word order into account, in the same way of n-gram model, while bag-of n-grams model would create a very high-dimensional representation that tends to generalize poorly.

The main contribution of this work is that, we integrate concepts into basic sentence embedding representation, and allow the resulting conceptual sentence embedding to model different meanings of a word under different concept domain. The experimental results demonstrate that this representation of semantic in sentence-level is robust. The outline of the paper is as follows. Section 2 surveys the related researches. Section 3 formally describes the proposed model of conceptual sentence embedding. Corresponding experimental results are shown in Sect. 4. Finally, we conclude the paper.

2 Related Works

Text classification and clustering play an important role in many applications, e.g., information retrieval, spam filtering. At the heart of these applications is machine learning algorithms such as logistic regression or K-means. These algorithms typically require the text input to be represented as a fixed-length vector. Conventionally, one-hot sentence representation has been widely used as the basis of bag-of-words (BOW) document model. However, it suffers from several challenges, the most critical one of which is it cannot take the semantic relationship between words into consideration.

Recently, much of the work with deep learning methods has involved learning sentence vector representations [2–6], most of which are inspired by word embedding [7]. [2] proposed the paragraph vector (PV), an unsupervised algorithm that learns fixed-length feature representations from variable-length pieces of texts. Their model represents each document by a dense vector which is trained to predict words in the document. In this paper, we based on PV to extend our models.

Aiming at enhancing discriminativeness for ubiquitous homonymy and polysemy, [3] employed latent topic models to assign topics for each word in the text corpus, and learn topical word embeddings (TWE) and sentence embeddings based on both words and their topics. Nevertheless, compared with traditional Skip-Gram, TWE models require more parameters to record more discriminative information for word representation. Besides, the window size is experientially set as 1, which may result in losing contextual information. When applying in short text (e.g., social media text or query), such topic model based algorithms usually do not perform well, because suffer extremely from the sparsity and noise. On the contrary, our models utilize text conceptualization models to discriminate sentence senses by considering all words and their contexts together, and provide a good performance on short text.

In addition, to combine deep learning with linguistic structures, many syntax-based algorithms have been proposed [9, 10] to utilize long-distance dependencies. However, for short text, syntax-based DNN faces the same problem. Since short texts usually do not observe the syntax of a written language, nor do they contain enough signals for statistical inference, we must derive more semantic signals from the input, e.g., concepts, which have been demonstrated effective in knowledge representation [11, 12].

3   Conceptual Sentence Embedding

We propose two models conceptual sentence embedding in this section. The first one is based on bag-of-word model (denoted as **CSE-1**) which have not taken word order into consideration. To overcome this drawback, we propose the extension model (denoted as **CSE-2**), which is based on Skip-gram model.

3.1   Preliminary

Firstly, we sketch previous methods for learning word vectors, which are the inspiration for our models. Continuous Bag-of-Words (CBOW) and Skip-Gram are well-known frameworks for learning word vectors [7].

In the framework of CBOW (Fig. 1(a)), each word is mapped to a unique vector, represented by a column in a word matrix $\mathbf{W}$. Then the vectors are concatenated or averaged to predict the next word in a context. Formally, Given a sentence $S = \{w_1, w_2, \ldots, w_l\}$, the objective of CBOW is to maximize the average log probability:

$$\mathcal{L}(S) = \frac{1}{l - 2k - 2} \sum_{t=k+1}^{l-k} \log \Pr(w_t | w_{t-k}, \ldots, w_{t+k}) \qquad (1)$$

Wherein, k is the contextual window size. Usually, the prediction task is done via a multiclass classifier, such as softmax as follows:

$$\Pr(w_t | w_{t-k}, \ldots, w_{t+k}) = \frac{e^{y_{w_t}}}{\sum_{w_i \in W} e^{y_{w_i}}} \qquad (2)$$

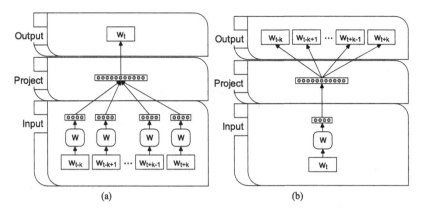

Fig. 1. (a) CBOW model and (b) Skip-Gram model.

Each of y_{w_t} is un-normalized log-probability for each output word w_t, computed as

$$y_{w_t} = \mathbf{U}h(w_{t-k}, \dots, w_{t+k}; \mathbf{W}) + b \tag{3}$$

Wherein, $\mathbf{U}$ and b are the softmax parameters. And $h(\cdot)$ is constructed by a concatenation or average of word vectors.

In contrast, Skip-gram (Fig. 1(b)) aims to predict context words given a target word in a sliding window. The vector of target word is used as features to predict the context words. Given a sentence $S = \{w_1, w_2, \dots, w_l\}$, the objective of Skip-Gram is to maximize the average log probability as follows:

$$\mathcal{L}(S) = \frac{1}{l} \sum_{t=1}^{l} \sum_{-k \le c \le k} \log \Pr(w_{t+c}|w_t) \tag{4}$$

Wherein, l is the context size of the target word w_t. Skip-Gram formulates the probability $\Pr(w_c|w_i)$ using a softmax function as follows:

$$\Pr(w_c|w_t) = \frac{e^{\overline{w_c} \cdot \overline{w_t}}}{\sum_{w_i \in W} e^{\overline{w_c} \cdot \overline{w_i}}}$$

Wherein, $\overline{w_t}$ and $\overline{w_c}$ are respectively the vector representations of target word w_t and context word w_c, and V represents the word vocabulary. In order to make the models efficient for learning, the techniques of hierarchical softmax and negative sampling are used when learning CBOW and Skip-Gram [13, 14]. In the proposed work, the structure of the hierarchical softmax is a binary Huffman tree [14], which is a good speedup trick because frequent words could be accessed quickly. Moreover, during training stage of CBOW or Skip-Gram, the word vectors are usually trained using stochastic gradient descent where the gradient is obtained via backpropagation [8].

The proposed conceptual sentence embedding model for learning sentence vector representation is inspired by the methods for learning the word vectors. The inspiration is that the word vectors are asked to contribute to a prediction task about the next word in the sentence.

3.2   CSE Based on CBOW

Following the inspiration of algorithms for word embeddings, wherein word vectors are asked to contribute to a prediction task about the next word in the sentence. Despite the fact that the word vectors are initialized randomly, they could eventually capture semantics as an indirect result of the prediction task [2]. In the proposed work, the sentence vectors are also asked to contribute to the prediction task of the next word given many contexts. In the framework of **CSE-1** (see Fig. 2(a)), there are three-layers: input layer, project layer and output layer. Wherein, words in text window $\{w_1, w_2, \ldots, w_l\}$, sentence ID and concept distribution θ_C corresponding to sentence are the inputs. Every sentence, denoted by sentence ID, is mapped to a unique vector $\bar{s}$, represented by a column in matrix $\mathbf{S}$ and every word is also mapped to a unique vector $\bar{w}_i$, represented by a column in matrix $\mathbf{W}$. The concept distribution θ_C is generated from a knowledge-based conceptualization algorithm [11], which will be described in later section. Besides, $\mathbf{C}$ is a fixed liner operator similar to the one used in [15] that converts the concept distribution θ_C to a concept vector (CV), denoted as $\bar{c}$. Note that, this is very different from the approach in [2] where no concept information is used. It is clear that **CSE-1** also does not take word order into consideration just like CBOW.

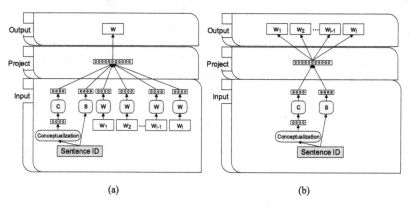

(a)   (b)

Fig. 2. CSE-1 model (a) and CSE-2 model (b). Green circles indicate word embeddings, blue circles indicate concept embeddings, and purple circles indicate sentence embeddings. Besides, orange circles indicate concept distribution generated by text conceptualization algorithm (Color figure online).

Afterward, the sentence vector $\bar{s}$, word vectors $\{\bar{w}_1, \bar{w}_2, \ldots \bar{w}_l\}$ and the concept vector $\bar{c}$ are concatenated or averaged to predict the next word in a context. Obviously, compared to the word vector framework, the only change in **CSE-1** is in Eq. 3, where $h(\cdot)$ is constructed from $\mathbf{W}$, $\mathbf{C}$ and $\mathbf{S}$. Note that, the sentence ID, which acts as a memory that remembers what is missing from the current context, could be thought of as another word. What's more, the contexts are fixed-length (length is l) and sampled from a sliding window over the sentence. The sentence vector is shared across all contexts generated from the same sentence but not across sentences. The word vector matrix $\mathbf{W}$, however, is shared across sentences [2].

In summary, **CSE-1** itself has following three key stages:

(1) **Conceptualization Stage**

For each sentence in corpus, we obtain the corresponding concepts about it by employing the idea in [11]. Following [11], we built a lexical knowledge base based on Probase[1] [16] to discover fine-grained semantic signals from the input. We organize words, concepts, and all relevant signals into a semantic graph (example shown in Fig. 3), and utilize an iterative random walk approach to determine concepts about the sentence. Given a sentence, once we have accessed the concepts and corresponding probabilities, we could obtain the concept distribution θ_C of this sentence.

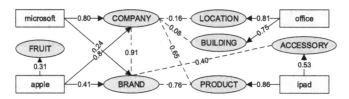

Fig. 3. Subgraph of example sentence "microsoft unveils office for apple's ipad". Ellipses indicate concept defined in Probase and rectangles indicate instances or attributes. Directed links indicate *isA* relationship between instances (attributes) and concepts (i.e., term–concept), and dashed lines indicate *correlation* relationship between two concepts (i.e., concept-concept). Moreover, the numerical values on the line is corresponding probability.

(2) **Training Stage**

In this stage, we aim at obtaining word matrix **W**, sentence matrix **S**, and softmax weights **U**, b on already observed sentences. **W** and **S** are trained using stochastic gradient descent, and the gradient is obtained via backpropagation. At every step of stochastic gradient descent, one could sample a fixed-length context from a random sentence, compute the error gradient from the network in Fig. 2, and then use the gradient to update the parameters in the proposed model.

(3) **Inference Stage**

At prediction time, we needs to perform an inference step to compute the sentence vector for a new sentence. More specially, we get sentence vectors for new sentences (unobserved before) by adding more columns in **S** and gradient descending on **S** while holding **W**, **U** and b fixed. Finally, the sentence matrix **S** could be used as features for the sentence, and we use **S** to make a prediction about some particular labels by using a standard classifier (e.g., logistic regression or K-means) in output layer.

3.3   CSE Based on Skip-Gram Model

The CBOW-based algorithm above loss information about word order. Furthermore, there exists another way for generating word vectors, which ignores the context words in the input, but force the model to predict words randomly sampled from the fix-length

---

[1] Probase data is available at http://probase.msra.cn/dataset.aspx.

contexts of sentence in the output. That is, only sentence vector $\bar{s}$ and concept vector $\bar{c}$ are used to predict the next word in a text window. The contextual words are no longer used as inputs, whereas they become what the output layer predict. Hence, this model is similar to the Skip-gram model in word embedding [14]. Typically, we sample a text window at each iteration of stochastic gradient descent, then sample a random word from the text window and form a classification task given the sentence vector. This technique is shown in Fig. 2(b). We name this version the **CSE-2**, as opposed to **CSE-1** in previous section. The scheme of **CSE-2** is similar to that of **CSE-1**.

As described before, concept distribution θ_C yields a considerable influence on conceptual sentence embedding (**CSE-1** and **CSE-2**). This is because, each dimensionality of this distribution denotes the probability of the concept (topic or category) this sentence is respect to. In other words, the concept distribution is a solid semantic representation of the sentence. Nevertheless, the information in each dimensionality of sentence (or word) vector makes no sense. Hence, there exists a linear operator in both **CSE-1** and **CSE-2**, which transmit the concept information of the sentence into word vector and sentence vector, as shown in Fig. 2.

4   Experiments and Results

To validate the performance of related models, we conduct experiments on two text understanding problems: text classification and information retrieval. The source codes and datasets of this paper are publicly available[2].

4.1   Dataset

For text classification, we use three datasets: **NewsTile**, **Twitter** and **TREC**. Dataset **Tweet11** is used for evaluation in information retrieval task.

NewsTitle: We extract news titles from a news corpus containing about one million articles searched from Web pages. Following [12], the news articles are classified into topics, and we select six topics, i.e., *company, health, entertainment, food, politician*, and *sports*, to evaluate different approaches. We randomly select 3,000 news articles in each topic, and only keep the title field. The average word count of titles is 9.41.

Tweet11: This is the tweet collections used in TREC Microblog Task 2011 and 2012 [17, 18]. Our local **Tweet11** collection has a sample of about 16 million tweets, and a set of 49 (TMB2011) and 60 (TMB2012) timestamped topics (i.e., query).

Twitter: Then we utilize previous dataset **Tweet11** to construct this dataset. By manually labeling, the dataset contains 12,456 tweets which are in five categories: *company* 3,144), *country* (2,918), *entertainment* (3,029), and *device* (3,365). The average length of the tweets is 13.16 words. Because of noise and sparsity, this dataset is challenging.

TREC: It is the corpus for question classification on TREC [19], which is widely used as benchmark in text classification task. The entire dataset of 5,952 sentences are

---

[2] http://hlipca.org/index.php/2014-12-09-02-55-58/2014-12-09-02-56-24/57-cse.

classified into the following 6 categories: *person, abbreviation, entity, description, location* and *numeric*.

4.2 Alternative Algorithms

We compare the proposed model with the following algorithms. We use the sentence vectors generated by each algorithm as features and run a linear classifier using Liblinear [20] for evaluation. We just provide the overviews of these algorithm, and the parameter settings will be described in latter section.

BOW: It represents each sentence as bag-of-words. A high weight will be given by a high word frequency in the given sentence and a low frequency of the word in the whole corpus. We compute TF-IDF scores [21] of the words in each sentence based on the given corpus and use the TF-IDF scores as features to generate sentence vector for clustering. Specially, **BOW** is employed as a basic baseline.

LDA: It represents each sentence as its inferred topic distribution [22], and the dimensions of the sentence vector of is number of topics as we presuppose.

PV: Paragraph Vector models are variable-length text embedding models proposed most recently, including the distributed memory model (**PV-DM**) and the distributed bag-of-words model (**PV-DBOW**). PV models are reported to achieve the state-of-the-art performance on sentiment classification [2].

TWE: By taking advantage of topic model, this Skip-Gram based algorithm overcomes ambiguity to some extent, and contributes to sentence embedding [3]. For **TWE** models, we learn topic models using latent dirichlet allocation [22] on the training set. It further learn topical word embeddings using the training set, then generate sentence embeddings for both training set and testing set. [3] proposed three models for topical word embedding, and we present the best results here.

4.3 Experiment Setup

The details about parameter settings of the comparative algorithms are described in this section, respectively. For **BOW** and **LDA**, It also tends to filter out common terms. For our datasets, we remove about 400 stop words (such as "the" "of" "good" etc.,) by using InQuery stopword list. For **BOW** method, we select top 50,000 words according to TF-IDF scores as features. That is, the dimensions of the sentence vector in **BOW** is 50,000. For **LDA**, in the text classification task, we set the topic number to be the cluster number or twice the cluster number, and report the better of the two. While in the information retrieval task, we experimented with a varying number of topics from 100 to 500 for **LDA**, which gives similar performance, and we report the final results of using 500 topics. For **TWE** models, we learn topic models using latent dirichlet allocation [22] on the training set by setting the number of topics as that in **LDA**. For **TWE, CSE-1** and **CSE-2**, we set the dimensions of word, topic and concept embeddings as 5,000, which is like the number of concept clusters in Probase [11, 16]. Usually, in project layer, the sentence vector and the word vectors could be averaged or concatenated to predict the next word in a context. In the experiments, we use concatenation as the method to combine the vectors.

In summary, we first generate sentence embedding based on different algorithms, and then regard sentence embedding vectors as features and train a linear classifier using Liblinear [20]. All the experiments are run on a PC with a 2.93 GHz Intel Core 2 Duo Processor and 64 GB memory.

4.4   Text Classification

Multi-class text classification is a well-studied problem in NLP. In this section, we run the experiments on the dataset **NewsTitle**, **Twitter**, and **TREC**. We report macro-averaging precision, recall and F-measure for comparison, as shown in Table 1.

Table 1. Evaluation results of multi-class text classification task.

Model	NewsTitle			Twitter			TREC		
	P	R	F	P	R	F	P	R	F
BOW	0.782	0.791	0.786	0.437	0.429	0.433	0.892	0.891	0.891
LDA	0.681	0.678	0.679	0.295	0.221	0.253	0.778	0.761	0.769
PV-DM	0.725	0.719	0.722	0.413	0.408	0.410	0.824	0.819	0.821
PV-DBOW	0.748	0.740	0.744	0.426	0.424	0.425	0.836	0.825	0.830
TWE	0.804	0.795	0.799	0.457	0.434	0.445	0.892	0.882	0.887
SCE-1	0.815	0.809	0.812	0.461	**0.449**	0.454	0.896	0.890	0.893
SCE-2	**0.827**	**0.817**	**0.822**	**0.475**	0.447	**0.462**	**0.901**	**0.895**	**0.898**

We could observe that **CSE-2** outperforms all baselines significantly. This indicates that the proposed model can capture more precise semantic information of sentence as compared to topic models and other embedding models. Because the concepts we obtained contribute significantly to the semantic representation of the sentence, meanwhile suffer slightly from text's noisy and sparsity. Moreover, as compared to the **BOW** model, the **CSE-1** and **CSE-2** models manage to reduce the feature space by 90 %.

From the results we can also see that, our proposed model **CSE-2** significantly outperforms **PV-DBOW**, the state-of-the-art model for sentence embedding. Moreover, **LDA** performs worst because it is trained on very sparse short texts, where there is no enough statistical information to infer word co-occurrence and word topics. **BOW** is a letter better, but it still underperforms the topical embedding methods (i.e., **TWE**) and conceptual embedding methods (i.e., **CSE-1** and **CSE-2**). As described in Sect. 3, **CSE-2** performs better than **CSE-1**, because the former one take the advantage of word order. In addition to being conceptually simple, **CSE-2** requires to store less data. We only need to store the softmax weights as opposed to both softmax weights and word vectors in **CSE-1**. Based on Skip-gram similarly, **CSE-2** outperforms **TWE** which is enhances sentence representation by using topic model. Obviously, neither parsing nor topic modeling works well because there are simply not enough signals in the input.

What's more, nearly all the algorithms perform better on **TREC** and **NewsTitle**, while perform badly on **Twitter**, especially **LDA** and **TWE**. This is mainly because that texts in **Twitter** are more challenging as short-texts are noisy, sparse, and ambiguous.

Certainly, we could also train **TWE** (even **LDA**) on a very large corpus, e.g., Wikipedia, and could expect much better results. However, training latent topic model on very large data set is very slow, although many fast algorithms of topic models are available [23, 24]. What's more, from the complexity analysis, we could conclude that, compared with **PV** and **CSE**, **TWE** models require more parameters to record more discriminative information for word embedding.

For the comparison of **BOW** and **LDA**, it is amazing that the simplest **BOW** achieves better performance than **LDA**. Such phenomenon indicates again that latent topic model suffer extremely from the sparsity of the short text, because the length of the text in experimental datasets is usually short (i.e., question and social media text). Moreover, the size of experimental dataset is to some extent not very large as discussed above, we guess **LDA** may achieve better performance given more data for learning. Besides, the number of topics also impacts the performance of **LDA**. In future, we may conduct more experiments to explore genuine reasons.

4.5   Information Retrieval

We also evaluate the proposed model on an information retrieval task, where the goal is to decide whether a sentence should be retrieved given a query. Specially, we mainly focus on short-text retrieval by utilizing official tweet collection **Tweet11**, which is the benchmark dataset for microblog retrieval [17, 18]. We index all tweets in this collection by using Indri toolkit[3], and then implement a general relevance-pseudo feedback algorithm. Given a query (with timestamp), we firstly obtain associated tweets, which are before query issue time, via preliminary retrieval as feedback documents. Afterwards, we generate the sentence vector of query and these feedback tweets by the alternative algorithms above. With such efforts, we could compute cosine scores between query vector and each tweet vector to measure the semantic similarity between the query and candidate tweets, and then re-rank the feedback tweets with descending cosine scores.

Experimental results for this task are shown in Table 2 using the official metric for the TREC Microblog track, i.e., Precision at 30 (P@30), and also on Mean Average Precision (MAP) for evaluating the ranking performance of different algorithms.

As shown in Table 2, the **SCE-2** significantly outperforms all these models, and exceeds the best baseline model (**TWE**) by 11.9 % in MAP and 4.5 % in P@30, which is a statistically significant improvement. As we pointed out before, such an improvement comes from the **CSE-2**'s ability to embed the contextual and semantic information of the sentences into a finite dimension vector. Just like phenomenon observed in previous experiments, topic model based algorithms (such as **LDA** and **TWE**) suffer extremely from the sparsity and noise of tweet collection. For the twitter data, since we are not able to find appropriate long texts, latent topic models are not performed.

---

[3] http://www.lemurproject.org/lemur.php.

Table 2. Evaluation results of information retrieval task.

Model	TMB2011		TMB2012	
	MAP	P@30	MAP	P@30
BOW	0.304	0.412	0.321	0.494
LDA	0.271	0.407	0.307	0.483
PV-DM	0.285	0.412	0.324	0.491
PV-DBOW	0.327	0.431	0.340	0.524
TWE	0.328	0.441	0.346	0.509
SCE-1	0.337	0.451	0.344	0.512
SCE-2	**0.367**	**0.461**	**0.350**	**0.517**

Originally, **BOW** is the most common fixed-length vector representation for texts used in information retrieval task, due to its simplicity, efficiency and often surprising accuracy. It's amazing that the simplest **BOW** provides sustainable results, even better than **LDA** and **PV-DM**. For the comparison of **BOW** and **LDA**, it is amazing that the simplest **BOW** achieves better performance than **LDA**. However, it is clear that the bag-of-words has many disadvantages. The word order is lost, and thus different sentences can have exactly the same representation, as long as the same words are used. Even though bag-of-n-grams considers the word order in short context, it suffers from data sparsity and high dimensionality. Moreover, bag-of-words have very little sense about the semantics of the words or more formally the distances between the words.

5   Conclusion

By inducing concept information, the proposed conceptual sentence embedding maintains and enhances the semantic information of word embedding, meanwhile it take into consideration the word-order as well as n-gram model. We compare it with different algorithms, including bag-of-word models, topic-model-based model and other state-of-the-art sentence embedding models. The experimental results demonstrate that the proposed method performs the best and shows improvement over the compared methods.

Acknowledgement. The work was supported by National Natural Science Foundation of China (Grant Nos. 61132009, 61201351), and National Hi-Tech Research & Development Program (863 Program, Grant No. 2015AA015404).

References

1. Harris, Z.S.: Distributional structure. Synth. Lang. Libr. **10**, 146–162 (1954)
2. Le, Q., V., Mikolov, T.: Distributed representations of sentences and documents (2014). arXiv preprint arXiv:1405.4053
3. Liu, Y., Liu, Z., Chua, T.S., et al.: Topical word embeddings. In: Twenty-Ninth AAAI Conference on Artificial Intelligence (2015)

4. Ma, M., Huang, L., Xiang, B., et al.: Dependency-based convolutional neural networks for sentence embedding (2015)
5. Palangi, H., Deng, L., Shen, Y., et al.: Deep sentence embedding using long short-term memory networks. Arxiv. **24**(4) 694–707 (2015)
6. Wieting, J., Bansal, M., Gimpel, K., et al.: Towards universal paraphrastic sentence embeddings (2015). arXiv preprint arXiv:1511.08198
7. Mikolov, T., Chen, K., Corrado, G., et al.: Efficient estimation of word representations in vector space (2013). arXiv preprint arXiv:1301.3781
8. Rumelhart, D.E., Hinton, G.E., Williams, R.J.: Learning representations by back-propagating errors. Nature **323**(6088), 533–536 (1986)
9. Wang, M., Lu, Z., Li, H., et al.: Syntax-based deep matching of short texts (2015). arXiv preprint arXiv:1503.02427
10. Severyn, A., Moschitti, A., Tsagkias, M., et al.: A syntax-aware re-ranker for microblog retrieval. In: 37th International ACM SIGIR Conference on Research & Development in Information Retrieval, pp 1067–1070. ACM Press (2014)
11. Wang, Z., Zhao, K., Wang, H., et al.: Query understanding through knowledge-based conceptualization. In: 24th International Joint Conference on Artificial Intelligence, pp. 3264–3270. AAAI Press (2015)
12. Song, Y., Wang, S., Wang, H.: Open domain short text conceptualization: a generative +descriptive modeling approach. In: 24th International Conference on Artificial Intelligence, pp. 3820–3826. AAAI Press (2015)
13. Morin, F., Bengio, Y.: Hierarchical probabilistic neural network language model. In: International Workshop on Artificial Intelligence and Statistics, pp. 246–252 (2005)
14. Mikolov, T., Sutskever, I., Chen, K., et al.: Distributed representations of words and phrases and their compositionality. In: Advances in Neural Information Processing Systems, pp. 3111–3119 (2013)
15. Huang, P.S., He, X., Gao, J., et al.: Learning deep structured semantic models for web search using clickthrough data. In: 22nd ACM International Conference on Conference on Information & Knowledge Management, pp. 2333–2338. ACM Press (2013)
16. Wu, W., Li, H., Wang, H., et al.: Probase: a probabilistic taxonomy for text understanding. In: 2012 ACM SIGMOD International Conference on Management of Data, pp. 481–492. ACM Press (2012)
17. Ounis, I., Macdonald, C., Lin, J., Soboroff, I.: Overview of the TREC-2011 microblog track. In: TREC 2011 (2011)
18. Soboroff, I., Ounis, I., Lin, J.: Overview of the TREC-2012 microblog track. In: TREC 2012 (2012)
19. Li, X., Roth, D.: Learning question classifiers. In: 19th International Conference on Computational Linguistics, pp. 1–7. Association for Computational Linguistics (2002)
20. Fan, R.E., Chang, K.W., Hsieh, C.J., et al.: LIBLINEAR: a library for large linear classification. J. Mach. Learn. Res. **9**, 1871–1874 (2008)
21. Salton, G., McGill, M.J.: Introduction to Modern Information Retrieval. McGraw-Hill, New York (1986)
22. Blei, D.M., Ng, A.Y., Jordan, M.I.: Latent dirichlet allocation. J. Mach. Learn. Res. **3**, 993–1022 (2003)
23. Smola, A., Narayanamurthy, S.: An architecture for parallel topic models. VLDB Endowment **3**(1–2), 703–710 (2010)
24. Ahmed, A., Aly, M., Gonzalez, J., et al.: Scalable inference in latent variable models. In: Fifth ACM International Conference on Web Search and Data Mining, pp. 123–132. ACM Press (2012)

Inferring Social Roles of Mobile Users Based on Communication Behaviors

Yipeng Chen[1,2], Hongyan Li[1,2(✉)], Jinbo Zhang[1,2], and Gaoshan Miao[2]

[1] Key Laboratory of Machine Perception (Peking University),
Ministry of Education, Beijing 100871, China
[2] School of Electronics Engineering and Computer Science,
Peking University, Beijing 100871, China
lihy@cis.pku.edu.cn

Abstract. Social roles of mobile users have widespread applications. However, most of users' social roles and other personal information are missing due to privacy and some other reasons, which makes it difficult to infer users' social roles precisely. Though mobile operators are lacking in information about users' social roles, they have mobile communication data which records users' communication behaviors. Since users with same social role have similar communication behaviors, it is possible to infer users' social roles based on their communication behaviors. This paper studies the problem of inferring social roles of mobile users from users' communication behaviors. A Mobile Communication Behaviors based framework (MCB) is proposed to infer social roles of mobile users. MCB solved the difficulties of inferring users' social roles with few labeled users, inaccurate label information, and few users' feature information. Our study is based on a real-world large mobile communication dataset and the experiment shows the accuracy and effectiveness of the method.

Keywords: Mobile communication network · Social role

1 Introduction

Mobile users have varieties of social roles which represent their roles played in social lives such as students, teachers and company employees. The information on users' social roles has extensive applications in many fields, especially on mobile industry. For example, student users have special consumption habits such as higher demand for Internet consumption. With information on users' social roles, mobile operators can identify student users and then promote targeted data plans to students, which will definitely enhance user experiences.

However, most users' social role information is missing due to privacy and some other reasons. The statics of International Telecommunications Union indicates that lots of people across the world use prepaid services, which allow users

This work was supported by Natural Science Foundation of China (No.61170003).

B. Cui et al. (Eds.): WAIM 2016, Part I, LNCS 9658, pp. 402–414, 2016.
DOI: 10.1007/978-3-319-39937-9_31

to be anonymous. About 95 % mobile users in India, 80 % in Latin America use this kind of services [1]. If mobile operators can infer the missing social role information precisely, based on which, they can then adjust sales strategies and further promote service level.

This paper studies the problem of inferring mobile users' social roles. Lots work have devoted to inferring users' information [1–9]. However, they can hardly adapt scenarios of inferring mobile users' social roles for several reasons:

1. **Few labeled users.** Social roles information is not mandatory in prepaid and some other services. Few users initiative offer their social role information considering privacy. The serious scarcity of social role information results in few labeled users in training set, which makes it difficult to accurately infer social roles of mobile users.
2. **Incorrectly labeled users.** People's social role information may turn into inconsistent with information they previously provided because of work change, school graduation or some other reasons. It generates incorrect label information which further diminishes the precision of previous inferring methods.
3. **Few user features.** Despite the lack of social role information, anonymous users are also lacking in other personal information such as age, interpersonal relationship, and economic background. Besides, text content of mobile communication is often encrypted according to privacy policy. All these reasons result in the lack of user features when inferring users' social roles.

Existing approaches require adequate label information and user features to maintain accuracy thus can not perform well concerning mobile communication dataset.

Mobile operators though are lacking in users' social role and feature information, they have capable fortunately of mobile communication data which records the users' communication behaviors. Based on these records, we can construct a mobile communication network similar to social network. Some research shows that Internet social network can reflect users' features [6,9]. Based on this idea, we exploit users' features from mobile communication network. A Mobile Communication Behaviors based framework (MCB) is proposed to infer social roles of mobile users with few label and feature information specifically.

2   Related Work

Traditionally, Lots of work on inferring information about users has done. They can be grouped into three categories based on: (1) methods using relational classifier. (2) methods using community detection. (3) methods using label propagation. (4) methods using probabilistic model.

Methods Using Relational Classifier. Hecht [3] used Multinomial Naive Bayes model to classify user locations from their tweets. Pennacchiotti [5] used decision trees model to infer user's labels like political orientation from profile features, linguistic content features, and social network features. These methods require text content features such as tweets, or users' personal information such as relationship and economic background which are unavailable in mobile user data. Thus, they cannot be applied to infer mobile users' social roles.

Methods Using Community Detection. These methods extracted communities from network and then decide users' attributions by voted on their communities. Backstrom [9] proposed a community detection based method to infer college and major of students. However, these methods required enough labeled users(at least 20 %) which mobile user dataset cannot provide, to achieve high accuracy. Besides, incorrectly labeled users will decrease the accuracy of these methods. Thus, they can not perform well in inferring mobile users' social roles.

Methods Using Label Propagation. These method used ties between users to propagate labels from the labeled users to the whole dataset. Zeng [7] used label propagation model to infer user affiliation based on social activities on the Internet. However, similar to methods using community detection, these methods require enough labeled users and accurate label information which mobile datasets cannot provide. Thus, they can not perform well in inferring mobile users' social roles either.

Methods Using Probabilistic Model. Dong [1] used a probabilistic model which integrated users' social properties and network features to infer users' age and from mobile social network. Li [6] proposed a unified discriminative influence model which integrated network and tweets to infer user's location. These methods rely on the distribution of particular attribute such as age, gender and location. The distribution of social roles of mobile users is different from the distributions of these attributes. Thus, existing methods that use probabilistic model cannot be directly applied to infer mobile users' social roles.

3   Problem Definition

This study aims at inferring mobile users' social roles. We note the set of users $U = U^L \cup U^N$, where $U^L(U^N)$ is the set of labeled(unlabeled) users respectively. $S = \{s_1, ..., s_a\}$ represents the set of a kinds of social roles. For any user u in U, $s^u \in S$ represents the social role of u. $K = \{k_1, ..., k_b\}$ is the set of b kinds of communication behaviors. The problem can be defined as follow: given a partially labeled(less than 20 %) mobile user set U, social role set S, and communication behavior set K, infer the social roles of all users in U^N. We propose a *Mobile Communication Behaviors based framework* (MCB) to solve this problem.

Firstly, MCB constructs a mobile communication network based on communication records. MCB chooses some network characteristics which can effectively distinguish different types of users as structure features according to the analysis of real-world datasets. Then MCB combines label propagation and homophily theory to get connection features which reflect users' relationship with different user groups. Next, MCB constructs feature vectors combining structure features and connection features. The classifier is trained with feature vectors of all users and label information of few users. Then MCB uses classifier to infer users' social roles and updates the dataset according to preliminary inferring results to improve connection features. Labeled users whose inferring results differ significantly from reality are removed from the training set. And unlabeled users whose inferring results have high confidence levels are added to training set. After updating training set, MCB repeats feature extraction, classifier training, social role inferring until most users are filtered or transformed into labeled users.

In MCB, structure features is irrelevant to users' label information, and connection features only require a little of label information. By that, MCB can perform well with few labeled users and users' feature information. Besides, users' connection features will be updated on every round, the incorrectly labeled users will be polished gradually which makes MCB can handle the scenarios where incorrectly labeled users exist.

4 Network Construction and Feature Extraction

As discussed in Sect. 1, few features can be directly extracted. Thus, we instead construct a network based on mobile communication records and explore the connection between network features and users' social roles.

4.1 Network Construction

Let $G = (U, E)$ denotes the undirected and weighted mobile communication networks, where U is a set of user nodes as defined in Sect. 3. Each $u_i \in U$ represents a user. E is a set of communication edges between users. If u_i contacts u_j, then an edge $e_{ij} \in E$ exists. Most mobile communication datasets record users' calling behaviors K_{CALL} and texting behaviors K_{SMS}. Each $m_{ij} \in K_{SMS}$ represents texting frequency of u_i towards u_j, $c_{ij} \in K_{CALL}$ represents calling frequency of u_i towards u_j, $d_{ij} \in K_{CALL}$ represents the calling duration of u_i towards u_j.

MCB extracts users' features from the mobile communication network in two ways: on one hand, network structure features of individuals which will not be affected by users' label information. On the other hand, connection features which reflect users' relationship with users of different social roles.

4.2 Structure Feature

Mobile communication network is an abstract of communication behaviors. Users' network structure can reflect their behaviors in real world. Degree centrality, average neighbor degree, local clustering coefficient, and embeddedness are the most common network characteristics, which show different aspects of network properties [1]. We analyze what these characteristics implies in mobile communication network.

Degree Centrality (DC). Degree centrality is the number of ties a node has [1]. On mobile communication network, it represents frequency of users' communication behaviors. For example, on CALL network, it represents calling frequency. Users with higher DC are more active than users with lower DC.

Dong [11] found that users of various age have different DC distributions. Similarly, users of various social roles have different DC distributions. Figure 1(a)(e) show the DC distributions of mobile users. It can be clearly observed that the curve for students shifts right compared with other users which means student users have higher DC than other users. This suggests that students contact with people more frequently than others. We also analyze the distribution of weighted degree centrality(use texting frequency, calling frequency, and calling duration as weight respectively). The distributions are similar to distribution of DC.

Average Neighbor Degree (AND). Average neighbor degree is The average degree of neighbors of a particular node, it reflects the connectivity of the node [8]. On mobile communication network, it shows how easily a users can be contacted. AND indicates users' communication activeness from another perspective. Users with higher AND are more easily reached and more active than users with lower AND.

Zhao [8] found that Internet users of various social roles have different AND distributions. Similarly, this is also the case on mobile network. Figure 1(b)(f) show the AND distributions of mobile users. We can observe from both figures that the curve for students shifts right compared with other users, which means student users have higher AND than other users. This suggests that student users can be more easily contacted than other users, which means mobile users of various social roles have different AND distributions.

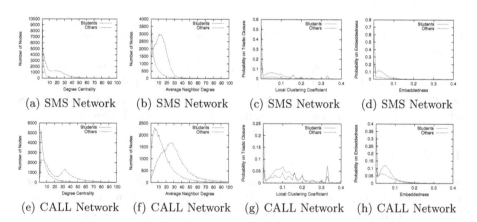

(a) SMS Network   (b) SMS Network   (c) SMS Network   (d) SMS Network

(e) CALL Network   (f) CALL Network   (g) CALL Network   (h) CALL Network

Fig. 1. Distribution of network characteristics. X-axis: number of nodes. Y-axis: value of specific network characteristic. The results are based on a real-world mobile user dataset which contains 77,577 student users and 77,577 other users. SMS and CALL networks are constructed based on users' texting and calling behaviors respectively.

Triadic Closure. Triadic closure is a property among three node A, B and C, such that if a strong tie exists between A-B and A-C, there is a weak or strong tie between B-C [12]. It can reflect the strength and density of social ties. Local clustering coefficient (LCC) can measure triadic closure [10]. LCC is defined as:

$$LCC_i = \frac{2 \cdot |\{e_{jk} : j, k \in N_{u_i}\}|}{|N_{u_i} \cdot (|N_{u_i} - 1|)|} \tag{1}$$

where u_i is a given user node, N_{u_i} is the set of u_i's neighbors, e_{jk} is the edge connects u_j and u_k.

On mobile communication network, LCC shows the phenomenon that the contacts of a user may contact with each other as well. Users with higher LCC have a more intensive and closer contact network.

Dong [1] showed that mobile users of different age and gender have different LCC distributions. Similar results can be found in mobile users of different social roles. Figure 1(c)(g) show the distribution of local clustering coefficient of mobile users. It shows that the curve for students shifts right compared with other users in both figures. This means that students have relative dense social ties compared with others. One thing to note that most users' LCC distributed on zero, especially on SMS network. Thus, LCC is not as efficient as DC and AND in distinguishing different types of users.

Embeddedness (EMB). Embeddedness measures the degree that nodes are enmeshed in networks [2]. EMB is defined as:

$$EMB_{u_i} = \frac{1}{|N_{u_i}|} \sum_{u_j \in N_{u_i}} \frac{|N_{u_i} \cap N_{u_j}|}{|N_{u_i} \cup N_{u_j}|} \tag{2}$$

where u_i is a given user node, N_{u_i} is the set of u_i's neighbors.

On mobile communication network, EMB shows the phenomenon that contact networks of users may be overlapping, it can reflect the de density of users' contact networks.

Dong [1] indicated that mobile users of different age and gender have different EMB distributions. Similar results can be found in mobile users of different social roles. Figure 1(d)(h) shows the distribution of embeddedness of mobile users. Figure 1(d)(h) shows that students tend to have a relative higher embeddedness score than others, which means students have relative stronger contact networks compared with others. Similar to LCC, most users' EMB distributed on zero. Thus EMB is not as effective as DC and AND on distinguishing different types of users.

In summary, network structure features can reveal the different communication activeness as well as contact networks on different social roles. Structure features exploit structure information of mobile communication network and don't need any label information of users. Structure features, especially degree centrality and average neighbor degree of individuals can reflect users' social roles. Most users' LCC and EMB of SMS network are distributed on zero, so we only use the LCC and EMB on CALL network, DC, and AND on both networks as users' structure features.

4.3 Connection Feature

Connection feature is designed to reflect users' relationship with different user groups. Unlike structure feature, connection feature can be affected by other users' social roles. One possible way to design connection feature is to use network characteristics related to nodes' labels such as the number of ties with some particular users. However, as discussed in Sect. 1, there are few labeled users on mobile dataset which makes this kind of connection feature can not actually reflect users' relationship with other users. For example, most users' ties with some particular users are zero because most users' neighbors are unlabeled in this situation.

To solve this problem, we use label propagation and homophily theory to extract connection feature. To began with, we discover the homophily on social role in mobile communication network. Homophily refers that people tend to be connected with those who are similar to them [13]. Birds of a feather flock together. Homophily has been extensively studied and confirmed in Internet social network and mobile social network. Zhao [8] discovered the homophily pattern on social roles and statuses in Internet social network. Dong [1] found the homophily on both gender and age on mobile social network.

We study the homophily on social role in mobile communication network. Figure 2(a) shows that the student users call more frequently than call other kinds of users. This implies that users are more likely to communicate with users with the same social role. Same results can be observed from Fig. 2(b)(c) in the aspects of texting frequency and calling duration. This suggests that people tend to have ties with those who have same social role.

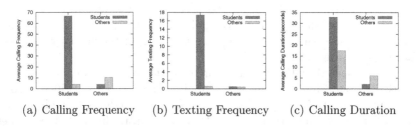

(a) Calling Frequency   (b) Texting Frequency   (c) Calling Duration

Fig. 2. Homophily on affiliation. X-axis: average calling frequency, average texting frequency, and average calling duration respectively. Y-axis: social roles of users. The results are based on a real-world mobile user dataset which contains 77,577 student users and 77,577 other users. SMS and CALL networks are constructed based on users' texting and calling behaviors respectively.

The observations above suggest that closer data points tend to have similar social roles. Thus, label propagation can be adopted here to propagate labels through dense unlabeled data regions, the value measures the probability a node belongs to a class [14]. However, methods which directly applied label propagation perform poor on this scenario for limited labeled users in dataset. Thus, we modified the label propagation in three ways: Firstly, we propagate labels for every class, and users' labels will not be decided during the propagation. Secondly, we get propagation values for every user as their connection features, while traditional label propagation methods directly assign label with the highest propagated value for every user. Lastly, users' values will propagate to itself with certain probability during propagation. Thus users' label value can be affected by its previous value and initial label information. The propagation value is computed as follow:

$$P_u^t = dP_u^{t-1} + (1-d) \sum_{v \in B(u)} w_{vu} P_v^{t-1} \qquad (3)$$

where P_u^t is the label propagation value of user u in step t, $B(u)$ is the set of users which point to user u, $0 < d < 1$, d is the probability of staying put. w_{ij} is the weight of e_{ij}. w_{ij} have different computing methods for different communication behaviors. For calling and texting behaviors, w_{ij} is computed as follow:

$$w_{ij}^m = \frac{m_{ij}}{\sum_{u_k \in U} m_{ik}}, w_{ij}^c = \frac{c_{ij}}{\sum_{u_k \in U} c_{ik}}, w_{ij}^d = \frac{d_{ij}}{\sum_{u_k \in U} d_{ik}} \qquad (4)$$

m_{ij}, c_{ij}, d_{ij} represent u_i's texting frequency, calling frequency, total calling duration towards u_j.

P_u^0 is the initial value of user, it is decided by users' label information. For social role $s_i \in S$, we set $P_u^0 = 1$ for users whose $s_u = s_i$. For other users(including unlabeled users), we set $P_u^0 = 0$. And after performing modified label propagation based on three kinds of weight, we can get three values(referred as sf_i, cf_i, cd_i). These values can reflect users' relationship toward labeled users whose social role are s_i.

Connection feature shows users' relationship with different user groups. We note that although we design the connection features based on homophily, connection features can capture not only relationships with users of same social role but also relationships with different user groups. In other words, connection features can reflect cross-social-roles relationship as well. By label propagation, most users can get a propagation value as their connection features. By removing incorrectly labeled users and adding labeled users, connection features can reflect users' relationship more precisely. Besides, from the formula of connection features, we can find that connection features take into consideration not only social roles of neighbors but also social roles of indirectly connected users. Thus connection features can reflect users' relationship with users of different social roles more thoroughly.

5   Inferring Social Roles

5.1   Feature Vector Construction

Firstly, MCB constructs feature vectors for every user based on their structure features and connection features extracted from mobile communication network.

Let dc, nd, lcc, emb represent users' degree centrality, average neighbor degree, local clustering coefficient, and embeddedness respectively. Suffix "sms" represents the feature is extracted from SMS network, suffix "call" represents the feature is extracted from CALL network. We use $(dc_{sms}, dc_{call}, nd_{sms}, nd_{call}, lcc_{call}, emb_{call})$ as users' structure features according to the analysis in Sect. 4.2. For social role $s_i \in S$, we calculate and denote (sf_i, cf_i, cd_i) as corresponding connection features according to the discussion in Sect. 4.3. Thus, for every $u_i \in U$, we can construct its feature vector as:

$$x_{u_i} = (dc_{sms}^{u_i}, dc_{call}^{u_i}, nd_{sms}^{u_i}, nd_{call}^{u_i}, lcc_{call}^{u_i}, emb_{call}^{u_i}, sf_1^{u_i}, cf_1^{u_i}, cd_1^{u_i}, ..., sf_m^{u_i}, cf_m^{u_i}, cd_m^{u_i})$$
(5)

5.2 Classifier Training

Logistic regression model is most widely used classifier model, we use logistic regression model for two reasons: First, logistic regression model is an effective classifier model, and we use parallelized trust region Newton method to compute its regression coefficients which makes it much faster and adapt large-scale mobile communication data. Second, logistic regression model can compute an estimated probability of users' belonging to given type. Based on estimated probability, we can further update the labeled users and improve the connection features of users. According to logistic regression model, the probability that $u_i \in U$ belongs to a given social role group $s_i \in S$ is computed as:

$$P(s^{u_i} = s_i|x_i) = \pi(x_i) = \frac{1}{1 + e^{-g(x_i)}}$$
(6)

where $g(x_i) = \beta \cdot x_i$, β is the regression coefficient, and x_i is feature vector of u_i[15].

Trust Region Newton Method (TRON) [15] is applied here to effectively compute the maximum likelihood of regression coefficient, which satisfies $\nabla f(\beta) = 0$, where $f(\beta)$ follows:

$$f(\beta) = \sum_{i=1}^{n} \{y_i \ln[(\pi(x_i))] + (1 - y_i) \ln[1 - \pi(x_i)]\}$$
(7)

where $y_i = 1$ if $s_{u_i} = s_i$ and $y_i = 0$ otherwise. $\{(x_i, y_i)\}, 1 \leq i \leq n$ is the training set.

5.3 Inferring and Updating Dataset

After training classifier, we can infer social roles of every user. As discussed in Sect. 1, there is some inaccurate social role information in the dataset which will result in bias to the classifier. And as mentioned in Sect. 4.3, updating dataset can improve users' connection features. Thus instead of outputting users' social roles directly, we update the dataset based on elementary computational results and repeat the whole process.

6   Experiments

6.1   Data Sets

Our data is extracted from a real-world mobile communication records of an anonymous school base in China which spans 3 months. It consists of 1.5 million users and 14 million calling records and 7.7 million text messaging records. Each record contains IDs of calling and called part, frequency of communication behaviors, and duration of communication behaviors. There are three kinds of users in the dataset, including 77 thousand students users, 77 thousand users who are not students, and other users whose social roles are unknown.

6.2   Baseline

To demonstrate the effectiveness of MCB, we compare our method against a number of baselines:

KNN. As discussed in Sect. 4.3, there is homophily on users' social roles. We can infer users' social roles based on the information of their neighbors. Thus KNN can be used as a baseline.

Community Detection. Community detection is often used to infer users' information. We use the method of [16] to infer users' social roles.

Label Diffusion. Label Diffusion is often used to infer users' information. We use the method of [17] to infer users' social roles.

MCB-dc. Connection features are removed from MCB to evaluate the effects of structure features.

MCB-ds. Structure features are removed from MCB to evaluate the effects of connection features.

6.3   Experimental Result

Inferring Performance. We conduct 10 fold cross validation on each method. The averaged precision, recall, and F-score results are illustrated in Fig. 3.

From Fig. 3, we can observe that MCB method outperforms other methods on all the measures. MCB improves the results by $0.1 \sim 0.2$ compared with other methods and MCB can infer 88 % of users' social roles, which shows the effectiveness of MCB.

We further analyze features' contributions for inferring by removing structure features and connection features respectively. We can see that the performance when removing connection features drops more than when removing structure features, which indicates a stronger contribution of connection features.

Sensitivity Analysis Results. As introduced in Sect. 1, there are very few labeled users in dataset, it is valuable to test the effectiveness of the proposed method over different setting of labeled users. Therefore, we test the performance of MCB and baselines by varying the fractions that users are labeled in mobile network.

From Fig. 4, we can see that as the fraction of labeled users declines, the accuracy of baseline methods falls significantly because they need enough training samples to ensure accuracy. By contrast, the accuracy of MCB doesn't change much because MCB takes use of structure features, which are irrelevant to users' label information, and connection features, which require few labeled users. It shows that even when only 0.001 of users are labeled, MCB can achieve about 0.88 precision. It shows that the MCB can effectively infer users' social roles even under few labeled users setting.

Robustness Analysis Results. As introduced in Sect. 1, there are incorrectly labeled users in mobile dataset, To verify the robustness of MCB, we transform part of labeled users into incorrectly labeled users and then test the performance of all methods. From Fig. 5, we can observe that as the fraction of incorrectly labeled users increases, the accuracy of MCB falls little than other methods mainly because MCB can find out and remove incorrectly labeled users and iteratively improve its connection features It shows that MCB can handle the situation where incorrectly labeled users exist.

Attribute Evaluator. To analyze how much each factor influences the inferring result. We compute the information gain of each attribute.

From Table. 1, we can see that connection features have higher information gain than structure features. This is in accordance with the results that MCB-ds has higher F-scores than MCB-dc. The most significant attribute of network characteristic is degree centrality. And the average neighbor degree also has relative high information gain than local clustering coefficient and embeddedness. This is line with our observations on the distribution of these network characteristics in Sect. 4.2.

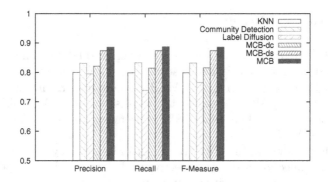

Fig. 3. Inferring performance.

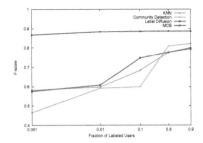 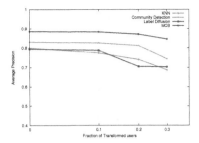

Fig. 4. Sensitivity analysis results.   **Fig. 5.** Robustness analysis results.

Table 1. Information gain of each feature. $cd_1(cd_2)$, $cf_1(cf_2)$, $sf_1(sf_2)$ are connection features calculated by propagating from student(other) users

Attribute	Information gain	Attribute	Information gain
cd_1	0.4884	dc_{sms}	0.4255
cf_1	0.481	nd_{sms}	0.3877
sf_2	0.4534	dc_{call}	0.1709
sf_1	0.442	lcc_{call}	0.1537
cd_2	0.4287	nd_{call}	0.1345
cf_2	0.4262	emb_{call}	0.0714

7 Conclusion

In this paper, we construct a mobile communication network based on real-world large mobile communication data and analyze which factors can reflect users' social roles. Then we extract from mobile communication network the connection feature and structure feature of users, which reflects users' social relationships and network structure properties respectively. Furthermore, we provided a method to infer users' social roles. We evaluated our method by comparing with some baselines and testing sensitivity and robustness under different setting of labeled users and incorrectly labeled users. The experimental results show that our method is effective, and can adapt to the real-world situation where users' label information and feature information are severe lack of and even incorrect.

In the future, we plan to infer other attributions from mobile communication records and analyze their relation to users' social roles.

References

1. Dong, Y., Yang, Y., Tang, J., Yang, Y., Chawla, N.V.: Inferring user demographics and social strategies in mobile social networks. In: Proceedings of the 20th ACM SIGKDD International Conference on Knowledge Discovery and Data Mining, pp. 15–24. ACM (2014)

2. De Choudhury, M., Mason, W.A., Hofman, J.M., Watts, D.J.: Inferring relevant social networks from interpersonal communication. In: Proceedings of the 19th International Conference on World Wide Web, pp. 301–310. ACM (2010)

3. Hecht, B., Hong, L., Suh, B., Chi, E.H.: Tweets from Justin Bieber's heart: the dynamics of the location field in user profiles. In: Proceedings of the SIGCHI Conference on Human Factors in Computing Systems, pp. 237–246. ACM (2011)

4. Rao, D., Yarowsky, D., Shreevats, A., Gupta, M.: Classifying latent user attributes in twitter. In: Proceedings of the 2nd International Workshop on Search and Mining User-Generated Contents, pp. 37–44. ACM (2010)

5. Pennacchiotti, M., Popescu, A.M.: Democrats, republicans and starbucks afficionados: user classification in twitter. In: Proceedings of the 17th ACM SIGKDD International Conference on Knowledge Discovery and Data Mining, pp. 430–438. ACM (2011)

6. Li, R., Wang, S., Deng, H., Wang, R., Chang, K.C.C.: Towards social user profiling: unified and discriminative influence model for inferring home locations. In: Proceedings of the 18th ACM SIGKDD International Conference on Knowledge Discovery and Data Mining, pp. 1023–1031. ACM (2012)

7. Zeng, G., Luo, P., Chen, E., Wang, M.: From social user activities to people affiliation. In: 2013 IEEE 13th International Conference on Data Mining (ICDM), pp. 1277–1282. IEEE (2013)

8. Zhao, Y., Wang, G., Yu, P.S., Liu, S., Zhang, S.: Inferring social roles and statuses in social networks. In: Proceedings of the 19th ACM SIGKDD International Conference on Knowledge Discovery and Data Mining, pp. 695–703. ACM (2013)

9. Mislove, A., Viswanath, B., Gummadi, K.P., Druschel, P.: You are who you know: inferring user profiles in online social networks. In: Proceedings of the Third ACM International Conference on Web Search and Data Mining, pp. 251–260. ACM (2010)

10. Kossinets, G., Watts, D.J.: Empirical analysis of an evolving social network. Science 311(5757), 88–90 (2006)

11. Dong, Z.B., Song, G.J., Xie, K.Q., Wang, J.Y.: An experimental study of large-scale mobile social network. In: Proceedings of the 18th International Conference on World Wide Web, pp. 1175–1176. ACM (2009)

12. Watts, D.J.: Six Degrees: The Science of a Connected Age. WW Norton & Company, New York (2004)

13. McPherson, M., Smith-Lovin, L., Cook, J.M.: Birds of a feather: homophily in social networks. Ann. Rev. Sociol. 27(1), 415–444 (2001)

14. Zhu, X., Ghahramani, Z.: Learning from labeled and unlabeled data with label propagation. Technical report, Citeseer (2002)

15. Lin, C.J., Weng, R.C., Keerthi, S.S.: Trust region Newton method for logistic regression. J. Mach. Learn. Res. 9, 627–650 (2008)

16. Raghavan, U.N., Albert, R., Kumara, S.: Near linear time algorithm to detect community structures in large-scale networks. Phys. Rev. E 76(3), 036106 (2007)

17. Zhou, D., Bousquet, O., Lal, T.N., Weston, J., Schölkopf, B.: Learning with local and global consistency. Adv. Neural Inf. Process. Syst. 16(16), 321–328 (2004)

Sparse Topical Coding with Sparse Groups

Min Peng, Qianqian Xie[✉], Jiajia Huang, Jiahui Zhu,
Shuang Ouyang, Jimin Huang, and Gang Tian

School of Computer, Wuhan University, Wuhan, China
{pengm,xieq,huangjj,zhujiahui,ouyangshuang,
huangjimin,tiang2008}@whu.edu.cn

Abstract. Learning a latent semantic representing from a large number of short text corpora makes a profound practical significance in research and engineering. However, it is difficult to use standard topic models in microblogging environments since microblogs have short length, large amount, snarled noise and irregular modality characters, which prevent topic models from using full information of microblogs. In this paper, we propose a novel non-probabilistic topic model called *sparse topical coding with sparse groups* (STCSG), which is capable of discovering sparse latent semantic representations of large short text corpora. STCSG relaxes the normalization constraint of the inferred representations with sparse group lasso, a sparsity-inducing regularizer, which is convenient to directly control the sparsity of document, topic and word codes. Furthermore, the relaxed non-probabilistic STCSG can be effectively learned with *alternating direction method of multipliers* (ADMM). Our experimental results on Twitter dataset demonstrate that STCSG performs well in finding meaningful latent representations of short documents. Therefore, it can substantially improve the accuracy and efficiency of document classification.

Keywords: Document representation · Topic model · Sparse coding · Sparse group lasso

1 Introduction

In social media, short texts have been an important form of information carrier, which are generated in large quantities at all times. Discovering latent semantic representing of short text corpora is an effective means to acquire valuable information from overwhelming amount of noisy short texts. Topic models, a kind of statistical models for analyzing a great quantity of texts, have received increasing attention in recent years.

This research is supported by the Natural Science Foundation of China No.61472291, and the Natural Science Foundation of Hubei Province No. ZRY2014000901.

B. Cui et al. (Eds.): WAIM 2016, Part I, LNCS 9658, pp. 415–426, 2016.
DOI: 10.1007/978-3-319-39937-9_32

LDA-based probabilistic topic models (PTMs) [1,2,4], are based on the thought that documents are mixtures of topics, where a topic is a probability distribution over words. These LDA-based probabilistic topic models have demonstrated great successes on long documents, but not on short texts. Because short and sparse texts do not have abundant co-occurrence information that can be utilized by probabilistic topic models for precisely learning the proportions.

Standard topic models have trouble learning a sparse representation in microblogging environments. There are two main strategies to deal with this sparsity problem. First is to use a sparse priorinto LDA-based models, which can indirectly control a sparsity bias over the posterior representations [14–17]. These models usually fail to achieve sparse posterior representations in a real sense, because the sparse prior results in an worse effect on smoothing. The second method is based on non-probabilistic coding, in which coding is used to represent the coefficients of corresponding topical basis in topic space. These models drop out solutions to the sparsity problem by imposing various sparsity constraints like l_1 norm [9,11,12]. However, there are still some challenges to solve the sparsity problem based on all of the above methods. That is, it is difficult to effectively yield truly fully sparse posterior representations, which significantly decrease consuming in memory and time in some large scale of text mining tasks.

In our work, to effectively overcome the challenge described above, we present the sparse topical coding with sparse groups (STCSG), which can conclusively learn sparse latent document, topic and word representations. In our model: (1) by imposing sparse group lasso regularizers, STCSG can directly control and achieve fully sparsity of inferred representations, (2) inferring problems of STCSG can be efficiently solved by ADMM [5] algorithm. The main contributions in this paper are listed as follow:

1. We design a novel non-probabilistic topic model STCSG. In this model, document, topic and word-level sparsity can all be directly controlled with sparse group lasso like NPMs, and STCSG is able to derive document - topic proportions like PTMs.
2. We incorporate the ADMM to handle the biconvex learning problems, which results in closed-form iteration patterns.
3. Experimental results on the Twitters dataset show that STCSG yields higher word code sparsity, topic code sparsity and document code sparsity than other baselines. Therefore, STCSG achieves higher classification accuracy and time efficiency.

The rest of this paper is organized as follows. Section 2 introduces the related work. Section 3 presents the details of the proposed STCSG. Section 4 shows the experimental results. Finally, we draw our conclusions in Sect. 5.

2   Related Work

2.1   Probabilistic Topic Models

Typical probabilistic topic models like probabilistic latent semantic analysis (PLSA) [18] and latent Dirichlet allocation (LDA) [1] have been widely used in text mining. These topic models do not require any prior annotations or labeling of documents. On the basis of this fundamental works, DTM [2], TOT [4] and OLDA [3] were proposed with considering time factor into LDA to deal with large scale of texts and analyze the topic evolution over time. In order to solve the sparsity problem further, Wang et al. [14] proposed sparseTM, which decouples sparsity and smoothness in the component distributions with selecting variables to determine which terms could appear in the topic. However, all of these models lack the ability of directly controlling the posterior sparsity of learned representations.

2.2 Non-probabilistic Topic Models

Contrary to the probabilistic topic models, the non-probabilistic topic models (NPMs) can directly control the sparsity by imposing l_1 norm or other composite regularizer on codes rather than using a sparse prior or introducing auxiliary variables, such as NMF [9], STC [12], GSTC [11]. Although, all the above models can solve the sparsity problem in some extent, they still fail to achieve fully sparse.

2.3 Sparse Techniques

The fundamental sparse regularizer, lasso [13], has been widely used in machine learning. Lasso is an estimation technique which gives interpretable models and has the capacity of variable selection. However, the lasso lacks a mechanism to control group sparsity. To discover some special sparsity patterns, Yuan et al. [6] proposed group lasso, to achieve group sparsity. Namely, if a group parameter is non-zero, then the parameters of this group will all be non-zero. However, the group lasso is just able to select group variables. Then, Simon et al. [7] presented the sparse group lasso via combining the penalty function of lasso with group lasso's. The sparse group lasso has the capacity of variable and group variables selection. In our work, we employ the sparse group lasso regularizer on word codes to achieve fully sparsity.

3 Sparse Topical Coding with Sparse Groups

Let $V = \{1, ..., N\}$ be a vocabulary with N words, $D = \{d_1, ..., d_m\}$ be a document collection with size M, where $w_d = \{w_1^d, ..., w_{|I|_d}^d\}$ is a vector of terms to represent a document d, M is the index set of words that appear in document d, w_n^d ($n \in I_d$) counts the appearances of word n in document d. Let $\beta \in \mathbb{R}^{K \times N}$ be a dictionary with K basis, where each row $\beta_{k.}$ in the directory is a topic basis that has a unigram distribution over vocabulary.

3.1 Probabilistic Generative Process for STCSG

For simplicity, we let $s \in \mathbb{R}^K$ be the word code of word n in document d, and assume that the observed word counts are independent from each other. The core thought of STCSG is to impose the sparse group lasso on word codes to induce document, topic and word sparsity. We start with describing a probabilistic generative procedure. Each document d in D is generated by the following process:

For each document $D = \{d_1, ..., d_m\}$:

1. Sample the word code $s \sim p(\lambda_1)$;
2. For each topic $k \in \{1, ..., K\}$:
   (a) Sample the word code vectors $s_{d,.k} \sim p(\lambda_2, s_{d,.})$;
   (b) For each observed word $n \in I$:
       i. Sample the latent word count $w_n \sim p(w_n | s_{d,.k}, \beta_{.k})$:

In this process, we assume that: (1) the observed word counts w_n can be reconstructed by the linear combination of word code vectors and dictionary, (2) word codes $s_{d,.}$ are derived via the prior $p(\lambda_1)$, (3) and the word code vectors $s_{d,n}$ are obtained via the prior $p(\lambda_2)$. Based on the previous work [12] [11], we think that the document codes are the average aggregation of the word code vectors. STCSG is depicted in Fig. 1.

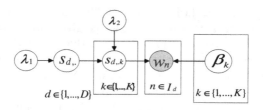

Fig. 1. The graphical model of STCSG.

3.2 Formulation of STCSG

According to the above generation process, the joint distribution of STCSG can be defined as follow:

$$p(s, w | \beta) = \prod_{d \in D} p(s_{d,.} | \lambda_1) \prod_{k \in K} p(s_{d,.k} | \lambda_2) \prod_{n \in I_d} p(w_n | s_{d,.k}, \beta_{.k}), \qquad (1)$$

In this study, to achieve sparse word and topic codes which are defined as $\theta_{d,k} = \sum_{d=1}^{M} \sum_{n=1}^{I_d} s_{d,nk} \beta_{kn} / \sum_{d=1}^{M} \sum_{n=1}^{I_d} \sum_{k=1}^{K} s_{d,nk} \beta_{kn}$, we use the Laplacian prior. The Laplacian prior and the lasso play an equivalent affect under a MAP-solution [12], which is defined as:

$$Laplace(s_{d,.} | 0, \lambda_1^{-1}) \propto \exp(-\lambda_1 \|s_{d,.}\|_1), \qquad (2)$$

Similarly, to achieve sparse document and topic coding, we assume that the word codes under the same topic are in group structure. In considering of the above factors, we choose the Multi-Laplacian distribution [8] $p(s_{d,.k} | \lambda_2)$ which is defined as:

$$M - Laplace(s_{d,.k} | 0, \lambda_2^{-1}) \propto \lambda_2^{N/2} \exp(- \lambda_2 \| s_{d,.k} \|_2), \tag{3}$$

Suppose the distribution of w_n satisfies Poisson distribution [12]:

$$p(w_n | s_{d,.k}, \beta_{.k}) \propto Possion(w_n; s_{n.}^T \beta_{.n}), \tag{4}$$

With defining $S = \{s_d\}$ and $W = \{w_d\}$, STCSG solves the following optimization problem:

$$\min_{S,\beta} L(S, \beta) = l(S, \beta) + \lambda_1 \sum_{d,k \in I_d} \| s_{d,.k} \|_2^2 + \lambda_2 \|S\|_1$$

$$s.t. : S \geq 0, \forall d; \sum_{n=1}^{I_d} \beta_{kn} = 1, \forall k; \lambda_1 + \lambda_2 = 1; \tag{5}$$

In Eq. (5), the parameters (λ_1, λ_2) are non-negative, which can be selected via cross-validation. The first part of the objective function is a loss function, which is equivalent to minimizing the unnormalized KL-divergence between the observed word counts and their reconstructions. The second and third part form the sparse group lasso, inducing fully sparse coding.

3.3 Optimization

The optimization Eq. (5) is a bi-convex problem, which can be solved by ADMM. ADMM [5] uses variable splitting, and is fit for decomposing our problem into easily solvable sub-problems. As it is usually difficult to directly minimize s and β on $l(s, \beta)$ independently, we introduce a new variable x, with the constraint $x_k = \beta_{k.} s_{.k}$. However, the optimization problem is still intractable, as the word codes are grouped in terms of topics and each group is independent. Therefore, we consider to optimize each group. According to the framework of ADMM, we optimize (x, s, β, u) separately according to the following steps: When $n \leq k$,

$$x^{m+1} := \arg\min_x (x_k^m + w_k \ln(x_k^m + A_k) + \frac{\rho}{2} \| x_k^m - \beta_{k.}^m s_{.k}^m + u_k^m \|_2), \tag{6}$$

When $n \geq k$,

$$\begin{cases} x_k^{m+1} := \arg\min_x (x_k^m + \frac{\rho}{2} \| x_k^m - \beta_{k.}^m s_{.k}^m + u_k^m \|_2) \\ \beta_{k.}^{m+1} := \arg\min_\beta (\frac{\rho}{2} \| x_k^{m+1} - \beta_{k.}^m s_{.k}^m + u_k^m \|_2) \\ s_{.k}^{m+1} := \arg\min_s (\lambda_1 \| s_{.k}^m \|_2^2 + \lambda_2 \| s_{.k}^m \|_1 + \frac{\rho}{2} \| x_k^{m+1} - \beta_{k.}^{m+1} s_{.k}^m + u_k^m \|_2) \\ u_k^{m+1} := u_k^m + (x_k^{m+1} - \beta_{k.}^{m+1} s_{.k}^{m+1}) \end{cases} \tag{7}$$

where (x, β, s) are primal variables and u is the dual variable, the parameter ρ is non-negative which can be selected via cross-validation like λ_1. As all of the above iterative process are in explicit expression, hence, we can analyze the issue comprehensively and critically.

Optimize over X: We learn it for Eq. (6) each document and each topic separately by optimizing the. When $n \leq k$, Eq. (7).1 is a convex quadratic problem, which can be efficiently solved with Quasi-Newton Methods, one of the most effective methods for solving nonlinear optimization problems.

When $n \geq k$, taking the subderivative of Eq. (7).1 with respect to s_k^m, we can obtain the solution for Eq. (7).1 by setting the first-order sub-gradient equivalent to zero. That is

$$x_k^{m+1} = \beta_{k.}^m \, s_{.k}^m - u_k^m - 1/\rho, \tag{8}$$

Optimize over β: We can also obtain the solution for Eq. (7).2 by setting the first-order sub-gradient equivalent to zero:

$$\beta_{kn}^{m+1} = (x_k^{m+1} + u_k^m - \sum_{i=1 \wedge i \neq n}^{I_d} \beta_{ki}^m \, s_{ik}^m) / s_{nk}^m, \tag{9}$$

Optimize over S: It is a sparse group lasso problem that has only one group and is non-convex. We also use the ADMM algorithm to solve Eq. (7).3 with introducing variables W and Λ. The learning algorithm of $s_{.k}$ can be summarized as Algorithm 1. Where:

Algorithm 1. ADMM algorithm for sparsity with group sparsity (SG-ADMM).

Input: $x_k^{m+1}, \beta_{k.}^{m+1}, s_{.k}^m, u_k^m$;
Output: $s_{.k}^{m+1}$;
 1: initialize: $W_0 = \Lambda_0 = 0$;
 2: **for** number of iterations **do**
 3:     $(I + (\beta_{k.}^{m+1})^T \beta_{k.}^{m+1}) (s_{.k}^m)_t = W_{t-1} - \Lambda_{t-1} + (\beta_{k.}^{m+1})^T (x_k^{m+1} + u_k^m)$;
 4:     $W_t = S_q(S_p((s_{.k}^m)_t, \lambda_1), \lambda_2)$;
 5:     $\Lambda_t = \Lambda_{t-1} + (s_{.k}^m)_t - W_t$;
 6: **end for**

$$S_p(x, \alpha) = x/\, ||x||_2 \max\{0, ||x||_2 - \alpha^{2-p} \, ||x||_2^{p-1}\} \tag{10}$$

The parameters (p, q) [10] are non-negative and can be selected via cross-validation also. Meanwhile, STCSG learning algorithm is presented in Algorithm 2.

4   Experiments

In this section, we evaluate the efficiency of the proposed STCSG by implementing experiments on the Twitter feeds, which contains 107120 tweets with

Algorithm 2. ADMM algorithm for STCSG.

Input: s, β, D;
Output: s, β;
1: **for** each $d \in \{1, ..., D\}$ **do**
2:    **for** each $k \in \{1, ..., K\}$ **do**
3:       **for** number of iterations **do**
4:          **if** $n \leq k$ **then**
5:             Quasi-Newton Methods to optimize s_k^{m+1};
6:          **else**
7:             $x_k^{m+1} = \beta_{k.}^m \, s_{.k}^m - u_k^m - 1/\rho$;
8:          **end if**
9:          **for** $n = \{1, ..., I_d\}$ **do**
10:             $\beta_{kn}^{m+1} = (x_k^{m+1} + u_k^m - \sum\limits_{i=1 \land i \neq n}^{I_d} \beta_{ki}^m \, s_{ik}^m)/ s_{nk}^m$;
11:          **end for**
12:          $s_{.k}^{m+1} = SG - ADMM(x, \beta, s, u)$;
13:          $u_k^{m+1} = u_k^m + (x_k^{m+1} - \beta_{k.}^{m+1} \, s_{.k}^{m+1})$;
14:       **end for**
15:    **end for**
16: **end for**

---

7 categories which are gathered by our web crawler[1]. It also has a vocabulary that contains 500 terms after removing stop words and words which have low frequency. All pre-processing work including word segmentation, stop words removal and vector space modeling are implemented by Python 2.7. We employ several state-of-the-art methods for comparison, including LDA[2], STC[3], and GSTC. The performance of this methods is evaluated from sparse ratio of latent representations of document, topic and word, classification accuracy of documents and time efficiency.

4.1 Evaluation of Sparse Latent Representations

We first show the average sparse ratio of word, topic and document codes.

Word code: We calculate the average word code for every word over all documents. We find that using of norm l_1 on s makes sparser word codes and topic codes, and using of l_1 / l_2 norm on s leads to sparser document codes. In Fig. 2(a), we show the average word sparse ratio of four models. We can see, (1) the word codes learned by our STCSG are much sparser than those learned by LDA and GSTC, this is mainly because the LDA and the GSTC do not have the sparse priors. (2) STCSG also has higher improvement on sparse ratio than STC. It is possibly because imposing sparse group lasso on word representations, makes a group of and in-group word code variables to sparse.

---

[1] http://sc.whu.edu.cn/.
[2] http://www.cs.princeton.edu/blei/lda-c/.
[3] http://bigml.cs.tsinghua.edu.cn/~jun/stc.shtml/.

Then, we analyze impact of different setting of alpha on the sparse ratio. Parameters λ_1 and λ_2 have the same impact as alpha on the sparse ratio, so we omit results of λ_1 and λ_2 here. From Fig. 2(d), we observe that the sparse ratio of document codes increases with α, but not for word and topic codes. The result manifests that λ_1 norm on s makes a sparser word code and topic code, whereas l_1 / l_2 norm on s leads to a sparser document code.

Figure 3 shows the average word representations of top-representation words learned by different models of first category in train document. We also omit the results of GSTC in order to save space. We can see that: When topic number is 10, there are little non-zero word code elements in STCSG, STC and LDA, this may because there are almost no more redundancy topics when topic number closes to the actual topic number. However, when topic number is 20, the word codes are sparse only in STCSG, in which a word interprets only one topic. In contrast, the STC and LDA achieve denser word codes, in which a word may explain several topics.

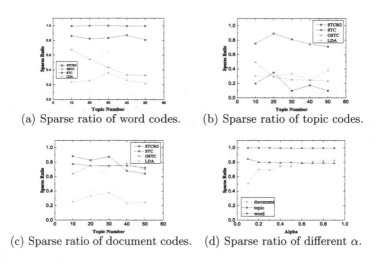

(a) Sparse ratio of word codes.   (b) Sparse ratio of topic codes.

(c) Sparse ratio of document codes.   (d) Sparse ratio of different α.

Fig. 2. The sparse ratio of a latent semantic representing with different topic number k and different α.

Topic code: We compare the topic sparse ratios learned by different models. Figure 2(b) shows the sparse ratio of topic codes under different topic number k. We can see that STCSG achieves much sparser topic codes than other three models because of the sparse topic regularizers. Figure 4 shows the average topic codes of first category in training documents. we can observe that: (1) The topic codes of LDA tend to have many small non-zero weights due to data rare, (2) On the contrary, the topic codes of STCSG have many zero weights, and the non-zero elements have significant weights. (3) The sparsity of topic codes of STC are between LDA and STCSG, the word codes of STCSG also has many non-zero elements.

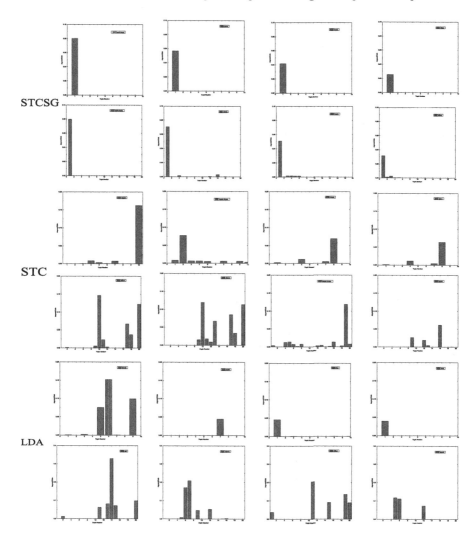

Fig. 3. The average word codes of top-4 word of first category in training document.

Document code: We further compare the document sparse ratios. Figure 2(c) presents the sparse ratio of document codes with setting different topic number k. We can see that: STC, GSTC, and STCSG all achieve much sparser document codes. But, LDA has a lower sparse ratio of document codes. It is because the LDA dose not have sparsity control mechanism. In addition, STCSG achieves a higher sparse ratio of document codes than STC and GSTC. It illustrates that the sparse group lasso can further improve sparse ratio of document codes. Figure 5 shows the average document codes of the first category in training document. We can find that: 1) STCSG can obtain more discriminative representations of documents compared with other two methods. 2) The results of

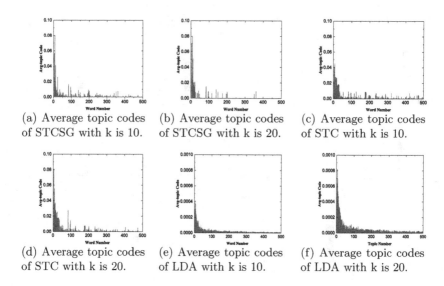

Fig. 4. The average topic codes of first category in training document.

STC and of LDA are worse than STCSG. It is also worth noting that, in STC, although each document code is sparse, the average document code is not. Fortunately, STCSG is opposite, because STCSG uses norm l_1 / l_2 on word code, which is endowed with group sparse property.

4.2 Evaluation of Document Classification

In order to further verify the quality of our model in learning meaningful sparse representations, we perform text classification tasks on tweets with using the sparse document representation of each document as the feature vector. Here, 8900 documents are used for training and 2000 documents are used for testing. We adopt the R package e1071[4] as classifier. From Fig. 5(a), We can see that: (1) STCSG performs better than other three methods all the time, and it performs the best when the topic number is setting to 40. (2) LDA, STC and GSTC performs similarly when topic number is setting to 10, but LDA performs worse with increasing topic number. (3) GSTC performs better than STC and LDA, mainly due to the using of group lasso on word codes. (4) STCSG has higher classification accuracy than GSTC. It is because STCSG can find more meaningful representations than GSTC by using the sparse group lasso.

4.3 Evaluation of Time Efficiency

Finally, we evaluate the time efficiency of different methods. In this paper, STC, GSTC and STCSG are all implemented by R and C++ under 2.33 GHZ intel processor. LDA is fully implemented in C++, so we omit the result of LDA

---
[4] https://cran.r-project.org/web/packages/e1071/.

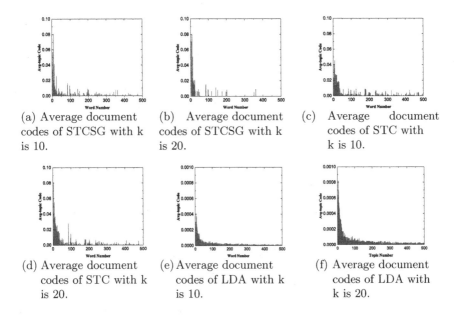

(a) Average document codes of STCSG with k is 10.

(b) Average document codes of STCSG with k is 20.

(c) Average document codes of STC with k is 10.

(d) Average document codes of STC with k is 20.

(e) Average document codes of LDA with k is 10.

(f) Average document codes of LDA with k is 20.

Fig. 5. The average document codes of first category in training document.

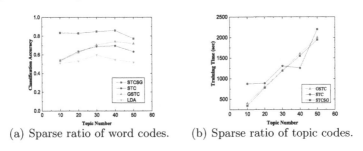

(a) Sparse ratio of word codes.   (b) Sparse ratio of topic codes.

Fig. 6. The classification accuracy of testing document and time efficiency of training process.

here. We conduct our experiments on training documents and compare the running time. Figure 5(b) shows the corresponding experimental results. Obviously, STCSG and GSTC are lower time consuming and STCSG is more efficient than GSTC. It is mainly because that there are many zero elements in the sparse representations at each iteration which potentially speeds up the learning process of STCSG (Fig. 6).

5   Conclusions

In this paper, to address the sparsity problem of short texts, we proposed a novel non-probabilistic topic model, named STCSG, with imposing the sparse group lasso on word codes. STCSG can effectively model the sparsity of word, topic and document codes, and directly discover these latent sparse representations. Moreover, ADMM was incorporated to learn STCSG effectively. Experimental results on Twitter dataset show that STCSG can discover fully sparse representations of short texts, and makes great improvement on document classification and running time. In the future, we plan to devise online learning algorithm in our model, as to deal with large scale document sets, and the complicated dictionary optimization.

References

1. Blei, D., Ng, A., Jordan, M.: Latent Dirichlet allocation. J. Mach. Learn. Res. **3**, 993–1022 (2003)
2. Blei, D., Lafferty, J.: Dynamic topic models. In: ICML 2006, pp. 113–120. ACM, Pittsburgh (2006)
3. AlSumait, L.,Barbar, D., Domeniconi, C.: Online LDA: adaptive topic models for mining text streams with applications to topic detection and tracking. In: ICDM 2008, pp. 3–12. IEEE, Pisa (2008)
4. Wang, X., McCallum, A.: Topics over time: a non-Markov continuous-time model of topical trends. In: KDD 2006, pp. 424–433. ACM, Philadelphia (2006)
5. Boyd, S., Parikh, N., Chu, E., et al.: Distributed optimization and statistical learning via the alternating direction method of multipliers. Found. Trends Mach. Learn. **3**, 1–122 (2011)
6. Yuan, M., Lin, Y.: Model selection and estimation in regression with grouped variables. J. Roy. Stat. Soc. **68**, 49–67 (2006)
7. Simon, N., Friedman, J., Hastie, T., et al.: A Sparse-Group Lasso. J. Comput. Graph. Stat. **22**, 231–245 (2013)
8. Eltoft, T., Kim, T., Lee, T.: On the multivariate Laplace distribution. IEEE Sig. Process. Lett. **13**, 300–303 (2006)
9. Heiler, M.: Learning sparse representations by non-negative matrix factorization and sequential cone programming. J. Mach. Learn. Res. **7**, 1385–1407 (2006)
10. Chartrand, R., Wohlberg, B.: A nonconvex ADMM algorithm for group sparsity-with sparse groups. In: ICASSP 2013, pp. 6009–6013. IEEE, Vancouver (2013)
11. Bai, L., Guo, J., Lan, Y., et al.: Group sparse topical coding: from code to topic. In: WSDM 2013, pp. 315–324. ACM, Rome (2013)
12. Zhu, J., Xing, E.: Sparse topical coding. In: UAI 2011, pp. 831–838. AUAI, Barcelona (2011)
13. Tibshirani, R.: Regression shrinkage and selection via the lasso. J. Roy. Stat. Soc. **58**, 267–288 (1996)
14. Wang, C., Blei, D.: Decoupling sparsity and smoothness in the discrete hierarchical dirichlet process. In: NIPS 2009, pp. 1982–1989. MIT Press, Vancouver (2009)
15. Lin, T., Tian, W., Mei, Q., et al.: The dual-sparse topic model: mining focused topics and focused terms in short text. In: WWW 2014, pp. 539–550. ACM, Seoul (2014)
16. Chien, J., Chang, Y.: Bayesian sparse topic model. J. Sig. Process. Syst. **74**, 375–389 (2014)
17. Than, K., Ho, T.B.: Fully sparse topic models. In: Flach, P.A., De Bie, T., Cristianini, N. (eds.) ECML PKDD 2012, Part I. LNCS, vol. 7523, pp. 490–505. Springer, Heidelberg (2012)
18. Hofmann, T.: Probabilistic latent semantic indexing. In: SIGIR 1999, pp. 50–57. ACM, Berkeley (1999)

Information Retrieval

Differential Evolution-Based Fusion for Results Diversification of Web Search

Chunlin Xu, Chunlan Huang, and Shengli Wu[✉]

School of Computer Science and Telecommunication Engineering,
Jiangsu University, Zhenjiang 212013, China
swu@ujs.edu.cn

Abstract. Results diversification has been a key research issue on web search in the last couple of years. Some recent research work suggests that data fusion, especially linear combination of multiple results, is a good option of dealing with this problem. However, there are many different ways of setting weights. In this paper, we propose a differential evolution-based method to find optimal weights in the weight space for the linear combination method. Experimental results show that the proposed method is effective compared with the state-of-the-art techniques.

1 Introduction

For an ambiguous query submitted by a user, web search engines try to provide diversified results [1,6,26] so as to let them satisfy the user's various kinds of information needs. Most of the time a two-step procedure is taken: first for a given query, a traditional search model is used to retrieve a ranked list of documents. At this step the search model only concerns relevance of the documents to the query. The next step is to use a re-ranking algorithm to re-rank all the documents generated at the first step. This time diversity needs to be considered along with relevance. Different kinds of diversity re-ranking algorithms have been proposed, and related experiments demonstrate that many of them are able to achieve certain improvement on diversity. However, there is room for further improvement.

On the other hand, data fusion is an effective approach for performance improvement in many retrieval-related tasks [2,25]. The general idea of data fusion is to combine results from different retrieval models/mechanisms that work with the same collection of documents. Quite a few data fusion methods, such as CombSUM [11], CombMNZ [11], linear combination (LC) [22], ClustFuse [15], LambdaMerge [21], optimization algorithms such as the genetic algorithm [12], and so on, have been proposed and investigated.

Although data fusion has been widely used in many tasks in information retrieval/web search, most of them focus on improving relevance-related performance. There are two pieces of work that address the issue of results diversification. The first piece of work undertaken by Liang et al. in [17] is a combination of different techniques: in the first part they combine a few results from different

© Springer International Publishing Switzerland 2016
B. Cui et al. (Eds.): WAIM 2016, Part I, LNCS 9658, pp. 429–440, 2016.
DOI: 10.1007/978-3-319-39937-9_33

search engines. The data fusion methods used are CombSUM and CombMNZ. In the second part they infer latent subtopics by topic modeling. The document set used is the output of the first step with full text for all the documents. Lastly, result diversification is performed by a typical result diversification method PM-2 [9].The second piece of work is undertaken by Wu and Huang in [24]. Linear combination is used for fusing results, while the weights are set by considering diversity-related performance of each search engine and similarity between component results. In short, Liang et al. [17] only uses very common data fusion methods CombSUM and CombMNZ, while Wu and Huang [24] uses a heuristic-based method to set weights for the linear combination method.

As a typical method in data fusion, linear combination is a straightforward and flexible approach because different weights can be assigned to cope with different situations accordingly. Besides, there are many different ways of assigning weights. In this paper, we move a step further from Wu and Huang's work in [24] for weight assignment. We apply an optimization method, Differential Evolution (DE) [10,19], for this. Experiments are carried out with three groups of data from TREC[1]. and experimental results demonstrate that the proposed method performs better than the state-of-the-art techniques.

The remainder of this paper is organized as follows: in Sect. 2 we discuss some related work. In Sect. 3 we introduce differential evolution-based fusion method and some other fusion methods as the baseline for comparison. Experimental settings and results are reported in Sect. 4. Section 5 concludes the paper.

2 Related Work

Researchers have investigated search results diversification through a variety of approaches. Generally the methods for search result diversification can be categorized into two types: the implicit approach and the explicit approach [20]. The explicit approach tries to work out all the sub-topics that are related to the query, and usually obtaining such information requires analyzing queries and visiting external resources; while the implicit approach does not consider any aspect of the query and does not need any other resources apart from the results, or the contents of all the documents involved.

A typical implicit approach is Maximal Marginal Relevance (MMR) [4]. It selects documents one by one iteratively from the original ranked list of documents. The first one selected by a typical implicit approach has the highest relevance score. Suppose S is the set of documents that have already been selected, then the next one is selected by a linear combination of the relevance score of the candidate document and the minimum dissimilarity score between any of the documents in S and the candidate document. PM-2 [9] and xQuAD [20] are two representative explicit approaches. Both of them assume that for every document retrieved in the original list, the search engine knows the relevance of

---

[1] TREC (Text Retrieval Conference) is an annual information retrieval evaluation event held by the National Institute of Standards and Technology, USA. Its web site is located at http://trec.nist.gov/.

the document to all the sub-topics of the query. And both of them use a similar selection process that MMR uses but apply different selection criteria. xQuAD estimates the combination of relevance of the candidate document to the original query and to all sub-topics of that query. PM-2 treats the problem as a proportional election of the documents for all sub-topics. It only considers relevance of the candidate document to all sub-topics without taking the original query into account.

Additionally, many learning-based methods have also been proposed, especially the methods that can directly optimize evaluation measures on the training data. These methods try to maximize the objective function value in training and the final evaluation measure in testing [26,28]. SVM-DIV [27] is a representative learning approach to search result diversification by using structural Support Vector Machines (SVMs) to optimize subtopic coverage. ClustFuse [15] is a clustering-based approach. LambdaMerge (λ-Merge) [21] directly optimizes a retrieval metric (such as NDCG or MAP) with a mixture of scoring functions.

Different from above-mentioned approaches, data fusion algorithms combine documents retrieved from multiple search engines for better performance. CombSUM, CombMNZ and a series of combination methods are some of the earliest data fusion methods proposed [11]. The linear combination method (LC) [22] re-ranks all the documents to form a new single list by a linear combination of the document's scores obtained from multiple component search engines. The major difference between CombSUM and linear combination is the way of weight assignment. For CombSUM, the weight assigned to each component search engine is always equal to 1, while for linear combination, each weight may be different from the others.

When only ranking information is available, various voting-based strategies may be used for fusion. Borda count [2], Condorset fusion [18], Reciprocal Rank Fusion (RRF) [8] and some other voting-based fusion methods merge documents based on their rank positions in different results.

Weight assignment is a major issue in linear combination, and many different strategies have been proposed. Some heuristic-based methods, such as performance-related weighting strategies, are straightforward. More sophisticated and expensive methods, such as conjugate gradient [3], multiple linear regression [23], expectation maximization [13], and optimization algorithms such as the genetic algorithm [12], have been utilized to weight assignment. However, in all above-mentioned research work, relevance is the only concern and diversity is not addressed.

As mentioned in Sect. 1, there are two pieces of work that address results diversification via the data fusion technique completely or partially. In Liang et al.'s work [17], they only used very common data fusion methods CombSUM and CombMNZ as part of their approach. In Wu and Huang's work [24], they proposed a heuristic-based method to assign weights for linear combination. The latter is the most relevant to our research reported in this paper. We propose a differential evolution-based weighting assignment method and empirical studies show that the proposed method performs better than all those heuristic-based methods.

3   Data Fusion Techniques

In data fusion, documents from multiple search engines are fused to form a new single list of ranked documents. In this section we first brief the data fusion methods that will be used later in this paper, which is followed by the differential evolution-based method.

3.1   Data Fusion Methods Served as Baseline

Suppose there are n component search engines E_1, E_2,...,E_n, each of which retrieves documents in the corpus to match the same query and returns a ranked list of them as retrieval results. Thus we obtain n ranked lists $L_1, L_2, ..., L_n$. For CombMNZ, it computes the sum of each document's initial scores, which is multiplied by the number of component ranked lists that contain the document. In the following equation, $m(d)$ denotes the number of search engines in which d is retrieved.

$$s(d) = m(d) * \sum_{i=1}^{n} s_i(d) \tag{1}$$

where $s_i(d)$ is the score of document d in the ranked list L_i $(1 \leq i \leq n)$. If document d is not retrieved by a search engine E_i, we may set a default value, such as 0, as its score (or $s_i(d) = 0$). According to [16], CombMNZ is a good data fusion method.

Linear combination is a fusion method that merges every document's scores from different component search engines, and each of which is assigned a weight.

$$s(d) = \sum_{i=1}^{n} w_i * s_i(d) \tag{2}$$

where w_i is the weight assigned to E_i. With different weighting assignment strategies, performance of linear combination may vary considerably. If all weights are the same, then linear combination becomes CombSum.

Performance-based weighting assignment strategy is applicable to linear combination [24]. It utilizes component search engines' retrieval performance to quantify their importance. The performance of a search engine is often measured by some evaluation metric, such as MAP, NDCG and so on, on a group of training queries, and then set as its weight for testing. More formally, let us assume that the performance of a group of search engines E_i $(1 \leq i \leq n)$ is P_i $(1 \leq i \leq n)$ by using some training data, and the weight w_i of E_i for combination can be set as P_i, P_i^2, and so on. For the dissimilarity-based weighting assignment strategy [24], we also need some training data, which are the results obtained from multiple search engines for a group of queries. Note that relevance judgment is not required for this purpose. We first calculate the ranking difference of all the documents for each pair of component results L_j and L_k $(j \neq k)$. Here we assume that the length of all ranked lists is l and $|L_j \cup L_k|$ documents appear in both ranked lists L_j and L_k. $p_j(d_i)$ and $p_k(d_i)$ denote the rank positions of d_i

in L_j and L_k, respectively. For those documents that only appear in one result, we further assume that they apprear one by one successively in the other result after position l whilst keeping the same relative order. Therefore, we have

$$v(L_j, L_k) = \frac{1}{l * |L_j \cup L_k|} \{ \sum_{d_i \in L_j \cap L_k} |p_j(d_i) - p_k(d_i)|$$
$$+ \sum_{d_i \in L_j \setminus L_k} |p_j(d_i) - (l + i)| \qquad (3)$$
$$+ \sum_{d_i \in L_k \setminus L_j} |p_k(d_i) - (l + i)|\}$$

Note that the weights calculated in Eq. 4 is slightly different from the weights calculated in [24]. We think the one used in this paper is better normalized. Based on Eq. 4, the dissimilarity weight of L_j $(1 \leq j \leq n)$ is defined as

$$diss_j = \frac{1}{n - 1} \sum_{k=1 k \neq j}^{n} v(L_j, L_k) \qquad (4)$$

Up to this point, we have calculated dissimilarity weight $diss_j$ of E_j $(1 \leq i \leq n)$ using the results of one query as training data. If multiple queries are involved in the training data, then we can define the average weight of them as the final weight for testing.

Learning-based algorithms ClustFuse [15] and λ-Merge [21] are not involved in the experiment because they are not comparable. These methods require a very different experimental setting compared with the methods reported in this paper. As a matter of factor, ClustFuse needs to know the contents of all the documents in the resultant lists to calculate their inter-document similarities; while λ-Merge needs to know multiple query-document features and query-reformulation features to train weights. Those features are not required/available in our case.

3.2 Differential Evolution-Based Linear Combination

In this paper, weight assignment is treated as an optimization problem to improve the diversity of fusion result directly. We use differential evolution to find the best weighting scheme so that the linear combination with such weights can lead to the optimal performance.

For an optimization problem, an objective function needs to be defined on all possible solutions. We try to find a best solution by searching the solution space. For linear combination with n component systems, its weighting scheme can be represented as a weight vector $W = [w_1, w_2, ..., w_n]$, and the weight space is $w_j \in [0, 1]$ $(1 \leq j \leq n)$ and $\sum_{j=1}^{n} w_j = 1$. The objective function f is defined as $f(W) = E(\mathcal{L}, W)$, where $\mathcal{L}$ is the set of n ranked lists, which are fused by linear combination with the weighting scheme W, and then the fused result is evaluated by operation E. Therefore, the task of differential evolution is to find the best solution W^* so that $f(W^*) \geq f(W)$ holds for all possible W $(W^* \neq W)$.

Any novelty or diversity-related metric, such as ERR-IA@20 [5], can be used in operation E to measure retrieval performance of the fused results.

There are four basic operation steps in DE [10]. Firstly, the operation of initialization is a setting of initial weight vectors. The initial population is composed of a certain number of points that are randomly selected in the weight space. Then, those vectors in the population will be mutated, combined, and selected iteratively. After several generations of reproduction, the population will fit the optimization objective better and better and a best vector is achievable finally. A stopping criterion for the evolution may be defined as the maximum number of generations G. The iteration does not stop until the number of generations equals to G.

Population Initialization. The population $P = \{W_1, W_2, ..., W_{NP}\}$ consists of NP weight vectors, and each element of vectors $w_{ij} = rand(0,1)$ ($1 \leq i \leq NP, 1 \leq j \leq n$), where $rand(0,1)$ is a function that can generate a random value in $[0,1]$. Additionally, all the weights should be normalized to guarantee $\sum_{j=1}^{n} w_j = 1$.

Difference-Vector Based Mutation. The mutation of DE is based on the differential strategy to generate donor vectors. By randomly selecting two weight vectors W_{i1} and W_{i2} in the current population, it merges the target vector W_i with the differential of the two selected vectors to get the donor vector for next generation: $V(g+1) = W_i(g) + F(W_{i1}(g) - W_{i2}(g))$, $i \neq i1 \neq i2$. F is a constant that controls the amplification of the differential variations.

Crossover. The crossover operation generates a new vector, trial vector, by mixing the target vector W_i and the donor vector V_i.

The trial vector U_i is defined as:

$$u_{ij}(g+1) = \begin{cases} v_{ij}(g+1) & \text{if } rand(0,1) < CR \text{ or } j = j_{rand} \\ w_{ij}(g) & \text{otherwise} \end{cases} \tag{5}$$

Here, j_{rand} is a random integer in $\{1, 2, ..., n\}$, and CR is a constant that can control the proportion of elements inherited from the donor vector to the trial vector.

Selection. The target vector W_i is compared with the trial vector U_i based on the objective function, then the better one is chosen to the next generation.

$$W_i(g+1) = \begin{cases} U_i(g+1) & \text{if } f(U_i(g+1)) \geq f(W_i(g)) \\ W_i(g) & \text{otherwise} \end{cases} \tag{6}$$

The algorithm for differential evolution-based weight assignment is presented in Algorithm 1. Let us assume that there are n component results and each comprises a ranked list of l documents, then linear combination can be completed in $O(n*l)$ time. Therefore, the complexity of the whole training process by using differential evolution is $O(NP * G * m * n * l)$.

Algorithm 1. The algorithm of weight assignment for linear combination by using differential evolution-based method

Input:
The training set including retrieval results from n component systems for m queries, $\mathcal{L}^q = L_1^q, L_2^q, \cdots, L_n^q, 1 \leq q \leq m$.
DE parameters: NP (size of the population), F (a constant), CR (a constant), G (maximum number of generations).

Output:
The best weight vector for all component systems $W_{best} = [w_1, w_2, \cdots, w_n]$.
1: Initialize the population $P = \{W_1, W_2, ..., W_{NP}\}$.
2: **for** each $k \in [1, NP]$ **do**
3:    **for** each $q \in [1, m]$ **do**
4:       Combine the ranked lists in $\mathcal{L}^q$ with W_k using Eq. 3.
5:       Evaluate the fused result by $f(W_k^q) = E(\mathcal{L}^q, W_k^q)$.
6:    **end for**
7:    Compute the average score of the $f(W_k)$ over m queries.
8: **end for**
9: Initialize W_{best} with the weight vector that has the highest average score.
10: **for** each $iter \in [1, G]$ **do**
11:    **for** each $k \in [1, NP]$ **do**
12:       Mutate to create the donor vector V_k.
13:       Cross to get the trial vector U_k using Eq. 6.
14:       Select the better performed vector: evaluate the target vector W_k and the trial vector U_k(same as Steps 3–6) and get the better one to update W_k using Eq. 7.
15:       Update the best weight vector W_{best} with the maximum score.
16:    **end for**
17: **end for**
Output the best weight vector W_{best}.

4   Experiment Setting and Results

In order to measure the performance of weight assignment using differential evolution, we work with the runs submitted to the Diversity Task in the TREC 2009, 2010, 2011 Web Track[2]. The data set used in these events is ClueWeb09[3]. The diversity task is a competition that encourages search engine participants to retrieve a wide coverage of all possible subtopics for any given query, while avoiding redundancy among the results [6].

50 ambiguous queries are used each year. For every query, each search engine retrieves a ranked list of documents. From all those runs submitted by participants, we select 8 of them. All of them are submitted by different participants and are top performers among all the submissions in that year. Detailed information about them is summarized in Table 1. For retrieval evaluation, we use

---

[2] Runs can be downloaded from TREC's web site http://trec.nist.gov.
[3] http://1boston.lti.cs.cmu.edu/Data/clueweb09

Table 1. Information summary of three groups of runs submitted to TREC (measured by ERR-IA@20 and α-NDCG@20; figures in bold indicate the best performance in that year group)

TREC 2009		TREC 2010		TREC 2011	
Run	ERR-IA α-NDCG	Run	ERR-IA α-NDCG	Run	ERR-IA α-NDCG
MSRAACSF	**0.2144** **0.3653**	msrsv3	**0.3473** **0.4909**	uogTrA45Nmx	**0.5284** **0.6298**
MSDiv3	0.2048 *0.3456*	THUIR10DvNov	0.3355 *0.4745*	msrsv2011d1	0.4994 *0.6079*
uogTrDYCcsB	0.1922 *0.3081*	ICTNETDV10R2	0.3222 *0.4637*	UWatMDSqltsr	0.4939 *0.5947*
UamsDancTFb1	0.1774 *0.2808*	uogTrB67xS	0.2981 *0.4178*	ICTNET11DVR3	0.4764 *0.5818*
mudvimp	0.1746 *0.2747*	UMd10IASF	0.2546 *0.3773*	UAmsM705tFLS	0.4377 *0.5222*
UCDSIFTdiv	0.1733 *0.2776*	cmuWi10D	0.2484 *0.3452*	uwBA	0.3986 *0.4918*
NeuDiv1	0.1705 *0.2781*	UAMSD10aSRfu	0.2423 *0.3413*	CWIcIA2t5b1	0.3487 *0.4316*
THUIR09AbClu	0.1665 *0.2782*	UCDSIFTDiv	0.2100 *0.3118*	liaQEWikiAnA	0.2287 *0.3161*
Average	0.1814 *0.3011*	Average	0.2823 *0.4028*	Average	0.4265 *0.5220*
Variance	0.0003 *0.0013*	Variance	0.0025 *0.0047*	Variance	0.0098 *0.0113*

two metrics ERR-IA@20 and α-NDCG@20 [7]. Both of them are widely used in literature on search result diversification.

The experiment is aimed at testing the effect on result diversification by differential evolution-based weights assignment. The parameters for DE are set as: $NP = 30$, $F = 0.5$, $CR = 0.5$, $G = 1000$. 0.5 is a typical value for both F and CR. The values of NP and G should be set in appropriate ranges. If they are too small, then the performance of the algorithm decreases significantly; if they are too large, then too much time is required to perform the algorithm. If this experiment, we find that the performance of the algorithm stabilizes when $NP \geq 20$ and $G \geq 800$. We use the 5-fold cross validation [14] to verify the performance: 50 queries are divided into five groups. Each time we choose one different group as testing data, and the four remaining groups as training data for learning weights by DE. In the end, every group is selected for testing once and we can get the results on all the queries. In our experiment, we randomly pick up queries without replacement to form query groups. The above process

Table 2. Fusion performance measured by ERR-IA@20 (Figures in parentheses indicate the improvement rates of the method over the best search engine)

Group	2009	2010	2011	Average
Best	0.2144	0.3473	0.5284	0.3634
CombMNZ	0.2493	0.4052	0.5368	0.3971
	(16.28 %)	(16.67 %)	(1.59 %)	(9.28 %)
LC-P	0.2513	0.3953	0.5397	0.3955
	(17.21 %)	(13.83 %)	(2.14 %)	(8.83 %)
LC-P^2	0.2514	0.3879	0.5343	0.3912
	(17.28 %)	(11.68 %)	(1.11 %)	(7.66 %)
LC-PD	0.2536	0.3933	0.5411	0.3960
	(18.30 %)	(13.25 %)	(2.41 %)	(8.99 %)
LC-DE	**0.2551**	**0.4067**	**0.5571**	**0.4063**
	(18.99 %)	(17.11 %)	(5.43 %)	(11.82 %)

Table 3. Fusion performance measured by α-NDCG@20 (Figures in parentheses indicate the improvement rates of the method over the best search engine)

Group	2009	2010	2011	Average
Best	0.3653	0.4909	0.6298	0.4953
CombMNZ	0.4111	0.5555	0.6471	0.5379
	(12.54 %)	(13.16 %)	(2.75 %)	(8.59 %)
LC-P	0.4107	0.5524	0.6513	0.5381
	(12.42 %)	(12.53 %)	(3.41 %)	(8.64 %)
LC-P^2	0.4111	0.5490	0.6468	0.5356
	(12.54 %)	(11.84 %)	(2.70 %)	(8.14 %)
LC-PD	**0.4123**	0.5515	0.6519	0.5386
	(12.87 %)	(12.35 %)	(3.51 %)	(8.73 %)
LC-DE	0.4117	**0.5619**	**0.6691**	**0.5476**
	(12.71 %)	(14.47 %)	(6.23 %)	(10.55 %)

is repeated 10 times so as to obtain more reliable results. The results reported in this paper is the average of them.

For the comparison of fusion performance, some existing data fusion methods are involved in our experiment as baseline. We have: the best component run, CombMNZ, Linear Combination with Performance level weighting (referred to as LC-P)and its power variation (referred to as LC-P^2), and Linear Combination with Performance and Dissimilarity weights (equal weight for performance and dissimilarity, referred to as LC-PD). For LC-P, LC-P^2, and LC-PD, they use the 5-fold cross validation as LC-DE does.

Table 4. An example of weighting scheme used in LC-DE and its fusion performance (the 2009 year group, measured by ERR-IA@20)

Components	1	2	3	4	5
MSRAACSF	0.1636	0.0427	0.1520	0.0823	0.1035
MSDiv3	0.1477	0.1934	0.1664	0.2262	0.1648
uogTrDYCcsB	0.1919	0.1947	0.2332	0.2152	0.2262
UamsDancTFb1	0.1063	0.1358	0.0308	0.1182	0.1037
mudvimp	0.2120	0.1722	0.1649	0.1428	0.1851
UCDSIFTdiv	0.1635	0.1464	0.1325	0.1679	0.1647
NeuDiv1	0.0150	0.1035	0.0548	0.0426	0.0472
THUIR09AbClu	0.0000	0.0113	0.0654	0.0049	0.0048
Performance on training set	0.2681	0.3059	0.2605	0.3004	0.2980
Performance on testing set	0.2375	0.2857	0.1878	0.2153	0.3537

Because it is very often that the original scores are not comparable across different search engines, score normalization is required as a necessary step for data fusion. In our experiment, a natural logarithm function is used to normalize all component results. The function is $s_i(d) = 1 - 0.2 * \ln(r_i(d) + 1)$, where $r_i(d)$ is the rank position of document d in a ranked list L_i.

Tables 2 and 3 show diversification performance of all the data fusion methods involved and their improvement rates over the best component run submitted. From these two tables, we can see that data fusion can achieve good performance on search result diversification. On average of three year groups, all data fusion methods perform better than the best component run by over 8 %. LC-DE performs the best, and the figures for improvement rates are 11.82 % and 10.55 %, for ERR-IA@20 and α-NDCG@20, respectively. The improvement rates for other data fusion methods range from 7.66 % to 9.28 %. In every year group, the linear combination with differential evolution-based weighting assignment performs best, with only one exception in the 2009 year group measured by α-NDCG@20 (where LC-PD achieves the best). Therefore, we conclude that LC-DE is effective compared with other weighting schemes.

In order to understand more about LC-DE, we also look at weights and fusion performance of it on the training data set and testing data set separately. We find that the weights of the same search engine varies considerably from one training group to another, although 30 out of 40 results are the same in any of these two training groups. It is also very common that performance of LC-DE varies across two groups (training and testing). Table 4 shows a typical example of it. Although on average performance of LC-DE on the training data set is 18 % more effective than that on the testing data set, the figures are not even for different groups, from as high as 40 % to as low as -16 %. We hypothesize that there is certain relationship between consistency of weights and the size of the training data set. It is possible that such a phenomenon is due to not large enough training data set (40 queries) and more reliable weights may be achievable if a larger training data set is used.

It is also notable that performance of a result is an important but not the only factor that affects the weight of the result. Similarity between results is another factor that affects weights of linear combination for achieving possible best performance. This can explain why MSRRAACSF consistently receives low weights than mudvimp in Table 4 although the former is more effective than the latter.

5   Conclusion

In this paper we have addressed the issue of search results diversification by data fusion. Especially we have investigated how to use differential evolution to learn weights for the linear combination method. Experiments with three groups of data from TREC show that differential evolution is a good option for such a task and it performs better than heuristic-based weighting schemes. Experiments also show that in general data fusion is an effective approach for performance improvement over component search engines.

As our future work, we plan to investigate reliability of the weights learned by differential evolution and the size of the training data set. This work needs a large data set with hundreds of or even thousands of queries. Another direction is to apply differential evolution to more tasks. One particular task in which we are interested is a task in NTCIR-12[4]. The task "Temporally Diversified Retrieval (TDR)" is a subtask of the Temporal Information Access (Temporalia) Task. This task seems more challenging because it requires that the search results are not only diversified but also satisfying temporal restrictions.

References

1. Agrawal, R., Gollapudi, S., Halverson, A., Ieong, S.: Diversifying search results. In: Proceedings of WSDM 2009, Barcelona, Spain, pp. 5–14 (2009)
2. Aslam, J.A., Montague, M.: Models for metasearch. In: Proceedings of ACM SIGIR 2001, New Orleans, Louisiana, USA, pp. 276–284 (2001)
3. Bartell, B.T., Cottrell, G.W., Belew, R.K.: Automatic combination of multiple ranked retrieval systems. In: Proceedings of ACM SIGIR 1994, Dublin, Ireland, pp. 173–184 (1994)
4. Jaime, G., Goldstein, C.J.: The use of MMR, diversity-based reranking for reordering documents and producing summaries. In: Proceedings of ACM SIGIR 1998, Melbourne, Australia, pp. 335–336 (1998)
5. Chapelle, O., Metlzer, D., Zhang, Y., Grinspan, P.: Expected reciprocal rank for graded relevance. In: Proceedings of ACM CIKM 2009, Hong Kong, China, pp. 621–630 (2009)
6. Charles, L., Clarke, A., Craswell, N., Soboroff, I.: Overview of the TREC web track. In: Proceedings of the Eighteenth Text REtrieval Conference, TREC 2009, Gaithersburg, MD, USA (2009)
7. Clarke, C.L.A., Kolla, M., Cormack, G.V., Vechtomova, O., Ashkan, A., Büttcher, S., MacKinnon, I.: Novelty and diversity in information retrieval evaluation. In: Proceedings of ACM SIGIR 2008, Singapore, pp. 659–666 (2008)

---

[4] Its web site is located at http://research.nii.ac.jp/ntcir/ntcir-12/index.html.

8. Cormack, G.V., Clarke, C.L.A., Büttcher, S.: Reciprocal rank fusion outperforms Condorcet and individual rank learning methods. In: Proceedings of ACM SIGIR 2009, Boston, MA, USA, pp. 758–759 (2009)

9. Van Dang, W., Bruce Croft, W.: Diversity by proportionality: an election-based approach to search result diversification. In: Proceedings of ACM SIGIR 2012, Portland, OR, USA, pp. 65–74 (2012)

10. Das, S., Suganthan, P.N.: Differential evolution: a survey of the state-of-the-art. IEEE Trans. Evol. Comput. 15(1), 4–31 (2011)

11. Fox, E.A., Shaw, J.A.: Combination of multiple searches. In: Proceedings of the Second Text REtrieval Conference, TREC 1993, Gaithersburg, MD, USA, pp. 243–252 (1993)

12. Ghosh, K., Parui, S.K., Majumder, P.: Learning combination weights in data fusion using genetic algorithms. Inf. Process. Manage. 51(3), 306–328 (2015)

13. Hong, D., Si, L.: Mixture model with multiple centralized retrieval algorithms for result merging in federated search. In: Proceedings of ACM SIGIR 2012, Portland, OR, USA, pp. 821–830 (2012)

14. Kohavi, R.: A study of cross-validation and bootstrap for accuracy estimation and model selection. In: Proceedings of IJCAI 1995, Montréal, Québec, Canada, pp. 1137–1145 (1995)

15. Kozorovitzky, A.K., Kurland, O.: Cluster-based fusion of retrieved lists. In: Proceeding of ACM SIGIR 2011, Beijing, China, pp. 893–902 (2011)

16. Lee, J.H.: Analysis of multiple evidence combination. In: Proceedings of ACM SIGIR 2007, Philadelphia, PA, USA, pp. 267–275 (1997)

17. Liang, S., Ren, Z., de Rijke, M.: Fusion helps diversification. In: Proceeding of ACM SIGIR 2014, Gold Coast, QLD, Australia, pp. 303–312 (2014)

18. Montague, M., Aslam, J.A.: Condorcet fusion for improved retrieval. In: Proceedings of ACM CIKM Conference, McLean, VA, USA, pp. 538–548 (2002)

19. Price, K., Storn, R.M., Lampinen, J.A.: Differential Evolution-A Practical Approach to Global Optimization. Natural Computing. Springer, Heidelberg (2005)

20. Rodrygo, L., Santos, T., Macdonald, C., Ounis, I.: Exploiting query reformulations for web search result diversification. In: Proceedings of WWW, Raleigh, North Carolina, USA, pp. 881–890 (2010)

21. Sheldon, D., Shokouhi, M., Szummer, M., Craswell, N.: Lambdamerge: merging the results of query reformulations. In: Proceedings of WSDM, Hong Kong, China, pp. 795–804 (2011)

22. Vogt, C.C., Cottrell, G.W.: Fusion via a linear combination of scores. Inf. Retrieval 1(3), 151–173 (1999)

23. Wu, S., Bi, Y., Zeng, X.: The linear combination data fusion method in information retrieval. In: Hameurlain, A., Liddle, S.W., Schewe, K.-D., Zhou, X. (eds.) DEXA 2011, Part II. LNCS, vol. 6861, pp. 219–233. Springer, Heidelberg (2011)

24. Wu, S., Huang, C.: Search result diversification via data fusion. In: Proceedings of ACM SIGIR 2014, Gold Coast, QLD, Australia, pp. 827–830 (2014)

25. Wu, S., McClean, S.I.: Performance prediction of data fusion for information retrieval. Inf. Process. Manage. 42(4), 899–915 (2006)

26. Xia, L., Xu, J., Lan, Y., Guo, J., Cheng, X.: Learning maximal marginal relevance model via directly optimizing diversity evaluation measures. In: Proceedings of ACM SIGIR 2015, Santiago, Chile, pp. 113–122 (2015)

27. Yue, Y., Joachims, T.: Predicting diverse subsets using structural SVMs. In: Machine Learning, Proceedings of ICML, Helsinki, Finland, pp. 1224–1231 (2008)

28. Zhu, Y., Lan, Y., Guo, J., Cheng, X., Niu, S.: Learning for search result diversification. In: Proceedings of ACM SIGIR 2014, pp. 293–302 (2014)

BMF: An Indexing Structure to Support Multi-element Check

Chenyang Xu$^{(\boxtimes)}$, Qin Liu, and Weixiong Rao

Department of Software Engineering, Tongji University, Shanghai, China
{1336321,qin.liu,wxrao}@tongji.edu.cn

Abstract. Standard Bloom Filter is an efficient structure to check element membership with low space cost and low false positive rate. However, the standard Bloom Filter assumes that all elements belong to a single set. When given multiple sets of elements, it cannot efficiently check whether or not multiple input elements belong to the same set, called *multi-element check*. To support the multi-element check, in this paper, we design a new data structure, namely Bloom Multi-filter (BMF). BMF maintains an array of integer numbers to support (1) the insertion of multiple sets of elements into BMF and (2) the lookup to answer multi-element check. We propose four techniques to improve the BMF and optimize the false positive rate. We conducted intensive experiments to study the tradeoff between BMF's space cost and lookup precision. Our experimental results indicate that BMF greatly outperforms the standard bloom filters with around 9.82 folds of lower false positive rate.

1 Introduction

Standard Bloom Filter (SBF) [1] is a powerful compact data structure to represent a set of elements and to check the membership of an element. With low space cost, SBF can encode a large amount of elements by using a bit array. Due to the high space-efficiency and low false positive rate, SBF has been widely used in many applications such as distributed databases and distribute systems [8].

Beyond the element-membership check above, nowadays many applications require more advanced membership check. For example, in keyword search applications, users input multiple terms to retrieve those web documents containing at least one of such input terms. Therefore, it is useful to check whether or not there exist documents containing multiple input terms, instead of a single term. Equivalently, we need to check whether or not multiple terms co-appear in the same documents. For simplicity, we call such a check to be *multi-element check*.

SBF cannot efficiently answer the multi-element check. More specifically, though SBF encodes a large amount of elements, it can only check whether or not an element is inside the bloom filter. Instead, we need to encode multiple

---

The original version of this chapter was revised. An erratum to this chapter can be found at 10.1007/978-3-319-39937-9_41

© Springer International Publishing Switzerland 2016
B. Cui et al. (Eds.): WAIM 2016, Part I, LNCS 9658, pp. 441–453, 2016.
DOI: 10.1007/978-3-319-39937-9_34

sets (or groups) of elements and want to answer the multi-element check. Unfortunately, existing literature work [2,5,6] cannot efficiently answer this question. Such works suffer from scalability issue: they can work only for a very small number of groups, but not for a large number of sets. For example, in the keyword search application above, we have a large amount of web documents, each of which represents a set of elements (document terms).

To tackle the performance issue above, in this paper, we design a novel data structure called Bloom Multifilter (in short BMF) to answer the multi-element check. Instead of a bit array in SBF, we use an array of integer numbers. To encode a group of sets of elements, we assign each set with a unique ID. The insertion of the elements in each group adopts the bitwise OR operation onto the IDs that are at the same entry, and the lookup of multi-element check instead adopts the bitwise AND operation onto the entries associated with the input elements. The result of such an AND operation next decides the lookup value. Beyond the basic idea, we further propose 4 techniques to optimize the false positive rate for higher precision. We conducted intensive experiments on a real data set and study the tradeoff between BMF's space cost, lookup precision and running time. Our experimental results indicate that BMF greatly outperforms the standard bloom filters with around 9.82 folds of lower false positive rate.

The reminder of this paper is organized as follows. Section 2 first introduces the preliminaries of our work including problem definition and related work. Section 3 describes the basic structure of BMF. Next, Sect. 4 proposes four techniques to optimize the basic BMF. After that, Sect. 5 evaluates the performance of BMF, and Sect. 6 finally concludes the paper.

2 Preliminaries

In this section, we first give the problem definition, next investigate related work, and finally give two baseline solutions.

2.1 Problem Definition

Multi-element Check: Consider a group S of sets S_i ($1 \leq i \leq n$). Each set S_i contains $|S_i|$ members $x_i^1...x_i^{|S_i|}$. Next, we consider a query q containing the number of $|q|$ elements $y_1...y_{|q|}$ (with $|q| \geq 1$). Given the group S and query q, the multi-element check needs to find whether or not there exist sets S_i containing all members y_i in q (equivalently $q \subseteq S_i$). If such sets S_i do exist in the group S, the multi-element check returns *true* and otherwise *false*.

For example, we have a group S containing three sets $S_1 = \{A, C\}$, $S_2 = \{A, B, D\}$ and $S_3 = \{B, C\}$. Then, for a given query $q = \{A, D\}$, the multi-element check of q in S returns *true*, and yet *false* for query $q' = \{A, B, C\}$.

Problem Definition: Given the above group S (containing n sets S_i with $1 \leq i \leq n$), each set S_i contains $|S_i|$ members $x_i^1...x_i^{|S_i|}$, we want to design a data structure to represent the group S. The structure should answer multi-element check with low false positives, high space efficiency and fast time response.

2.2   Related Work

The (single) element membership check problem has been widely studied for long time including standard bloom filters, counter bloom filter [1], scalable bloom filter [9], dynamic bloom filter [3,4], adaptive bloom filters [7], etc.

To the best of our knowledge, few work in literature have tackled the lookup of multi-element check, except the following three related works to design the variant of standard bloom filters [5]. The first approach [2] is to build a standard bloom filter for each set of members (resp. a document containing terms in the above example). We next have to check the standard bloom filter one by one for the multi-element membership check. When the number of such sets is large, this approach suffers from both high space cost and slow query time, both of which is linearly proportional to the number of sets.

Differing from the first approach above, the second approach [2,6] assigns each set with a unique ID, and represents the ID with some bits. Each bit of the set Id is with a standard bloom filter, such that $\log k$ bloom filters were used to represent k sets.

Differing from the above both approaches, [5] uses a different way. It assigns each set with a bit vector, and each bit of a position is with a set of hash functions. To insert each element of a set, this approach checks the bits of the set Id. For the bits equal to 1, the element was then encoded with the hash functions associated with such bits. To determine whether or not two members co-appear inside the same set, [5] checked their set Ids by using all hash functions to verify the bits equal to 1.

In addition, the very recent work [10] considered the Set queries including membership queries, association queries and multiplicity queries. The first one is the single element membership query, the second one is to identify which set(s) among a pair of sets contain a given element, and the last one is to check how many times an element appears in a multi-set. All such queries are with a single input element. Instead, the proposed multi-element check in this paper focuses on the query consisting multiple elements.

2.3   Baseline Solution

In order to answer the proposed problem, we probably first follow [5] to use standard bloom filters (SBF) as follows. For each set S_i, we build an associated SBF to encode the members $x_i^1...x_i^{|S_i|}$ in S_i. Next, to answer the multi-element check, we check the SBFs one by one. For each SBF associated with set S_i, we can successfully determine whether or not all $|q|$ members in q are encoded by the SBF (though with a low false positive rate). If yes, we then return $true$ and otherwise continue the check of remaining SBFs.

The baseline solution above can correctly answer the multi-element check but suffer from high space cost and slow running time. It requires n SBFs to encode each set S_i, and the running time is also at the complexity of $O(n)$. When the number of n is very large, the space and time cost is prohibitively high at the scale of $O(n)$.

To overcome the low running time of above naive solution, an alternative is to use a single SBF. That is, we first find all combinations of elements in each set S_i and insert each of such combinations to the SBF. After that, to answer the multi-element check of query q, we can follow the same combination rule as the insertion to find the combination result of q and then check the SBF to return the result. Unfortunately, due to the exponential number of combinations of set S_i, our experiment will verify that the lookup approach suffers from high false positive rate.

3   Basic Structure and Operations

The proposed Bloom Multifilter (BMF) is an extension of standard bloom filter. Specifically, BMF maintains an array $\mathcal{R}$ of integer numbers. Instead, SBF maintains an array of bits (or bit vector). The basic operations of BMF include (i) insertion of the elements in S_i and (ii) lookup of multi-element check to answer query q.

3.1   Insertion

To enable an insertion, we assume that each set $S_i \subseteq \mathcal{S}$ is associated with a unique ID Id_i. Before insertion, all entries in the array $\mathcal{R}$ of BMF are initialized to be null. Next, to insert set S_i and element x_i^j with $1 \leq j \leq |S_i|$, we first lookup the entry associated with $h(x_i^j)$ where $h(x_i^j)$ is the hashing result over x_i^j by a hashing function $h(\cdot)$. For convenience, we denote such an entry to be $E[h(x_i^j)]$. In case that the entry $E[h(x_i^j)]$ is null, we simply set $E[h(x_i^j)] = Id_i$. Otherwise, we reset the entry by the union $Id_i \cup E[h(x_i^j)]$, i.e., $E[h(x_i^j)] \leftarrow Id_i \cup E[h(x_i^j)]$.

In Algorithm 1, we give the insertion pseudocode. Here, suppose that we use L bits to represent the ID (i.e., $1 \leq Id_i \leq 2^L - 1$) of S_i. Thus, we can assign $2^L - 1$ IDs (we will give algorithms about how to tune the number L and assign a reasonable Id for set S_i). When K hash functions are used, we map each member x_i^j to K entries (line 1). For each of such K hash functions, we follow the above idea to insert the member x_i^j (lines 2–4).

Theorem 1. *Consider a set S_i with the Id_i, the result $(E[h(x_i^j)] \cap Id_i) = Id_i$ must hold for each member $x_i^j \in S_i$.*

3.2   Lookup

After all sets S_i in group $\mathcal{S}$ are inserted to a BMF, we next give the lookup algorithm to answer multi-element check. Specifically, for a query q containing $|q|$ members y_j (with $1 \leq j \leq |q|$), we map each member $y_j \in q$ to the entries $E[h(y_i))$. Next, we conduct an intersection operation over the numbers in such entries. If the intersection result is non-zero, we return *true* (indicating that at least a set S_i contains all the members in q) and otherwise *false*.

Algorithm 1. Basic BMF

```
 1  //Insert (Subset Sᵢ with Idᵢ)
 2  for 1 ≤ k ≤ K do
 3  |   for 1 ≤ j ≤ |Sᵢ| do
 4  |   |   if the entry E[hₖ(xⁱⱼ)] is null then E[hₖ(xⁱⱼ)] ← Idᵢ;
 5  |   └   else  E[hₖ(xⁱⱼ)] ← Idᵢ ∪ E[hₖ(xⁱⱼ)]
 6  //Lookup (Query q)
 7  flag ← null
 8  for 1 ≤ k ≤ K do
 9  |   for 1 ≤ j ≤ |q| do
10  |   |   if E[hₖ(yⱼ)] is null then return false;
11  |   |   if j == 1 then flag ← E[hₖ(y₁)];
12  |   └   else flag ← flag ∩E[hₖ(yⱼ)];
13  └   if flag == 0 then  return false;
14  return true;
```

In Algorithm 1, we give the lookup pseudocode. Similar to the insertion algorithm, we also conduct the same K hash functions to lookup the associated entries. If an associated entry is `null` (line 8), we easily find that no set contains the member $y_j \in q$ and return *false*. Otherwise, we conduct the intersection among the numbers in all entries associated with $|q|$ members $y_j \in q$ (lines 9–10). If the intersection result $flag$ is larger than zero, line 8 then returns *true* (line 12).

Theorem 2. *There is no false negative for Algorithm 1 to answer multi-element check.*

We use Fig. 1 as an example to illustrate the insertion and lookup of BMF as follows.

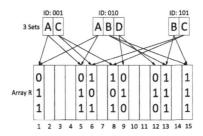

Fig. 1. Example of a BMF: 3 sets $S_1 = \{A, C\}$, $S_2 = \{A, B, D\}$ and $S_3 = \{B, C\}$; 2 hash functions $h_1(A) = 1$, $h_2(A) = 5$, $h_1(B) = 13$, $h_2(B) = 15$, $h_1(C) = 6$, $h_2(C) = 8$, $h_1(D) = 9$, $h_2(D) = 12$.

Example 1. Fig. 1 shows an example of the basic BMF. We have three sets with unique IDs 001, 010 and 101 (with $L = 3$) and two hash functions $h_1(\cdot)$ and $h_2(\cdot)$. The entries in the array $\mathcal{R}$ are shown after the insertion of such three sets.

In addition for set $S_2 = \{A, B, D\}$ with the ID 010, it is not hard to find that the intersection between $E[h_1(A)] = 011$ and the ID $Id_2 = 010$ satisfies the claim, i.e., $011 \cap 010 = 010$; similarly, $E[h_1(D)] \cap Id_2 = 010 \cap 010 = 010$ and $E[h_2(B)] \cap Id_2 = 111 \cap 010 = 010$ hold, too.

Consider query $q = \{A, D\}$. Following the lookup algorithm, we have $E[h_1(A)] = 011$, $E[h_1(D)] = 010$, $E[h_2(A)] = 011$, $E[h_2(D)] = 010$. Now, by intersection, we get $E[h_1(A)] \cap E[h_1(D)] = 010 > 0$ and $E[h_2(A)] \cap E[h_2(D)] = 010 > 0$. Then the lookup returns *true*, indicating that at least one set contains the both members $\{A, D\}$. Among the three sets in Fig. 1, we do have the 2nd set $\{A, B, D\}$ containing such members.

Next consider query $q = \{C, D\}$. We have $E[h_1(C)] = 101$ and $E[h_1(D)] = 010$. By intersection $E[h_1(C)] \cap E[h_1(D)] = 000$, we determine that no set contains the two members $\{C, D\}$, which is the correct answer to the query. However, consider the third query $q = \{A, B, C\}$; and we have false positive issue. Due to $E[h_1(A)] \cap E[h_1(B)] \cap E[h_1(C)] = 001 > 0$ and $E[h_2(A)] \cap E[h_2(B)] \cap E[h_2(C)] = 001 > 0$, we have a return value *true*, indicating that some sets contain such input members. However it's not the truth, leading to a false positive. ∎

Discussion: Based on the example above, we discuss the reason that leads to the false positive issue. First, when a specific element appears in more sets, the entries associated with the element have high chance to be reset to the numbers with more 1-bits during the insertion phase (due to the union $\cup$). In the worst case, the entries are all with L 1-bits leading to the value $2^L - 1$. In the lookup phase to use the intersection $\cap$ operation, such 1-bits lead to the false positive issue.

We are interested in the factors that incur a high false positive rate. (1) As shown above, the element frequency (i.e., the number of sets containing a given element) directly determines the number of 1-bits in the associated entries. In real applications such as keyword search, the frequency of terms in documents is typically very skewed (such as Zipf distribution). It significantly incurs high false positive rate in the lookup phase. (2) The parameters in the basic BMF also decides the false positive rate. For example, a larger size of array $\mathcal{R}$ and larger length L of bits in IDs obviously lead to lower false positive rate. In addition, (3) the number of hash functions is useful to reduce the false positive rate. By the above discussion, we will propose an improved BMF in the next section to optimize the false positive rate of lookup.

4   Improvement

In this section, we introduce four methods to optimize the false positive rate of BMF.

4.1   Bloom Multifilter Group

Our first technique partitions the sets in S into multiple groups of sets. Before giving the detail, we motivate the technique by plotting the term frequency distribution in a real data set. We use a keyword search query log from a real search engine. The log contains totally 3,500,000 queries. Figure 2 plots the top-15 highest frequency. The highest frequency 125,333 indicates that 125,333 queries in this query log contain such a term. If we simply use the basic BMF in Sect. 3 to encode the 3,500,000 sets (each of which represents a query consisting of at least one term), the false positive rate of the basic BMF to answer multi-element check is very high.

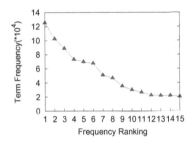

Fig. 2. Term frequency distribution

For a given term with very high frequency, our basic idea to tackle the scalability issue is to use multiple BMFs and partition those sets containing such a term evenly into such BMFs. For each BMF, the term frequency then becomes lower compared with the original one. After that, the lookup needs to check all BMFs one by one. If one of the BMFs returns *yes*, the final answer of multi-element check is *yes*. Otherwise, we continue the check on remaining BMFs. With the basic idea, we formally define the following problem.

Problem 1. Given a fixed size of space cost to maintain m BMFs, we want to design an efficient partitioning algorithm with the objective to minimize the overall false positive rate of multi-element check.

Due to the NP-hard problem, we give a greedy algorithm as follows. First, we calculate the frequency f_i of every term in the group S of sets. Next we create an array for partitioning sets $P[0...m-1]$, each of which maps to a BMF, such that we will use a BMF to encode the sets S_i inside the partitioning set $P[j]$ $(0 \leq j \leq m-1)$. After that, inside the **while** loop (lines 3–10), we process the terms one by one to select those still unprocessed terms by descending order of current term frequency. For a currently selected term t_i, we are interested in those sets S_i containing t_i and add some of such sets to $P[j]$. To this end, we compute an average frequency of t_i by a function ComputeAvg(). Here, for each partitioning set $P[j]$, we define an internal term frequency $f'_{i,j}$ of t_i based by the sets inside

$P[j]$ (instead of the global frequency f_i computed by the global sets $\mathcal{S}$). Then, ComputeAvg() does not need to consider those partitions $P[j]$ associated with the internal term frequency $f'_{i,j} > f_i/m$. That is, the *newAvg* is computed on the global sets $\mathcal{S}$ except those sets inside $P[j]$ with $f'_{i,j} > f_i/m$. After that, for each partitioning set $P[j]$, we determine whether or not the internal frequency $f'_{i,j}$ of t_i is larger than *newAvg*. If true, we randomly choose the amount $(newAvg - f'_{i,j})$ sets from G_i to $P[j]$. In this way, we partition those sets $S[i]$ containing high-frequency terms t_i evenly into partitioning sets $P[j]$.

Algorithm 2. Partitioning Algorithm (Group $\mathcal{S}$ of Sets $S_1, S_2...S_n$, m BMFs)

---

1 Compute the frequency f_i of each term t_i in $\mathcal{S}$;
2 Initiate an array of partitioning sets $P[0...m-1]$;
3 **while** *there still exists unprocessed t_i* **do**
4 | Let t_i to be the unprocessed term with the highest frequency;
5 | Let G_i to be those sets $S_i \subseteq \mathcal{S}$ containing t_i;
6 | int $newAvg = $ ComputeAvg(f_i/m, P);
7 | **for** *each partitioning set $P[j]$ with $0 \leq j \leq m-1$* **do**
8 | | **if** $f'_{i,j} < newAvg$ **then**   randomly choose $(newAvg - f'_{i,j})$ sets from G_i into $P[j]$;
9 | update the term frequency f_i and mark term t_i to be processed;

4.2   Use of the Number of 1-Bits in ID

As shown previously, the number of 1-bits in ID is important to optimize the false positive rate of BMF. It is not hard to find that more 1-bits in ID incur a higher false positive rate. Denote $Count(\cdot)$ to be the number of 1-bits in ID. Consider an example set $S_3 = \{B, C\}$ with $Count(Id_3) = Count(101) = 2$. After the insertion, we have the intersection $E[h(B)] \sqcup E[h(C)] > 0$ (i.e., the condition in the lookup of Algorithm 1). Beyond that, we further find that $Count(E[h(B)] \sqcup E[h(C)]) \geq Count(Id_3) = 2$.

The observation gives us a chance to improve the lookup algorithm by setting a stricter condition. When assigning ID to sets, we denote lb to be the minimal number of 1-bits in all assigned IDs. Then the lookup algorithm needs to satisfy the condition that the intersection result is with at least lb 1-bits. If the condition is dissatisfied, the lookup then returns *false*.

4.3   Single-Hash Insertion

Besides the technique above, we propose our third technique to optimize the false positive rate by directly reducing the number of 1-bits in the entries of array $\mathcal{R}$. One possible solution is to use a smaller $K' < K$ hash functions. It could reduce the amount of 1-bits in the entries of array $\mathcal{R}$. However, for the terms with very high frequency, the associated entries are still with more 1-bits.

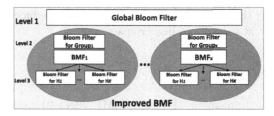

Fig. 3. Improved BMF

To this end, our basic idea is to randomly choose only one from K hash functions in the insertion phase, and still apply all K hash functions in the lookup phase. In more detail, to lookup a query $q = \{t_1...t_{|q|}\}$, we only require that one hash function to be passed. For example, when inserting set $S_1 = \{A, C\}$, we may choose a hash function, say $h_2()$, to insert the ID of S_1 to the two entries $E(h_2(A))$ and $E(h_2(C))$. Next, to lookup $q = \{A, C\}$, we check the entries of each term in q for two hash functions. If any of such two functions can pass the lookup and return *true*, the lookup directly returns *true*, regardless of the another hash function.

For a given set S_i with $|S_i|$ terms, the proposed technique inserts the ID of S_i into $O(|S_i|)$ entries, compared with $K \times |S_i|$ entries inserted by the basic BMF. Thus, the Single-Hash Insertion can dramatically decrease the number of 1-bits in the entries of $\mathcal{R}$. Due to a smaller number of 1-bits in the entries of $\mathcal{R}$, the lookup algorithm, though still with K hash functions, can achieve better false positive rate than the basic BMF.

4.4 Using Standard Bloom Filter

Lastly, we propose a three-level standard Bloom Filters to optimize the false positive rate of lookup. For simplicity, we call the overall structure including the standard bloom filters and BMF together Improved BMF (as shown in Fig. 3). In more detail, we first use a global standard bloom filter to encode all terms appearing in $\mathcal{S}$, with no consideration of the sets S_i that a particular term belongs to. Second, following the partitioning algorithm in Sect. 4.1, we divide each set $S_i \subseteq \mathcal{S}$ into a unique partitioning set $P[j]$, and again use a standard bloom filter to encode all elements appearing in $P[j]$. Thus, for each partitioning set $P[j]$, we maintain a standard bloom filter and a BMF. Finally, as shown in Sect. 4.3, each hash function is randomly selected to insert a set S_i. Thus, for a hash function h_k, we use a bloom filter to encode the elements appearing in those sets S_i that is associated with h_k. Given x groups (Sect. 4.1) and K hash functions (Sect. 4.3), we totally have $1 + x + x \times K$ standard bloom filters in the improved BMF.

Given the improved BMF, we still follow the lookup algorithm onto the BMF in each partitioning set $P[j]$ as before. Besides that, we can use the three level standard bloom filters to check (single) element membership. Our experiment will very soon verify that the improved BMF can greatly reduce the false positive ratio of lookup but at the cost of higher space cost.

5   Experiments

In this section, we implement the basic BMF (Sect. 3) and improved BMF (Sect. 4) and evaluate their performance in three aspects: *false positive ratio, space cost* and *time efficiency*. We use the same data set as Sect. 3. The data set is with 4,000,000 queries (or sets) and 81.32 MB. By splitting the data set into two parts, the first one (with 3,500,000 sets and 72.01 MB) is used for insertion, and the second one (with 500,000 sets and 9.31 MB) for lookup. Table 1 lists the parameters used in the experiment and associated default values.

Table 1. Used parameters

Parameter	Default value
x: Number of BMFs	150
K: Number of hash functions	18
$|\mathcal{R}|$: size of array $\mathcal{R}$	8000000
L: number of bits in ID	20
Level-1 bloom filter	1 KB
Level-2 bloom filter	2.5 KB
Level-3 bloom filter	4 KB

5.1   False Positive Rate

By varying the values of key parameters, we study the effect of false positive rate of BMF, compare the effectiveness between BMF and improved BMF and plot the result in Fig. 4.

Size of Array $\mathcal{R}$: In Fig. 4(a), when the size of $\mathcal{R}$ grows from $6*10^6$ to $18*10^6$, the false positive rate of BMF becomes significantly lower from 31.2 % to 7.1 %. However, the space cost of BMF grows from 17.01 MB to 50.39 MB. This result is obvious and consistent with standard bloom filters: when the size of $\mathcal{R}$ in BMF grows, fewer hash collisions leads to lower false positive rate.

Num. L of bits in ID: Fig. 4(b) shows the relation between L and false positive rate of two different ID assignment approaches. The first approach proposed by Sect. 4.2 is to use a fixed number of 1-bits in ID, and second one is to incrementally increase the value of IDs without any optimization. By varying L from 17 to 38, we have higher space cost from 23.71 MB to 43.74 MB, and lower false positive rate from 29.3 % to 9.1 % for the proposed approach. Yet for the incremental ID, the false positive rate is rather high and slight decreased from 63.2 %

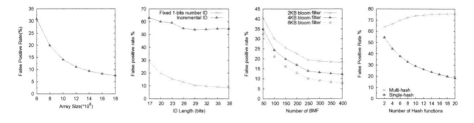

Fig. 4. Effect of false positive rate (from left to right): (a) size of array $\mathcal{R}$; (b) num. L of bits in ID; (c) num. x of BMFs; (d) num. of hash functions

to 55.8 %. This result indicates the benefits of the proposed ID assignment approach by fixing the 1-bits in ID.

Num. x of BMFs: Fig. 4(c) plots the figure of false positive rate with various numbers x of BMFs to show the experiment result of BMF Group proposed by Sect. 4.1. In addition, we vary the space cost of level-3 bloom filters by 2 KB, 4 KB and 8 KB. As shown in this figure, more BMFs lead to smaller false positive rate. Meanwhile, in terms of level-3 bloom filter, when the associated space cost becomes higher, we have chance to achieve lower false positive rate.

Num. K of Hash Functions: Fig. 4(d) shows the relation between K and false positive rate. For comparison, we also implement a variance of BMF by using the same K hash functions for insertion and lookup. Instead, the improved BMF uses the Single-Hash function solution for insertion and K-Hash function for lookup. As shown in this figure, the improved BMF (using single-hash insertion) always outperforms the variance BMF (using multi-hash insertion). Moreover, a larger K leads to lower false positive ratio for our improved MBF and instead higher false positive ratio for the variance BMF. The result clearly indicates the advantage of the improved BMF by using the single-hash insertion.

5.2 Comparison Between BMF and Standard Bloom Filter

In this experiment, we compare the improved BMF with standard bloom filters. In the standard bloom filter, for each set S_i, we find all combinations of the terms in S_i and insert each of such combinations into the standard bloom filters. For fairness, both the standard bloom filters and improved BMF use the same space cost.

In Fig. 5(a), we vary the used space cost by standard bloom filters and improved BMF, and plot the false positive rate of both approaches. This figure clearly indicates that the improved BMF greatly outperforms standard bloom filters when given the same space cost. For example with the space cost 45 MB, improved BM can achieve 9.82 folds of lower false positive rate than standard bloom filters. The higher false positive rate of standard bloom filters is due to the very large amount of elements caused by exponential combinations.

In addition, consider that the improved BMF uses both BMF and three level standard bloom filters. To study the effect of standard bloom filters, we vary

Table 2. Bloom filters and false positive rate

Level 1 bloom filter								
Space cost	0 MB	0.2 MB	0.4 MB	0.6 MB	0.8 MB	1 MB	1.2 MB	1.4 MB
FPR	32.02 %	26.14 %	22.36 %	20.80 %	20.14 %	19.85 %	19.82 %	19.72 %
Level 2 bloom filter								
Space cost	0	0.5 MB	1 MB	1.5 MB	2 MB	2.5 MB	3 MB	3.5 MB
FPR	27.04 %	23.35 %	22.06 %	21.16 %	20.35 %	19.88 %	19.60 %	19.34 %
Level 3 bloom filter								
Space cost	0	1 MB	2 MB	3 MB	4 MB	5 MB	6 MB	7 MB
FPR	36.44 %	30.29 %	25.71 %	22.23 %	19.90 %	18.32 %	17.26 %	16.66 %

the space cost of each level standard bloom filter (and fix the space cost of the other two level standard bloom filters). Table 2 gives the associated false positive rate. For each level standard bloom filters, the more space bloom filter used, the lower false positive rate of the improved BMF has. When comparing the false positive rate of three different level standard bloom filters, we find that level-3 standard bloom filters use the largest space cost than the other two levels in order to achieve the roughly same false positive rate. For example, level 1 uses 1 MB to reach 19.85 %, level-2 uses 2.5 MB to reach 19.88 %, and level 3 uses 4 MB to reach 19.90 %. The main reason is because the overall number of level-3 standard bloom filters is $m \times K$ and the level-1 is only one.

5.3   Time Efficiency

Finally, in Fig. 5(b and c), we plot the running time of lookup of various K hash functions and x BMFs. When either the number K of hash functions or the number of x of BMF numbers grows, the lookup becomes slower. It is not hard to explore the behind reason: a larger K indicates more probes of the array $\mathcal{R}$; more x indicate that the lookup has to check more BMFs. Both obviously lead to higher running time.

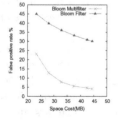 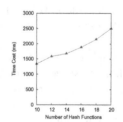 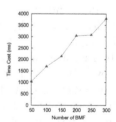

Fig. 5. From left to right: (a) Comparison study btw. BMF and standard BF; Effect of running time by (b) num. of hash functions and (c) num. of BMFs

6   Conclusion

In this paper we proposed a novel data structure BMF to answer the multi-element check with the tradeoff between low space cost, high precision and fast lookup speed. We studied the factors which cause the change in false positive rate, and proposed the optimization techniques for the enhancement of false positive rate. Our experiments verified that the improved BMF can greatly outperform the literature work with much better precision.

As the future work, we continue the better optimization of BMF with lower space cost, higher precision and faster lookup speed. In addition, we are planning the application of BMF onto real applications, such as how to speedup keyword search (such as top-k aggregation queries) to prune irrelevant results.

References

1. Broder, A.Z., Mitzenmacher, M.: Survey: network applications of bloom filters: a survey. Int. Math. **1**(4), 485–509 (2003)
2. Chang, F., Li, K., Feng, W.: Approximate caches for packet classification. In: Proceedings of the 23rd Annual Joint Conference of the IEEE Computer and Communications Societies, IEEE INFOCOM, Hong Kong, China, 7–11 March 2004
3. Guo, D., Wu, J., Chen, H., Luo, X.: Theory and network applications of dynamic bloom filters. In: 25th IEEE International Conference on Computer Communications, Joint Conference of the IEEE Computer and Communications Societies, INFOCOM 2006, Barcelona, Catalunya, Spain, 23–29 April 2006
4. Guo, D., Wu, J., Chen, H., Yuan, Y., Luo, X.: The dynamic bloom filters. IEEE Trans. Knowl. Data Eng. **22**(1), 120–133 (2010)
5. Hao, F., Kodialam, M.S., Lakshman, T.V., Song, H.: Fast multiset membership testing using combinatorial bloom filters. In: 28th IEEE International Conference on Computer Communications, Joint Conference of the IEEE Computer and Communications Societies, INFOCOM 2009, Rio de Janeiro, Brazil, 19–25 April 2009, pp. 513–521 (2009)
6. Lu, Y., Prabhakar, B., Bonomi, F.: Bloom filters: design innovations and novel applications. In: Proceedings of the Forty-Third Annual Allerton Conference (2005)
7. Tapolcai, J., Bíró, J., Babarczi, P., Gulyás, A., Heszberger, Z., Trossen, D.: Optimal false-positive-free bloom filter design for scalable multicast forwarding. IEEE/ACM Trans. Netw. **23**(6), 1832–1845 (2015)
8. Tarkoma, S., Rothenberg, C.E., Lagerspetz, E.: Theory and practice of bloom filters for distributed systems. IEEE Commun. Surv. Tutorials **14**(1), 131–155 (2012)
9. Xie, K., Min, Y., Zhang, D., Wen, J., Xie, G.: A scalable bloom filter for membership queries. In: Proceedings of the Global Communications Conference, GLOBECOM 2007, Washington, DC, USA, 26–30 November 2007, pp. 543–547 (2007)
10. Yang, T., Liu, A.X., Shahzad, M., Zhong, Y., Fu, Q., Li, Z., Xie, G., Li, X.: A shifting bloom filter framework for set queries. PVLDB **9**(5), 408–419 (2016)

Efficient Unique Column Combinations Discovery Based on Data Distribution

Chao Wang, Shupeng Han, Xiangrui Cai, Haiwei Zhang[✉],
and Yanlong Wen

College of Computer and Control Engineering, Nankai University,
38 Tongyan Road, Tianjin 300350, People's Republic of China
{wangc,hansp,caixr,zhanghw,wenyl}@dbis.nankai.edu.cn

Abstract. Discovering all unique column combinations in a relation is a fundamental research problem for modern data management and knowledge discovery applications. With the rapid growth of data volume and popularity of distributed platform, some algorithms are trying to discover uniques in large-scale datasets. However, the performance is not always satisfactory for some datasets which have few unique values in each column. This paper proposes a parallel algorithm to discover unique column combinations in large-scale datasets on Hadoop. We first construct a prefix tree to depict all unique candidates. Then we parallelize the verification of candidates in the same layer of the prefix tree. Two parallel strategies can be chosen: one is parallelizing across all subtrees, the other is parallelizing only in a single subtree. The parallel strategies and pruning methods are self-adaptive based on the data distribution. Eventually, experimental results demonstrate the advantages of the method we proposed.

Keywords: Unique column combinations · Hadoop · Data distribution · Pruning

1 Introduction

Unique column combinations play a fundamental role in understanding the structure of data. In any datasets, column combination whose value is unique can identify a unique row in the collection. The identification of uniques is crucial in many areas of modern data management, including data modeling, query optimization, indexing and data integration. Without the uniques, the datasets will be in a mass, developer and DBAs can't establish an efficient data management system.

Unfortunately, there are many complex and large-scale datasets with incomplete unique column combinations, and the research and industrial communities have paid relatively little attention to finding uniques in large datasets. Most previous work can solve such problem in normal size datasets or samples. However, modern applications produce large-scale data sets are not only in respect of numbers of rows, but also numbers of columns. Discovering all uniques in large data sets is quite difficult, because combination number increases exponentially as column increases. In recent years, some algorithms try to find all uniques in large-scale datasets in parallel on

B. Cui et al. (Eds.): WAIM 2016, Part I, LNCS 9658, pp. 454–466, 2016.
DOI: 10.1007/978-3-319-39937-9_35

distributed platform, but they don't always work well in different datasets even when all datasets have the same size. For example, given a dataset showed in Table 1. Candidates need to be verified as unique by scanning all rows. The column combination {*Name, Major*} would be detected in a new set created by row-based pruning which is widely used in existing algorithms to reduce calculation after verifying {*Name*}. The pruning rules are related to the rows. In these rows, the count of identical tuples projected on {*Name*} is 1 are ignored when creating new set. So {*Name, Major*} would only be verified on row 1 and row 4, so row-based pruning can reduce 50 % of the amount of calculation. But in another scenario, given a dataset showed in Table 2 which has the same size as Table 1, none of the rows can be deleted when pruning because there are no uniques in {*Name*}. {*Name, Major*} must scans all rows from row 1 to row 4, which not only makes the decrease of calculation by row-based pruning ZERO, but also brings extra IO to create new input. Datasets whose rows are similar to each other are very common in finance, medical and biology. Existing parallel algorithms have a poor efficiency dealing with datasets in such data distribution, sometimes it might be even worse than serial algorithms. What we need is a pervasive method that discovers uniques in large-scale datasets.

Table 1. Data sample with unique values

Name	Major	Grade
Lucy	English	Freshman
John	Math	Freshman
Jack	Computer	Junior
Lucy	Computer	Freshman

Table 2. Data sample with no unique values

Name	Major	Grade
Lucy	English	Freshman
John	English	Junior
John	Computer	Junior
Lucy	Computer	Freshman

This paper proposes a parallel algorithm *HUD* (Hadoop based Unique Column Combination Detection) on Hadoop to discover non-redundant uniques. Firstly, we construct a prefix tree to depict all unique candidates, and *HUD* verifies unique candidates in layer sequence in prefix tree. Then, *HUD* calculates the data distribution of given dataset D by a selectivity factor function $T(D)$, the range of $T(D)$ is between 0 and 1, D has more duplicates in each columns when $T(D)$ get closer to 0, on the contrary, D has more unique values in each columns. *HUD* has two different parallel strategies to verify candidates. We set a threshold for $T(D)$ to specify which one of the parallel strategies to be used: when $T(D)$ is greater than threshold, *HUD* verifies unique candidates which have the one and same father node in single subtree simultaneously, which called Parallel Subtree Strategy (*PSS*), when $T(D)$ is smaller than threshold, all column combinations in the same layer in prefix tree are detected simultaneously, which called Parallel Layer Strategy (*PLS*). Furthermore, we use pruning techniques in two parallel strategies to prune candidates. In particular, before non-unique based pruning, *HUD* run our *Non-Unique Discovery Module* which combines *GORDIAN* with *MapReduce* scheme to find the maximal non-unique set which is a parameter in non-unique based pruning.

In general, we make three contributions.

1. We find that the data distribution will influence the efficiency of non-redundant unique discovery, besides the scale of datasets.
2. We propose *HUD*, an efficient parallel algorithm for discovering unique column combinations on Hadoop. *HUD*'s parallel strategies and pruning methods are self-adaptive based on the data distribution of given dataset.
3. We demonstrate the efficiency of *HUD* on a real dataset.

The remainder of this paper is organized as follows. In Sect. 2, we review the related work. Then we discuss pruning techniques in Sect. 3. Section 4 contains the details of HUD, including non-unique discovery module, selectivity factor function and parallel strategies. Section 5 shows the results of an empirical evaluation of *HUD*. At last, we conclude this paper in Sect. 6.

2   Related Work

Currently, only a few known approaches discover all uniques of a table, most of them can only be applied to small data or detect uniques of single column due to the limitation of CPU and memory capacity. [1–4] can be used in large-scale data sets but it can only detect single column uniques. *GORDIAN* [5] is a row-based algorithm. It formulate dataset as a prefix tree and find the maximal non-uniques without scanning all rows in data table. But *GORDIAN* needs to load the entire prefix tree into the memory. The high memory consumption makes it hard to be applied on large-scale datasets. There are some methods based on *GORDIAN*, one of the best is *HCA* [6]. It applies an efficient candidate generation, consideration of column statistics as well as ad-hoc inference and uses FDs to improve the efficiency. However, *HCA* needs a large memory to maintain a large number of statistical data and has an exponential runtime as the number of column grows. Every algorithm we talked above is serial algorithm. As a parallel algorithms, *MRUCC* [7] combines column-based pruning and row-based pruning to improve efficiency. However, as we talked in Sect. 1, row-based pruning has negative effect when a dataset has many duplicates.

All existing works only consider the scale of datasets and do not realize that data distribution can greatly affect the efficiency. Some algorithms and pruning methods are extremely dependent on the number of unique values in each columns. For datasets which have inadaptable data distribution, such as dataset has overwhelming duplicates, these methods have low efficiency dealing with them. *HUD* has obvious advantages, the parallel strategies and pruning methods in *HUD* are self-adaptive based on the data distribution, all methods in *HUD* work efficiently and have no negative effect.

3   Pruning Techniques

In this section, we define the basic concepts of unique column combinations formally and proceed to introduce our pruning techniques.

3.1 Preliminaries

Given a relation R with schema S and instance r, a *unique column combination* is set of one or more columns (attributes) whose projection has only unique rows while *non-unique column combination* has at least one duplicate row. The unique and non-unique are defined as follows [8, 9].

Definition 1. Unique. *A column combination* $K \subseteq S$ *is a unique for* R, *iff* $\forall$ $r_i, r_j \in R$ *and* $i \neq j : r_i [K] \neq r_j [K]$.

Definition 2. Non-Unique. *A column combination* $K \subseteq S$ *is a non-unique for* R, *iff* $\exists$ $r_i, r_j \in R$ *and* $i \neq j : r_i [K] = r_j [K]$.

Definition 3. Minimal Unique. *A column combination* $K \subseteq S$ *is a minimal unique for* R, *iff* $\forall$ $K' \subset K : (\exists$ $r_i, r_j \in R$ *and* $i \neq j : r_i [K'] = r_j [K'])$.

Definition 4. Maximal Non-Unique. *A column combination* $K \subseteq S$ *is a maximal non-unique for* R, *iff* $\forall$ $K' \supset K: (\forall$ $r_i, r_j \in R$ *and* $i \neq j : r_i [K'] \neq r_j [K'])$.

Discovering all minimal unique column combinations in relation R with n columns needs to check all $2^n - 1$ column combinations, each check involves scanning over all rows to search for duplicates. That is the only way to check if the column combination is a unique. We model all column combinations as a prefix tree, for example, given a relation R with schema (a, b, c, d), the prefix tree for R is showed in Fig. 1. The complexity of traversing all column combinations in prefix tree is $O(2^n \times Rows)$ which shows the brute force is impossible. So we use pruning techniques here to promote efficiency.

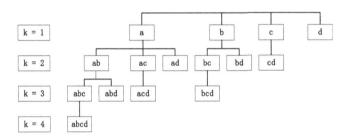

Fig. 1. Prefix tree. We build the prefix tree by lexicographical order. K in each layer means the number of columns. For 4 columns (attributes in a relation), there are 15 column combinations to be checked

3.2 Non-Unique Based Pruning

Non-unique based pruning is based on definition 4 to avoid scanning all rows to search for duplicates. Before non-unique based pruning, we first execute the *Non-Unique Discovery Module* which is introduced in Sect. 4 to find the maximal non-unique set S_{nu} in a given dataset. In pruning process, if a column combination K under verification is included in any subset of S_{nu}, it is a redundant non-unique.

For example, we verify unique candidates for a given relation R showed in Fig. 1. If column combination $\{abc\}$ is a non-unique detected by non-unique discovery module, the column combination $\{a\}$, $\{b\}$ and $\{c\}$ do not need to be checked when $K = 1$. So do the column combination $\{ab\}$, $\{ac\}$ and $\{bc\}$ when $K = 2$ and $\{abc\}$ when $K = 3$.

3.3   Functional Dependency Based Pruning

Functional dependency is a very important concept in database. Given a relation R, a set of attributes A in R is said to functionally determine another set of attributes B (written $A \rightarrow B$) if, and only if, each A value is associated with precisely one B value. In this paper, we focus on functional dependency of single attribute, that is to say, both A and B are single column.

We discover FDs by counting the number of different values projected on each columns:

$$A \rightarrow B, if\ Uc(AB) = Uc(A)$$

where $Uc\ (X)$ is the number of different values projected on column combination $\{X\}$.

For verification based on FDs, given a relation R with attributes A, B, C and a FD $A \rightarrow B$: The column combination $\{AC\}$ is a unique when $\{BC\}$ is a unique while $\{BC\}$ is a non-unique when $\{AC\}$ is a non-unique. For example, given a relation shown in Fig. 1. We get the FD $b \rightarrow a$ when checking $\{ab\}$ and detect $\{ac\}$ is a unique after checking $\{ac\}$.When $\{bc\}$ is checked, it can be directly marked as unique by FD without scanning all rows.

3.4   Cartesian Product Based Pruning

Cartesian Product Based Pruning is a very sample idea based on statistics. Given a unique candidate K consisted of column combinations K_1 and K_2, assuming the number of different values are n_1 and n_2 respectively projected on K_1 and K_2. If $n_1 * n_2$ is less than the number of rows of the dataset, then K is a non-unique. In particular, K can not be regarded as a unique even $n_1 * n_2 > rows$. To avoid the exponential time complexity, we only divide K into two uncorrelated column combinations which appear in the past path in prefix tree.

3.5   Unique-Based Pruning and Row-Based Pruning

Unique-based pruning is also called column-based pruning and row-based pruning have been widely used in many previous works. Unique-based pruning is based on Minimal Unique (Definition 3). If unique candidate K includes any elements in unique set S_u, then K is a redundant unique.

Row-based pruning improves efficiency by reducing the size of input. The main idea behind it is to create a new input by removing the tuples which appear only once in the projection of a column combination K for verifying the K's descendants.

4 HUD Algorithm

In this section, we introduce our unique discovery algorithm HUD. It takes large-scale dataset as input and uses different parallel strategies (PLS and PSS) based on data distribution to find all uniques. We first introduce the *Non-Unique Discovery Module* for non-unique based pruning. Then, we give the details about the parallel strategies PLS and PSS in HUD. At last, we introduce a formula based on statistics to describe the data distribution and determine which strategy to use for given datasets.

4.1 Non-Unique Discovery Module

Non-Unique Discovery Module is crucial point for non-unique based pruning. It is a parallel method to discover maximal non-unique set based on $GORDIAN$.

To solve the time crisis in large-scale datasets, we improve $GORDIAN$ with *MapReduce* model in Hadoop, the whole model is showed in Fig. 2. Firstly, the Hadoop distributed file system HDFS divide the dataset into fixed size. Then each sub-dataset can execute the *Map* function independently. In *Mapper*, we use GOR-$DIAN$ to discover non-uniques in sub-dataset. In *Reducer*, non-uniques from all *Maps* are sent to *Reduce* node to remove the redundant non-uniques. The output of *Reducer* is the maximal non-unique set.

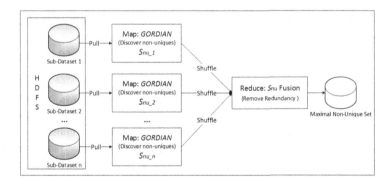

Fig. 2. Non-Unique Discovery Module. *Map* pull Sub-dataset from HDFS. The non-uniques set discovered in each *Map* is written as S_{nu}. Shuffle is the transmission mechanism for data transmission from *Map* to *Reduce*

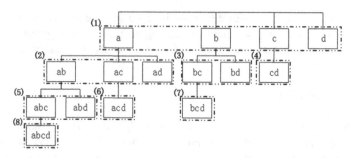

Fig. 3. Verification order of *PSS*. Column combinations in grid would be checked simultaneously. The number in top left corner of each grid shows the order of verification

4.2 Parallel Subtree Strategy

Parallel Subtree Strategy (*PSS*) verifies all unique candidates by traversing prefix tree in order of layer and all candidates that belong to the same father node would be checked simultaneously. Figure 3 illustrates the verification order of prefix tree showed in Fig. 1 in *PSS*.

The procedure of *PSS* (Algorithm 1) can be divided into three phases.

(1) First of all, we generate unique candidates for each verification. *PSS* uses queue to store non-uniques, we pop the head of queue each time and generate its descendant nodes in prefix tree as the unique candidates (lines 2–3). Initially, we push the root of prefix tree into queue, the descendant nodes of root are all single column combinations.

(2) For each candidate in unique candidates, we first try to detect it to be unique or non-unique by pruning techniques before checking all rows (lines 4–14). In *PSS*, there are four pruning methods to be used (line 5): unique-based pruning, non-unique based pruning, FDs-based pruning and Cartesian product based pruning. Once candidate K is detected to be unique, it would be put into the unique set (line 7). On the contrary, it would be put into the end of queue when K is a non-unique (line 10).

(3) After pruning, there may be some candidates without result and we will check these candidates simultaneously on Hadoop by scanning all rows. *MRCheckDP* (line 19) is a unique verification method that contains a *Mapper* and a *Reducer*. The former is responsible for the projection of each column combination, and the latter is in charge of aggregation. If the number is equal to the row number, the candidate is a unique. The data from *Mapper* to *Reducer* are formed as <Key-Values>. We use row-based pruning to materialize the new table in order to accelerate the verification of subtree (line 25), and find possible FDs when candidate is consisted of two columns (line 24).

Algorithm 1. Parallel Subtree Strategy

```
Input : Dataset, NonUniques
Output: Uniques
 1: while Queues is not empty do
 2:     currentNonUnique = GetandRemoveFirst(Queues)
 3:     Candidates = PSSGenerate(currentNonUnique)
 4:     for each candidate in Candidates do
 5:         if PSSCheckByPruningTechs(candidate)
 6:             if candidate is a unique
 7:                 Uniques.add(candidate)
 8:             end if
 9:             if candidate is a non-unique
10:                 Queues.add(candidate)
11:             end if
12:             Candidates.remove(candidate)
13:         end if
14:     end for
15:     if Candidates is empty
16:         continue
17:     end if
18:     Get new input created by row-based pruning
19:     MRCheckDP(newInput, Candidates)
20:     for each candidate in Candidates do
21:         if candidate is unique
22:             Uniques.add(candidate)
23:         end if
24:         DiscoverFDs(candidate)
25:         RowPruning(candidate, DataSet)
26:     end for
27: end while
28: return Uniques
```

4.3   Parallel Layer Strategy

Parallel Layer Strategy (Algorithm 2) has different idea with *PSS*. *PLS* goes across all subtrees and verifies all unique candidates in the same layer in prefix tree simultaneously. Figure 4 illustrates the verification order in *PLS*, we can see the difference between *PLS* and *PSS* clearly.

Although *PLS* has completely different strategy with *PSS*, *PLS* has approximately the same procedure of *PSS*. For simplicity, we only introduce the difference in the procedure, which can be divided into three parts.

(1) The maximum number of loops is the height of prefix tree which is equal to the number of single columns in dataset. We keep all non-uniques in a container, when the container is empty, *PLS* can break the loop to end the algorithm (lines 23–25).

(2) We use all pruning methods in *PSS* except FDs-based pruning (line 4). Because candidates across the subtree are verified simultaneously and FDs-based pruning only works when detecting candidates in the same layer in another subtree.

(3) *MRCheckNDP* (line 17) has the same mechanism with *MRCheckDP*, but we only form data as *<Key>*, because we don't use row-based pruning in *PLS* to create new input, abandoning values could reduce memory consumption.

Algorithm 2. Parallel Layer Strategy

```
Input  : Dataset, NonUniques
Output : Uniques
 1: while i = 1 to columns
 2:     Candidates = PLSGenerate(NonUniques, i)
 3:     for each candidate in Candidates do
 4:     if PLSCheckByPruningTechs(candidate)
 5:         if candidate is a unique
 6:            Uniques.add(candidate)
 7:         end if
 8:         if candidate is a non-unique
 9:            NonUniques.add(candidate)
10:         end if
11:         Candidates.remove(candidate)
12:     end if
13:     end for
14:     if Candidates is empty
15:         continue
16:     end if
17:     MRCheckNDP(Candidates)
18:     for each candidate in Candidates do
19:         if candidate is unique
20:            Uniques.add(candidate)
21:         end if
22:     end for
23:     if NonUniques is empty
24:         break
25:     end if
26: end while
27: return Uniques
```

4.4 Parallel Strategy Selection

As we talked above, there is no algorithm works well in all datasets. For different datasets, *HUD* (Algorithm 3) first discover maximal non-unique set by *Non-Unique Discovery Module*, then it uses proper method, *PSS* or *PLS*, which depends on the given dataset's data distribution to minimize the time consumption of finding all unique column combinations. We propose a simple selectivity factor function $T(D)$ based on statistics to describe the data distribution as followed.

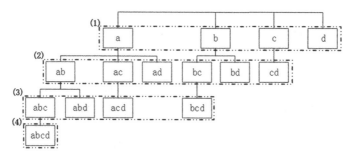

Fig. 4. Verification order of *PLS*. Column combinations in grid would be checked simultaneously. The number in top left corner of each grid shows the order of verification

$$T(D) = \frac{\sum_{i=1}^{N} U(C_i)}{N \times M} \tag{1}$$

where $U(C_i)$ denotes the number of different values projected on single column C_i. N is the number of columns and M is the number of rows in dataset.

$T(D)$ which range between 0 and 1 indicates the characteristic of given dataset D. The experiment results in Sect. 5 show that, *PLS* is better for D when $T(D) < 0.33$ and *PSS* is better when $T(D) > 0.33$. So we set the threshold to 0.33.

Algorithm 3. HUD

```
Input  : Dataset
Output : Uniques
1: NonUniques=NonUniquesFinder(Dataset)
2: Calculate Cardinality from Counters
3: if Cardinality ≥ Threshold
4:    Uniques = PSS(DataSet, NonUniques)
5: else
6:    Uniques = PLS(DataSet, NonUniques)
7: return Uniques
```

---

5   Experiment

In this section, we use a series of experiments to evaluate the efficiency of our method. In unique discovery field, accuracy must be 100 %, so we only evaluate runtime. Our Hadoop environment has totally 4 nodes: one master and 3 slaves.

5.1   Data Sets

We use one synthetic and one real-world datasets in our experiments: TPC-H is a benchmark database which consists of a suite of business oriented ad-hoc queries and concurrent data modifications; delivery is a real-world database of information management system from the insurance field. The statistics of each data set are shown in Table 3.

Table 3. Statistics of data sets

DataSet	Table	Avg col	Max col	Avg row	Max row
TPC-H	8	7	16	1082655	6001215
Delivery	92	6	75	66110	1786228

5.2   Evaluation of Efficiency

We first evaluate our algorithm with regard to increasing number of rows. We select 10 tables from Delivery with 12 columns differing in the row-count ranging from 5000 to 50000, and the step is 5000. We compare our algorithm with *GORDIAN, MRUCC, HCA, PLS* without *Non-Unique Discovery Module (PLS-NG)* and *PSS* without *Non-Unique Discovery Module (PSS-NG)*. Figure 5 illustrates the runtime of all algorithms. The experimental results show *HUD* is a linear growth method which has a better performance and high scalability. In addition, *Non-Unique Discovery Module* gets more complete maximal non-unique set with increasing number of rows. It will greatly reduce candidates later.

To analyze the effect of column number, we compare our method with *GORDIAN, MRUCC, HCA, PLS-NG* and *PSS-NG* in tables selected in Delivery with 15000 rows differing in the column-count ranging from 6 to 12, the step is 1. Figure 6 illustrates the runtime of all algorithms. It is easy to see that *MRUCC* and *HCA* are exponential growth with column grows, *PLS* and *PSS* in *HUD* grows slowly and smoothly, In conclusion, *HUD* algorithm is more suitable for dealing with large data sets.

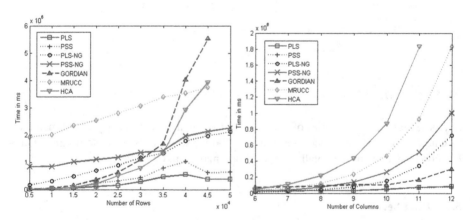

Fig. 5. Row number effect   **Fig. 6.** Column number effect

5.3   Selectivity Factor

In above experiments, we use both *PLS* and *PSS* in *HUD* to evaluate efficiency, in fact, only one strategy can be used in each dataset. Function $T(D)$ indicate the data distribution of dataset D that the range is between 0 to 1. We randomly select 9 tables in TPC-H and Delivery, 5 from TPC-H and 4 from Delivery. All tables are in same size, but have different $T(D)$. The minimum $T(D)$ is 0.03 and the maximum is 0.857. We evaluate the runtime in both *PLS* and *PSS* and show the results in Fig. 7.

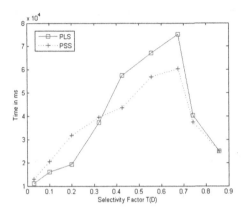

Fig. 7. Runtime with respect to increasing $T(D)$

The experimental results show that *PSS* and *PLS* are applicable to different data distributions, respectively. When function $T(D)$ is between 0.33 and 0.857 *PSS* has better performance than *PLS*. Because there are more unique values in each columns with the $T(D)$ grows, FDs-based pruning can reduce more candidates and row-based pruning generates smaller input to reduce the time cost of scanning all rows greatly. When $T(D)$ is between 0.03 and 0.33, *PLS* is better than *PSS*. Because there are more duplicates in each column, the new input created by row-based pruning not only may almost has the same size as the original dataset but also brings extra IO.

In particular, when $T(D)$ approaches 0 and 1, *PLS* has almost the same performance as *PSS*. $T(D)$ is 1 means that every single column is a unique, when $T(D)$ gets closer to 1 ($T(D) \geq 0.857$), unique-based pruning can prune almost all candidates in second layer of prefix tree both in *PLS* and *PSS*. When $T(D)$ approaches 0, complete maximal non-unique set becomes easier for *Non-Unique Discovery Module* to get, non-unique based pruning can prune almost all candidates without verification. In both situation, *PLS* and *PSS* have few calculation and almost the same runtime. To simplify the algorithm implementation, we set the threshold to 0.33, when $0 \leq T(D) < 0.33$, *HUD* uses *PLS* and when $0.33 \leq T(D) \leq 1$, *HUD* uses *PSS*.

6 Conclusion

A unique is the basis of understanding a data table, but it is often incomplete in large data set. This paper proposes a distributed non-redundant unique discovery algorithm named *HUD* on Hadoop. Additionally, we use different parallel strategies which depends on data distribution and pruning techniques to improve efficiency. One interesting future work is implementing *HUD* in Spark and improving the selectivity factor function to describe the data distribution more precisely.

Acknowledgments. This work is partially supported by National 863 Project of China under Grant No. 2015AA015401, and Research Foundation of The Ministry of Education and China Mobile under Grant No. MCM20150507. This work is also partially supported by Tianjin Municipal Science and Technology Commission under Grant Nos. 13ZCZDGX01098, 14JCQNJC00200 and 14JCYBJC15500.

References

1. Brown, P., Haas, P.J., Myllymaki, J., Pirahesh, H., Reinwald, B., Sismanis, Y.: Toward automated large-scale information integration and discovery. In: Härder, T., Lehner, W. (eds.) Data Management in a Connected World. LNCS, vol. 3551, pp. 161–180. Springer, Heidelberg (2005)
2. Bell, S., Brockhausen, P.: Discovery of constraints and data dependencies in relational databases. In: Lavrač, N., Wrobel, S. (eds.) ECML 1995. LNCS, vol. 912. Springer, Heidelberg (1995)
3. Kivinen, J., Mannila, H.: Approximate dependency inference from relations. Theoret. Comput. Sci. **149**, 129–149 (1995)
4. Petit, J.-M., Toumani, F., Boulicaut, J.-F., Kouloumdjian, J.: Towards the reverse engineering of renormalized relational databases. In: Proceedings of the ICDE, pp. 218–227 (1996)
5. Sismanis, Y., et al.: GORDIAN: efficient and scalable discovery of composite keys. In: Proceedings of the 32nd International Conference on Very Large Data Bases. VLDB Endowment (2006)
6. Abedjan, Z., Naumann, F.: Advancing the discovery of unique column combinations. In: Proceedings of the 20th ACM International Conference on Information and Knowledge Management. ACM (2011)
7. Han, S., Cai, X., Wang, C., Zhang, H., Wen, Y.: Discovery of unique column combinations with hadoop. In: Chen, L., Jia, Y., Sellis, T., Liu, G. (eds.) APWeb 2014. LNCS, vol. 8709, pp. 533–541. Springer, Heidelberg (2014)
8. Heise, A., Quiané-Ruiz, J.A., Abedjan, Z., et al.: Scalable discovery of unique column combinations. Proc. VLDB Endowment **7**(4), 301–312 (2013)

Event Related Document Retrieval
Based on Bipartite Graph

Wenjing Yang, Rui Li$^{(\boxtimes)}$, Peng Li, Meilin Zhou, and Bin Wang

Institute of Information Engineering, Chinese Academy of Sciences, Beijing, China
`lirui@iie.ac.cn`

Abstract. Given a short event name, event retrieval is a process of retrieving event related documents from a document collection. The existing approaches employ the-state-of-art retrieval models to retrieve relevant documents, however, these methods only regard the input query as several keywords instead of an event, thus the special aspects of the event are not considered in the models. Aiming at this problem, we first propose a novel bipartite graph model to describe an event, where one bipartition represents event type and the other represents the event specific information. Each edge between two bipartitions issues co-occurrence relationship. Then we model an event with a unigram language model estimated through the corresponding bipartite graph. Based on KL-divergence retrieval framework, event model is integrated into the query model for more accurate query representation. Experiments on publicly available TREC datasets show that our method can improve the precision@N metric of event retrieval.

Keywords: Event retrieval · Bipartite graph · Event model · Query model

1 Introduction

Many public events (natural disasters, protests, accidents, etc.) once happen, an explosive growth of online reports causes information overload for individual users. A method that can retrieve the relevant documents about the specific event from the massive collection is essential. Event Retrieval (ER) is a task which takes a short event name as input and returns event related documents to satisfy the information need about the specific event. As a useful fundamental task, ER is now becoming increasingly important and has been used in many different application domains, such as event summarization, sub-event detecting, event monitoring, and so on. The rapid rise of the practical scenarios concentrating upon events has generated much interest in ER research.

The traditional retrieval methods assume that a document associating with higher similarity with query is more relevant. As shown in Fig. 1(a), these approaches score the documents via keywords queries and rank by $sim(q,d)$. While the goal of ER is to gather event-related documents, a document which

© Springer International Publishing Switzerland 2016
B. Cui et al. (Eds.): WAIM 2016, Part I, LNCS 9658, pp. 467–478, 2016.
DOI: 10.1007/978-3-319-39937-9_36

describes the actuality or history of the specific event either in general outline
or in much detail is markable and greatly desired. The relevance between an
event and a document is $sim(e,d)$, showed in Fig. 1(b). Especially, we note that
$sim(e,d)$ is not equivalent to $sim(q,d)$: high $sim(q,d)$ value does not mean high
$sim(e,d)$ value and vice versa. E.g., given an event query q "*Queensland flood*",
though a finance article that says the stocks and bonds are affected by Queensland flood, obtains high $sim(q,d)$ value, in fact it should be neglected due to
its low $sim(e,d)$. On the other hand, with high $sim(e,d)$ score, an article that
depicts the floods in Bundaberg, Brisbane (cities in central and southern Queensland) might be belittled in $sim(q,d)$ for the absence of keyword "*Queensland*".
The inequivalence between $sim(e,d)$ and $sim(q,d)$ may lead unsatisfied retrieval
performance by applying traditional methods directly to ER task.

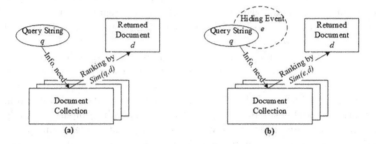

Fig. 1. The difference between query-related and event-related retrieval

The differences between $sim(e,d)$ and $sim(q,d)$ lies in the fact that the short
event query string q can hardly preserve the rich information of the hiding event
e. It indeed presents a major challenge to improve the retrieval effectiveness
in ER task. So the motivation of our work is to bridge the gap between the
event query string q and the hiding event e. The key point of our approach is to
model the event directly, then incorporate the event model into the query model
for a more accurate query representation. Next, we focus on following research
questions: (1) How to model an event? (2) How to apply the event model in ER
process?

The remainder of this paper is structured as follows. Section 2 states the
related work. In Sect. 3, we discuss how to model an event via bipartite graph,
leveraging the event type information from external resources. Section 4 details
how to estimate the event query model for retrieval. In Sect. 5, we present the
experimental setup and results. Finally, in Sect. 6, we summarize our method
and present the future work.

2   Related Work

To the best of our knowledge, event based retrieval has been studied well. The
computational cost and inflexibility make existing approaches tend to be insignif-
icant or at least not so significant. The common retrieval methods, keyword query

with pseudo feedback, despite its simplicity, have been remained to be the most popular ways in most event retrieval cases.

Existing event-based retrieval methods either consider a proper event representation or propose a method to measure the event-document similarity. Typically, in [1], to understand the event semantically, natural language processing tools are applied to parse each sentence within textual documents into three elements: subject, object and predicates, which can subsequently be used for event retrieval. While in [2], the authors structure both queries and documents as graphs of event mentions and employ graph kernels to measure $sim(q, d)$. These two methods are both sophisticated and time-consuming. Furthermore, they greatly depend on natural language processing results which are not good enough till now. Some others [3,4] are focus on the event search application, which are not the concerns of this paper.

So far at least, in most event-based retrieval cases, people still regard the event query as several keywords rather than an event object, and use pseudo-relevance feedback (PRF) to expand the query and re-rank the documents. Reference [6] indicates that a language model incorporating feedback generally gives the best performance, compared with traditional models(e.g. Rocchio for the vector space model). A new family of models called probabilistic distance retrieval model is standing out for better accommodating feedback. These models score a document with respect to a query based on the distance between the corresponding documents language model and the query language model. Feedback can be flexibly casted as to improve the estimation of the query language model. Based on KL-divergence retrieval model, representative methods for estimating query models are Relevance Model [7], Divergence Minimization Model [8], Mixture Model [8], Markov Chain [5], etc. Some existing work has also attempted to further extend PRF in language model by incorporating additional evidences(e.g.Term proximity for Relevance Model [9], maximum entropy for Divergence Minimization Model [10]).

Generally, the role of the query model is to incorporate knowledge of information need clearly. However, in ER task, the re-estimated query model by above methods can hardly give prominence to the characteristics of the target event. The motivation of our work is to reconcile the query model with the target event, as accurate query model is more likely to achieve high retrieval effectiveness.

3   Model the Event via Bipartite Graph

A standard definition of event does not yet exist. "Something that happens at a particular time and place" [11] is a possible definition which is generally accepted. Though it is hard to define event, it is easier to recognize parts of event identity, the properties that make two events the same.

With the aid of event identity, we have first got to motivate an event representation method, so that a certain event could be exclusively determined. We draw inspirations from two observations about event identity: (1) events that belonged to the same category classified by event type share common information in aspects of event venation and event context; (2) The specific information of an event (the location, the time, the core persons, etc.) distinguishes it from broader classes of events. E.g., "Russia protests against legislative election" and "Egyptian protests against then-President Morsi" are events whereas "protests" is the more general class containing them. The two stories share the properties "protests": demonstration by crowd in favor of political or other cause. Further additional specific information (election, Putin, Russia, etc.) facilitates narrowing the target event to the Russia protests caused by fraud election.

In light of above observations, the cognition process of an event can be considered in two-fold:(1) considering the type of event, which helps to depict the outline and context of the event, then (2) considering the unique information of the specific event, which differentiates it from similar events that belonged to a same event category. Thus, we propose to consturct a bipartite graph to identify an event from these two aspects. Next, we first introduce several related concepts (Sect. 3.1) and detail bigraph representation for an event (Sect. 3.2), then, propose the notion of event model (Sect. 3.3).

3.1 Several Notions

We state several important notions which will be involved later:

Bipartite Graph: A bipartite graph (or **bigraph**) is a graph whose vertices can be divided into two disjoint sets U and V. Every edge e connects a vertex in U to one in V. Formally it is denoted by $G =< U, V, E >$.

One-Mode Projection: The one-mode projection onto U (**U projection** for short) means a network containing only U nodes, where two U nodes are connected when they have at least one common neighboring V node.

Resource-Allocation Weighting: Zhou et al. [12] proposed this method to determine the edge weight in U projection. It assumes that a certain amount of resource is associated with each node in the projection and the directional weight w_{ij} represents the proportion of the resource node j would like to distribute to node i. The resource allocation process is based on the bipartite graph, involving *equal distribution* across neighbors and consisting of two steps: first from U to V, and then back.

Event Model: We suppose that there is an accurate **unigram language model** of the specific event, named event model θ_E. All the event-related documents and event query are randomly sampled from the underlying θ_E.

3.2 Event Representation via Bipartite Graph

Given an event with several event-related documents, we intend to represent the unique event by combining information from two facets: event type (**ET**) and event specific information (**ES**). The ET word set is denoted as $ET = \{t_1, t_2, \cdots, t_n\}$ and the ES word set is denoted as $ES = \{s_1, s_2, \cdots, s_m\}$. ET and ES words interact in event snippets. Reflected onto the bigraph, the event can be formulated as $Event =< ET, ES, Connections >$. An example is shown in Fig. 2(a). Algorithm 1 implements the construction of the graph.

ET: In bigraph, $ET = \{t_1, t_2, \cdots, t_n\}$ consists of salient words of a certain event type. Given an event (event type is given beforehand), sufficient auxiliary information can be obtained from external sources, by counting the words in a collection of events with the same event type.

ES: To simply the problem slightly, we generally assume that word w in descriptive event snippets is sampled from either ET or ES set. Excluding ET words, the remaining words in event snippets are added to the ES set.

Connections: The co-occurrence issues the relationship between ES and ET. *Connections* can be fully described by an $n \times m$ matrix $\{a_{ij}\}$, $a_{ij} = k$, $k > 0$, if s_j has already been co-occured with t_i in same sentences for k times in event-related documents.

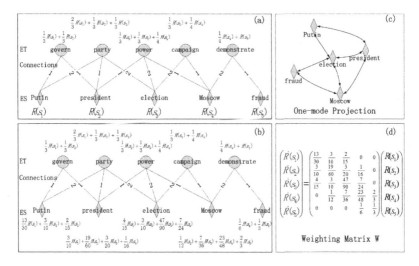

Fig. 2. An example of event bigraph representation. (a) and (b) show two-step modified resource-allocation weighting scheme, from ES to ET, then back. (c) represents the one-mode projection onto ES. (d) shows the corresponding weighting matrix W.

3.3   Event Model

As stressed before, an event model is a unigram language model. If we have available event description snippets enough, we could estimate $p(w|\theta_E)$ by counting the number of occurrences of w in the event-relevant documents and smoothing the counts. However, the main obstacle is that no labeled training data is available to get the probability distribution $p(w|\theta_E)$.

To approximate the event model, we assume that event model is a mixture model. The words in event-related documents are drawn from two models, event type model $p(w|\theta_{ET})$ and event specific model $p(w|\theta_{ES})$. It can be written as:

$$p(w|\theta_E) = \alpha p(w|\theta_{ET}) + (1-\alpha)p(w|\theta_{ES}) \qquad (1)$$

Algorithm 1. Event bigraph constructing algorithm

Input:
    target event e, event type T, event-related(e) documents $\mathbb{D}$
Output:
    A bigraph $G =< U, V, E >$ to represent e
1: $G, U, V, E \leftarrow \emptyset$
2: $Eset \leftarrow \{eve|eve$ is an event, $eve.event\_type = T\}$ by searching in Wikipedia
3: kowledge base $\mathbb{KB} \leftarrow \{page|page = Article(eve), \forall eve \in Eset\}$
4: $\phi \leftarrow$ words distribution with using MLE on $\mathbb{KB}$
5: $ET = \{w|p(w) \in \phi, p(w) > THRESTHOLD\}$ or cut-off with top N words.
6: $U \leftarrow ET$
7: **for** each $d \in \mathbb{D}$ **do**
8:     **for** each $sentence \in d$ **do**
9:         $et_s \leftarrow ET \cap words(sentence)$
10:        $es_s \leftarrow words(sentece) - et_s$
11:        **if** $et_s, es_s \neq \emptyset$ **then**
12:           $V \leftarrow V \cup es_s$, $Edges \leftarrow \{(a, b, 1)|a \in et_s, b \in es_s\}$
13:           $E \leftarrow E \cup Edges$
14:        **end if**
15:     **end for**
16: **end for**
17: $V \leftarrow V$ is pruned by degree centrality (dc_{thres})
18: **return** $G =< U, V, E >$

Estimating $p(w|\theta_{ET})$: It is mentioned ahead events within the same event category (T) are involved to estimate $p(w|\theta_{ET})$.

$$p(w|\theta_{ET}) = \frac{count(w, \mathbb{KB})}{|\mathbb{KB}|} \tag{2}$$

where $\mathbb{KB}$ is the external knowledge base for a certain event type. $count(w, \mathbb{KB})$ is the count of word w in the $\mathbb{KB}$, and $|\mathbb{KB}|$ is the total number of words in $\mathbb{KB}$.

Estimating $p(w|\theta_{ES})$: The difficult stage is estimating $p(w|\theta_{ES})$ for the lack of sufficient specific event data. In bigraph, if a node in ES set is tightly and widely linked to the nodes in ET set, it is worthwhile to emphasize this particular node, as it is more likely to be the key elements. In spirit of this, we estimate $p(w|\theta_{ES})$ based on one-mode projection to compress the connections onto ES set with modified Resource-allocation Weighting method (*ratio distribution*, instead of *equal distribution*) to retain the original information better.

Now, consider the event bigraph $Event =< ET, ES, Connections >$. For convenience, we write $Connections$ as $A = \{a_{ij}\}_{n \times m}$. We are looking for a matrix $W = \{w_{ij}\}_{m \times m}$ representing the weighted ES projection. Two steps will be involved in resource-allocation process. Assuming that the initial resource located on the ith ES node is $R(s_i)$. After the first step, showed in Fig. 2(a), all the resource in ES flow to ET, and the resource located on the lth ET node reads

$$R(t_l) = \sum_{i=1}^{m} a'_{li} R(s_i) \tag{3}$$

where $A' = \{a'_{ij}\}_{n \times m}$, $a'_{ij} = \frac{a_{ij}}{\sum_{j=1}^{m} a_{ij}}$, namely, row by row normalization of matrix A. In the next step, all the resource flows back to ES, as shown in Fig. 2(b), and the final resource located on s_i reads

$$R'(s_i) = \sum_{l=1}^{n} a''_{li} R(t_l) = \sum_{l=1}^{n} a''_{li} \sum_{j=1}^{m} a'_{lj} R(s_j) \tag{4}$$

where $A'' = \{a''_{ij}\}_{n \times m}$, $a''_{ij} = \frac{a_{ij}}{\sum_{i=1}^{n} a_{ij}}$, A is normalized by column. Above equation can be rewritten as

$$R'(s_i) = \sum_{j=1}^{m} w_{ij} R(s_j) \tag{5}$$

where

$$w_{ij} = \sum_{l=1}^{n} a'_{li} a''_{lj} \tag{6}$$

It indicates that two nodes get closer when they share more resources through the two-step paths. It also can be written in matrix form as $W = A'^{T} A''$. w_{ij} can be considered as the importance of node i in j's sense in ES set, and it is generally not equal to w_{ji}, as expressed in Fig. 2(d).

We wish to estimate $p(w|\theta_{ES})$, which measures the words' relative importance in ES set. W records the nodes' relations in ES. Normalized by column, $W' = \{w'_{ij}\}_{m \times m}, w'_{ij} = \frac{w_{ij}}{\sum_{j=1}^{m}}$, can be regard as the transition matrix of a Markov chain. The centrality vector $\boldsymbol{p}$ corresponds to the stationary distribution W'. Borrowing the idea of PageRank, we weight each word in ES by considering where the votes come from and taking the centrality of the voting words into account. The formulation can be expressed by the matrix form equation

$$\boldsymbol{p} = [d\boldsymbol{U} + (1 - d)W']^{T} \boldsymbol{p} \tag{7}$$

where $\boldsymbol{U}$ is a square matrix with all elements being equal to $\frac{1}{m}$. A walker on this Markov chain chooses one of the adjacent states of the current state with probability $1 - d$, or jumps to any state in the graph with probability d. We assign the stationary distribution $\boldsymbol{p}$ on ES to $p(w|\theta_{ES})$.

4   Event Retrieval Using Event Model

In previous section, a graph-based method was proposed to optimize the event representation and used for estimating the event model θ_E. But, it offers no answer to such a question: how to improve the event retrieval accuracy by leveraging event model. Next, we first introduce a probabilistic distance retrieval framework, then we explore incorporating event model into language query model under this framework.

4.1 KL-Divergence Retrieval Model

The KL-divergence retrieval model represents the state-of-the-art of the language modeling approaches to retrieval. It assumes that the query is a sample observed from a query language model θ_Q, while the document is a sample from a document language model θ_D. KL-divergence distance of these two models is used to measure how close they are to each other and the documents are ranked by the distance (indeed, negative distance). Formally, the score of a document D w.r.t a query Q is given by:

$$Score(D, Q) = -D(\theta_Q \parallel \theta_D) = -\sum_{w \in V} p(w|\theta_Q) log \frac{p(w|\theta_Q)}{p(w|\theta_D)} \tag{8}$$

Generally, the estimation of θ_Q offers interesting opportunities of leveraging feedback information or other user information to improve retrieval accuracy. In event retrieval cases, a major deficiency of other query model estimation methods is that it cannot easily achieve full benefit of feedback due to the limited consideration of event object. Thus, it is worth exploring that whether the performance will be improved if event model is accommodated into the query model.

4.2 Event Query Model

In event retrieval, ideally, an accurate event model would be wished to cast to query model for clear event representation. An event model θ_E referred in Sect. 3 can be approximately estimated with the pseudo feedback documents, when the event type is identified beforehand.

Event query model is a unigram language model, which combines the event model with the original query model θ_Q, where $p(w|\theta_Q) = \frac{count(w,Q)}{|Q|}$, $count(w, Q)$ is the count of word w in the query Q. Like most other estimation methods, the re-estimated query model can be interpolated with the original one to improve the performance. So, the event query model θ_Q' is

$$p(w|\theta_Q') = \lambda p(w|\theta_Q) + (1 - \lambda)\hat{p}(w|\theta_E) \tag{9}$$
$$= \lambda p(w|\theta_Q) + (1 - \lambda)[\alpha p(w|\theta_{ET}) + (1 - \alpha)p(w|\theta_{ES})]$$

With above interpretations, we conclude the procedure of the event retrieval via bipartite graph in Algorithm 2.

5 Experiments

To the best of our knowledge, there is no standard test collection available for event retrieval task that we could use to evaluate our method directly, and several traditional standard TREC datasets are invalid for the lack of enough event-related data. To evaluate our approach, we use two datasets for TREC Temporal Summarization (TREC-TS) task[1], respectively called TREC-TS 2013

---

[1] http://www.trec-ts.org/.

Algorithm 2. Event retrieval process via bigraph

Input:
    target event e, event type T, a collection of documents $\mathbb{C}$
Output:
    A ranked list of event-related documents Ret
1: $FD \leftarrow$ first retrieval result
2: $G \leftarrow$ call algorithm 1 "event bigraph constructing algorithm" with FD
3: $\theta_E \leftarrow$ estimating with G
4: $\theta'_Q \leftarrow \lambda p(w|\theta_Q) + (1 - \lambda)\hat{p}(w|\theta_E)$
5: $Ret \leftarrow$ ranking by $-D(\theta'_Q \parallel \theta_D)$ with respect to $\mathbb{C}$

---

and TREC-TS 2014. They consist of a set of timestamped documents from a variety of news and social media sources. TREC-TS 2013 contains 10 topics representing events, along with an event query q also event type T, while TREC-TS 2014 contains 15 events. The event type is one of the followings: accident, bombing, hostage, protests, riot, shooting, earthquake and storm.

Setup: The corpura are indexed by indri. Stopword removal and Porter stemming is applied. We have completed 5 groups of experiments. Bigragh model (BIG) is our bigraph-based method, Relevance Model (RM3, RM4) and Divergence Minimization Model (DivMin) are both based on the language modeling retrieval model, the remaining one is non-language model, okapi BM25 with feedback. In BM25, we simply set $k_1 = 1.2$, $b = 0.75$, $k_3 = 7$. Our method shares following common parameters with RM and DivMin: Dirichlet smoothing ($\mu = 1000$) for estimation the document language models, the number of feedback documents ($fbDocs = 10$), the number of terms in feedback model ($fbTerms = 30$), interpolating parameter ($\lambda = 0.5$) for controlling the amount of feedback. In particular, in order to construct ET part of bigraph and estimate $p(w|\theta_{ET})$, for each event type T, we have collected 20 events by searching "list of T"(e.g.,"list of earthquake") in Wikipedia to build the external knowledge base. After word counting and stopword removing, we cut top 200 terms to be ET words for each event type.

Parameter Selection: Importantly, in BIG, the parameter α makes a trade off between ES and ET. In Algorithm 1, line 17 indicates that the number of ES nodes N_{ES} is controlled by dc_{thres}. $dc_{t_i} = \frac{\sum_j^m A_{ij}}{sum(A)}$ and $dc_{s_j} = \frac{\sum_i^n A_{ij}}{sum(A)}$ are respectively for ET and ES words, where $sum(A) = \sum_i^n \sum_j^m A_{ij}$. We have observed rough corresponding non-linear relation between dc_{thres} and N_{ES}, showed in abscissa axis in right subplot of Fig. 3. To learn the optimal parameter setting, we use TREC-TS 2013 for the training usage. Linear search is performed for α, when dc_{thres} is fixed as 0.004. From Fig. 3(a), we are aware of that the $P@20$, $P@30$ are stable when α ranges but $P@10$ gradually slips. Comprehensively, $\alpha = 0.4$ is optimal. What can be draw from the observation from right subplot of Fig. 3 is that N_{ES} within a range 20 to 50 is suitable. So we set $\alpha = 0.4$, $dc_{thres} = 0.005$ in follow-up experiments.

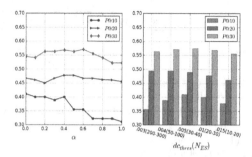

Fig. 3. Parameter selection for α and dc_{thres} on TREC-TS 2013 (Color figure online).

Experimental Result: To provide a high-quality evaluation setting, pooling technique is adopted, and the judgments are made by two sets of assessors and any conflicts were reconciled by a third user. The instruction of evaluation considers the density and quality of event information, preferring to an article providing sufficient details. Table 1 reports the performance of our proposed approach on TREC-TS 2014 measured by average $P@N$ (N=10,20,30). We observe that our approach wins other methods, with a commanding lead over its closest competitors DivMin (13.3 %), and BM25(17.9 %) on $P@10$. The number of test queries in our experiments is limited, so we make a supplementary analysis about why our method is better than others.

Table 1. Retrieval precision($P@N$) on TREC-TS 2014

	BM25 + FB	DivMin	RM3	RM4	BIG
$avg.P@10$	0.390	0.406	0.386	0.340	**0.460**
$avg.P@20$	0.403	0.430	0.416	0.406	**0.473**
$avg.P@30$	0.434	0.442	0.444	0.422	**0.470**

Further Analysis: The intrinsic part of BIG is to re-estimate a query model which might describe the event appropriately. To verify the inner property of our re-estimated query model, we have conducted a case study towards the event query "Costa Concordia"[2], representing a shipwreck accident happened in Italy. For comparison usage, we also generate a query model using RM1. Figure 3 makes a simple visualization. The upper subplot is $logp(w|\theta_{BG})$, representing the collection background model, The bottom subplot shows two expanded query models (without interpolation with original query model). From Fig. 3, two observations are evident. (1) The query model generated by our method (the yellow line) fluctuates more violently than the model produced by RM1 (the blue line), also a larger divergence against the background model. From the perspective of query

---

[2] https://en.wikipedia.org/wiki/Costa_Concordia_disaster.

clarity score [13], if a query model is smoother and more like the background model, it is more ambiguous. It means that BIG has *lower degree of ambiguity* than RM1. (2) The gap points $(logp(w|\theta_{BIG}) - logp(w|\theta_{RM1}) > 3)$ confirm that our method assigns more sufficient significance on representative event words than RM1. Namely, BIG shows the *property of event-prominence*. A high coherence query model (unusually large probabilities for a small number of topical terms), apparently, will return the topic related articles consistently higher than the ordinaries in the ranked list Fig. 4.

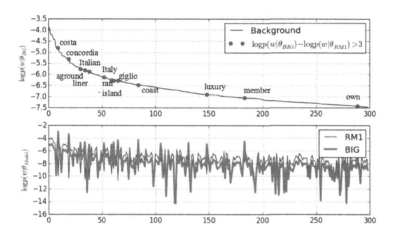

Fig. 4. Query models generated by RM1 and BIG for event query "Costa Concordia". Top 300 terms are plotted in descending order by background probability $p(w|\theta_{BG})$. The points interspersed sporadically throughout the background distribution are the salient terms on which BIG emphasizes more than RM1.

Further more, we conclude that the prior information of event type introduced in bigraph offers great help in recognizing important event specific words, and interaction between ET and ES contributes significantly to a better description of a certain event. These valid information brings above two excellent properties to BIG query model, so as to improve event retrieval performance.

6   Conclusion and Future Work

In this paper, we have focused on two research questions: how to model an event and how to apply the event model in the event retrieval task. Experiments show that the event model (unigram language model) estimated from the bigraph event representation, could express the event's characteristics clearly, and it is also feasible to improve the event retrieval effectiveness by incorporating event model into original query model. A major deficiency of our approach is that the event type of query must be known in advance, meanwhile the corresponding event type model has been estimated already. Future explorations can be done to complete or improve our work: (1) automatic event type detection for adaptive retrieval, e.g. building a preposing classifier to identify the event type of input queries; (2) establishing accurate event type models is potential for improving performance; (3) trials on modeling events from other perspectives.

Acknowledgments. We would like to thank the anonymous reviewers for their valuable comments and suggestions. This work is supported by the National Natural Science Foundation of China (grant No. 61572494), the Strategic Priority Research Program of the Chinese Academy of Sciences (grant No. XDA06030200), and the National Key Technology R&D Program (grant No. 2012BAH46B03).

References

1. Lin, C.-H., Yen, C.-W., Hong, J.-S., Cruz-Lara, S.: Event-based textual document retrieval by using semantic role labeling and coreference resolution. IADIS international conference WWW/Internet (2007)
2. Glavaš, G., Šnajder, J.: Event-centered information retrieval using kernels on event graphs. In: Graph-Based Methods for Natural Language Processing (2013)
3. Kawai, H., Jatowt, A., Tanaka, K., Kunieda, K., Yamada, K.: ChronoSeeker: Search engine for future and past events. In: Proceedings of the 4th International Conference on Uniquitous Information Management and Communication, p. 25 (2010)
4. Shan, D., Zhao, W.X., Chen, R., Shu, B., Wang, Z., Yao, J., Yan, H., Li, X.: Eventsearch: a system for event discovery and retrieval on multi-type historical data. In: SIGKDD, pp. 1564–1567 (2012)
5. Lafferty, J., Zhai, C.: Document language models, query models, and risk minimization for information retrieval. In: SIGIR, pp. 111–119 (2001)
6. Lv, Y., Zhai, C.: A comparative study of methods for estimating query language models with pseudo feedback. In: CIKM, pp. 1895–1898 (2009)
7. Lavrenko, V., Croft, W.B.: Relevance based language models. In: SIGIR, pp. 120–127 (2001)
8. Zhai, C., Lafferty, J.: Model-based feedback in the language modeling approach to information retrieval. In: CIKM, pp. 403–410 (2001)
9. Lv, Y., Zhai, C.: Positional relevance model for pseudo-relevance feedback. In: SIGIR, pp. 579–586 (2010)
10. Lv, Y., Zhai, C.: Revisiting the Divergence Minimization Feedback Model. In: CIKM, pp. 1863–1866 (2014)
11. Allan, J., Papka, R., Lavrenko, V.: On-line new event detection and tracking. In: SIGIR, pp. 37–45 (1998)
12. Zhou, T., Ren, J., Medo, M., Zhang, Y.-C.: Bipartite network projection and personal recommendation. Phys. Rev. E **76**(4), 046115 (2007)
13. Cronen-Townsend, S., Zhou, Y., Croft, W.B.: Predicting query performance. In: SIGIR, pp. 299–306 (2002)

SPedia: A Semantics Based Repository of Scientific Publications Data

Muhammad Ahtisham Aslam$^{(\boxtimes)}$ and Naif Radi Aljohani

Faculty of Computing and Information Technology,
King Abdulaziz University, Jeddah, Saudi Arabia
{maaslam,nraljohani}@kau.edu.sa

Abstract. There is a noticeable increase in the number of scientific publications. These publications are being published by different publishers. *Springer* is one of those publishers which has published more than nine million scientific documents. *SpringerLink* is the portal providing the gateway to searching and accessing these published scientific documents. The structure, as well as the way, the contents are presented on the portal, provides valuable information about documents metadata such as author, ISBN, references, articles, chapters. However, this metadata is understandable by human in such a way that it facilitates the keyword-based searches through *SpringerLink* portal. At the same time this huge data about scientific documents is in silence as it is neither open nor linked to other datasets. To address these issues, we have created a semantics based repository called *SPedia* which consists of semantically enriched data about documents published by *Springer*. Currently, *SPedia* datasets consist of more than 300 million RDF triples. In this paper we describe *SPedia* and examine the quality of its extracted data by performing semantically enriched queries. The results show that *SPedia* facilitates the users to put sophisticated queries by employing semantic Web techniques instead of keyword-based searches. In addition, *SPedia* datasets can be utilized to link to other datasets available in the Linked Open Data (LOD) cloud.

1 Introduction

With the current deluge of data, we live in the fourth paradigm of scientific research which called data intensive research. The data is available in high volume about different domains such as educational data, government data, geo data etc. *Semantic Web* and *Linked Open Data (LOD)* communities have been working for more than a decade to produce semantically enriched data that is *open* as well as *linked* (i.e. *Linked Open Data (LOD)*). The purpose is to bring data on a global scale (which is currently dispersed, accessible on limited scale and available in data chunks) and to interlink different datasets so that semantically enriched queries could be made for sophisticated query answering purposes.

Many approaches have been proposed so far to extract and produce the structured and semantically enriched data from different existing data resources.

© Springer International Publishing Switzerland 2016
B. Cui et al. (Eds.): WAIM 2016, Part I, LNCS 9658, pp. 479–490, 2016.
DOI: 10.1007/978-3-319-39937-9_37

For example [2,3,6,14] focuses on extracting structured knowledge from Wikipedia contents, such that Semantic Web techniques can be used to ask sophisticated queries against Wikipedia. Similarly in [7,15], authors propose different methodologies to extract structured information from different unstructured data sources belonging to different domains. Another approach to extract semantically enriched data from three most popular Chinese encyclopedia sites and to link extracted data entities as Chinese LOD (*CLOD*) has been describe in [8]. In SwetoDblp project [1], authors propose a framework to create dataset in RDF form from XML document containing DBLP information.

When it comes to scientific publications such as journals, articles, books, chapters etc., there are many publishers (e.g. *Springer, IEEE, Amazon, Pearson, Thomson Reuters* etc.) that are publishing scientific documents. *Springer* is one of the leading global scientific publisher and *SpringerLink* is the worlds most comprehensive online collection of scientific documents [12]. In this paper we focus on extracting structured and semantically enriched datasets from *Springer-Link* so that semantically enriched queries can be put to huge datasets of scientific documents published by *Springer*. A very little initial work has been completed by way of extracting *LOD* only from *Springer* conferences on computer science [11]. This provides information about conferences\proceedings in the computer science domain only, leaving untouched a huge amount of remaining data. Our extracted datasets offer far more coverage of the disciplines and documents data.

We have extracted the structured data from the *SpringerLink* (as source of data) and produced a huge repository (i.e. *SPedia*) of semantically enriched data of scientific documents published by *Springer*. *SPedia* datasets consist of more than 300 million RDF triples and are available to download and experiment with at local level (in *N-Triple* format)[1]. We can use these datasets to formulate sophisticated queries about the documents published by *Springer* rather than relying on keyword-based searches through the portal. We ran SPARQL queries, visualizing document information and browsing the knowledge by using *Semantic Web* browsers (e.g. *Gruff* [5]). We also created a *SPARQL Endpoint* to pose semantically enriched queries from the extracted datasets.

The rest of the paper is organized as follows: Related work is discussed in Sect. 2. In Sect. 3 we describe extraction of structured information from *Springer-Link*. We next briefly describe the resulting datasets in Sect. 4, while Sect. 5 describes different ways to browse and query resulting datasets. Finally we conclude and discuss the future potential of our work in Sect. 6.

2   Related Work

To date, many approaches have been adopted to extract structured as well as semantically enriched data from existing sources so that users can pose semantically enriched queries. For example, in [1], the authors presented an approach

---

[1] http://wo.kau.edu.sa/Pages-SPedia.aspx.

to extract and produce an RDF version of the publications and author information contained in DBLP. The authors also describe their approach about crawling, parsing, and extracting structured data from source DBLP XML file. The main limitation of both source and resulting RDF datasets is that they provide only bibliographic information about documents without establishing relations between them.

The DBPedia project focused on the task of converting Wikipedia content into structured knowledge, so that Semantic Web techniques can be employed to pose sophisticated queries from Wikipedia [2,3,6]. Another project (i.e. YAGO [14]) also extracted structured information from Wikipedia. However, in [14], the authors proposed the extraction of data of 14 relationship types, ignoring extraction from Wikipedia infobox templates, in contrast to the DBPedia data extraction process. Continuing towards Wikipedia as a source, an entity based extraction approach has also been presented in [4] to extract entity based structured data from Wikipedia.

In [10], the authors proposed an approach for extraction of structured data from *The Cancer Genome Atlas (TCGA)* [15], organized as text archives in a set of directories. The approach made it easier for domain experts and bioinformatics applications to process and analyze cancer-related data by RDFizing the data. The resulting structured data is available to the public for remote querying and analysis of the structured domain knowledge.

An approach to transforming the data of the *Financial Transparency System (FTS)* of the European Commission is described in [7]. The authors give examples of queries to the resulting datasets not possible using existing HTML and XML formats of the data.

An effort to produce large-scale Chinese semantic data and to link these together as *Chinese LOD (CLOD)* is presented in [8]. In this work the authors proposed a method and strategy to extract structured data from the three largest Chinese encyclopedia sites and then link them together. They also established SPARQL Endpoint to query the extracted datasets using semantic web-based techniques.

In a similar fashion, *Springer Developer APIs* [13] provide sets of APIs to access programmatically the metadata of articles published by *Springer*. Some major limitations of these APIs are: Firstly, they provide the results in formats (e.g., mostly XML or JSON) that need to be parsed, to extract the required information, and RDFized. Secondly, they work over the traditional HTTP Get request/response method, making performance almost the same as by accessing the content online through the portal. Thirdly, these APIs provide access to 7 million articles, resulting in the straight forward absence of metadata from over 2 million published articles. Fourthly, by using these APIs structural information defining the relationships between data entities might be missed that could be detected if we access the contents through the *SpringerLink* portal. Our approach made full use of structural as well metadata content on the *SpringerLink* portal to extract the maximum number of data items and the relationships between them.

3   Extraction of Structured Information from SpringerLink

SpringerLink is the portal that we used as our source of data, so in this section we first present the content and structure of its existing data about scientific documents. We next present the *SPedia* knowledge extraction process we developed to crawl, parse, and extract the required data and to produce semantic data in RDF triples format.

3.1   SpringerLink Templates

The data source (i.e., *SpringerLink*) used in our approach contains information about the scientific documents in typical templates. These have the metadata as well as the relational information between the various documents (as shown in Fig. 1). All this information is available in HTML pages, so these were processed to extract the required structured and relational information. We extracted documents from 24 disciplines, as explained below:

Type and Discipline. As discussed above, in the data source are various types of documents (e.g., book, chapter, journal, article, etc.) and 24 disciplines (e.g., computer science, engineering, and medicine) to which a document may belong. Thus, both document type and discipline are represented by *rdf:type*.

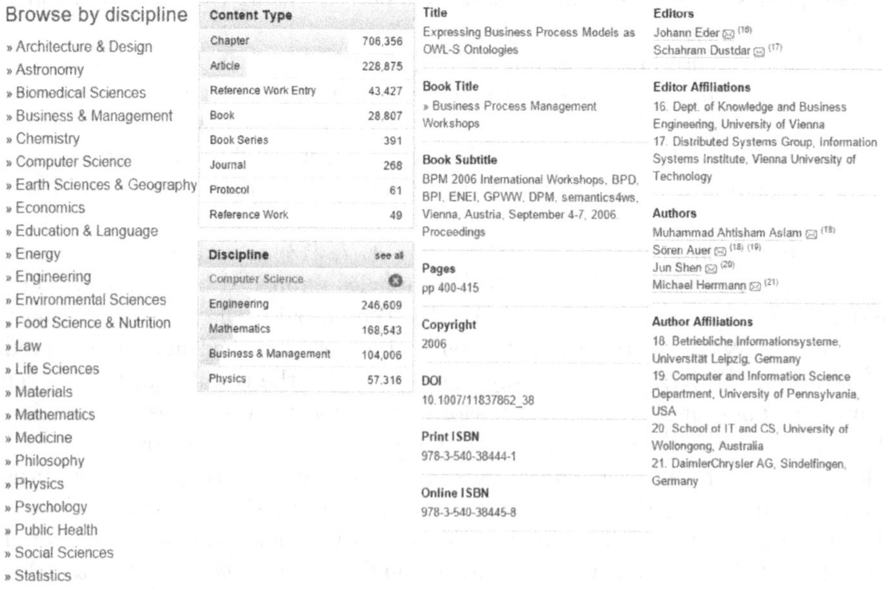

Fig. 1. Different views (combined together) of *SpringerLink* template to present relational as well as metadata information of documents.

Document Metadata. Metadata properties (e.g., title, abstract, copyright, doi, print ISBN, online ISBN, print ISSN, online ISSN, publisher, volume, issue, coverage, pdf link, references, pages, publication year) about documents are presented by using properties *publication:has_Title, has_Abstract, has_Copyrigth, has_DOI, has_Print_ISBN, has_Online_ISBN, has_print_ISSN, has_Online_ISSN, has_Publisher, has_Volume, has_Issue, has_Coverage, has_PDF_Link, has_Reference, has_Pages, has_Publication_Year* respectively.

Authors, Editors and Affiliation. Every document has one or more author/s, editor/s, editor-in-chief and affiliation/s. Therefore, author/s, editor/s editor-in-chief and their affiliation/s data were extracted and presented by *publication:has_Author, has_Editor, has_Editor_In_Chief and has_Affiliation* properties.

Relations Between Documents. Documents have relationships with documents that are very important, especially when we work in domain of scientific documents. For example, a *chapter* has relation to a *book*, an *article* has relation to a *Journal* and a *reference work entry* has relation to a published *reference work* and vice versa. These relations are described by using *publication:has_Book_Chapter, is_Book_Chapter_Of, has_Article, is_Article_Of, has_Reference_Work_Entry, is_Reference_Work_Entry_Of* properties respectively.

3.2   Extraction Process

The extraction process consisted of four main steps, each of which is interlinked in a recursive manner (as shown in Fig. 2). The Fig. 2 shows that the information extraction process starts by crawling through *SpringerLink* for a particular discipline, then crawling and extracting information for a particular content type, and then for sub-content types, using a recursive approach. Here we describe each step of extraction process:

SPedia Crawler. Information about documents on *SpringerLink* is available in such a way that crawlers can be written to start crawling through documents

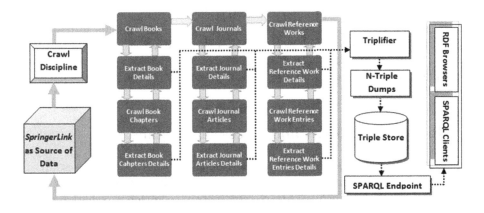

Fig. 2. Overview of the *SPedia* knowledge extraction process.

of any discipline from the first document to the last. Therefore, this module starts crawling through the *SpringerLink* portal from the first discipline and continue through every discipline by using a recursive approach (as shown in Fig. 2). In fact, crawling through a particular discipline is the starting point for further crawling through the various types of document in a particular discipline (e.g., book, journal, reference work, etc.). Also, the relations between disciplines and documents are saved during the crawling process for use in describing the relational properties (e.g., "Book" publisher:has_Book_Chapter "Book_Chapter" etc.) in later stages.

Information Parser and Extractor. The information parsing and extraction takes place at every stage of the crawling process, as well as during particular stages at which metadata of documents or relational information become available. For example, when a stage comes during the extraction process where a document (e.g., Book) exists, the information parser and extractor module gets activated. It parses the existing page and extracts the relational information (e.g., books relation to a particular discipline, or book chapters relation to a particular book, etc.) as well as metadata information (e.g., a book ISBN, pdf link, etc.). As soon as the information parsing and extraction step is complete, the process lets the crawler go on to the next item so information from every last document and its related information in the whole chain is extracted and utilized.

Triplifier. The information and metadata of documents extracted at any stage during the crawling process is triplified and saved to the data model. Every document (i.e. book, chapter, article etc.) is triplified as a resource, and every resource is triplified for their properties (e.g., title, abstract, ISBN, etc.). Relations (e.g., book *has_book_chapter*, journal *has_journal_article* etc.) are also triplified a this stage. Authors, editors, and affiliated organizations are identified as individual resources (resulting in large dataset of scientific authors and organizations). The triplifying process, together with the information parser and extraction process, also takes care of the data types of values of extracted properties.

Datasets Generator. This is the last step of the information extraction process which generates *SPedia* datasets. Data models of the information in triplicate are taken as inputs at this stage, inconsistencies in URIs are resolved and the final datasets in N-Triple format are generated. All datasets are generated at two different levels (e.g., property level, document level (further explained in Sect. 4.1)).

4   Extracted Dataset and Necessary Details

The *SPedia* datasets provide information on more than 9 million resources, including over 3.9 million journal articles, 3.1 million chapters, and 0.65 million reference work entries. Extracted datasets also provide information about more than 3100 journals, 181,000 books, and 650 reference works. Information

Table 1. *SPedia* datasets statistics.

Resource	Description	Numbers	Triples
Book	Provides information about book metadata as well as chapters of the book	181K	5.3M
Chapter	Provides information about book chapter metadata and the book in which chapter is published	3.1M	103M
Journal	Contains information about journal, its metadata as well as its sub resources i.e. articles published in a journal	3K	4M
Article	Provides information about article, the parent journal and all related metadata	3.9M	126M
Reference work	Resources of the type reference work provide information about reference works, their metadata and corresponding entries	858	0.7M
Reference work Entry	Contains information about metadata of reference work entries and their parent reference works	0.65M	11.9M
Person	Persons contains information about people as author, editor, editor-in-chief etc. with their contact information	31M	104M
Organization	Provides information about names of organization as affiliation of persons	12M	24M

about other resources, including 31 million people (e.g., authors, editors, editors-in-chief, etc.) and 12 million organizations is part of the extracted datasets. The resulting datasets consist of more than 300 million triples, including (approximately) 126 million triples on journal articles, 103 million triples on book chapters, and 11.9 million triples on reference work entries. A summary of the statistics on resources and extracted triples is provided in Table 1.

4.1 Different Levels of Extracted Datasets

SPedia datasets are available for download (in *.nt* format) at two different levels. Users can download these datasets to their requirements, starting with property-level datasets (smaller in size, but distributed in large number of property level .nt files) through to discipline level datasets (bigger in size, but with all data about documents relating to a single discipline in one .nt file). These different levels of data sets are further explained below:

I **Property Level.** Property-level datasets provide information about every property of documents in N-Triple format. For example, the *book type.nt* dataset contains information in (triple format) about the types of documents (e.g., a document type is *Book* and *Computer Science*). Similarly, the *book* *title.nt* dataset contains titles of all books published in a particular discipline. In the same way, other property level datasets (e.g., *journal hasPrint-ISBN.nt*, *article pdf link.nt* etc. provide data about the different properties of documents). Some common metadata properties of scientific documents is described in Table 2.

II **Document Type Level.** Datasets at document type level provide complete information about a particular type of document. For example, the type of document can be Book, Chapter, Article, etc. Therefore, document type-level datasets provide complete information about a particular document type belonging to a particular discipline in one dataset (e.g., the *book.nt* dataset provides all information about all books published in a particular discipline). Similarly, the *chapter.nt* dataset provides all information about all chapters published in particular discipline.

Table 2. Description of some common properties extracted for different document types.

Property	Description
type	Provides data about the document type (e.g. *type* can describe that a document is a *Chapter* in discipline *Computer Science*)
has_Title	Provides the title of the document
has_DOI	Provides the DOI of the document
has_Print_ISBN	Contains the print ISBN number of the document
has_Author	Contains the information about the authors of the document which then further contains his name, email and affiliation
has_Book_Chapter	Links every book with the every chapter publisher in that book
is_Book_Chapter_Of	This property links every chapter with the book in which chapter is published
has_Editor	Links every type of document (e.g. book, chapter, article etc.) with person (as editor)
is_Editor_Of	Links a person (as editor) with any type of document

4.2 Inconsistencies in Source Data

Since, the source of extracted datasets is the *SpringerLink* portal, many inconsistencies in the source data were noted during the extraction process. For example, sometimes information is not as per the template and heading of the topic e.g. the portal provides information about a single editor by using the heading Editors. This heading leads the parsing and extraction process to extract more than one editor, although there is actually just one. Likewise, ISBNs are sometimes available in a number format and sometimes as a string (by adding the - character, which cannot be typecasted and is thus treated as an integer). Moreover, our approach uses the title of the document to create a URI for the extracted resource and sometimes this contains special characters (e.g., mathematical signs, language-based special characters, etc.) that cannot be used as part of the standard URI of a resource. In this case, such characters are treated as illegal characters and replaced by the most appropriate option, following best practice for publishing RDF vocabularies [9].

5 Accessing and Querying SPedia

SPedia datasets can be accessed from the project website and may also be queried from SPARQL Endpoint. Any third party application and semantic Web browser (e.g., Gruff [5]) can be used to query the datasets through SPARQL Endpoint, and to browse the datasets as well as to visualize the resulting data. We discuss below these scenarios of accessing and querying the resulting datasets.

5.1 Querying Through SPARQL Endpoint

The most common way to query the datasets is through SPARQL Endpoint. We have created a SPARQL Endpoint for *SPedia* datasets. This can be used to put sophisticated queries to the SPedia. Here, we give some sample queries and the results obtained from our extracted datasets.

We asked to search all articles in a journal that is published in the Mathematics discipline with the title OPSEARCH.

Example 1 (Find a Journal in Mathematics whose title is OPSEARCH and find all Articles published in that Journal.).
    PREFIX spedia:<http://www.kau.edu.sa/fcit/ontology/2015/3/v1.8# >
    select ?articles where {
        ?document rdf:type spedia:Journal.
        ?document rdf:type spedia:Mathematics.
        ?document spedia:has_Title"OPSEARCH"^^xsd:string.
        ?document spedia:has_Article ?articles.
}

The results that we obtained from this query are shown in the Fig. 3(a) indicating that the Journal with title *OPSEARCH* has 245 articles.

Then we further filter our results by querying for only those articles which are published in Volume 49, Issue 1 of the *OPSEARCH Journal*. We also queried for additional information (i.e. *PDF Link*) of these articles. Figure 3(b) shows six articles (with their *PDF Links*) published in Volume 49, Issue 1.

Example 2 (Find articles published in a particular volume 49 and issue 1 of the "OPSEARCH" Journal.).

PREFIX spedia:<http://www.kau.edu.sa/fcit/ontology/2015/3/v1.8# >
select ?articles ?PDF_Link where {
    ?document rdf:type spedia:Journal.
    ?document rdf:type spedia:Mathematics.
    ?document spedia:has_Title"OPSEARCH"^^xsd:string.
    ?document spedia:has_Article ?articles.
    ?articles spedia:is_In_Volume "49"^^xsd:string.
    ?articles spedia:is_In_Issue "1"^^xsd:string.
    ?articles spedia:has_PDF_Link ?PDF_Link.
}

245 Results
?articles
On solving multi level multi objective linear programming problems through fu
Flow constrained minimum cost flow problem
Nonsmooth multiobjective optimization involving generalized univex functions
Analysis of M X G 1 feedback queuewith two phase service compulsory server
Stochastic behaviour analysis of power generating unit in thermal power plant
Batch service queue with change over times and Bernoulli schedule vacation i
Optimum configuration selection in Reconfigurable Manufacturing System inv

(a)

6 Results	
?articles	?PDF_Link
A new approach for solving the network problems	http://link.springer.com/article/10.1007/s12597-012-0063-8.pdf
How effective is aggregation for solving 0–1 models?	http://link.springer.com/article/10.1007/s12597-011-0062-1.pdf
On constrained optimization by interval arithmetic and interval order relations	http://link.springer.com/article/10.1007/s12597-011-0061-2.pdf
Time dependent analysis of an MM1N queue with catastrophes and a repairable server	http://link.springer.com/article/10.1007/s12597-012-0065-6.pdf
Location models with stochastic demand points	http://link.springer.com/article/10.1007/s12597-012-0066-5.pdf
Observations on some heuristic methods for the capacitated facility location problem	http://link.springer.com/article/10.1007/s12597-012-0064-7.pdf

(b)

Fig. 3. Results of example queries 1 and 2.

5.2   Browsing via Semantic Web Browsers

The semantic Web and LOD community has developed various tools and client applications that could be used to browse RDF data, either by loading the RDF triples directly or by connecting to SPARQL Endpoint. Following the same trend, *SPedia* datasets can be browsed by using the semantic Web browsers in tabular, as well as in visual form (as shown in Figs. 3 and 4). Figure 3 shows the results of our queries in a semantic Web browser (connected to SPARQL Endpoint) in tabular form. These can be further explored as long as linked RDF data is available or we reach a literal value. Also, as discussed before, some semantic Web browsers facilitate visualization of the datasets as well visual browsing. Figure 4 provides an example for some sample RDF records for a book and its related metadata.

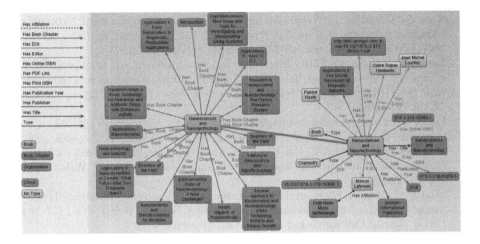

Fig. 4. Visual representation of extracted data.

6   Conclusion and Future Work

In this paper we presented *SPedia*: a semantics based repository of scientific publications data. *SPedia* aims to facilitate the semantic based search in order to allow users to ask sophisticated queries to overcome the limitation of the traditional keyword-based search. To achieve this goal, data about nine million scientific documents consisting of about 300 million RDF triples were extracted and produced as a machine-processable data. Furthermore, we created a SPARQL Endpoint to query *SPedia* datasets. The results showed that *SPedia* can play a key role in allowing users to put sophisticated queries by employing semantic Web techniques instead of keyword-based searches. In addition, *SPedia* datasets can be utilized to link to other datasets available in the Linked Open Data (LOD) cloud.

For the future work, we are aiming to extract semantically enriched structured data from publicly available resources of other publishers such as *Pearson, Elsevier, Amazon* and interlink them to explore the value of linked open scientific publications data. We are also working on increasing the coverage of the entities in *SPedia* datasets. Similarly, we are working to treat references as resources rather than strings, ultimately helping semantic Web-based agents to crawl through the interlinked data.

References

1. Aleman-Meza, B., Hakimpour, F., Budak Arpinar, I., Sheth, A.P.: Swetodblp ontology of computer science publications. Web Semant. Sci. Serv. Agents World Wide Web **5**(3), 151–155 (2007)
2. Auer, S., Bizer, C., Kobilarov, G., Lehmann, J., Cyganiak, R., Ives, Z.G.: DBpedia: a nucleus for a web of open data. In: Aberer, K., et al. (eds.) ASWC 2007 and ISWC 2007. LNCS, vol. 4825, pp. 722–735. Springer, Heidelberg (2007)

3. Auer, S., Lehmann, J.: What have Innsbruck and Leipzig in common? Extracting semantics from wiki content. In: Franconi, E., Kifer, M., May, W. (eds.) ESWC 2007. LNCS, vol. 4519, pp. 503–517. Springer, Heidelberg (2007)
4. Exner, P., Nugues, P.: Entity extraction: from unstructured text to dbpedia RDF triples. In: Proceedings of the Web of Linked Entities Workshop in Conjuction with the 11th International Semantic Web Conference (ISWC 2012), pp. 58–69. CEUR-WS (2012)
5. Franz: Gruff: A grapher-based triple-store browser for allegrograph (2015)
6. Lehmann, J., Isele, R., Jakob, M., Jentzsch, A., Kontokostas, D., Mendes, P.N., Hellmann, S., Morsey, M., van Kleef, P., Auer, S., Bizer, C.: DBpedia - a large-scale, multilingual knowledge base extracted from wikipedia. Semant. Web J. 6(2), 167–195 (2015)
7. Martin, M., Stadler, C., Frischmuth, P., Lehmann, J.: Increasing the financial transparency of European commission project funding. Semant. Web J. 5(2), 157–164 (2013). Special Call for Linked Dataset Descriptions
8. Niu, X., Sun, X., Wang, H., Rong, S., Qi, G., Yu, Y.: Zhishi.me - weaving Chinese linking open data. In: Aroyo, L., Welty, C., Alani, H., Taylor, J., Bernstein, A., Kagal, L., Noy, N., Blomqvist, E. (eds.) ISWC 2011, Part II. LNCS, vol. 7032, pp. 205–220. Springer, Heidelberg (2011)
9. Berrueta, D., Phipps, J.: Best practice recipes for publishing RDF vocabularies. w3c working group note (2008). http://www.w3.org/TR/2008/NOTE-swbp-vocab-pub-20080828
10. Saleem, M., Shanmukha, S., Ngonga, A.-C., Almeida, J.S., Decker, S., Deus, H.F.: Linked cancer genome atlas database. In: Proceedings of the 9th International Conference on Semantic Systems, I-SEMANTICS 2013, pp. 129–134. ACM, New York (2013)
11. Springer: Lod for conferences in computer science (2015). http://lod.springer.com/wiki/bin/view/Linked+Open+Data/About
12. Springer: Facts and figures. Springer Science+Business Media (2015). http://resource-cms.springer.com/springer-cms/rest/v1/content/20616/data/v11/Facts+and+Figures+April+2015
13. Springer: Springer—biomed central API portal (2015). https://dev.springer.com/
14. Suchanek, F.M., Kasneci, G., Weikum, G.: Yago: a core of semantic knowledge. In: Proceedings of WWW 2007, pp. 697–706 (2007)
15. Tomczak, P.C., Katarzyna, M.W.: The cancer genome atlas (TCGA): an immeasurable source of knowledge. Contemp. Oncol. 19(1A), A68–A77 (2015)

A Set-Based Training Query Classification Approach for Twitter Search

Qingli Ma, Ben He$^{(\boxtimes)}$, Jungang Xu$^{(\boxtimes)}$, and Bin Wang

University of Chinese Academy of Sciences, Beijing, China
maqingli13@mails.ucas.ac.cn, {benhe,xujg}@ucas.ac.cn, wangbin@iie.ac.cn

Abstract. Learning to rank is a popular technique of building a ranking model for Twitter search by utilizing a rich list of features. As most learning to rank algorithms are supervised, their effectiveness is heavily affected by the quality of labeled training data. Selecting training queries with high quality is an important means to improving the effectiveness of ranking model for Twitter search. Existing approach for this problem learns a query quality classifier, which estimates the training query quality on a per query basis, but ignores the dependence between queries. This paper proposes a set-based training query classification approach that estimates a training query's quality by taking its usefulness in combination with other training queries into consideration. Evaluation on standard TREC Microblog track test collection shows effective retrieval performance brought by the proposed approach.

Keywords: Training query classification · Learning to rank · Information retrieval

1 Introduction

With the rapid development of online social networks, Twitter, one of the most popular microblogging services, is attracting more and more attention from Internet users [1]. On Twitter, there is a large quantity of updates every day, or even every minute, created by its hundreds of millions of users. The overwhelming amount of updates on Twitter lead to the difficulty in finding the interesting messages, which are fresh and relevant to the given query, whereby a user's information need is represented by a query issued at a specific time. However, Twitter's search engine only sorts the tweets containing the query terms in chronological order. This mechanism cannot guarantee that the most interesting tweets are top-ranked [2].

To achieve an effective real-time Twitter search, there have been efforts in applying learning to rank to Twitter search by integrating various sources of evidence of relevance. Learning to rank (LTR), as known as machine-learned ranking, is a family of algorithms and techniques that automatically learn a ranking model from training data, where explicit or implicit evidence of relevance are usually represented by a feature vector. Compared to the classical retrieval

© Springer International Publishing Switzerland 2016
B. Cui et al. (Eds.): WAIM 2016, Part I, LNCS 9658, pp. 491–503, 2016.
DOI: 10.1007/978-3-319-39937-9_38

models, learning to rank has the advantage of combining the predefined features to construct a ranking model. As shown by the evaluation forum in the Text REtrieval Conference (TREC) Microblog track [3,4], most top runs apply LTR algorithms.

Most LTR algorithms are supervised approaches and their effectiveness is heavily affected by the quality of labeled training data [5,6]. As the low-quality training queries may lead to degraded retrieval performance, it is necessary to select the high quality ones before the deployment of LTR [7], so does for its application to Twitter search [8,9].

Existing LTR approaches may be categorized into three groups, the point-wise, pairwise, and listwise algorithms [5]. Among them, the pairwise and listwise algorithms are popular for their superior effectiveness [5]. For the application of the pairwise algorithms to Twitter search, a recent approach by Li et al. proposes a query quality classification algorithm, denoted by TQC_{Single}, to select training queries on a per query basis [9]. Their approach learns a query quality classifier using the retrieval performance gain as the measurement of the quality of individual training queries, based on a list of query features such as the relevance scores and tweet length [9]. Despite the effectiveness brought by TQC_{Single}, its application to Twitter search also suffers from the following problems. First, this approach does not consider the dependence and overlap between training queries. As a consequence, it may end up with selecting a highly redundant training query set and result in low retrieval effectiveness. Second, using the listwise algorithms, the number of training examples, namely the queries, is much less than that of the pairwise algorithms, namely the document pairs. Therefore, it is difficult to apply TQC_{Single} to learn a training query classifier for the listwise LTR algorithms due to the small amount of training examples.

To this end, this paper proposes a novel training query classification approach for LTR by considering the dependency between training queries. Specifically, the proposed approach estimates a training query's quality by its usefulness in combination with other training queries. In this way, the dependency and redundancy among training queries are taken into consideration, and the number of training examples can be increased to a sufficient level by using pairs or sets of training queries, instead of the individual queries. Evaluation on the standard TREC microblog track dataset shows the statistically significant retrieval performance improvement of our proposed approach.

2   Related Work

2.1   Existing Twitter Search Approaches

Due to the convenience in combining various sources of features for the relevance weighting, learning to rank is a popular approach to Twitter search [10–13]. Xu et al. adopt eight different learning to rank algorithms in the re-ranking phase [11]. Magdy et al. employ six learning to rank algorithms as the candidate rankers and try to combine the ranking scores of these candidate rankers by *weighed Condorce-fuse* [13,14]. Because learning to rank is a data-driven approach which integrates a number of features, many participants try to extract more features of tweets and adopt more external evidence[1]. Lv et al. extract 240 features from four different query sets on three different corpora [10]. Li et al. learn a binary classifier to separate relevant document from non-relevant ones in the retrieved document pool [15], besides they utilize the feedback documents and classifier-judged documents to update and formulate the queries. All in all, few groups try to select high-quality queries to further enhance the retrieval performance, until recently Li et al. propose a query selection algorithm by learning a query quality classifier, which automatically selects the training data on a per-query basis [9].

2.2 Training Query Selection

There have been quite a few research efforts in selecting training queries to improve the retrieval effectiveness of LTR. For example, Long et al. propose an active learning approach, Expected Loss Optimization (ELO), to select the most informative examples for the learning by optimizing the expected Discounted Cumulative Gain (DCG) loss in order to reduce the dependence on the number of labeled documents in the training set for Web search [7].

As for the application to Twitter search, recently, Li et al. propose a simple but effective training query classification approach based on the retrieval performance gain using a number of pre-defined query features [9]. The main idea is to establish a linear relationship of a set of query features and the estimated retrieval performance gain to select training queries with high quality from the candidate training set. Specifically, the quality of a training query is given by the retrieval performance of the ranking model learned using the query and its associated documents as the training set. Their experimental results on the Tweets 2013 collection show that this approach outperforms the conventional application of learning to rank that learns the ranking model on all training queries available.

Despite the above merits of the approach in [9], it suffers from the issues mentioned in Sect. 1, and therefore is not applicable to the listwise LTR algorithms. To this end, this paper proposes a set-based approach in the next section that not only works well with the listwise LTR algorithms, but also improves the retrieval effectiveness over [9].

3   Training Query Classification: The Proposed Approach

In this section, we propose a set-based training query classification approach. The main idea of training query classification is to establish a linear relationship of a set of query features and the estimated retrieval performance gain to select training queries with high quality from the original training set.

A brief description of the steps involved in the training query classification approach is given in Fig. 1. Similar to the approach in [9], we assume the

---

[1] https://github.com/lintool/twitter-tools/wiki/TREC-2014-Track-Guidelines.

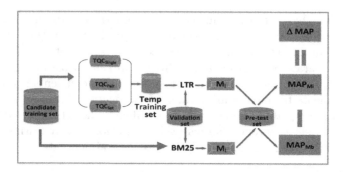

Fig. 1. The flowchart of the three training query classification algorithms: TQC_{Single}, TQC_{Pair}, TQC_{Set}.

availability of a candidate training set from which the final training queries are selected, a validation set for the parameter tuning, and a pre-test set for estimating the retrieval performance gain of the a given candidate training query. It firstly obtains a baseline estimation of the retrieval performance using the classical BM25 weighting function, given by mean average precision [16] on the pre-test set (MAP_{M_b}). Next, for each sample of the candidate queries, a ranking model M_i is learned using a given LTR algorithm. The composition of the sample depends on the training query classification algorithm used. Two training query classification algorithms are proposed based on different granularities of the samples from the candidate set. The algorithm proposed by Li et al. [9] is denoted as TQC_{Single}, which uses each single candidate query to learn the ranking model. Instead, our proposed algorithms use pairs of candidate queries (denoted as TQC_{Pair}), or sets of candidate queries (denoted as TQC_{Set}). The effectiveness of the learned ranking model M_i is also evaluated on the pre-test set by MAP. The difference between the above two MAP values, ΔMAP, is considered to be an estimation of the quality of the sample candidate set. A linear classifier is learned using Logistic Regression based on a list of query features as given in Sect. 4, where the target value is ΔMAP. Finally, for a given new set of candidate training queries, the classifier is applied to select the high-quality ones for learning to rank, and those with predicted ΔMAP larger than 0 are selected as high-quality queries. The dataset used and the pre-defined features are introduced in details in Sect. 4.

Next, we introduce the two proposed algorithms in details. Figure 2 presents the steps of the first proposed algorithm, TQC_{Pair}. It firstly obtains a baseline estimation of the retrieval performance using BM25. Next, for each $pair(q_i, q_i)$ of the candidate queries, a ranking model M_{ij} is learned using a given LTR algorithm. The effectiveness of this ranking model is also evaluated on the pre-test set by MAP, and the query quality estimation ΔMAP is then obtained. The training query classifier is learned over pairs of candidate queries using ΔMAP as the target. Finally, for a given new set of candidate training set, it adds pairs of high-quality queries to the training set using the learned classifier. An obvious

---

Strategy 2: TQC_{Pair}

Train the basic retrieval model (BM25) on the candidate training set, denoted as M_b;
Tune the parameter on validation set;
Apply M_b to the pre-test set, obtained the MAP, denoted as MAP_{M_b};

for query pair $(q_i, q_j) \in candidate\ training\ set$ **do**
    Use the query $pair(q_i, q_j)$ as temporal training set to train the LTR model, denoted as M_{ij};
      % Pairwise and listwise algorithms are applicable;
    Tune parameter of M_{ij} on validation set;
    Obtain $MAP_{M_{ij}}$ on pre-test set with ranking model M_{ij};
    Calculate the retrieval performance gain:
      $\Delta MAP = MAP_{M_{ij}} - MAP_{M_b}$
      % ΔMAP is the retrieval performance gain for each query pair (q_i, q_j)
    Extract features for each query, $(f_1, f_2, f_3 \dots f_n)$, and calculate the mean feature value
    F=$(\overline{f_{i1}f_{j1}}, \overline{f_{i2}f_{j2}}, \overline{f_{i3}f_{j3}} \dots \overline{f_{in}f_{jn}})$, we obtain the tuples($\Delta MAP$, F);
      % $\overline{f_{i1}f_{j1}}=(f_{i1}+f_{j1})/2$, $\overline{f_{i2}f_{j2}}=(f_{i2}+f_{j2})/2 \dots \overline{f_{in}f_{jn}}=(f_{in}+f_{jn})/2$
end for

Learn a linear function using logistic regression with (ΔMAP, F);

Fig. 2. TQC_{Pair}: Training query classification with pairs of queries' retrieval performance gain.

Strategy 3: TQC_{Set}

Train the basic retrieval model (BM25) on the candidate training set, denoted as M_b;
Tune the parameter on validation set;
Apply M_b to the pre-test set, obtained the MAP, denoted as MAP_{M_b};

for each $q_i \in candidate\ training\ set$ **do**
    Select another k queries from (candidate training set - q_i) randomly;
    Combine the q_i and selected k queries as the temporal training set to train the LTR model;
    Repeat the previous two steps 3 times, to obtain three LTR models, denoted as $M_{i_1}, M_{i_2}, M_{i_3}$;
      % Pairwise and listwise algorithms are applicable;
      % k $\geqslant$ 2 (k=4 in our experiment), if k=2 then a temporal training set containing 3 queries;
    Tune parameter of $M_{i_1}, M_{i_2}, M_{i_3}$ on validation set;
    Obtain $MAP_{M_{i1}}, MAP_{M_{i2}}, MAP_{M_{i3}}$ on pre-test set, to calculate the average value of the
    three MAP, denoted as MAP_{M_i};
    Calculate the retrieval performance gain of each query q_i:
      $\Delta MAP = MAP_{M_i} - MAP_{M_b}$
    Extract features for each one single query, F = $(f_1, f_2, f_3 \dots f_n)$, obtain the tuples($\Delta MAP$, F);
end for

Learn a linear function using logistic regression with (ΔMAP, F);

Fig. 3. TQC_{set}: Training query classification with sets of queries' retrieval performance gain.

advantage of using pairs of candidate queries, instead of the single queries as in [9], is the enlarged set of samples from the candidate set, such that the training queries classifier can be learned with more labeled examples, and hopefully lead to a better selection of the training queries.

Our second proposed algorithm, called TQC_{Set}, estimates the quality of a candidate training query by the retrieval performance gain when combined with other training queries. By doing so, it takes the dependency among the candidate training queries into consideration. As described in Fig. 3, TQC_{Set} estimates the quality of a candidate training query q_i by its usefulness in combining with other training queries. Specifically, for each candidate training query q_i, TQC_{Set} learns a ranking model using a temporal training set composed of q_i and k randomly selected candidate training queries, and obtains an estimate of ΔMAP on the pre-test set. Such a process is repeated for 3 times, and the quality of q_i is given by the average of the 3 estimates of ΔMAP. Similar to TQC_{Single}, a training query classifier is learned over individual candidate training queries using the average ΔMAP as the target.

4   Experimental Setup

We experiment on Tweets2013, a standard test collection used in the TREC 2013 and TREC 2014 Microblog track[2]. It contains 243 million tweets gathered from the public Twitter stream from February 1, 2013 to March 31, 2013 (inclusive, UTC time). We fetch up to 10,000 tweets for each query via the track API with the official token provided by the official organizer to the participants.

There are 60 topics in TREC 2013 and 55 topics in TREC 2014. We use one year of the TREC topics as the training data and the dataset of the other year as the test data. For data of each year of TREC, we divide them into three equal-size subsets. One subset is used as the candidate training set, one as the validation set, and the last one as the pre-test set. There are no overlap between any two subsets. The learned ranking model using the training queries selection algorithm is evaluated on the other year of the test data.

We preprocess the tweet documents to filter the pure retransmission (beginning with the 'RT'), repeat tweets and non-English tweets. Then we extract features from the tweet documents and queries. In this study, we use 20 features to represent the tweet documents and 12 features to represent the queries. According to the guidelines of the TREC Microblog track, we return top 1000 tweets published prior to and including the query time defined by the topic.

Table 1 introduces the features of tweet we adopted. There are five types of tweet features, i.e. the tweet content, content richness, author authority, tweet recency and Twitter specific features. These types of features are adopted by many participants of Microblog tracks [10–12] and their meanings are introduced in the Table 1.

There are three types of query features for inferring the benefit brought by a given training query for learning to rank, i.e. content-based relevance scores, the Normalized Query Commitment (NQC) [20], and the Twitter specific features. Table 2 presents all the query features exploited in this paper. And these query features are all used in the training query classifier phase.

---

[2] https://github.com/lintool/twitter-tools/wiki/TREC-2013-Track-Guidelines.

Table 1. Pre-defined tweet features.

Feature type	Feature ID	Description
Content relevance	PL2	Retrieval performance of PL2 [17] model measured by MAP
	PL2QE	Retrieval performance of PL2 [17] model with query expansion measured by MAP
	BM25	Retrieval performance of BM25 [18] model measured by MAP
	BM25QE	Retrieval performance of BM25 [18] model with query expansion measured by MAP
	TF*IDF	Term frequency multiply by inversed document frequency [16]
	LMDir	Retrieval performance of the KL-divergence language model with Dirichlet smoothing [19] measured by MAP
Content richness	Content Length	Length of tweet
	Word Count	Number of the word
	Mean Word Length	Mean Word Length
	URL Count	Number of URL contained in the Tweet
	OOV	The ratio of unique word
Author Authority	Retweet Count	Number of the Retweets
	Followers Count	Followers count of the author
	Statuses Count	Total number of the statuses published by the author
	Tweet in Topic	Tweet count in the same topic
Tweet recency	Query Day Dif	Time difference between tweet posttime and query posttime
	Burst Day Dif	Time difference between topic bursttime and tweet posttime
Twitter-specific	AT Count	Count of @ in the tweet
	RT	Whether the tweet is a retweet
	HASHTAG Count	Count of # in the tweet

Four different learning to rank algorithms are applied in our experiment, which are the pairwise RankSVM [8,21], and three listwise algorithms, namely Listnet, Coordinate Ascent (CA for short in tables), and Lamdamart [5,6,22]. We tune the parameters of algorithms on the validation set and the parameter values that lead to the best MAP on validation set are applied to the test set.

The baseline is the recent TQC_{Single} algorithm that has shown effective for Twitter search [9]. As introduced in Sect. 1, TQC_{Single} is not suitable for Listwise Algorithms, so only RankSVM is applied in baseline TQC_{Single}. Mean average

Table 2. Pre-defined query features.

Feature type	Feature ID	Description
Content relevance	PL2	Mean PL2 [17] score of the top-5 tweets
	PL2QE	Mean PL2 [17] score of the top-5 tweets with query expansion
	BM25	Mean BM25 [18] score of the top-5 tweets
	BM25QE	Mean BM25 [18] score of the top-5 tweets with query expansion
	TF*IDF	Mean TF*IDF [16] score of the top-5 tweets
	LMDir	Mean score of top 5 tweet given by the KL-divergence language model with Dirichlet smoothing [19]
NQC	PL2NQC	NQC of relevance score given by PL2 with query expansion
	BM25NQC	NQC of relevance score given by BM25 with query expansion
	TF*IDF-NQC	NQC of relevance score given by TF*IDF
	LMDirNQC	NQC of relevance score given by KL-divergence language model with Dirichlet smoothing
Twitter-specific	URL Count	The percentage of the top 10 tweet's with URLs in their content
	Followers Count	Average number of followers of the top 10 tweet's authors

precision (MAP), precision at 30 (P@30) and R-precision (R-prec) are reported in the next section.

5   Results and Analysis

This section presents the evaluation results. In Tables 3, 4 and 5, a star indicates a statistically significant improvement over the baseline according to the Wilcoxon matched-pairs signed-rank test at the 0.05 level.

Table 3 is the evaluation results compared with baseline on both TREC 2013 and 2014 test queries when evaluation metric is MAP. From this table, we can see the two proposed query classification algorithms perform better than the baseline algorithm no matter the learning to rank is pairwise (RankSVM) or listwise (Listnet, Coordinate Ascent, Lamdamart) algorithms. As for the final test set is TREC 2013, we obtain the best MAP, 0.3642, when we employ the TQC_{Set} and the Lamdamart algorithm, improved by 5.50 % than the baseline 0.3452. When we employ RankSVM, the same LTR model with the baseline algorithm, the result is improved by 3.33 % and 3.82 % with TQC_{Pair} and TQC_{Set}, respectively. When the final test set is TREC 2014, we obtain the best MAP, 0.5031, with

Table 3. Evaluation results measured by MAP.

TREC 2013	Baseline	Listnet	CA	LM	RankSVM
TQC_{Pair}	0.3452	0.3581	0.3581	0.3621	0.3567
Improve (%)		3.74 % ⋆	3.74 % ⋆	4.90 % ⋆	3.33 % ⋆
TQC_{Set}		0.3587	0.3585	0.3642	0.3584
Improve (%)		3.91 % ⋆	3.85 % ⋆	5.50 % ⋆	3.82 % ⋆
TREC 2014	Baseline	Listnet	CA	LM	RankSVM
TQC_{Pair}	0.4748	0.4881	0.4918	0.5001	0.4904
Improve (%)		2.80 % ⋆	3.58 % ⋆	5.33 % ⋆	3.29 % ⋆
TQC_{Set}		0.4859	0.4939	0.5031	0.4911
Improve (%)		2.34 %	4.02 % ⋆	5.96 % ⋆	3.43 % ⋆

Table 4. Evaluation results measured by P@30.

TREC 2013	Baseline	Listnet	CA	LM	RankSVM
TQC_{Pair}	0.5587	0.5743	0.5789	0.5813	0.5752
Improve (%)		2.79 % ⋆	3.62 % ⋆	4.05 % ⋆	2.95 % ⋆
TQC_{Set}		0.5789	0.5805	0.5807	0.5773
Improve (%)		3.62 % ⋆	3.90 % ⋆	3.94 % ⋆	3.33 % ⋆
TREC 2014	Baseline	Listnet	CA	LM	RankSVM
TQC_{Pair}	0.6648	0.6701	0.6834	0.689	0.6829
Improve (%)		0.80 %	2.80 % ⋆	3.64 % ⋆	2.72 % ⋆
TQC_{Set}		0.6756	0.6899	0.6875	0.6912
Improve (%)		1.62 %	3.78 % ⋆	3.41 % ⋆	3.97 % ⋆

TQC_{Set} and the Lamdamart algorithm, improved by 5.96 % than the baseline 0.4748. When we employ RankSVM, the result is improved by 3.29 % and 3.43 % with TQC_{Pair} and TQC_{Set} respectively.

Tables 4 and 5 are the evaluation results compared with the baseline on both TREC 2013 and 2014 test queries when evaluation metrics are P@30 and R-prec, respectively. The results demonstrate our algorithms perform better than the baseline algorithm. When the metric is P@30, TQC_{Pair} with Lamdamart algorithm perform best, improved by 4.05 % than the baseline for TREC 2013. The P@30 result of TQC_{Pair} and TQC_{Set} with RankSVM is increasing by 2.95 % and 3.33 % than the baseline respectively. For TREC 2014, TQC_{Pair} with Lamdamart algorithm perform best, increasing by 3.64 % than the baseline. The P@30 result of TQC_{Pair} and TQC_{Set} with RankSVM is improved by 2.72 % and 3.97 % over the baseline respectively.

When the metric is R-prec, TQC_{Set} with Lamdamart algorithm perform best, improved by 4.10 % than the baseline for TREC 2013. The R-prec result of

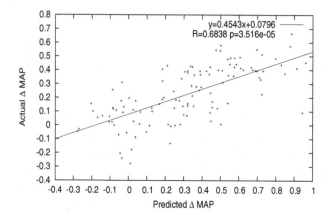

Fig. 4. The predicted ΔMAP against the actual ΔMAP of TQC_{Single}.

Table 5. Evaluation results measured by R-prec.

TREC 2013	Baseline	Listnet	CA	LM	RankSVM
TQC_{Pair}	0.3513	0.3551	0.3603	0.3634	0.356
Improve (%)		1.08 %	2.56 % ⋆	3.44 % ⋆	1.34 %
TQC_{Set}		0.3545	0.3595	0.3657	0.3579
Improve (%)		0.91 %	2.33 %	4.10 % ⋆	1.88 %
TREC 2014	Baseline	Listnet	CA	LM	RankSVM
TQC_{Pair}	0.4684	0.4856	0.478	0.4823	0.4801
Improve (%)		3.67 % ⋆	2.05 %	2.97 % ⋆	2.50 % ⋆
TQC_{Set}		0.487	0.4769	0.4906	0.4831
Improve (%)		3.97 % ⋆	1.81 %	4.74 % ⋆	3.14 % ⋆

ᵃCA: Coordinate Ascent Algorithm.
ᵇLM: Lamdamart Algorithm.

TQC_{Pair} and TQC_{Set} with RankSVM is improved by 1.34 % and 1.88 % than the baseline. For TREC 2014, TQC_{Set} with Lamdamart algorithm perform best, improved by 4.74 %. The R-prec result of TQC_{Pair} and TQC_{Set} with RankSVM is improved by 2.50 % and 3.14 % respectively.

Our proposed approach assumes a linear relationship between the training query quality, estimated by ΔMAP, and the effectiveness of the learned ranking model. Figures 4, 5 and 6 plot the predicted ΔMAP, estimated by the training query classifier, against the actual ΔMAP obtained on the test queries. These figures combine the results of all queries from both TREC 2013 and 2014 using RankSVM. From these figures, we can see that the ΔMAP predicted by the logistic regression has indeed a moderate linear correlation with the actual ΔMAP. The correlation coefficients are R = 0.6838, R = 0.5716, R = 0.5641 and the P-values are P = 3.516e-05, P = 0.0342, P = 7.649e-06 for three TQC

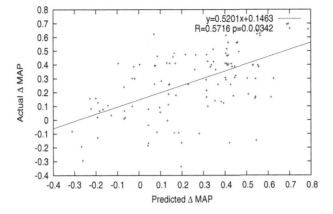

Fig. 5. The predicted ΔMAP against the actual ΔMAP of TQC_{Pair}.

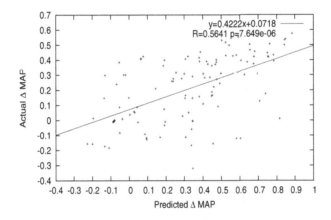

Fig. 6. The predicted ΔMAP against the actual ΔMAP of TQC_{Set}.

algorithms, respectively, indicating that the strength of the linear relationship is moderate but statistically significant.

6   Conclusions and Future Work

This paper has proposed a set-based training query classification approach by taken the dependency and redundancy among candidate training queries into account in order to improve the retrieval effectiveness of Twitter search. Two different algorithms, TQC_{Pair} and TQC_{Set}, based on different granularities of the samples from the candidate set are proposed. According to the evaluation results on the standard Tweets2013 collection with two years of the TREC Microblog track topics, our proposed algorithms outperform the TQC_{Single} baseline that estimates the training query quality for individual queries, instead of pairs (as by TQC_{Pair}) or sets (as by TQC_{Set}) of queries. Application of LTR to Twitter search can be benefit from our proposed approach from the following two aspects. First, the retrieval effectiveness can be enhanced by selecting the high-quality training queries for learning the ranking model. Second, the relevance assessments of the documents is only necessary for the selected training queries, instead of all those in the candidate set, such that the human efforts and the related costs are largely reduced.

The proposed approach in this paper is not only applicable to Twitter search, but also a potentially general solution for the training query selection of learning to rank. In the future, we plan to investigate the scalability and effectiveness of the proposed approach for other applications such as ad-hoc retrieval and Web search. Public datasets such LETOR can be used in the related study.

Acknowledgments. This work is supported in part by the National Natural Science Foundation of China (61472391), and Beijing Natural Science Foundation (4142050).

References

1. Kwak, H., Lee, C., Park, H., Moon, S.: What is twitter, a social network or a news media? In: Proceedings of WWW, pp. 591–600 (2010)
2. Duan, Y., Jiang, L., Qin, T., Zhou, M., Shum, H.Y.: An empirical study on learning to rank of tweets. In: Proceedings of COLING, pp. 295–303 (2010)
3. Lin, J., Efron, M.: Overview of the TREC 2013 microblog track. In: TREC (2013)
4. Lin, J., Efron, M.: Overview of the TREC 2014 microblog track. In: TREC (2014)
5. Liu, T.Y.: Learning to rank for information retrieval. Found. Trends Inf. Retr. **3**(3), 225–331 (2009)
6. Cao, Z., Qin, T., Liu, T.Y., Tsai, M.F., Li, H.: Learning to rank: from pairwise approach to listwise approach. In: Proceedings of ICML, pp. 129–136 (2007)
7. Long, B., Chapelle, O., Zhang, Y., Chang, Y., Zheng, Z., Tseng, B.: Active learning for ranking through expected loss optimization. In: Proceedings of SIGIR, pp. 267–274 (2010)
8. Zhang, X., He, B., Luo, T., Li, D., Xu, J.: Clustering-based transduction for learning a ranking model with limited human labels. In: Proceedings of CIKM, pp. 1777–1782 (2013)
9. Li, D., He, B., Luo, T., Zhang, X.: Selecting training data for learning-based twitter search. In: Hanbury, A., Kazai, G., Rauber, A., Fuhr, N. (eds.) ECIR 2015. LNCS, vol. 9022, pp. 501–506. Springer, Heidelberg (2015)
10. Lv, C., Fan, F., Qiang, R., Fei, Y., Yang, J.: Pkuicst at TREC 2014 microblog track: feature extraction for effective microblog search and adaptive clustering algorithms for TTG. In: TREC (2014)
11. Xu, T., Oard, D.W., McNamee, P.: HLTCOE at TREC 2014: microblog and clinical decision support. In: TREC (2014)
12. Zhang, Z., Lan, M.: Estimating semantic similarity between expanded query and tweet content for microblog retrieval. In: TREC (2014)
13. Magdy, W., Gao, W., El-Ganainy, T., Wei, Z.: QCRI at TREC 2014: applying the KISS principle for the TTG task in the microblog track. In: TREC (2014)
14. Montague, M., Aslam, J.A.: Condorcet fusion for improved retrieval. In: Proceedings of CIKM, pp. 538–548 (2002)

15. Li, C., Wang, Y., Mei, Q.: A user-in-the-loop process for investigational search: foreseer in TREC 2013 microblog track. In: TREC (2013)
16. Baeza-Yates, R.A., Ribeiro-Neto, B.: Modern Information Retrieval. Addison-Wesley Longman Publishing Co. Inc., Boston (1999)
17. Amati, G., Van Rijsbergen, C.J.: Probabilistic models of information retrieval based on measuring the divergence from randomness. ACM Trans. Inf. Syst. **20**(4), 357–389 (2002)
18. Robertson, S., Zaragoza, H.: The probabilistic relevance framework: Bm25 and beyond. Found. Trends Inf. Retr. **3**(4), 333–389 (2009)
19. Zhai, C., Lafferty, J.: A study of smoothing methods for language models applied to ad hoc information retrieval. In: Proceedings of SIGIR, pp. 334–342 (2001)
20. Shtok, A., Kurland, O., Carmel, D., Raiber, F., Markovits, G.: Predicting query performance by query-drift estimation. ACM Trans. Inf. Syst. **30**(2), 11:1–11:35 (2012)
21. Cao, Y., Xu, J., Liu, T.Y., Li, H., Huang, Y., Hon, H.W.: Adapting ranking SVM to document retrieval. In: Proceedings of SIGIR, pp. 186–193 (2006)
22. Xia, F., Liu, T.Y., Wang, J., Zhang, W., Li, H.: Listwise approach to learning to rank: theory and algorithm. In: Proceedings of ICML, pp. 1192–1199 (2008)

NBLucene: Flexible and Efficient Open Source Search Engine

Zhaohua Zhang, Benjun Ye, Jiayi Huang, Rebecca Stones, Gang Wang$^{(\boxtimes)}$,
and Xiaoguang Liu$^{(\boxtimes)}$

College of Computer and Control Engineering, Nankai University, Tianjin, China
{zhangzhaohua,yebj,huangjy,becky,wgzwp,liuxg}@nbjl.nankai.edu.cn

Abstract. The most popular open source projects for text searching have been designed to support many features. These projects are well-written in Java for cross-platform using. But when conducting research, the execution efficiency of program should be more essential, which is a problem for applications written in Java. It is also difficult for Java to use parallel mechanisms in the modern computer system like SIMD and GPUs. To this end, we expand an open source text searching project written in C++ for research purpose.

Our approach is to define a flexible and efficient search engine architecture which consists of extensible application programming interfaces. We aim to provide a flexible architecture to enable researchers to readily implement and modify search engine algorithms and strategies. Moreover, we integrate one generic mathematical encoding library which can be used to compress inverted index. We also implement an integral framework for result summarization, including snippet generation and cache strategies. Experiment results show that the new architecture makes a significant improvement versus original work.

1 Introduction

Due to the complex requirements, more and more open source projects for text searching have been well designed [11]. These projects support full-featured methods including text indexing, query processing and result presentation. For cross-platform and high performance purposes, most of them were written in Java, such as Lucene[1] and Nutch[2]. Because of their excellent design, these projects have been widely used in academic and commercial situations.

When conducting research, however, the projects which are written in Java face efficiency problems. Java programs have to been executed on a JVM (Java Virtual Machine), which allows application programs to be run on any platform without having to be rewritten or recompiled for each individual platform. This runtime environment does not directly support some parallelism in modern computer system, like SIMD (Single Instruction, Multiple Data) and GPUs. Both of

---

[1] https://lucene.apache.org/.
[2] http://nutch.apache.org/.

© Springer International Publishing Switzerland 2016
B. Cui et al. (Eds.): WAIM 2016, Part I, LNCS 9658, pp. 504–516, 2016.
DOI: 10.1007/978-3-319-39937-9_39

them have been utilized to boost query processing in [3,8,10,13,14,18], typically written in C/C++. The other problem of using Java is memory management. As the garbage collection and memory reallocation is fully controlled by the JVM, it is difficult for programmers to design the memory layout and control reallocation as needed. In some specific situations, such as cache strategy experimentation, it is a serious problem that the memory space can not be fully controlled by designers.

This paper is instead dedicated to design an efficient open source library for text searching written in C++. The core library should be designed to have good flexibility, which can be expanded for different specific algorithms and strategies expediently. As a result, we make a choice to expand the CLucene[3], which is a C++ migration of Lucene, with our experimental interfaces. We call this expansion of CLucene *NBLucene*.

The new experimental interfaces in NBLucene involve different stages during text searching, including text preprocessing, index compression, query processing and result summarization. In text preprocessing, we provide full support to analyze TREC web format and Web Archive format, which is the standard format for GOV2 and ClueWeb09 web collection, respectively. This work has not been implemented in the original CLucene project. For index compression and query processing, we implement several typical mathematical encoding methods with both scalar and SIMD versions. These methods can be used to compress posting lists and position information in the inverted index. Besides these, we also design a full-featured cache framework for result summarization, including snippet generation and cache strategies. All these interfaces can be easily expanded for other specific methods as needed. To the best of our knowledge, there is no previous open source project that provides all of the above features.

2   Related Work

2.1   Open Source Search Engines

Many text search engines have been made open source for researching and commercial use. Table 1 concludes the most popular projects for full-text searching. Lucene has become a top-level Apache project since 2005 and has been extended to a series of Lucence-based search engine, e.g. Nutch and Solr[4]. Lucene is not a complete search engine, but provides the core API library for full-text searching. CLucene is an existing port of Java Lucene written in C++. The latest version is converted from Lucene 2.3.2 and has not updated since March, 2011. Because our previous work has been integrated in Java Lucene, we refer to expand the CLucene as our baseline. Besides CLucene, there are also many other open source search engines written in C/C++, e.g. Indri [15] and Zettair [22].

---

[3] http://clucene.sourceforge.net/.
[4] http://lucene.apache.org/solr/.

Table 1. Comparison of the related open source search engines

Project	License	Language	Latest update
CLucene	Apache LGPLv2	C++	March, 2011
Galago	BSD	Java	January, 2016
Indri	BSD	C/C++	January, 2016
Lucene	Apache	Java	January, 2016
Nutch	Apache	Java	January, 2016
Solr	Apache	Java	January, 2016
Sphinx	GPLv2	C++	September, 2015
Terrier	MPL	Java	April, 2015
Zettair	BSD	C	March, 2009

2.2 Index Compression

In most search engines, an inverted index is the central data structure which maps terms to the documents that contain them. Given a collection of N documents, each document is represented by a unique document identifier, *docID*, between 0 and $N - 1$. An inverted index contains posting lists for all distinct terms in the collection. As the posting lists often take large fraction of storage, many previous studies focus on index compression techniques [1,2,5,19,20].

Besides compression ratio, decompression speed is also a crucial indicator since the compressed posting lists relative to a query have to be completely or partially decompressed during query processing. By using SIMD instructions, several papers have reported significant improvement on decoding speed. Willhalm [18] proposed a SIMD approach to execute the vectorized value decompression with very short latency. Stepanov [14] concentrated on their SIMD-based method, called *varint-G8IU*, and outperformed the classic variable byte coding methods. Zhang [21] reported a compression framework with a novel storage layout format and made very competitive performance. Lemire [9] found that compressing integers in large blocks of integers with minimal branching would be faster than previous algorithms.

2.3 Snippet Generation

Nowadays, most search engines will return a document summarization with each top-ranked result. These summarized text fragments will help users to judge the relevance before clicking the link to get full document. Previous work on document summarization can be categorized as either query-dependent summarization or query-independent summarization. In this paper, we focus on query-dependent method, which is known as snippet generation, i.e. to rank and select most relative text fragments for each query. This method has been studied in respective papers [4,16,17].

3   Architecture

CLucene already provides a basic API for text indexing and ranked search for common query models. By keeping the standard modular architecture, we define a series of modular interfaces which would be essential for relevant experiments. These scalable modules involve the different stages in the text searching, which consist of four major components: text preprocess, index compression, query process and result summarization. Figure 1 illustrates the overall architecture of NBLucene. The rectangles with grey background are the modular interfaces we have added to the original CLuence.

The text preprocess module would parse origin HTML pages and generate formatted documents for *Indexer*. With these documents, the index compression module will build an inverted index and use *Encoder* for compression. During query processing, *Decoder* will be called to decompress posting list blocks and corresponding position lists. For each top ranked document, the result summarization module will return the snippet directly when it is in the cache. Otherwise it will generate snippet and update cache as needed.

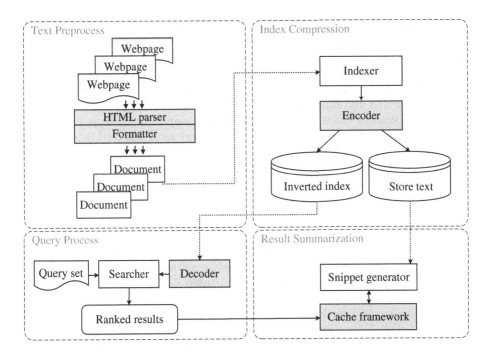

Fig. 1. Search engine architecture of NBLucene

3.1   Text Preprocess

Since CLucene does not provide any toolkit for HTML parsing, we have added an independent module for text preprocessing. The specific method for a given text collection file type needs to be implemented to compose independent documents. Each document has been stripped of HTML tokens and JavaScript, and only keeps the readable content of the original webpage. We implement the function for analysis of the TREC web format and Web Archive format corpus, and it is written in Java with JSoup[5].

In our experiments, we define four basic field types for webpage: URL, title, anchor text and content. In the text preprocess module, *Formatter* can split different fields into fixed positions. For instance, URL text, title words and anchor words are stored on the first three lines for each document, respectively. The content text will be kept next. For word stemming and stop words filtering, we would leave the option for users to determine whether they need them.

3.2   Index Compression

After index building, applications based on NBLucene may need to select different mathematical encoding methods to yield different trade-offs between space and time. The basic list compression method in CLucene is VByte [6]. We integrate a standard integer list compression library[6] in NBLucene to support other mathematical encoding algorithms. The methods of compression are implemented from unified interfaces. As instances, we also implement several common encoding and decoding methods, including both scalar and SIMD versions. For query processing efficiency, NBLucene utilizes skip-table to support random access on posting lists and position lists. For compression, inverted lists are also compressed in blocks consisting of a fixed number of postings in order to align to skip-table entries.

3.3   Query Processing and Result Summarization

To support result summarization, we integrate the standard query processing with an extensible snippet generation and cache framework. After result ranking, the most related documents, called Top-K results, will be returned. A most widely used relevance score, i.e. BM25 [12], is used in CLucene. Since CLucene has already implemented a snippet generation method which is called *Highlighter* in Lucene, we use it directly to extract the most relative text fragments for each result document. To improve query processing throughput, we utilize caching to store snippet results for future duplicate queries. In our experiment, we implement two basic cache types: results cache and snippet cache. The results cache stores queries and the corresponding result snippet for each Top-K document, and the snippet cache keeps snippet for each pair of query and result document.

---

[5] http://jsoup.org/.

[6] https://github.com/lemire/FastPFor.

We also implement two typical cache strategies, i.e. LRU and LFU. The caching framework is also easy to be extended for other types of cache, e.g. document cache and fragment cache, and other cache strategies.

4   Index Encoding

4.1   Posting and Position List Compression

In CLucene, the inverted index stores the relationship information between terms and documents. As CLucene supports incremental indexing, the full index is often composed of multiple sub-indexes, which are also called segments. In each segment, the index is organized into files storing distinct content. Among these files, the *term frequency file* stores the posting lists. While the *term proximity file* stores the position information of terms in each document. In this section, we will focus on these two types of files.

The term frequency file contains the lists of postings that each list is corresponding to one specific term. Each term list is composed of two parts, the posting list itself and the skip table. Figure 2 shows the layout of term frequency files. Each posting list is composed of postings and each posting is made up of a docID and the term frequency. Every docID and its term frequency will follow one special rule: if the term frequency is equal to one, its value will be omitted. In this case, the docID will be shifted left by one bit and the least significant bit will be set to 1. Otherwise, term frequency is greater than one, and we store it next to its docID. The docID will be also shifted left by one bit, but leave the least significant bit to be 0. Whether or not the frequency values are omitted, a compression method will encode them consistently.

After encoding, the term frequency file almost keeps the same format. Each term list still consists of a posting list and a skip table. In CLucene, the option *SkipInterval* is the interval between consecutive entries of the skip table. In other words, it is equal to the size of one block in the posting list. In our implementation, we compress each block independently. The first docID of each block will be stored in the corresponding skip table entry. After compression, the offset of each code words block will be updated in skip table.

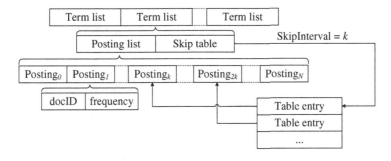

Fig. 2. Original file format of the term frequency file

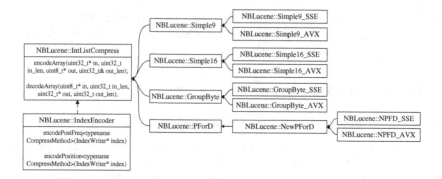

Fig. 3. Class inheritance and interfaces of list compression

Similar to the term frequency file, the term proximity file stores the lists of positions in which the term occurs in documents. We compress the position lists within each block of corresponding posting list. After compression, the offset of position list code words in the skip table entries will also be updated.

Figure 3 shows the interfaces and the class hierarchy of list compression. The base class of list compressors is *IntListCompress*, which has two main interfaces: *encodeArray* and *decodeArray*. The derived classes, such as Simple9 in the figure, or other methods designed by the users, have to be implemented for the specific method to encode original integer lists and decode code words, respectively. For encodeArray, there are four arguments: the origin integer list *in* with its length *in_len* and the output byte sequence *out* with its length *out_len*, while for decodeArray the arguments are similar. *IndexEncoder* is a C++ template which based on the derived class of IntListCompress, and it makes use of one specific encoding method to compress posting lists and position information.

4.2   Encoding Algorithm

Due to the lack of generic mathematical encoding methods in CLucene, we implement several typical algorithms to compress the index. In contrast to other previous mathematical encoding libraries, we have also provided relative SIMD implementation for different decoding algorithm, including Simple 9 [2], Simple 16 [20], NewPForD [19] and GroupByte [7]. We make use of the FastP-For and varint-G8IU as SIMD implementations for NewPForD and GroupByte, as given in [9] and [14], respectively.

For Simple 9 and Simple 16, we design a parallel decoding method. To illustrate, Fig. 4 shows the decoding procedure for an example. Suppose there are four consecutive integers b_0, b_1, b_2, b_3 which are compressed into one Simple 9 code word. That means each integer can be represented within seven bits. The parallel decoding method will first make four copies C_0, C_1, C_2, C_3 of the code word in a SIMD register as Fig. 4(a) shows. In Fig. 4(b), by utilizing parallel instructions, we can shift b_0, b_1, b_2, b_3 to the last position in C_0, C_1, C_2, C_3, respectively. Finally, we use the parallel bitwise AND instruction to "clear" other bits

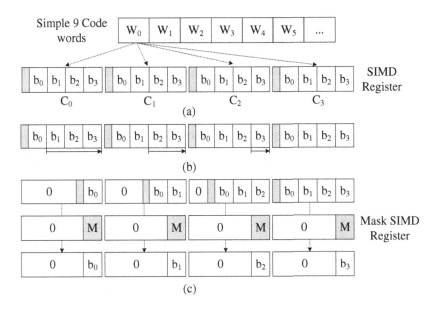

Fig. 4. Decoding procedure of Simple 9 in SIMD scheme

in C_0, C_1, C_2, C_3 leaving b_0, b_1, b_2, b_3. In Fig. 4(c), each rectangle with a grey background in Mask SIMD Register contains seven consecutive bits 1. The content in the last register is the final result. In our experiment, we implement these SIMD algorithms on an Intel platform with SSE and AVX instructions.

5   Cache Framework

CLucene provides one straightforward snippet generation method which we extend with cache strategies. We implement the cache framework and provide the interfaces for different types of cache. The kernel data structure of a cache is a *cache table*, and an entry in this table is a *cache node*, which is defined by a key-value pair. For instance, the key of a cache node in the results cache is query string and the value is the arrays of snippets of the top-ranked documents. When receiving a new query, the search engine will first lookup the cache table to find the results. If it succeeds, the result will be returned directly. Otherwise, it will search the top-ranked documents and generate snippets from them. According to its cache policy, the cache framework will decide whether or not to store a new snippet result, and which cache entries are replaced if available space is insufficient.

In addition to the kernel data structure and standard operations, we also define the interface for the cache replacement strategy for users to implement their own cache algorithms. The specific cache strategy is independent from cache type. We implement some basic cache replacement algorithms, including LRU and LFU.

1. LRU (Least Recently Used). We keep a list of pointers to record each cache node in the cache table. When a new cache node is inserted, or an existing cache node is hit, the corresponding element in the list will be moved to the head. When cache replacement occurs, the cache nodes will be removed from the tail of the list in reverse order. So the least recently used cache nodes will be evicted.
2. LFU (Least Frequently Used). For each cache node, we use a counter to record the access frequency. The frequency will increase if the cache node is hit. There is a *MinHeap* to maintain the frequency of all cache nodes. When cache replacement happens, the top element in the heap, i.e. the least frequently used one, will be deleted.

The process of query processing with cache is composed of two parts: the training phase and testing phase. Depending on whether or not the cache can be modified in testing phase, the cache mechanism can be described as one of two types: static cache or dynamic cache. In a static cache, the training phase will fill the cache space by following a specific cache strategy. The content of the cache can not be modified during the testing phase. While in a dynamic cache, the cache can always be updated whatever the phase is.

6   Experiments

6.1   Experimental Setup

In our experiments, we use two real-world data sets as our text collection: GOV2 and ClueWeb09 (English pages in Category A). Collections GOV2 and ClueWeb09 contain approximately 25 million and 503 million documents, respectively. We analyze and format the web pages from these two raw text collections with our text preprocessing module. The stop words are filtered, and other words are stemmed with the Porter algorithm. The sizes of the two indexes are roughly 124 GB and 2.6 TB, respectively. The full index contains the posting lists file, position lists file, store fields file (containing all of the readable content of each page) and other auxiliary files. The value of option *SkipInterval* is set to 256, which means there are 256 docIDs in each posting list block (typically except the last block of a list).

In addition, we use the TREC Terabyte Track (100,000 queries from year 2006), and an AOL query log (100,000 queries) which was randomly extracted from 14.4 million queries log of 650,000 AOL users. We carry out all experiments on a PC server with a quad-core of a 3.40 GHz Intel(R) Core(TM) CPU and 32 GB of memory, running Centos 6.5. All components of the compression and cache framework are implemented in C++ and are compiled with g++ 4.8.2, with optimization flag -O2.

6.2   Encoding and Decoding

In this section, we compare the compression ratio and decompression speed for different compression methods. The compression ratio is measured as the average

Table 2. Space usage (bits per integer) and decoding speed (thousand integers per millisecond) for different encoding and decoding methods

Method	GOV2		ClueWeb09	
	Space	Speed	Space	Speed
Vbyte	6.32	587	8.50	341
GByte	6.72	793	9.24	557
GByte-SSE	6.80	819	9.41	572
GByte-AVX	6.91	802	9.75	562
S9	5.81	613	7.54	342
S9-SSE	5.81	633	7.54	360
S9-AVX	5.81	631	7.54	359
S16	5.67	599	6.87	289
S16-SSE	5.67	598	6.87	292
S16-AVX	5.67	600	6.87	292
NPFD	5.57	842	7.93	442
NPFD-SSE	5.54	901	7.94	521
NPFD-AVX	5.54	914	7.95	544

number of bits used per integer. The integers here refer to the elements of posting lists in NBLucene term frequency file, including docIDs and term frequencies. To compare the decompression speed, we execute the TREC and AOL query logs on the GOV2 and ClueWeb09 collections, respectively. During query processing, we count the number of integers actually decompressed and the time cost for decompression from code words to the original integers.

Table 2 shows that except from GByte methods, the other methods get better compression (GByte refers to GroupByte). GByte-SSE refers to varint-G8IU while GByte-AVX is a variant of varint-G8IU implemented using AVX instructions. S9, S16 and NPFD refers to the scalar Simple 9, Simple 16 and NewP-ForD. On the GOV2 collection, the best encoding method is NPFD-SSE and NPFD-AVX. They yield an enhancement of 12 % over the baseline VByte. On the ClueWeb09 collection, the best encoding method is S16 and its SIMD variants with an improvement is of 19 %. The highest decompression throughput of two datasets is achieved by NPFD-AVX and GByte-SSE. They are 56 % and 68 % faster than the baseline, respectively. We also find that the SIMD variants can get faster decompression speed than scalar version, and the improvement is up to 23 %. Although we only compare several methods in our experiment, the index compression module in NBLucene can be easily extended for other encoding methods as needed.

Table 3. Cache hit ratio and average query response time (milliseconds) for different results caches and cache sizes (MB)

Cache size	Cache type	GOV2		ClueWeb09	
		Ratio	Time	Ratio	Time
100	S-LRU	0.7 %	211	2.2 %	365
	S-LFU	0.5 %	211	1.9 %	370
	D-LRU	1.1 %	208	3.0 %	351
	D-LFU	1.0 %	210	2.4 %	358
200	S-LRU	0.8 %	209	2.4 %	355
	S-LFU	0.7 %	209	2.0 %	358
	D-LRU	1.8 %	206	3.1 %	344
	D-LFU	1.4 %	207	2.4 %	349
300	S-LRU	0.8 %	209	2.4 %	355
	S-LFU	0.7 %	209	2.0 %	357
	D-LRU	2.1 %	205	3.5 %	339
	D-LFU	1.5 %	207	2.6 %	345

6.3   Cache Framework

For the cache framework, we compare the cache hit ratio and average query response time with different results cache setups. The result of snippet cache is similar. For the ClueWeb09 collection, we extract the first 50 million documents to build a new index for running queries. We execute the TREC and AOL query logs on the whole GOV2 index and the new ClueWeb09 index, respectively. In each query set, we extract the first 50,000 queries for training and the other 50,000 queries for testing. We compare the static result cache and dynamic result cache with LRU and LFU strategies under different cache sizes.

In Table 3, we can see that the cache hit ratio improves and average query response time decreases with cache size increasing. The prefix used for the cache type name indicates the cache mechanism. For example, S-LRU indicates using a static cache with the LRU strategy while D-LFU indicating using a dynamic cache with the LFU strategy. We find that the dynamic caches outperform the static caches with the same configuration. Also, the caches with the LRU strategy have a higher hit ratio than the ones using the LFU strategy. Moreover, by designing the specific types of cache and replacement strategies, users can easily extend the result summarization module to implement their own cache experiments.

7   Conclusions and Future Work

In this paper, we design an efficient open source library for text searching. The most important contribution of our work is to provide a flexible architecture to enable researchers to readily implement and modify search engine algorithms and strategies. For this purpose, we extend the CLucene by defining a series of typical experimental interfaces which involve the different stages during text searching. In addition to these interfaces, we also implement the parser for TREC web format and Web Archive format, the extensible mathematical encoding library and the cache framework with basic replacement strategies. In our experiments, we have compared the compression methods on two real-world datasets, GOV2 and ClueWeb09. We also compare the different cache strategies in our cache framework.

As future work, we plan to extend our core library to support more query operations, especially to support WAND queries and early termination in scored queries. Also, we aim to implement an architecture which could support query processing on GPU platforms.

References

1. Anh, V.N., Moffat, A.: Index compression using fixed binary codewords. In: Proceedings of 15th Australasian Database Conference, vol. 27, pp. 61–67 (2004)
2. Anh, V.N., Moffat, A.: Inverted index compression using word-aligned binary codes. Inf. Retrieval **8**, 151–166 (2005)
3. Ao, N., Zhang, F., Wu, D., Stones, D.S., Wang, G., Liu, X., Liu, J., Lin, S.: Efficient parallel lists intersection and index compression algorithms using graphics processing units. Proc. VLDB Endowment **4**(8), 470–481 (2011)
4. Bast, H., Celikik, M.: Efficient index-based snippet generation. ACM Trans. Inf. Syst. **32**(2), 6 (2014)
5. Büttcher, S., Clarke, C.L.A., Cormack, G.V.: Information Retrieval: Implementing and Evaluating Search Engines. MIT Press, Cambridge (2010)
6. Cutting, D., Pedersen, J.: Optimization for dynamic inverted index maintenance. In: Proceedings of 13th Annual International ACM SIGIR Conference on Research and Development in Information Retrieval, pp. 405–411 (1989)
7. Dean, J.: Challenges in building large-scale information retrieval systems. In: Proceedings of 2nd ACM International Conference on Web Search and Web Data Mining, WSDM 2009, p. 1. ACM (2009)
8. Ding, S., He, J., Yan, H., Suel, T.: Using graphics processors for high performance IR query processing. In: Proceedings of 18th International Conference on World Wide Web, pp. 421–430 (2009)
9. Lemire, D., Boytsov, L.: Decoding billions of integers per second through vectorization. Softw. Pract. Exp. **45**, 1–29 (2015)
10. Lemire, D., Boytsov, L., Kurz, N.: SIMD compression and the intersection of sorted integers. CoRR, abs/1401.6399 (2014)
11. Middleton, C., Baeza-Yates, R.: A comparison of open source search engines. Technical report, Department of Technologies, Universitat Pompeu Fabra (2007)
12. Robertson, S.E., Jones, K.S.: Relevance weighting of search terms. J. Am. Soc. Inform. Sci. Technol. **27**(3), 129–146 (1976)
13. Schlegel, B., Willhalm, T., Lehner, W.: Fast sorted-set intersection using SIMD instructions. In: International Workshop on Accelerating Data Management Systems Using Modern Processor and Storage Architectures, pp. 1–8 (2011)

14. Stepanov, A.A., Gangolli, A.R., Rose, D.E., Ernst, R.J., Oberoi, P.S.: SIMD-based decoding of posting lists. In: Proceedings of 20th ACM International Conference on Information and Knowledge Management, pp. 317–326 (2011)
15. Strohman, T., Metzler, D., Turtle, H., Croft, W.B.: Indri: a language model-based search engine for complex queries. In: Proceedings of International Conference on Intelligent Analysis, vol. 2, pp. 2–6 (2005)
16. Turpin, A., Tsegay, Y., Hawking, D., Williams, H.E.: Fast generation of result snippets in web search. In: Proceedings of the 30th Annual International ACM SIGIR Conference on Research and Development in Information Retrieval, pp. 127–134 (2007)
17. Varadarajan, R., Hristidis, V.: A system for query-specific document summarization. In: Proceedings of 15th ACM International Conference on Information and Knowledge Management, pp. 622–631 (2006)
18. Willhalm, T., Popovici, N., Boshmaf, Y., Plattner, H., Zeier, A., Schaffner, J.: SIMD-scan: ultra fast in-memory table scan using on-chip vector processing units. Proc. VLDB Endowment **2**, 385–394 (2009)
19. Yan, H., Ding, S., Suel, T.: Inverted index compression and query processing with optimized document ordering. In: Proceedings of 18th International Conference on World Wide Web, pp. 401–410 (2009)
20. Zhang, J., Long, X., Suel, T.: Performance of compressed inverted list caching in search engines. In: Proceedings of 17th International Conference on World Wide Web, pp. 387–396 (2008)
21. Zhang, X., Zhao, W.X., Shan, D., Yan, H.: Group-scheme: SIMD-based compression algorithms for web text data. In: Proceedings of 2013 IEEE International Conference on Big Data, pp. 525–530 (2013)
22. Zobel, J., Williams, H., Scholer, F., Yiannis, J., Hein, S.: The Zettair Search Engine. Search Engine Group, RMIT University, Melbourne (2004)

Context-Aware Entity Summarization

Jihong Yan[1,2], Yanhua Wang[1], Ming Gao[1(✉)], and Aoying Zhou[3]

[1] Institute for Data Science and Engineering,
East China Normal University, Shanghai 200062, China
mgao@sei.ecnu.edu.cn
[2] Economic Management Institute, Shanghai Second Polytechnic University,
Shanghai 201209, China
[3] Shanghai Key Lab for Trustworthy Computing,
East China Normal University, Shanghai 200062, China

Abstract. Entity summarization aims at selecting a small subset of attribute-value pairs of an entity from a knowledge graph, which provides users with concrete information given an entity-related query. However, previous approaches focus on the "goodness" of the attribute-value pairs, paying little attention to user preference towards them.

In this paper, we formalize the task of context-aware entity summarization, and propose an algorithm to solve this problem. We model user interest by mining the latent topics in a query log dataset. A modified Personalized PageRank algorithm is utilized to rank attribute-value pairs by leveraging three elements: relevance, informativeness and topic coherence. We evaluate our approach on real-world datasets and show that it outperforms the state-of-the-art approaches.

Keywords: Entity summarization · Query log · Personalized Page Rank · Topic model

1 Introduction

With the popularity of knowledge graphs, a collection of entity attribute-value pairs (AVP) can be presented to users when they issue an entity-related query. An emerging problem is that with the explosion of Linked Open Data (LOD), users are faced with a long list of AVPs of the same entity, causing the problem of information overload. For example, the latest version of DBpedia has 4.58 million entities and more than 3 billion facts [1]. In average, there are over 650 AVPs per entity, which hinders users from understanding the entity with a glance.

Entity summarization is a technique to identify salient features (i.e., AVPs) representing the entity, which focuses on summarizing entity-relation subgraphs from the entire knowledge graph [6,19] or selecting the top-k AVPs to describe the entity [3,16,17]. However, we notice that these methods focus on the summarization of knowledge graphs alone, paying little attention to user's own preference towards the attributes. Nowadays, user generated content, such as query log, comments and adopted information, can help us to improve the accuracy of knowledge graph summarization.

© Springer International Publishing Switzerland 2016
B. Cui et al. (Eds.): WAIM 2016, Part I, LNCS 9658, pp. 517–529, 2016.
DOI: 10.1007/978-3-319-39937-9_40

In this paper, we propose the problem of Context-aware Entity Summarization (CES). Because the user preference is usually expressed implicitly, we take the query context to model the user preference and generate the entity summary by ranking the AVPs. For example, for the query "IBM Waston", a TV viewer may be interested in its play in Jeopardy, and prefer to see AVPs such as "<isDefeated, Brad Rutter>", while a scientist may prefer to see the AVP "<isDeveloping, Power 7>". Therefore, CES is capable of recommending AVPs of interest in application scenarios such as semantic Web search, knowledge graph serving, etc.

In this paper, we present a graph-based ranking method to solve this problem. We model the behavior of each user as a collection of query sessions, detected from a query log dataset. The user profile is represented as a topic distribution, mined by the topic model LDA. For entity AVP ranking, we propose a modified Personalized PageRank algorithm to calculate the ranks of all the AVPs. Three measures are introduced to guide the ranking process, namely relevance, informativeness and topic coherence. Finally, an entity summary is generated by selecting k AVPs of the entity with top-k highest ranking values.

This paper makes the following major contributions.

- We propose and formalize the CES problem. User preference is modeled by mining the latent topics of the query context.
- A ranking algorithm based on Personalized PageRank is presented to select top-k AVPs for each entity as the summary. We employ three measures in ranking, namely, informativeness, relevance and topic coherence.
- We conduct extensive experiments and compare CES with state-of-the-art approaches to illustrate the effectiveness of our approach.

The rest of this paper is organized as follows. Section 2 summarizes the related work. We define the CES problem formally in Sect. 3. Details of the CES system and the ranking algorithms are introduced in Sect. 4. Experiments on real-life datasets and comparison with baselines are presented in Sect. 5. Finally, we conclude our paper in Sect. 6.

2   Related Work

Early researches on entity summarization mainly focus on selecting top-k sentences given a collection of documents related to a target entity. Statistical measures such as word frequency and distribution, phrase frequencies and key phrases, and sentence patterns are employed in the summary generation process. With the rise of machine learning techniques in the field of NLP, machine learning models are employed to produce summaries by extracting key sentences. [11] incorporates a Naive Bayes classifier in textual summarization. In addition, other learning methods are used such as Hidden Markov Model, Log-Linear Models and Neural Networks [15], etc. This kind of summary technique reformulates the original information and aims to produce a better representation of the original documents. There are other summarization paradigms that focus on summarizing a particular topic in the documents, which is termed as *topic-driven summarization* [5]. Yan et al. [20] propose a fine-grained event summarization algorithm by studying multiple topics in a document collection. Although our approach CES focuses on summarizing AVPs given an entity, it can be regarded as a type of topic-driven summarization at the entity level.

Due to the enthusiasm in construction and application of large-scale knowledge graphs, entity summarization in knowledge graphs has attracted a lot of attention. For summarization from entity-relation graphs, Cheng et al. [3] first give the definition of entity summarization and propose a summarization system RELIN, to select typical AVPs that can best describe the entity. They adopt a variant of random surfer model considering relativeness and informativeness of entities. Thalhammer and Rettinger [17] propose the system SUMMARUM to generate summaries for entities in DBpedia, which is a structured knowledge base derived from Wikipedia. Gunaratna et al. [8] present the Cobweb model to group facts by conceptually clustering, and then choose the representative AVPs from each cluster to form summarization. This work is the closest to our approach, but they do not take the preferences of different users into account. Thalhammer et al. [18] leverage usage data to rank movie AVPs according to the importance of a particular entity. They compute k-nearest neighbors for an entity by using log-likelihood score ratio and calculate the weight by the tf-idf method. However, it requires that each user must have usage data available and is not suitable for large complex knowledge graphs. In this paper, we propose a novel model that supports generating entity summaries with the help of knowledge graphs and query logs.

3   Problem Statement

For the CES problem, we represent entities in the form of AVP collections. In this section, we formally define the task of CES and give some basic concepts.

3.1   Entity Summarization

The task of CES can be directly derived from entity summarization. For completeness, we first introduce the task of entity summarization.

Definition 1 *(**Knowledge Graph**). A knowledge graph is a finite collection of triples $KG = \{(e, a, v) | e \in E \wedge a \in A \wedge v \in E \cup L\}$, where*

- *E and L are finite collections of entities and literals;*
- *A is a finite collection of attributes, where each $a \in A$ has a source $Src(a) \in E$ and a target $Tgt(a) \in E \cup L$.*

Note that in our definition, "attributes" of an entity can refer to both relations between two entities and attributes between entities and literals. For ease of simplicity, we will use the term "attribute" in this paper. For $a \in A$ and $v \in E \cup L$, pair $t = (a, v)$ is an **attribute-value pair (AVP)**.

Definition 2 (Entity AVP Set). *An entity AVP set w.r.t. an entity e is a collection of AVPs $EAS(e) = \{(a, v)|(a, v) \in EAS(e) \wedge Src(a) = e\}$.*

Therefore, the task of entity summarization is defined based on entity AVP sets extracted from the knowledge graph, shown as follows:

Definition 3 (Entity Summarization). *Given the entity AVP set $EAS(e)$ and a positive integer $k < |EAS(e)|$, the task of entity summarization is to generate the summary $Summ(e) \subset EAS(e)$ of entity e such that $|Summ(e)| = k$.*

3.2 Context-Aware Entity Summarization

The above dentition does not consider the query context of users. Because a query is usually short and contains little background information. In this paper, we regard a series of queries issued by a certain user as the context. The task of context-aware entity summarization is defined based on query context:

Definition 4 (Context-Aware Entity Summarization). *Given a collection of queries $\{q_i|1 \leq i \leq N_u\}$ (N_u is the number of queries issued by u), denoted as Q_u, the entity AVP set $EAS(e)$ and a positive integer $k < |EAS(e)|$, the task of context-aware entity summarization is to generate the summary $CSumm(e) \subset EAS(e)$ of entity e such that $|CSumm(e)| = k$.*

The major difference between traditional entity summarization and the CES proposed in this paper is that we consider user preference by taking the query context into account. In previous approach, given an entity (e.g. "IBM Waston"), the algorithm generates the same summary regardless of the query context Q_u. In contrast, CES may generate different summaries for different users.

4 Proposed Approach

In this section, we present our approach to generate CES in detail. We introduce our framework in general and give algorithmic descriptions of all the modules.

4.1 General Framework

The general framework of CES is presented in Fig. 1. The input is a query corpus and the knowledge graph. User and attribute profiles are represented as topic distributions of the contents of user query sessions and attributes, generated by the LDA model. A Personalized PageRank based algorithm is proposed to generate top-k AVPs as entity summaries for all the entities in a query.

4.2 Profile Generation

User Profile. All the queries in Q_u of user u can be employed to build user profile by mining the latent interest of u. However, the queries for the same user are usually related to different topical tasks [10]. In this paper, we regard a query session as a single unit and generate user profiles based on query sessions.

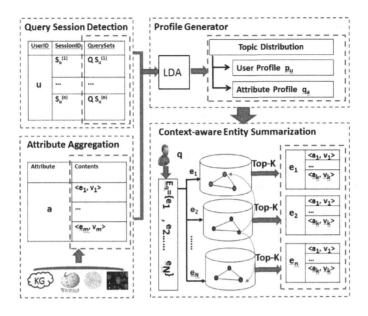

Fig. 1. The Framework of CES

More specifically, the query set Q_u can be divided by k_u query sessions:

$$Q_u = \bigcup_{i=1}^{k_u} QS_u^{(i)} \tag{1}$$

where $QS_u^{(i)} \cap QS_u^{(j)} = \emptyset$ for $\forall i, j (1 \le i \ne j \le k_u)$. We take all the contents of queries in $QS_u^{(i)}$ as the document for $QS_u^{(i)}$. Denote $\theta_{u,i}$ as the topic distribution of the query session $QS_u^{(i)}$. Then, we employ a simple heuristic rule. The profile p_u of user u is represented as the average of topic distributions, shown as follows:

$$p_u = \frac{1}{k_u} \sum_{i=1}^{k_u} \theta_{u,i} \tag{2}$$

The task of query session detection has been addressed in previous research [7]. In our research, we consider temporal clues and key word similarity factors for session boundary detection. The algorithm is presented in [4]. Due to space limitation, we omit the details here.

Attribute Profile. For attribute $a \in A$, we also calculate an attribute profile, represented as a topic distribution. To gather contents for these attributes, we employ an aggregation method. For simplicity, we first define the concept of attribute content set:

Definition 5 (Attribute Content Set). *An attribute content set w.r.t. an attribute a is a collection of tuples $ACS(a) = \{(e, v)|e \in E, v \in E \cup L\}$ such that for each tuple $(e, a, v) \in KG$.*

Thus, we can treat the content of an attribute a as a document which consisting of words in $ACS(a)$. We directly take the topic distribution of the attribute document (denoted as θ_a) as the profile for attribute a, defined as $q_a(= \theta_a)$.

Topic Analysis. To generate topic distributions for user and attribute profiles, we employ LDA model to compute the distributions in a uniform dimensionality. LDA [2] associates each document with a topic distribution, and models document-topic and topic-word distributions as multinomial distributions with Dirichlet prior.

In this paper, the document corpus for LDA contains two parts: user documents and property documents. As we mention above, for each user u, we generate k_u documents where the content of each document is the search queries in one query session $QS_u^{(i)}$. For each attribute a, we generate a single document whose content is the contents of $ACS(a)$. Therefore, we obtain $\sum_{i=i}^{|U|} k_{u_i} + |A|$ documents where U is the collection of distinct users in the query corpus. After the document set is genreated, we employ Gibbs sampling to obtain the topic distribution of each document in LDA.

In this way, the user and attribute profiles are both modeled as topic distributions of the same dimensionality. We need to note that in our research, the profiles are obtained offline by learning the parameters in the LDA model. In real-life applications, if there is a new user/attribute without a profile, our approach can be easily extended by inferencing the respect topic distribution(s) using the existing LDA model. See [2] for the inference algorithm.

4.3   Ranking Algorithm

The problem of entity summarization can be conceived as a ranking problem. We select top-k AVPs which can represent the characteristics of the entity and serve as the entity summary.

For an entity e and its AVP set $EAS(e)$, we construct a complete graph G_e where the nodes are all the AVPs in $EAS(e)$. Denote R_0 as the prior rank vector for $EAS(e)$, R_n as the rank vector in the nth iteration, and M as the transition matrix. The update rule of Personalized PageRank is represented as follows:

$$R_n = \lambda \cdot M \cdot R_{n-1} + (1 - \lambda) \cdot R_0 \tag{3}$$

where λ is a damping factor, typically set to 0.85 [9]. In the next part, we will discuss in detail about the prior rank R_0 and the transition matrix M.

Prior Rank. For CES, the prior rank R_0 encodes the prior importance of AVPs given the query context. Thus we employ the proximity between user and attribute profiles to assign the prior rank. Given a user u and an AVP t with attribute a, the prior rank is defined as the Jansen-Shannon divergence between the respective profiles, shown as follows:

$$r_0(u, t) = \frac{1}{Z} \cdot \frac{D_{KL}(p_u \| \frac{p_u + q_a}{2}) + D_{KL}(q_a \| \frac{p_u + q_a}{2})}{2} \tag{4}$$

where Z is a normalization constant and D_{KL} denotes the KL divergence between probabilistic distributions.

Transition Probability. Given two AVPs t_p and t_q in G_e, the transition probability $M_{p,q}$ defines the behaviors of random surfers on the graph. In the previous work, Cheng et al. [3] define two metrics to measure the "closeness" of two AVPs, namely *relevance* and *informativeness*. We notice that, there are more than one named entity in some user queries. These entities tend to share the same latent topic, therefore, the respective entity summaries should be topically coherent. As a result, we propose a novel measure named *topic coherence* to model the influence of rank propagation across different entity summaries in the same query. For completeness, we formally present the three measures as follows:

Relevance. The relevance of two AVPs t_p and t_q measures the degree that they have related attribute names and values. In [3], it is defined as a combination of the Pointwise Mutual Information (PMI) between their attributes and the PMI between their values:

$$Rel(t_p, t_q) = \sqrt{PMI(a_p, a_q) \cdot PMI(v_p, v_q)} \tag{5}$$

The PMI between two text strings x and y are calculated as follows:

$$PMI(x, y) = \log \frac{p(x, y)}{p(x) \cdot p(y)} \tag{6}$$

where $p(x)$ is the probability that x occurs in a large text corpus and $p(x, y)$ is the joint probability that x and y co-occur. According to [3], these probabilities can be estimated by measuring the number of documents returned by the Google search engine that match the text string.

Informativeness. The informativeness of t_p and t_q is measured by the amount of new information that the random surfer obtains by transferring from t_p to t_q [3]. According to information theory, it is defined as the self information of t_q given t_p:

$$Info(t_p, t_q) = -\log p(t_q | t_p) \tag{7}$$

where $p(t_q | t_p)$ is the probability that t_q belongs to an AVP set given that t_p is in the same one. The conditional probability is estimated by analyzing the knowledge graph statistically:

$$p(t_q | t_p) = \frac{|\{e \in E | t_p, t_q \in EAS(e)\}|}{|\{e \in E | t_p \in EAS(e)\}|} \tag{8}$$

Topic Coherence. For a query q issued by a user u, denote $E_q = \{e\}$ as the collection of entities that q contains. The topic coherence gives high scores to AVPs of entity e, if the attributes of these AVPs have high ranks for other entities in E_q. In this way, we guarantee that entity summaries (i.e., top-k AVPs) are topically coherent. Denote $r(t_p, e)$ as the rank value of AVP t_p of entity e, we define the topic coherence between t_p and t_p of entity e as follows:

$$TC(t_p, t_q) = \frac{\sum_{e' \in E_q \setminus \{e\}} r(t_q, e')}{\sum_{e' \in E_q \setminus \{e\}} r(t_p, e') + \sum_{e' \in E_q \setminus \{e\}} r(t_q, e')} \tag{9}$$

However, there are still one issue that need to be addressed in the above equation. We need to know $r(t_p, e')$ in advance before we calculate $r(t_p, e)$ where $e' \in E_q \setminus \{e\}$. It imposes a "chicken-and-egg" problem such that if we want to compute the AVP ranks of an entity, we must know AVP ranks of other entities. To solve this problem, we employ a heuristic to estimate $r(t_p, e')$. In previous work, Cheng et al. [3] propose a random surfer model for entity summarization based on relevance and informativeness. This method can generate approximate results compared to ours if we do not consider the query context. Denote the $r^*(t_p, e)$ as the rank value of AVP t_p of entity e calculated by [3]. The topic coherence is computed as follows:

$$TC(t_p, t_q) \approx \frac{\sum_{e' \in E_q \setminus \{e\}} r^*(t_q, e')}{\sum_{e' \in E_q \setminus \{e\}} r^*(t_p, e') + \sum_{e' \in E_q \setminus \{e\}} r^*(t_q, e')} \tag{10}$$

Therefore, by using topic coherence, we can rank AVPs on a query level if multiple entities exist in the same query.

Fusion of Measures. Finally, we fuse the three measures mentioned above to compute the transition probability via a linear combination, illustrated as follows:

$$M_{p,q} = \frac{1}{Z'} \cdot (\alpha \cdot Rel(t_p, t_q) + \beta \cdot Info(t_p, t_q) + \gamma \cdot TC(t_p, t_q)) \tag{11}$$

where α, β and γ are non-negative weight parameters with the constraint $\alpha + \beta + \gamma = 1$. Z' is a normalization constant that keeps the sum of all the transition probabilities given t_q to be 1. We tune the values of these parameters in the experiments. After the ranking procedure converges, we take the top-k AVPs as the entity summary.

5   Experiments

In the section, we present our experiments to demonstrate the effectiveness of the proposed approach. We also compare our method with existing baselines to make the convincing conclusion.

5.1 Datasets

We use two datasets in the experiments: the query log dataset and the knowledge graph dataset, introduced as follows:

- **Query Log:** We use a real-world query log set, which is released by AOL on August 2006, provided by Microsoft [13]. For each query, there are three attributes: AnonID (an anonymous user ID), QueryTime (the time at which the query was submitted) and Query (the query issued by the user). We employ query session detection techniques to identify query sessions of each user. In the log, 657,426 users submit 10,154,742 queries in 5,507,464 sessions.
- **Knowledge Graph:** We employ the YAGO [14] knowledge graph in the experiments, which is a huge semantic knowledge graph in which the knowledge is extracted from Wikipedia and other sources. We download the entire YAGO (version 3.0.2) from the official website[1]. Currently, YAGO contains over 10 million entities and 120 million facts about these entities.

5.2 Baselines

To illustrate the effectiveness of our method, we compare ours (i.e., CES) against four baselines, namely, RandomRank, PopularityRank, PageRank [12] and RELIN [3]. We introduce these baselines in brief:

- **RandomRank:** This is the most simple approach. We randomly select k AVPs from $EAS(e)$ as the summary for entity e.
- **PopularRank:** We first rank each attribute $a \in A$ according to the popularity, defined as:

$$Pop(a) = \frac{\{(e^{'}, a^{'}, v^{'}) \in KG | a^{'} = a\}}{|KG|} \times 100\% \qquad (12)$$

  Next, we select top-k AVPs from $EAS(e)$ with k highest attribute popularity values.
- **PageRank:** We employ classical PageRank algorithm to rank AVPs in G_e to select top-k AVPs as entity summary. The informativeness measure is employed to define the transition probability.
- **RELIN:** It is the entity summarization algorithm proposed by Cheng et al. [3]. It employs a random surfer model to rank AVPs, considering the relevance and informativeness.

5.3 Evaluation Method

To evaluate the effectiveness of our approach and compare it with baseline, we follow the evaluation method proposed by Cheng et al. [3] for entity summarization. For each entity e, we invite 15 participants majoring in computer science

---

[1] http://www.mpi-inf.mpg.de/departments/databases-and-information-systems/ research/yago-naga/yago/.

to manually select a summary from $EAS(e)$ as the ideal summary, denoted as $Summ_i^I(e)$. Denote the summary generated by machines as $Summ_i(e)$. The quality of $Summ_i(e)$ is defined as:

$$Quality(Summ(e)) = \frac{1}{15} \sum_{i=1}^{15} \frac{|Summ(e) \cap Summ_i^I(e)|}{|Summ_i(e)|} \qquad (13)$$

To better fit the task of CES, we define two metrics based on quality, namely *Average Entity-level Quality* (AEQ) and *Average Query-level Quality* (AQQ). For a collection of entities E_{Test}, average entity-level quality is defined as:

$$AEQ(E_{Test}) = \frac{1}{|E_{Test}|} \sum_{e \in E_{Test}} Quality(Summ(e)) \qquad (14)$$

For a collection of queries Q_{Test}, average quality-level quality is defined as:

$$AQQ(Q_{Test}) = \frac{1}{|Q_{Test}|} \sum_{q \in Q_{Test}} \frac{1}{|E_q|} \sum_{e \in E_q} Quality(Summ(e)) \qquad (15)$$

5.4 Results and Analysis

In the experiments, we use the Stanford NLP toolkit[2] to recognize named entities in the query log dataset. We randomly sample 100 queries with multiple entities, and present them to human labelers to generate top-5 and top-10 AVPs as entity summaries. We first tune the parameters α and β, whose default values are 0.2 and 0.4, respectively. The results are shown in Fig. 2.

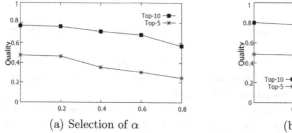
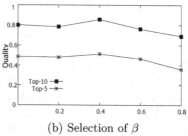

(a) Selection of α (b) Selection of β

Fig. 2. Tuning α and β for CES

Figure 2(a) illustrates the impact of weight for relevance via varying α from 0 to 0.8. We find that the quality of entity summarization declines smoothly when the value of α increases, i.e., the topic coherence is more important than relevance in the integrating measure. Similarly, Fig. 2(b) elaborates the impact

---

[2] http://stanfordnlp.github.io/CoreNLP/.

Table 1. Comparison with baselines

Method	Top-5 AEQ	Top-10 AEQ	Top-5 AQQ	Top-10 AQQ
RandomRank	0.67	1.23	0.41	1.41
PopularRank	1.65	3.23	1.83	3.12
PageRank	1.33	3.14	2.12	2.98
RELIN	1.79	3.56	2.45	3.23
CES	**2.56**	**4.29**	**3.16**	**4.54**
Min(CES % ↑)	**43 %**	**21 %**	**29 %**	**41 %**

of weight for informativeness via changing β from 0 to 0.8. We observe that CES achieves the optimal performance. As a result, we should fix the value of β and set the value of α as small as possible to obtain a better performance for CES.

We compare our method with baselines. The results are presented in Table 1, in which the minimum percentage increase is showed in last line. From the experiments, we can see that our approach outperforms all the baselines. Hence, we can conclude that the context information is helpful to improve the performance of entity summarization.

Table 2. Case studies

User	**Q1:** *the beatles* norwegian
Music-lover	<created, Norwegian Wood>
	<influences, Andrew Crayford>
	<wroteMusicFor, Help!>
	<hasWonPrize, Brit Awards>
	<isMarriedTo, Freddy Moore>
User	**Q2:** *nazi germany* empire
Military fan	<isCreatedOn, 1931-##-##>
	<hasCurrency, Reichsmark>
	<owns, Wolf's Lair>
	<participatedIn, Battle of Maastricht>
	<isLocatedIn, Europe>

5.5 Case Studies

To further illustrate the effective of CES, we present two examples of entity summarization. The results are shown in Table 2. A music-lover and a military fan submit queries *the beatles norwegian* and *nazi germany empire*, respectively. CES selects 5 AVPs out of 277 for the entity "the Beatles" and 5 AVPs out of

528   J. Yan et al.

285 for the entity "Nazi German". Given the user context, we can observe that all returning AVPs are very meaningful.

6 Conclusion

In this paper, we propose and formalize the problem of context-aware entity summarization. We mine user preferences from the query log by topic modeling. We build a random walk based on the concepts of relevance, informativeness and topic coherence. Our proposed algorithm returns the top-k AVPs, which are generated by the random walk-based ranking algorithm. The experiments show that the proposed algorithm outperforms the comparable baselines.

Acknowledgements. This work is supported by the National Basic Research Program (973) of China (No. 2012CB316203) and NSFC under Grant Nos. U1401256, 61402177, 61402180 and 61232002. This work is also supported by CCF-Tecent Research Program of China (No. AGR20150114) and NSF of Shanghai (No. 14ZR1412600).

References

1. Biega, J., Kuzey, E., Suchanek, F.M.: Inside yago2s: a transparent information extraction architecture. In: WWW, pp. 325–328 (2013)
2. Blei, D.M., Ng, A.Y., Jordan, M.I.: Latent Dirichlet allocation. J. Mach. Learn. Res. **3**, 993–1022 (2003)
3. Cheng, G., Tran, T., Qu, Y.: RELIN: relatedness and informativeness-based centrality for entity summarization. In: Aroyo, L., Welty, C., Alani, H., Taylor, J., Bernstein, A., Kagal, L., Noy, N., Blomqvist, E. (eds.) ISWC 2011, Part I. LNCS, vol. 7031, pp. 114–129. Springer, Heidelberg (2011)
4. Chien, S., Immorlica, N.: Semantic similarity between search engine queries using temporal correlation. In: WWW, pp. 2–11. ACM (2005)
5. Das, D., Martins, A.F.: A survey on automatic text summarization. In: Literature Survey for the Language and Statistics II Course at CMU, vol. 4, pp. 192–195 (2007)
6. Faloutsos, C., McCurley, K.S., Tomkins, A.: Fast discovery of connection subgraphs. In: KDD, pp. 118–127 (2004)
7. Gayo-Avello, D.: A survey on session detection methods in query logs and a proposal for future evaluation. Inf. Sci. **179**(12), 1822–1843 (2009)
8. Gunaratna, K., Thirunarayan, K., Sheth, A.P.: Faces: diversity-aware entity summarization using incremental hierarchical conceptual clustering (2015)
9. Haveliwala, T.H.: Topic-sensitive pagerank. In: WWW, pp. 517–526 (2002)
10. Hua, W., Song, Y., Wang, H., Zhou, X.: Identifying users' topical tasks in web search. In: WSDM, pp. 93–102 (2013)
11. Larsen, B., Aone, C.: Fast and effective text mining using linear-time document clustering. In: KDD, pp. 16–22 (1999)
12. Page, L., Brin, S., Motwani, R., Winograd, T.: The pagerank citation ranking: bringing order to the web (1999)

13. Pass, G., Chowdhury, A., Torgeson, C.: A picture of search. In: InfoScale, vol. 152, p. 1 (2006)
14. Suchanek, F.M., Kasneci, G., Weikum, G.: Yago: a core of semantic knowledge. In: WWW, pp. 697–706 (2007)
15. Svore, K.M., Vanderwende, L., Burges, C.J.: Enhancing single-document summarization by combining ranknet and third-party sources. In: EMNLP-CoNLL, pp. 448–457 (2007)
16. Sydow, M., Pikuła, M., Schenkel, R.: The notion of diversity in graphical entity summarisation on semantic knowledge graphs. J. Intell. Inf. Syst. **41**(2), 109–149 (2013)
17. Thalhammer, A., Rettinger, A.: Browsing DBpedia entities with summaries. In: Presutti, V., Blomqvist, E., Troncy, R., Sack, H., Papadakis, I., Tordai, A. (eds.) ESWC Satellite Events 2014. LNCS, vol. 8798, pp. 511–515. Springer, Heidelberg (2014)
18. Thalhammer, A., Toma, I., Roa-Valverde, A., Fensel, D.: Leveraging usage data for linked data movie entity summarization (2012). arXiv:1204.2718
19. Tong, H., Faloutsos, C.: Center-piece subgraphs: problem definition and fast solutions. In: KDD, pp. 404–413 (2006)
20. Yan, J., Cheng, W., Wang, C., Liu, J., Gao, M., Zhou, A.: Optimizing word set coverage for multi-event summarization. J. Comb. Optim. **30**(4), 996–1015 (2015)

Erratum to: BMF: An Indexing Structure to Support Multi-element Check

Chenyang Xu[✉], Qin Liu, and Weixiong Rao

Department of Software Engineering, Tongji University, Shanghai, China
{1336321,qin.liu,wxrao}@tongji.edu.cn

Erratum to:
Chapter 34: B. Cui et al. (Eds.)
Web-Age Information Management
DOI:10.1007/978-3-319-39937-9_34

In an older version of the paper starting on page 441 of this volume, the name and email address of the second author (Qin Liu) were missing. This has been corrected.

The updated original online version for this chapter can be found at 10.1007/978-3-319-39937-9_34

© Springer International Publishing Switzerland 2016
B. Cui et al. (Eds.): WAIM 2016, Part I, LNCS 9658, p. E1, 2016.
DOI: 10.1007/978-3-319-39937-9_41

Author Index

Printed in the United States
By Bookmasters